To
the memory of my father,
a lifelong practical biologist

Biology
Principles and Issues

William C. Schefler
College at Buffalo, State University of New York

Addison-Wesley Publishing Company
Reading, Massachusetts • Menlo Park, California
London • Amsterdam • Don Mills, Ontario • Sydney

This book is in the
ADDISON-WESLEY SERIES IN THE LIFE SCIENCES

Consulting Editor
Johns W. Hopkins III

The following figures originally appeared in Jeffery J. W. Baker and Garland E. Allen, *The Study of Biology,* 2nd ed. Reading, Mass.: Addison-Wesley, 1971: Figures 2.4, 3.2, 14.14 through 14.16, 14.20, 16.12, 16.18, and 16.19.

The following figures originally appeared in John W. Kimball, *Biology,* 3rd ed. Reading, Mass.: Addison-Wesley, 1974: Figures 3.9, 4.11, 4.14, 4.16, 8.11, 10.3, 10.4, 12.2, 12.8 through 12.10, 14.2, and 14.9.

The following figures originally appeared in Derry D. Koob and William E. Boggs, *The Nature of Life.* Reading, Mass.: Addison-Wesley, 1972: Figures 14.8 and 16.17.

The following figures originally appeared in Eric T. Pengelley, *Sex and Human Life.* Reading, Mass.: Addison-Wesley, 1974: Figures 5.3 through 5.14, 5.17 through 5.22, 6.1 through 6.6, and 9.10.

Copyright © 1976 by Addison-Wesley Publishing Company, Inc. Philippines copyright 1976 by Addison-Wesley Publishing Company, Inc.

All rights reserved. No part of this publication may be reproduced, stored in a retrieval system, or transmitted, in any form or by any means, electronic, mechanical, photocopying, recording, or otherwise, without the prior written permission of the publisher. Printed in the United States of America. Published simultaneously in Canada. Library of Congress Catalog Card No. 75-28725.

ISBN 0-201-06764-1
BCDEFGHIJ-HA-79876

Preface

The idea that a course ought to be designed to meet the special needs of its students is something of a novelty in higher education. In the dim but not too distant past, the groves of académé were noted mainly as a tranquil haven where the scholar could pursue his specialty through reflection and research. Unfortunately, he was also apt to take this opportunity to inflict his special interests on the helpless captives in his classroom, often without regard to differences in their backgrounds, interests, and objectives.

The idea of a biology course that is tailored to the interests, objectives, and needs of the English, social science, or art major has been frowned upon as lacking in "academic respectability," a term that even its users are unable to define but which is often related to (1) the degree of incomprehensibility and (2) what the professional biologist-specialist deems important.

It would be premature to say that this situation has radically changed. But there are some indications that more thought is being given to the needs of the student who is not a biology major, who more than likely "hates science" with a passion, and who signed up for biology because of some degree requirement or because someone told him that it is easier than physics.

These signs are welcome and long overdue, but there are those who fear with some justification that the academic world may be too quick to worship at the altar of a new god called *Relevance*. In some quarters the rush to be relevant has resulted in a course which appears to be a blatant appeal for student attendance and student approval. A course in "environmental biology" in which the instructor proclaims to a supposedly surprised class that the Hudson (Niagara, Mississippi, Ohio, etc.) River is polluted, and spends a semester exhorting the students to chase the evil-doers from the land, is not a biology course. All too often it becomes a propaganda session in which various aspects of the obvious are stated and restated.

This book is based on a one-semester course which the author teaches to nonmajors. Like the course, it attempts to promote an appreciation of biology as a science which is indeed relevant to many of the important issues that involve the future of man and of the planet that we call home. This book is firmly based, however, on the premise that a student can best arrive at an appreciation of the larger issues related to biology through a study of related basic principles. It seems unlikely, for example, that one can intelligently approach environmental problems without some awareness of at least the basic principles of ecology. Similar statements can be made about other issues related to genetics, abortion, population growth, racial differences, and so on.

The style, content, and level of sophistication are frankly directed toward the nonmajor. On the other hand, an emphasis on ease of comprehension should not prevent a consideration of some of the traditionally "difficult" topics such as those related to the general field of molecular biology. If such topics as DNA and energy transfer are approached with care, the nonmajor can gain a useful degree of understanding as well as a grasp of the significance of the principles involved.

It may seem trite to point out the need for an "enlightened citizenry," but we can no longer ignore the fact that the problems of man and his planet make it doubly important that people who are not biologists have an appreciation of major biological concepts. The major objective of this book, therefore, is to help the student attain a degree of biological sophistication through the study of a few selected basic principles, with applications to the larger issues of which they are a part.

There has been an attempt to organize this book on a logical sequential plan so that principles learned in earlier chapters will be useful to understanding those concepts taken up in later chapters. A special chapter on chemistry is not included; instead, a review of necessary chemical principles is brought in only where they are needed as part of a biological concept.

Buffalo, New York   W.C.S.
January 1976

Acknowledgements

A textbook writer owes a debt of gratitude to many people; not the least of these are the countless unnamed scientists and writers whose work has developed the body of knowledge that makes such a book possible. I must also thank all those who supplied photographs and illustrations, with special thanks to E. T. Pengelley and J. W. Kimball. I also want to thank Mrs. Marie Gale for the many hours she spent typing the manuscript. To the staff at Addison-Wesley goes my deep appreciation for their patient and always competent help. Finally, I wish to acknowledge with special gratitude my wife, Wilma, for her help in gathering materials, but most of all for her unfailing patience and support during the long hours of writing.

Contents

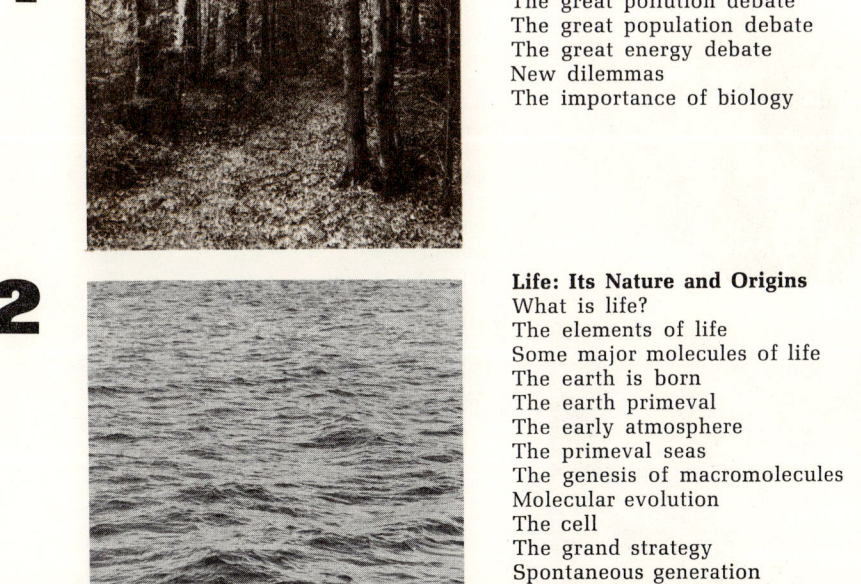

1 Introduction
The science of life	1
The great pollution debate	2
The great population debate	2
The great energy debate	3
New dilemmas	4
The importance of biology	5

2 Life: Its Nature and Origins
What is life?	7
The elements of life	9
Some major molecules of life	11
The earth is born	14
The earth primeval	16
The early atmosphere	16
The primeval seas	18
The genesis of macromolecules	20
Molecular evolution	24
The cell	24
The grand strategy	26
Spontaneous generation	27
Summary	28
Questions	29

ix

3 The Unit of Life
The cell	31
Cell structures	33
Energy	37
The energy scheme	38
Photosynthesis	39
Harvesting the sun's energy	42
The flow of cellular materials	45
Questions	49

4 The Cell Nucleus
Chromosomes	51
Cell division	55
Reproduction	62
Asexual reproduction	62
Sexual reproduction	65
Questions	65

5 "Be Fruitful, and Multiply..."
Human sexuality	67
The master controls	68
The male reproductive system	70
The testes	71
The female reproductive system	76
Sexual intercourse	78
The menstrual cycle	80
Fertilization	84
Pregnancy	86
Birth	93
Lactation	94
Summary	95
Questions	95

6 Birth Control
The need for birth control	97
Methods of birth control	99
Questions	111

7 The Legacy—DNA

The importance of the gene	113
The energy of chemical reactions	115
DNA	119
Replication of DNA	122
Ribonucleic acid	123
Protein synthesis	124
Sickle-cell anemia	126
Metabolic block disorders	126
Carrier identification	130
Questions	131

8 The Genetic Dice

The rules of the game	133
The determination of sex	135
Simple gene combinations	137
Gene frequencies	141
Inheritance of blood groups	149
Sex-linked genes	152
Polygenic inheritance	154
Summary	156
Questions	156

9 A Problem of Choice

Nondisjunction	159
Chromosome translocations	167
Chromosome deletions	169
Environmentally produced birth defects	170
Amniocentesis—a problem of choice	171
Questions	174

10 Human Nutrition

The importance of food	177
Classes of nutrients	178
Digestion and absorption	182
More on the metabolic mill	186
Earth—the hungry planet	189
Nutrition and the brain	191
Nutrition and aging	192
Nutrition and the pill	193
Eating for two	193
Too much food	195
Summary	197
Questions	197

11 "... And Replenish the Earth"

The problem	199
The arithmetic of population growth	201
Why explosive growth?	203
Consequences of population growth	208
What about the future?	210
Questions	211

12 Psychobiology

Introduction	213
The neuron—the basic unit of behavior	214
The nerve impulse	216
The neuron synapse—the switch	217
The command post—the brain	219
The control of behavior	229
Drugs and the brain	232
Why drug use?	237
Questions	237

13 Noah's Ark

Order out of chaos	239
Modern classification	241
Procaryotes: the Monera	242
Eucaryotes	244
Summary	261
Questions	261

14 The Ecosystem

The sun—the ultimate source	263
Ecology	265
The ecosystem	267
Energy flow in an ecosystem	270
Environmental factors in an ecosystem	273
The delicate web	274
Symbiosis	276
Circulation of materials	277
Ecosystem development	279
Summary	282
Questions	285

Contents xiii

15

"Subdue the Earth, and Have Dominion..."
Introduction 287
Water pollution 288
Air pollution 294
Pesticides 298
Energy problems—the machines 301
Energy problems—the people 309
Questions 311

16

Evolution
The changing earth 313
Darwinism 321
The process of evolution 326
Mutation 328
Selection 328
The formation of a species 331
The arrival of man 336
The races of man 340
Questions 345

17

Epilogue 347

Glossary 351

Annotated Reading List 361

Index 365

1

Introduction

The Science of Life

Too many of us recall our first encounter with the study of biology as a parade of rusty scalpels, impossibly huge earthworms, ancient skeletons, and long-dead frogs reeking of formaldehyde. It is indeed regrettable that so many former students of "the science of life" associate the study of biology with dead things and with long hours spent peering through a microscope at structures that bore only a fanciful resemblance to the beautiful diagrams in the book.

In point of fact, no other science is as lively as modern biology, and certainly none bears more immediate relevance to many of the important issues currently affecting planet Earth and the society of mankind. By way of introduction, let us briefly review some of the major problems in which biology plays an important role.

The Great Pollution Debate

It might be said with some justification that Rachel Carson started it all with her book *Silent Spring*, published in 1962. Although conservationists had been continually calling for an end to man's thoughtless exploitation of the natural environment of which he is a part, it was Miss Carson's eloquent imagery and her careful documentation of the effects of insecticides that first attracted significant public attention.

During the 1960's, the voices of protest grew louder and more numerous, and as the decade neared its end the "ecology movement" was in full swing. The more vocal and militant environmentalists were condemned by some as being "ecology nuts" and "doomsayers" who advocated laws that would close the factories and wreck the economy. The environmentalists, in turn, accused the polluters of sacrificing generations yet unborn on the altar of profit.

In point of fact, it is useless to question the evidence of the pollution which has spread to the air that we breathe, to the water that we drink, and to the soil on which we depend for our subsistence. Automobiles pour onto the streets and freeways in ever-increasing numbers, adding to the pollution that gushes from the stacks of the factories on which the livelihood of millions depends. Our lakes and streams once abounded with game fish and welcomed swimmers on a Sunday afternoon. Now they are often little better than cesspools, teeming with sewage and industrial wastes and fit for neither fish nor human swimmer. Meanwhile, the "Beach Closed" sign is a terse comment on man's careless exploitation of the environment.

Federal and state legislatures have taken belated steps to provide funds and laws for environmental action. Unfortunately, there is a serious question as to whether man's technology can restore that which took the forces of nature millions of years to create.

The no-return bottles, rustless cans, and plastic containers continue to pile up. The population shifts toward the great northern cities of the United States hoping for streets paved with gold and finding them paved with garbage. Is it somehow symbolic of man's character that the first men on the moon left behind on that virgin satellite an assorted pile of discarded junk? It is significant that this same space triumph first demonstrated vividly and in living color that the earth is not limitless, that it is indeed a spaceship of sorts, having life-support systems that are by no means inexhaustible.

Like any important problem, the environmental crisis will not yield to simple solutions. We must not delude ourselves that the environment can be cleaned up simply by removing phosphate detergents from the grocery shelves or by picking up empty bottles in the park on a Sunday afternoon. Significant and effective pollution abatement programs must be comprehensive and very expensive, and not all industries and municipalities are willing, or even able, to provide the vast sums of money that will be required.

The Great Population Debate

Some experts believe that efforts to control pollution merely attack a symptom, while the basic disease—uncontrolled population growth—remains unchecked. It seems impossible to deny the existence of a population problem when the cold mathematics of the population experts yield projections of an earth teeming with billions of people by the end of the present century. We hear warnings of impending disaster, that plague and famine will "solve" the problem for us if we do not take effective steps to curb our fertility.

Human misery already abounds in parts of the world where the energy demands of masses of human beings have outrun the resources, and the spectre of starvation hovers over half the people on earth today.

This most unpleasant picture was demonstrated vividly by Paul Ehrlich in *The Population Bomb,*

> By the time you have read these words, four people will have died of starvation, most of them children.

Yet, in spite of the figures and in spite of the impressive credentials of those who warn us against the ultimate tragedy of overpopulation, not everyone is convinced. In fact, it is probably more accurate to say that on a worldwide basis, very few are convinced! Those who oppose population control, voluntarily or otherwise, insist that plenty of room remains on the good earth to accommodate the exponential increase in humanity. They often point hopefully to the so-called "green revolution" and to the possibility of harvesting food from the seas. (These concepts will be discussed in a later chapter.)

Some object on religious grounds, some object on cultural grounds, and some suspect that those who urge population control are, in reality, advocating the suppression of specific racial and ethnic groups. Still others fear that population control efforts may ultimately lead to a loss of the very basic and personal freedom to have as many children as preference dictates. To all of this we must add that an appallingly large percentage of the world's population is illiterate, does not understand the nature of the problem, and lacks the knowledge and resources necessary to limit family size.

In Chapter 5 we will examine the basic physiological aspects of human fertility, and in Chapter 6 we will relate these basic principles to common methods for controlling fertility. Then in Chapter 11 we will take a closer look at the facts, figures, and social factors which underlie the population controversy.

The Great Energy Debate

In 1973 the Arab oil-producing countries placed an embargo on shipments to the United States and the great "energy crisis" burst upon the scene. Long lines of disgruntled motorists formed at the gas pumps and friendly service-station operators lost their friendliness and forgot to wipe your windshield. People stored firewood because of fear that there would not be enough fuel to heat their homes during the winter months. Factories cut back production and unemployment increased. The Nixon Administration promised the impossible—that the United States would be self-sufficient in terms of energy supplies by 1980! Meanwhile, the costs of energy rose rapidly and remained at higher levels after the worst of the crisis had passed. Most experts feel that we will probably never again see the cheap energy of former times.

While it lasted, the energy crisis made us realize some important truths. First, the affluence enjoyed by the industrialized nations is based on economic growth, which in turn depends upon adequate supplies of cheap energy. Secondly, the earth is not a horn of plenty capable of producing inexhaustible supplies of energy. And third, there are no quick and easy solutions to the energy problems that face us now and are sure to face us in the future.

The promises of self-sufficiency by 1980 were quickly scaled down, and with good reasons. What, for example, do we substitute for oil and natural gas as basic and relatively clean sources of energy? Is nuclear power the perfect solution? Can we take comfort from the fact that there is enough coal in the United States to last for several hundred years? Is the breakthrough in harnessing solar energy just around the corner?

In Chapter 15 we will see that none of the foregoing involves easy solutions; in fact, they tend to raise new problems! What, for example, will be the environmental consequences of using nuclear power and a return to widespread use of coal? In order to solve the energy problem, must we retreat from the relatively small environmental gains that have been made so far?

In Chapter 15 we will also examine the growing

problem of human energy. Although the reader of this book is much more likely to worry about an empty gas tank than an empty stomach, sooner or later we all must be called to account for the creeping malnutrition that affects over half of the growing billions that make up the earth's human population.

New Dilemmas

In the early 1950's, Nobel Prize winners James D. Watson and Francis E. Crick first described the structure and function of the gene. Their discovery set off a virtual explosion of knowledge in the field of genetics, and man may now be on the verge of attaining an awesome degree of control over his own destiny. Some scientists believe that it is only a matter of time until techniques of "genetic engineering" will permit us to manipulate the genetic code that determines our individuality to such a large degree. Would you like to arrange for your child to have a super I.Q.? Or would you prefer a musical genius? Before you decide that these seemingly farfetched possibilities are attractive, be sure to read Aldous Huxley's *Brave New World* if you have not already done so.

Science has a habit of handing us new problems along with the benefits it provides. The fact that the gift of nuclear energy is tainted with the potential horrors of nuclear warfare or nuclear "accidents" is a familiar example, but there are others which are more subtle and less well known. For example, we now know a good deal about the genetic processes that produce children with Down's syndrome, commonly known as mongoloid idiocy. In an earlier time, the unfortunate couple that produced a mongoloid child may have been assured by friends and clergy that their misfortune was God's will, and that such severely mentally and physically retarded children were placed on earth as part of the Almighty's master plan. While this explanation may have provided scant comfort to the afflicted family, it was at least better than nothing in a situation that offered no choice and no hope.

A technique is now available that permits the detection of certain types of genetic defects in the fetus during early pregnancy. The growing list of defects that are detectable by this method, called amniocentesis, includes mongolism. Thus it would seem that science has provided a choice that was denied to earlier generations. The way is now clear—if a pregnant woman learns that she will have a mongoloid child, she simply seeks an abortion. The problem is solved. Or is it?

Look closer and you will see that for many women this deceptively simple choice may be in reality a heart-breaking dilemma. Should the defective fetus be aborted and "God's will" be in effect denied? Or should human life in any form and at any stage be held sacred, not to be terminated by the hand of man for any reason.

Is abortion morally and ethically wrong under all circumstances? Or do we need to change our traditional attitudes toward abortion when failure to terminate a pregnancy may *knowingly* permit the birth of a child so retarded that it will live out its life scarcely aware of its own humanity?

Should the parents, who must bear the terrible emotional and financial burdens that come with a severely retarded child, have a legal right to make the decision? Perhaps of even greater significance is the question of whether society, through its laws, should have the right to make the decision for them one way or the other.

As we shall see in a later chapter, the foregoing is most emphatically not a hypothetical problem which might arise some time in the future. Prenatal examination of fetal chromosomes will soon be commonplace; soul-searching decisions of this kind will be required with greater frequency. For this and other reasons, the question of abortion continues to be an important public issue.

The Importance of Biology

Perhaps by now you are at least partially convinced that the study of biology does indeed involve more than frogs, rats, and nature walks in search of a glimpse of the great horned owl. It should be obvious that biology deals directly with some of the most critical problems that will face Spaceship Earth and its passengers during the last quarter of the twentieth century.

It has been said that war is too important to be left to the generals; it is equally true that biology is too important to be left exclusively to the biologists. Some problems are too complex for white-frocked scientists working in their laboratories. The complicated social, legal, economic, and moral ramifications of these problems require more than ever the presence of an enlightened citizenry, especially since many of these issues involve action by representative political bodies such as national, state, and local legislatures.

It is therefore one very important objective of this book to help you approach biology-related issues with a certain degree of sophistication. It is unfortunate, for example, that in recent years young people have gone about shouting "ecology" slogans without having taken the time or trouble to learn the meaning of the word, much less the basic principles involved. Progress toward solutions to environmental and other problems will not be made by bandwagon-riding sloganeers, but by thoughtful people who are conversant with the principles and concepts that are basic to the problem.

A second objective deals with your appreciation of biology as an inherently fascinating subject which is well worth your time and mental energy. Surely there is no tale more engrossing, even to the English major who may dutifully "hate science," than the story of the events that long ago brought forth on the primitive earth that ephemeral thing called "life." And one does not have to be a scientist to appreciate the wonderful simplicity of the genetic code or the processes of evolution which ultimately produced a creature with the capacity to study biology!

In our approach to biology, we will emphasize basic principles and concepts. This is as it should be, but we will relate these principles, where appropriate, to the larger issues and problems in which they play a significant role.

Then let us begin, appropriately at the beginning, by considering the nature of life and the way by which it probably originated on the young earth.

2

Life: Its Nature and Origins

What Is Life?

An old dog lies sleeping near the fireplace. Her arthritic joints no longer let her roam the fields in search of rabbits to chase, and she must be content with longer naps by the fire. She sleeps peacefully because her limited dog's brain keeps her mercifully unaware that her life is coming to a close. Unlike her, her master has the gift of reason, and with it comes the certain knowledge of the inevitability of her death —and his own. And so, like his primitive ancestors in the dim past, he wonders at the event that will shortly turn this living, breathing, canine personality into an object as unresponsive as a stone.

At some undefined point on the long road of evolution, man became the only animal that knows he must someday die. This unique awareness of his own mortality produced an age-old preoccupation with the mystery of death and endless speculations on the possibilities of life beyond the grave. In recent times, however, as man has grown more sophisticated, he has increasingly turned his efforts toward learning the secrets of life itself.

What is life? What constitutes that dreadfully final difference between a human being going about his daily activities and a cold, inanimate corpse? What distinguishes a turtle from the rock on which it rests at the water's edge? It is easy to understand how primitive man came to think in terms of a "spark of life," and to regard death as a point when the spirit or soul left the body to dwell forevermore in some far-better place. This view was given lasting emphasis by the biblical story in which God fashioned Adam from the dust of the earth and "breathed into his nostrils the breath of life."

This biblical "spark of life" was basic to the doctrine of *vitalism, which held that living matter was somehow different from that which was classified as nonliving.* The vitalists therefore believed that substances associated strictly with living organisms could only be produced with the aid of a mysterious vital force. Such substances were classified as *organic, that is, "produced by organs"* of the living body, and it was considered useless to attempt their synthesis in the laboratory because the vital force would be missing. Quite obviously, as long as this belief persisted, it presented a strong obstacle to the kinds of research that might have led sooner to a deeper understanding of life processes. If, for example, enough people of authoritative reputation were to keep assuring you that performing a certain task was impossible, you would very likely turn your attention elsewhere. Even today, many people assume that some mysterious, divinely inspired force or "spirit" separates the turtle from the rock, the living from the dead. It would certainly be difficult, however, to find a modern biologist who holds that view.

Vitalism suffered its first setback as long ago as 1828. Friedrich Wöhler, a German chemist, was heating a substance called ammonium cyanate and was astonished to discover that this simple laboratory procedure resulted in the formation of urea. Now, urea is a constituent of urine and was therefore regarded as organic; that is, it could supposedly be produced only by that mysterious force found solely in living tissue. Wöhler's discovery was therefore the first recorded instance of the production of a "life substance" in a test tube, an environment from which the vital force was obviously absent! This example led other scientists to laboratory production of an extensive list of substances that were supposedly impossible to create without the supernatural force that was thought to reside in the cell. Even so, the theory of vitalism endured until the infant science of biochemistry began to point the way toward a *mechanistic* explanation of life processes.

How does the mechanistic view differ from that of vitalism? Let us use the automobile as a rough analogy that may help explain this difference. A modern automobile is likely to be regarded by its owner as close to a living thing, fully eligible for all the tender loving care that is lavished upon it. Indeed, a fine automobile does create the illusion of having life as it twists and turns through traffic or hurtles at high speed down a highway. And yet, what driver has not been greeted by an ominous—indeed deathly—silence when the starter is engaged? And who has not gazed in frustration at a stalled car suddenly transformed into a useless mass of metal? Is the engine "dead" when it is stalled and "alive" when it is running? Did some automotive engineer at the factory endow it with a "breath of life"? The answer is obvious when we consider that practically any twelve-year-old boy can explain the operation of the automobile engine in great detail, and *in terms of basic laws of physics and chemistry!*

At this point you may feel vaguely uncomfortable with our imperfect analogy because you suspect, and rightly so, that an automobile, complicated though it may be, is hardly in the same league with a human being, or with a cockroach, or even with the lowly amoeba. Even so, our analogy helps to illustrate the point, and that is that the *mechanist* rejects the notion that a mysterious, unexplainable force is basic to life. He believes instead that all life processes can be explained in terms of physical principles.

How then can we define life? To attempt an easy definition is to enter a quagmire from which many a writer has emerged considerably more muddy than when he began. Some try to define it in terms of function, pointing to such supposedly unique characteristics as growth, reproduction, or irritability. But they inevitably find themselves apologizing for the many exceptions that appear to cross the fuzzy boundary between the worlds of the living and the nonliving. Perhaps it might be better to simply admit that complex concepts such as life and love do not lend themselves to neat and orderly textbook definitions.

Still, in keeping with our automobile analogy and the mechanistic view of life, we can regard a living organism in terms of complex chemical reactions taking place at a specific level of organization of matter. Taking this as a starting point, we might then explain the difference between the turtle and the rock in terms of chemistry, pointing out that the matter comprising the turtle is much more highly organized than that making up the rock. Certainly this is a greatly oversimplified approach, but it has some merit as a beginning. If living things differ from inorganic, inanimate objects according to the *level of organization* of the matter of which they are composed, then the "nonliving" world blends imperceptibly into the "living" world. The problem arises when we attempt to establish a well-defined boundary.

The Elements of Life

All forms of matter—the "stuff" of the universe—is composed of one or more of the elemental substances that presumably were present at the dawn of creation. This is consistent with the mechanistic concept since we can say that the stars, the planets, our own Earth, and the living things that originated from it are all composed of various combinations of the same basic elements. Water, for example, consists of the elements hydrogen and oxygen. The rock that you pick up in a field may be composed of calcium, carbon, and oxygen. The air we breathe contains oxygen and nitrogen in addition to smaller amounts of other substances. But it is also true that carbon, oxygen, hydrogen, and nitrogen are the principal elements that make up living tissue and participate in life processes. In fact, about 95 percent of your body tissues consists of various combinations of these four substances.

In addition, there are other common elements that play important roles in the living organism. Calcium, for example, is a known component of bones and teeth. Phosphorus is essential to the production of energy and it is also a part of the gene. Magnesium is a basic ingredient of chlorophyll; this element is therefore essential to the very existence of life on earth because of the importance of photosynthesis in tapping the sun's energy.

Figure 2.1 shows schematic diagrams of the atoms of the four principal elements of living tissue. Keep in mind that these are simplified versions of atomic structures and are used primarily to remind you of how atoms are put together to form compounds. For our purposes, therefore, we will use the familiar "solar system" model of the atom which shows negative particles, *electrons,* moving in orbits around the "sun." The nucleus, you will recall, consists of positive *protons* and neutral *neutrons,* except in the case of the most common form of hydrogen which has only a single proton in its nucleus.

Primarily, it is the number of electrons in the *outermost* orbit of an atom that will determine that atom's bonding behavior, that is, how that atom will combine with the atoms of other elements. Note that carbon has four electrons in its outer orbit, oxygen has six, nitrogen has five, and hydrogen has one elec-

10 Life: its nature and origins

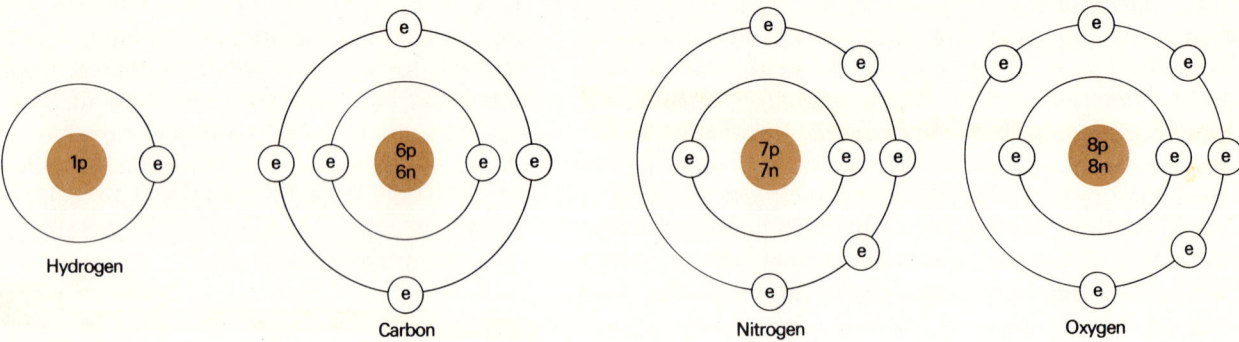

Figure 2.1
Structures of the four atoms that make up about 95 percent of the material of living organisms.

tron in its outermost and only orbit. Now look carefully at the structure represented in Figure 2.2, in which the ×'s represent electrons contributed by the carbon atom and the ○'s represent electrons provided by the hydrogen atoms. Between the carbon atom and each of the hydrogen atoms, you will note there is a pair of electrons consisting of one contributed by the carbon and one contributed by the hydrogen. Carbon and hydrogen have therefore entered into a sharing agreement, whereby each contributes an electron to a bond which exists between them. Thus, carbon contributes one of its four available electrons to an "electron-pair" bond, while each hydrogen atom contributes its one and only electron to the same electron pair. Each pair of shared electrons represents a single covalent bond.

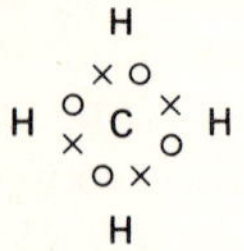

We now have a structural formula that represents a *molecule* of methane (CH$_4$). Note that for each electron-pair bond, we have now substituted a more convenient dash (—).

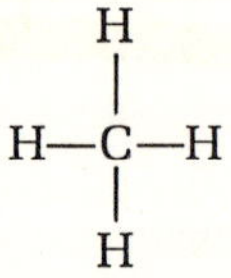

$$\begin{array}{c} H \\ | \\ H-C-H \\ | \\ H \end{array}$$

The above version of methane's structural formula is limited because it is all that can be shown on a flat piece of paper. To show more of the features of the methane molecule would require the use of a three-dimensional model.

As a further illustration, let us consider the formation of another compound called ethylene. Ethylene has the formula C$_2$H$_4$ and is bonded in the fashion shown below.

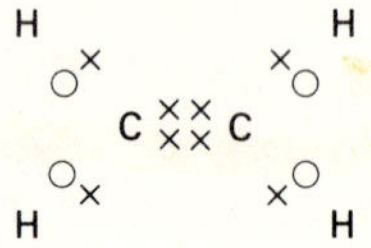

Figure 2.2
Structure of the methane molecule (CH$_4$). Note that the carbon atom contributes one electron (×) and each hydrogen atom provides its only electron (○) to each bond pair.

This time you can see that two carbon atoms and four hydrogen atoms are involved. A bond pair is formed by a sharing process between each carbon and hydrogen as before, but now each of the carbons has two of its original four available electrons left over. What now? In this case, each of the two carbons contributes its two remaining electrons to a bond between the carbons, but since this produces *two pairs of shared electrons, it represents a double bond.* The structural formula for ethylene is therefore written as follows:

$$\begin{array}{c} H \\ \diagdown \\ C=C \\ \diagup \diagdown \\ H H \end{array}$$

It is evident that when covalence is involved, the bonding power of an element depends on the number of electrons in its outermost energy level. We can therefore show the usual bonding behavior of the four principal elements of the living cell by:

$$H— \qquad —\overset{|}{\underset{|}{C}}— \qquad —\overset{|}{N}— \qquad —O—$$

Now we can briefly consider some of the major types of molecules that are specifically involved with life processes. When looking at their general structures, be sure to recall the bonding power of the four major elements shown above.

Some Major Molecules of Life

Carbohydrates

Carbohydrates are compounds that are composed of carbon, hydrogen, and oxygen. We should also note that the hydrogen and oxygen found in carbohydrates are almost always in a proportion of two to one, respectively.

One of the most important carbohydrates is *glucose*, and in Chapter 10 we will see how this compound is essential to the production of life energy. Glucose has the composition $C_6H_{12}O_6$, and its structural formula may be written as follows.

$$\begin{array}{c} H OH \\ \diagdown \diagup \\ C \\ | \\ H—C—OH \\ | \\ HO—C—H O \\ | \\ H—C—OH \\ | \\ H—C \\ | \\ H—C—OH \\ | \\ H \end{array}$$

Many of the bonds are left out of the above formula for the sake of convenience and space, but you should be able to supply them if you recall the "bonding powers" of carbon, hydrogen, and oxygen.

Glucose is regarded as a simple sugar or *monosaccharide*. Other common simple sugars include *fructose*, or fruit sugar, and *galactose*. The combination of two monosaccharide molecules forms a *disaccharide*. For example,

and

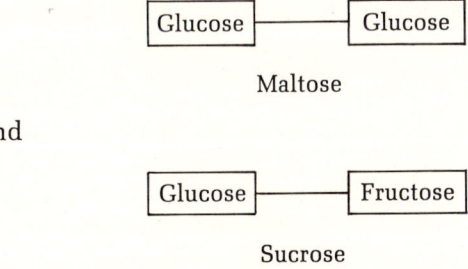

Sucrose is cane sugar, the kind we have on the table. Galactose and glucose combine to produce lactose, which is milk sugar.

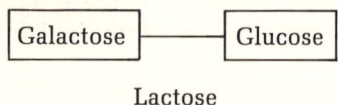

Lactose

When a large number of monosaccharide molecules are linked together, the result is a large molecule called a *polysaccharide*. Some important polysaccharides include starches, glycogen, cellulose, and chitin. Cellulose is an essential constituent of plant structure; glycogen is the form in which carbohydrate is stored in the animal body, and chitin is well known to lobster lovers because it is the material of which that animal's exoskeleton ("shell") is made.

Lipids

The category of compounds known as *lipids* includes materials such as fats, oils, and waxes. Fat is well known as the material to which extra food is converted for storage in the body, much to the disgust of the lobster lover who must watch his weight. Fats are excellent sources of energy, and pound for pound they produce more energy than any other food material.

The typical fat molecule consists of three molecules of fatty acids and one molecule of glycerine. You may have heard that glycerine, or glycerol, was a by-product of the soapmaking process which our ancestors carried on and in which fats were chief among the ingredients. The molecular building blocks of fats are shown at the bottom of the page, using stearic acid as an example of a fatty acid. When we combine three molecules of stearic acid with a molecule of glycerine, we will have what appears to be a very large molecule, but it is relatively small and simple when compared to many other types of molecules found in living systems.

Proteins

If any single kind of substance can be described as most fundamentally important to life, that substance is certainly *protein*. Even the word itself can be translated to mean "of first importance." Proteins are important to body structure. They also serve as enzymes, the significance of which will be discussed in Chapter 7.

The major building blocks of protein molecules are called *amino acids*, and there are approximately 20 such amino acids that are important to various forms of life. Now, with twenty *different* kinds of building blocks, it is obvious that proteins can be in-

Stearic acid: $C_{18}H_{36}O_2$

Glycerine: $C_3H_5(OH)_3$

finitely more complex than polysaccharides, which consist of chains of only one or two basic kinds of units.

The basic structure of amino acids may be illustrated by the following general formula.

$$H_2N-CHR-COOH$$

If you look at the left-hand portion of the molecule, you will see an NH₂ group—this is called an *amine group*. Looking now at the right-hand portion, you will find a COOH group—this is called a *carboxyl radical* and is typical in organic acids. Hence the term *amino acid*. Now, attached to the other carbon atom you will see the letter "R." This symbol stands for the side chain of the amino acid, and it is what makes one amino acid different from another. This in turn contributes to the tremendous variability in the structures of proteins that is fundamental to their essential, multipurpose role in the body. Consider, for example, the structures of the amino acids glycine, alanine, and glutamic acid. You can easily see that these three amino acids have different structures because of the kinds, numbers, and arrangements of the atoms making up their side chains.

Now let us consider the type of bond that links two or more amino acids together. The formation of that bond between glycine and alanine is shown below.

Glycine   Alanine

The dotted rectangle shows how the OH from glycine combines with the H from alanine to form H₂O. This molecule of water is released, and the process is therefore appropriately called *dehydration synthesis*, meaning "formation with removal of water."

Glycine

Alanine

Glutamic acid

This now leaves the two amino acids linked together as shown below.

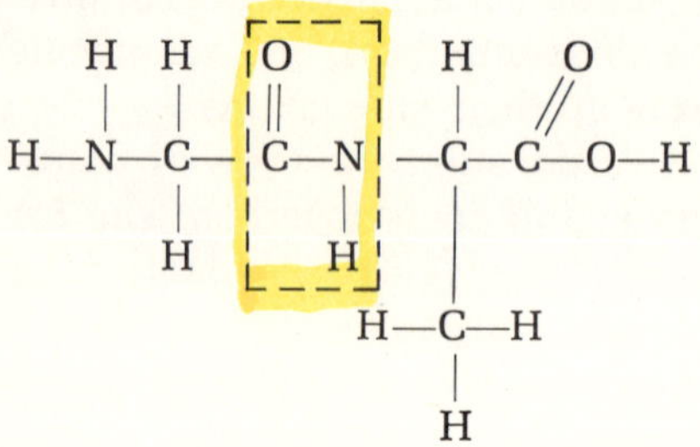

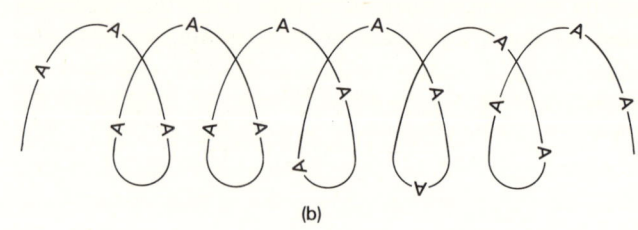

The COHN group outlined by the dotted rectangle is called a *peptide linkage,* and it is a fundamental step in the building of protein molecules. The two amino acids shown above represent a *dipeptide* chain, but if several or, as is usually the case, many amino acids are joined together by peptide linkages, we then have a *polypeptide* chain.

Now, if we can visualize a molecule composed of several polypeptide chains, each chain consisting of a large number of various combinations of amino acids, and if we can further imagine the chains folded in a variety of ways, then we can appreciate how infinitely complex and varied protein molecules can be (Figure 2.3). These complex possibilities of protein structure will be important to our discussion of the gene in Chapter 7.

The molecules discussed in this section are important architectural units of life, but they represent only a glimpse of the complex chemical organization of the living system. You can therefore see that it is helpful to have at least a nodding acquaintance with chemistry in order to appreciate the significance of modern biology. We will therefore continue to review certain chemical principles as the need arises throughout this book.

So far in this chapter, we have emphasized the idea that life is a function of the degree and kind of organization of those same elements that compose the universe in general. Keeping this concept in mind, we can now consider the processes by which life probably originated. Since, according to modern theory, the earth itself gave birth to living organisms, we will begin with the origin of our home planet.

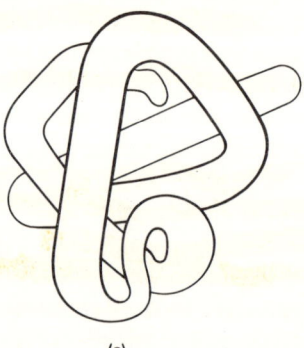

Figure 2.3
Primary (a), secondary (b), and tertiary (c) aspects of protein structure. Note how the primary polypeptide chain forms a springlike helix which then folds in various ways.

The Earth Is Born

The planet Earth was born nearly five billion years ago. No witnesses were there in the cold silence of space to record that cosmic blessed event, and the details will probably be forever obscured by an incredible expanse of time.

Yet man has always been fascinated by the origins of things, and practically every culture has evolved its own legendary version of creation. The Western world, being largely under the influence of the Judeo-Christian tradition, is most familiar with the story found in the Book of Genesis. There we are told that God created heaven and earth, the sun, the moon, the stars, and all life, ending His labors with the creation of man.

Such a concept of special creation by a supernatural force is outside the realm of science because it is simply not possible to test its validity by scientific methods. A belief in special creation at the hands of a deity is therefore a matter of one's personal philosophy and faith. The scientist, on the other hand, must seek a rational explanation that will be consistent with what is known about natural forces and with established scientific principles.

So far, science has been able to offer only guesses —educated guesses, to be sure—concerning the origin of the earth and the solar system of which it is a part. As a start, we can probably agree that our planet did have an origin by some means or other, since the idea that it has existed forever offends our sense of logic.

It would be wrong to suggest that any one version of the earth's origin enjoys total acceptance at this time, but scientists generally believe that the sun and its family of planets originated from a gigantic dust-gas cloud. Similar clouds of cosmic material are observable today in other parts of the universe, so the idea that the solar system could have originated from such a phenomenon is at least plausible.

This dust-gas cloud, so the story goes, was slowly rotating in space, and as pressure exerted by starlight and the force of gravity existing between the cloud particles brought them closer together, the cloud began to shrink. As it grew smaller, its speed of rotation increased. Small areas of turbulence appeared which were destined to form the nuclei of the planets. The major portion of the cloud material contracted to the center of the mass, creating enormous amounts of heat and pressure. Under these conditions, hydrogen atoms fused to form helium, releasing huge amounts of energy. This central core, essentially an incredibly powerful hydrogen bomb, was the newborn sun.

As time passed, the planetary nuclei gathered stray materials from the cloud and they increased in size by a process called *accretion*. That is, they grew simply by adding to their bulk material gained from the dust-gas cloud environment. One of these planets, the third from the sun, was to develop into Earth.

Once again it should be emphasized that the foregoing *dust-gas cloud hypothesis* should not be regarded as a final pronouncement concerning our planet's origin. In its most basic form, however, it is generally considered to be a logical explanation of the origin of the solar system, though not all experts agree on the details. In general, it appears to be consistent with known physical principles; furthermore, it fits our modern concept of a gradual, natural evolution as opposed to a sudden, cataclysmic creation brought about by supernatural forces.

The dust-gas cloud hypothesis also carries the rather startling implication that a similar process, occurring in other parts of the universe, has formed other suns with their own families of planets. In fact, many scientists think that it is not only possible but highly probable that literally millions of solar systems exist in the universe. Therefore, it is also highly probable that millions of planets similar to our own exist somewhere in the vastness of space.

Now, if we accept as plausible the theme which we will develop in the rest of this chapter, which is that life originated as a normal, *inevitable* process involving basic principles of chemical reaction, then we must entertain the logical if perhaps vaguely disturbing possibility that millions of life-supporting planets similar to our own exist somewhere in the universe. Throughout history, man has considered himself and his planet to have resulted from special creation by a lonely deity. Are we indeed special, or is life a common phenomenon of the cosmos and earth merely one of many islands of life in the incredible vastness of space? Perhaps someday we will know. Meanwhile it is at least interesting to speculate on the possibility that we are not alone.

The Earth Primeval

Most of us have given little thought to the remarkably hospitable environment which is provided by the planet that we call home. Notable exceptions have been astronauts who, while walking on the alien surface of the moon, have exclaimed at the beauty of their home planet as it rose above the lunar horizon. While absence makes the hearts of astronauts grow fonder from the vantagepoint of 250,000 miles away in space, the rest of us tend to take for granted the earth's green fields and lush forests, its life-sustaining atmosphere, and the familiar sounds of flourishing life.

If, on the other hand, we could step into a time-travel machine and be transported back in time to when the earth was young, we would find ourselves in an unbelievably forbidding place. We would be uncomfortably hot, the blazing sun would bombard us with intense ultraviolet radiation, and a few breaths of the primitive atmosphere would soon render us as lifeless as the rocks beneath our feet. Even if some magic potion were provided to protect us from the onslaught of this strange environment, we would still be struck by the eerie silence, broken only by some nearby volcanic activity or by the sound of primeval surf striking a rocky shoreline. No familiar birdsong, no cricket call, no rustling leaves, not even the unwelcome whine of a hungry mosquito would break the silence. We would quickly return to our time machine because the earth is not yet ready for us. We must wait a few billion years before appearing on stage.

Still, on that barren earth there would exist an abundance of the basic ingredients from which life would someday be formed. The rocks contain the sulfur, phosphorus, magnesium, and other elements so important to the evolution of living systems. The atmosphere and the seas will furnish the carbon, oxygen, hydrogen, and nitrogen that will comprise 95 percent of earth-dwelling organisms, and ultraviolet radiation together with other energy sources will drive the life-producing chemical reactions.

The Early Atmosphere

Our knowledge of the atmosphere that surrounded the primitive earth is necessarily a matter of speculation, and there is considerable disagreement on the details. One theory, for example, proposes that the ancient atmosphere consisted of carbon dioxide (CO_2), nitrogen (N_2), and water (H_2O). Another theory, however, suggests that methane (CH_4), ammonia (NH_3), water (H_2O), nitrogen (N_2), and possibly some free hydrogen (H_2) were the major ingredients. Advocates of the second theory point out that the most abundant elements—nitrogen, hydrogen, oxygen, and carbon—were heavily concentrated in the materials of the new earth and that hydrogen must have combined with the others to form methane, ammonia, and water vapor. It is also argued that atmospheres similar to this still exist on other planets that are farther from the sun, such as Jupiter.

Neither side has been able to present conclusive arguments, but the details of the early atmosphere are really not that important. What is important is the fact that the primitive atmosphere had to be a *reducing* atmosphere if life was to originate in the way that we think it did. On this, at least, everyone appears to agree. This, then, brings up the terms *oxidation* and *reduction,* which in their broadest meanings involve somewhat complicated chemistry. But for our purposes they can be explained as follows.

Most chemical elements have a tendency to combine with hydrogen or oxygen depending on which one is present in greater concentration. For example, in an atmosphere rich in oxygen, carbon will react with oxygen to produce carbon dioxide (CO_2). The carbon is therefore said to be *oxidized* and an oxygen-rich atmosphere is therefore an *oxidizing* atmosphere. On the other hand, if carbon is exposed to a hydrogen-rich atmosphere, it will react with the hydrogen to produce methane (CH_4). In that case, we say that the carbon is *reduced* and an atmosphere that is rich in hydrogen is a *reducing* atmosphere. What happens, for example, when hydrogen reacts with oxygen to form water? We then have the familiar reaction

$$2H_2 + O_2 \rightarrow 2H_2O$$

and we can see that hydrogen is oxidized while oxygen is reduced. Keeping in mind what we have seen so far, what would be the different fates of nitrogen in oxidizing and reducing atmospheres? In an oxidizing atmosphere, nitrogen would form nitrogen pentoxide (N_2O_5), while in a reducing atmosphere nitrogen would react with the hydrogen to produce ammonia (NH_3).

As we said earlier, everyone agrees that the early atmosphere must have been a reducing, not an oxidizing, atmosphere. Otherwise life could not have originated in the way we think it did. This is based on the fact that organic molecules are unstable in an oxidizing atmosphere and would therefore have been destroyed as soon as they were formed. In fact, it is considered highly unlikely that life could originate on earth today, partly because of our oxygen-rich atmosphere. Incidentally, the assumption of a reducing atmosphere on the primitive earth is supported by the fact that the most ancient rocks contain only ferrous iron compounds. Ferrous iron would have been formed in a reducing atmosphere, while the ferric variety would indicate an oxidizing atmosphere.

There is another reason why it is important to note that little or no free oxygen was present in the early atmosphere. The absence of free oxygen molecules also implies the absence of a special compound of oxygen known as *ozone*. Ozone is called an *allotropic* form of oxygen, because unlike the more common variety it consists of molecules formed by three atoms rather than two. Our present atmosphere contains a layer of ozone that effectively screens out a large portion of the sun's ultraviolet radiation which would otherwise saturate the earth's surface. Ultraviolet is that part of solar radiation that produces a tan—or a sunburn. If you recall your last sunburn, you can appreciate how intolerable it would be if the protective ozone screen were to suddenly disappear. Excessive vitamin D would be produced in the deeper layers of the skin, resulting in brittle bones overloaded with calcium. Perhaps most importantly, the intense ultraviolet would break down the hereditary material that makes up our genes. It is doubtful that we would long survive unless we resorted immediately to some kind of astronaut-on-the-moon existence.

Without a sheltering umbrella of ozone, the surface of the primeval earth must have been bombarded with intense ultraviolet radiation. In addition, it is believed that the heat generated by radioactive materials in the crust induced intensive volcanic activity far beyond that which we see today. Also, as the water vapor cooled and condensed, clouds formed, bringing torrential rains while thunder crashed and lightning illuminated the still barren landscape below.

In the preceding paragraph, we have actually suggested several sources of energy that could have been available to drive the chemical reactions that were to lead to the production of organic molecules. The most likely possibilities were (1) the intense ultraviolet radiation that penetrated the ancient atmosphere, (2) electrical energy derived from lightning, (3) heat produced by volcanic activity, and (4) shock waves originating from claps of thunder.

The formation of hydrogen cyanide (HCN) is a good example of how such energy sources could have worked. First of all, atmospheric nitrogen was broken down into free atomic nitrogen (N) by lightning:

$$N_2 \xrightarrow{lightning} 2N$$

Now, if our assumption that methane (CH_4) was an important ingredient of the ancient atmosphere is correct, we can further assume that the very active atomic nitrogen produced above would have reacted with methane as follows,

$$\underset{\text{Atomic nitrogen}}{2N} + \underset{\text{Methane}}{2CH_4} \rightarrow \underset{\text{Hydrogen cyanide}}{2HCN} + \underset{\text{Hydrogen}}{3H_2}$$

forming hydrogen cyanide, a simple but critical compound that can be the starting point in the synthesis of a number of important organic compounds.

Processes such as the foregoing continued in the ancient atmosphere over a period of time so vast that we find it hard to comprehend. Meanwhile methane, ammonia, and hydrogen cyanide were carried by the falling rain to the earth below, where they mixed with minerals that washed from the barren rocks into the ever-growing lakes and oceans.

The Primeval Seas

Life was born in the ancient seas and was cradled there for the millions of years that were to pass before the first living organisms ventured onto land. Even then, the colonists of the land took part of their watery birthplace with them, and water still remains one of the most important and abundant compounds in living tissue. Your own body, for example, is about two-thirds water by weight, and delicate physiological mechanisms have evolved that ensure that an appropriate amount of water is maintained in your tissues.

This ancient association of water with life processes is related to the unusual properties that make it one of the most versatile of all substances. For one thing, water is one of the most broadly effective of known solvents, and since chemical reactions tend to take place in solution, the solubility of a broad range of substances in water gives it an important role in the thousands of chemical reactions that take place in the cell.

The power of water as a solvent is also basic to its role in the transportation of a variety of substances to and from the cells. Transportation, for example, is the chief function of the blood, which is largely water and which carries food, wastes, antibodies, and other substances in solution to and from the cells in all parts of the body. The blood, in fact, is an important part of that watery environment that we still carry with us. In addition, our cells are bathed in lymph, also mostly water, without which materials would not be able to pass into and out of the cell through its membrane. In order to do so, these materials must be in solution.

Water has one of the highest specific heats of all common substances. This means that it can absorb or release considerable amounts of heat with relatively little change in temperature. Therefore, its presence in the living organism acts as a temperature buffer. Since living tissue is largely water, rapid temperature changes are avoided and a relatively steady body temperature is maintained even in so-called "cold-blooded" animals, such as reptiles, in which the body temperature tends to fluctuate with the temperature of the environment.

As water ran off the primeval rocks and fell into the lakes and oceans, it continued to bring into the ancient seas the raw materials that were to be fashioned into living organisms. Solutions of salts containing phosphorus, magnesium, iron, sulfur, and other compounds added to the stores of dissolved cyanide, methane, and ammonia. The time was ripe for the first stages of molecular genesis.

The production of the molecules that were to be the first level in the chemical organization of life was again mandated by known principles of chemistry. In other words, given (1) the presence of suitable concentrations of the necessary raw materials and (2) sources of sufficient energy, the origin of life in the primeval seas was probably not an accident, but an inevitability.

One of the first serious proposals that life originated from nonliving, inorganic materials came from A. I. Oparin, a Russian scientist. In 1936, Oparin stated his hypothesis in a book called *Origin of Life on Earth*, in which he proposed that the primitive atmosphere contained the necessary materials, and that these materials became part of the "warm, dilute soup." This term was used by another scientist, J. B. S. Haldane, to describe the ancient seas.

Oparin's ingenius hypothesis was regarded as merely interesting until 1953 when S. L. Miller devised and carried out an experiment which was to take its place as one of the classic experiments in biology. Miller, working at The University of Chicago, circulated a mixture containing methane, ammonia, hydrogen, and steam from boiling water past an electrode which provided energy in the form of an electrical discharge (Figure 2.4). You can see that Miller used a mixture of gases similar to that which many scientists believe existed in the ancient atmosphere, and the electrical discharge imitated lightning as an energy source. At the end of a week, the electrical discharge was turned off and the solution was carefully analyzed. As Oparin had predicted, the mixture contained varieties of amino acids, acetic acid, urea, and lactic acid! It was apparent that these *organic* substances, usually associated with living organisms, had indeed been produced by conditions similar to those which we believe existed on the primeval earth.

Similar controlled laboratory experiments have since been performed by Miller, Sidney Fox, and others, and the increasingly regular and predictable formation of "molecules of life" from primordial raw materials lends increasing support to the major points

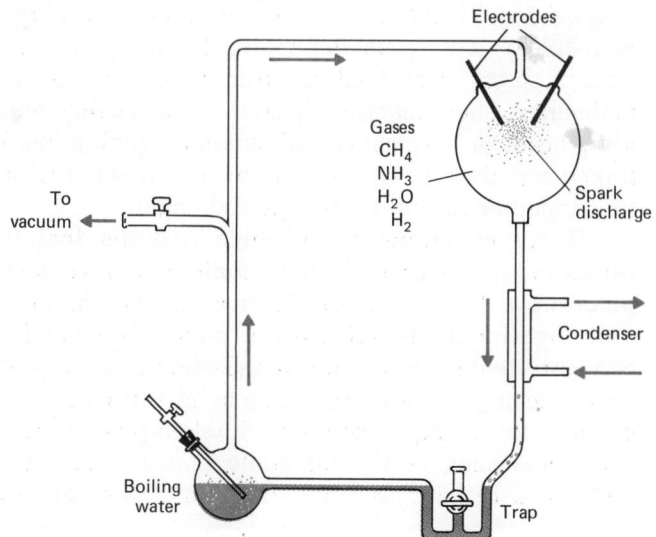

Figure 2.4
Apparatus used in S. L. Miller's experiments on the formation of organic molecules under conditions supposedly similar to those of the ancient earth.

of Oparin's hypothesis. Not only have all the essential amino acids been synthesized, but other fundamental organic molecules as well. For example, it has been shown that a reaction between hydrogen cyanide and ammonia produces not only certain amino acids, but also a compound called *adenine*. Adenine is a critically important constituent of the gene and is also part of a molecule known as *adenosine triphosphate* (ATP), which we will later see plays a fundamental role in energy relationships.

The details of the drama that was played out in those ancient seas are still a matter of speculation, but evidence provided by the experiments of Fox, Miller, and others allows us at least an educated guess as to what really happened. Certainly there will always be doubts and disagreement on details, since it will never be possible to perform an experiment that would exactly duplicate *all* the conditions that existed during the birth of our planet and during the subsequent origin of life.

Even so, if we disregard the controversy over details, we are probably safe in assuming that the "warm, dilute soup" of the primeval seas contained simple organic materials that awaited organization into the larger, fundamental molecules of life. It is likely that we can further assume that the energy needed to drive the building processes was supplied by the ultraviolet radiation from the unshielded sun, by the lightning that laced the sky during thunderstorms, and to a lesser extent by other sources such as volcanic heat and shock waves. Such conditions as these must have virtually dictated that further reactions would continue the chemical organization processes that would ultimately produce life.

The Genesis of Macromolecules

At this point in the story, our best educated guess suggests that Haldane's "warm, dilute soup" must have contained a variety of building blocks from

Adenine   Guanine

(a) Purines

Figure 2.5(a)

which the first major molecules of life could be synthesized. These fundamental materials certainly must have included inorganic salts, simple sugars, amino acids, and two extremely important classes of compounds known as *purines* and *pyrimidines* (Figure 2.5).

It is assumed that the next step toward the eventual origin of life had to be the production of *macromolecules*; that is, smaller molecules that formed in earlier stages combined to form large, or "macro" molecules. These macromolecules are also called *polymers* because they consist of smaller building blocks linked together, as in the case of amino acids which link together to build a polypeptide chain.

Direct experimental evidence suggests that the formation of macromolecules could not have taken place in a dilute solution. It is assumed, therefore, that the simpler building-block materials were first concentrated by some mechanism such as evaporation, freezing, or by gathering on solid surfaces. Evaporation, for example, could have taken place in shallow pools such as the tidepools that are left when high tides recede. Freezing could have been another

Figure 2.5
The molecular structures of (a) two kinds of purines and (b) two kinds of pyrimidines.

Figure 2.5(b)

important concentration mechanism. When water solutions freeze, the dissolved materials tend to separate out in more concentrated form. As a case in point, if the temperature of a mixture of water and alcohol is lowered far enough, the water will freeze, leaving highly concentrated alcohol behind. Still another theory suggests that building blocks of macromolecules could have been concentrated as a result of adhering to the surfaces of clay particles (Figure 2.6) or droplets scattered throughout the solution. Such droplets are spherical in shape and occur in *colloidal* solutions such as milk, which consists of fat droplets dispersed in a water medium. Oparin called these spherical droplets *coacervates*, and he suggested that they may have been an important factor in the necessary concentration of the simpler organic materials that were synthesized in the primeval seas.

Now that the simpler building blocks were brought into closer contact and that the necessary energy sources were present, the stage was set for the generation of forerunners of life in the form of important macromolecules. In this next process, simple sugars combined to produce complex sugars, and amino acids joined to form polypeptide chains in the first step toward the synthesis of complex proteins. Perhaps an even more fundamental event, however, was the combination of purines and pyrimidines with sugar and phosphate to form *nucleotides* (Figure 2.7).

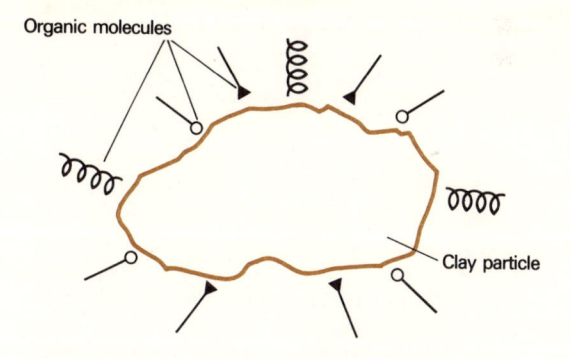

Figure 2.6
Concentration of organic molecules by adsorption on the surface of a clay particle.

Then, as these nucleotides linked together to form *polynucleotide chains,* they eventually produced very special macromolecules called *nucleic acids* (Figure 2.8). The special importance of the events that produced these first nucleic acids becomes obvious when we consider that nucleic acids form the structures that today we call genes. In a later chapter we will discuss nucleic-acid structure and function in more detail as we consider its role in heredity.

Returning to the primeval seas, however, the fundamental significance of nucleic-acid molecules to the formation of life lay in their remarkable and lifelike ability to duplicate themselves. We are tempted here to use the term "lifelike" because, as we all know, reproduction is an extremely important characteristic of *living* organisms. It might be appropriate, depending on one's point of view, to say that nucleic acids represented a borderline between the nonliving and the living, but it would be stretching things a bit to say that the nucleic acids represented a level of organization that would permit us to describe them as "alive" in the traditional sense.

We can now imagine the newly formed nucleic acids drifting in the primordial seas, partially protected by a shield of water from destruction by the intense ultraviolet radiation that continued to bombard the earth. As they drifted they combined with

Figure 2.7
The formation of nucleotides by the combination of a simple sugar, a phosphate group, and either a purine or a pyrimidine.

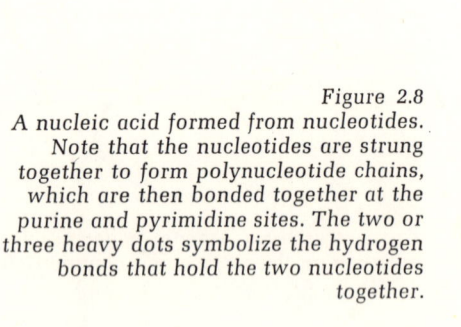

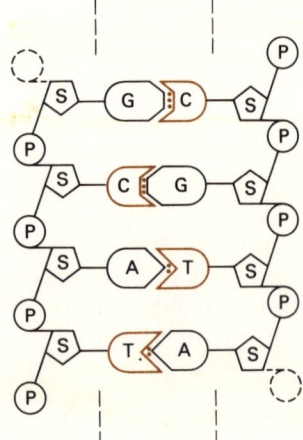

Figure 2.8
A nucleic acid formed from nucleotides. Note that the nucleotides are strung together to form polynucleotide chains, which are then bonded together at the purine and pyrimidine sites. The two or three heavy dots symbolize the hydrogen bonds that hold the two nucleotides together.

proteins, eventually reaching the stage where they *directed the specific structures of proteins* just as your nucleic-acid genes are doing right now. Meanwhile, they continued to duplicate themselves by utilizing the simpler building-block materials that abounded in the "dilute soup" that made up their environment (Figure 2.9). At this point we might throw caution to the winds and even suggest that these raw materials were "food" for the nucleic acids.

Meanwhile, a complication had entered the picture. The energy of ultraviolet radiation had been slowly causing atmospheric water molecules to split, releasing free hydrogen and oxygen. The hydrogen, being light, rapidly escaped the earth's gravitational

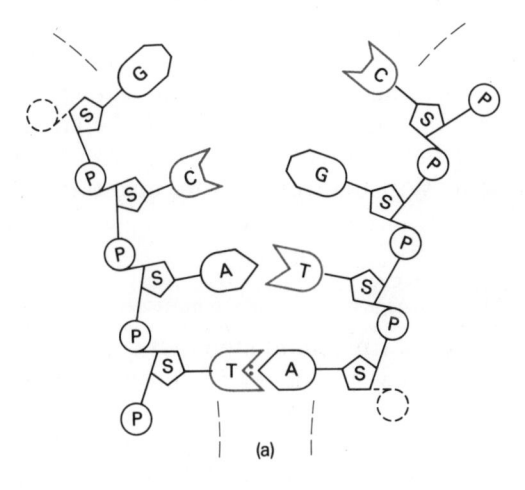

Figure 2.9
Duplication of a nucleic acid. Note how the molecule "unzips" (a). Each polynucleotide chain forms a new partner (b) by using materials from its environment. Step (c) shows the duplicated nucleic acid molecules.

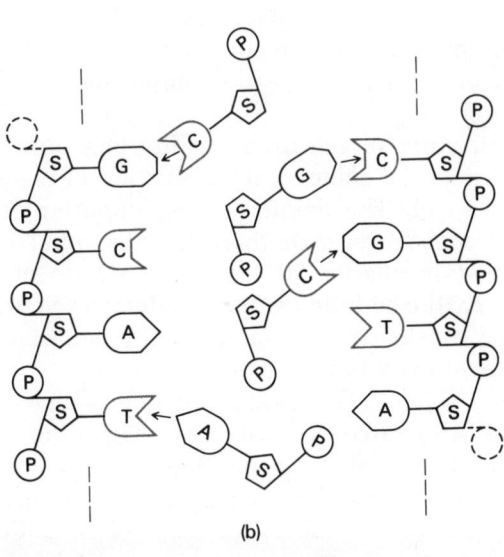

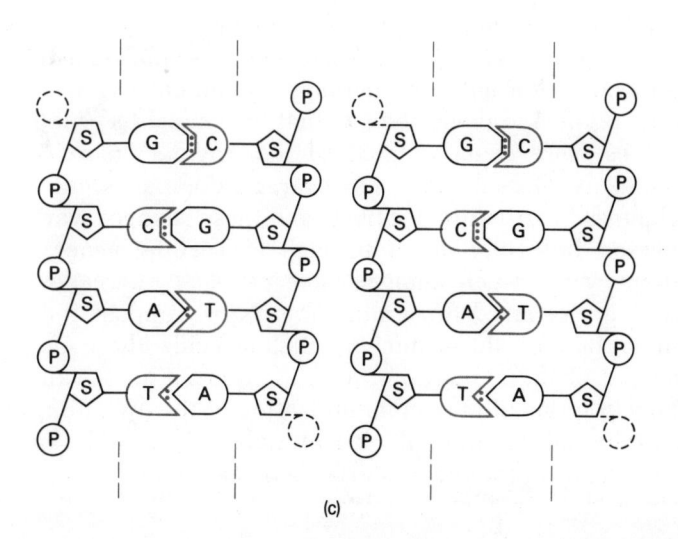

field and was lost to space, but some of the free oxygen was converted to ozone. Slowly a layer of ozone was built, forming an increasingly effective shield against the sun's ultraviolet radiation. Ultraviolet radiation, then, caused its own undoing by building an ozone umbrella that drastically reduced its intensity at the earth's surface. This reduction in ultraviolet intensity undoubtedly must have had two very important consequences: (1) the chemical consistency of nucleic-acid structure was much more safe from destruction by the ultraviolet radiation, just as the nucleic acids of your genes are today, and (2) the reduction of ultraviolet energy for the reactions that produced purines, pyrimidines, sugars, and amino acids resulted in a gradual decrease in the "food" molecules needed by the nucleic acids to carry on self-duplication.

The earth was facing its first famine.

Molecular Evolution

Like overabundant sheep in a poor pasture, the nucleic acids found themselves competing for the dwindling supply of organic molecules that made up their "food supply." At that point, a kind of molecular evolution process apparently began in earnest.

Like the genes in our own cells, the nucleic-acid molecules normally, or perhaps we should say usually, replicated themselves so that the "daughter" molecules were exact replicas of the original "parent" molecule. Occasionally, however, random, accidental changes in structure occurred, and these changes were passed faithfully down through succeeding generations. When such changes take place in our genes, they are referred to as *mutations*. Now, suppose that in certain of those ancient nucleic acids there occurred chance mutations which conferred an advantage in terms of maximum utilization of the dwindling supplies of building-block materials. We would say that such mutations conferred a *selective advantage*, since they enabled the mutant nucleic acids to better cope with the reduced availability of suitable organic molecules. Thus better able to survive, they would thereby produce whole populations having the new and more advantageous structure. In a later chapter we will see that this concept of *natural selection* is one of the prime forces behind the process of evolution and, indeed, accounts for how man himself appeared on earth.

One possibility, for example, is that certain "families" of nucleic acids developed an ability to direct the formation of polypeptides that acted as primitive catalysts, helping the nucleic acids gather the necessary building-block molecules from the environment. Others probably directed the development of a coat of protein. Viruslike forms were also born, their descendants remaining on earth to plague us still.

The Cell

The big step, a truly giant step that crossed that fuzzy border between nonlife and life, was yet to come. Eventually, through the slow process of evolutionary change, aggregates of nucleic acids developed a system which let them carry with them a part of the environment, all neatly packaged within a membrane that permitted selective communication with the outside world. These were the rudiments of the living cell.

Our story of the origin of life bogs down somewhat when we come to the transition of macromolecules to cell. The details of this important step are highly speculative and there is substantial information that is missing at this point. Quite obviously, however, the cell *did evolve*, producing the organized cellular existence that was critically important to the continued evolution of life.

In his book *The Origins of Life,* which is found in the Annotated Reading List beginning on page 361, L. E. Orgel suggests that nucleic acids may have first colonized by adhering to the surfaces of clay particles or coacervate droplets, or both. Along with this, Orgel raises an interesting point that the first "organisms" may have resembled pincushions (refer back to Figure 2.8).

The development of a true cell, however, required the evolution of a membrane that would surround and concentrate the macromolecules. Furthermore, this membrane had to be *selectively permeable*; in other words, unlike a sheet of glass, it had to be a living membrane that could allow certain materials to cross its boundary but still keep other materials from passing through. In this way, useful molecules would be retained inside the cell and necessary materials such as amino acids, purines, pyrimidines, and sugars could enter the cell from the outside environment.

Just how such a cell membrane evolved is another one of those matters of speculation. One of the more well-known educated guesses involves Oparin's coacervates. Figure 2.10 illustrates the development of a complex coacervate system, showing how electrically charged protein molecules might have attracted water particles to form a concentration of macromolecules surrounded by a primitive membrane. Whether or not this type of system did indeed develop is speculative. The method of formation of the complex living membrane of the cell is an unanswered question and is likely to remain so, but it is at least a possible explanation.

In the evolution of the cell, an efficient system was created in which nucleic acids could cooperate in directing the production of catalysts and thereby control a large number of chemical reactions. In Chapter 7 we will consider the details of how the nucleic acids of our genes control the formation of protein catalysts. We will also see how such catalysts, or enzymes, operate to increase the speed of chemical reactions.

Assisted by these enzymes, the newly evolved cell was able to carry on chemical activities that included the use of organic molecules obtained from the environment as building blocks for growth and repair. Also of major importance was the cell's ability to derive energy from the breakdown of organic molecules such as sugars. By this process, called *respiration*, the cell was able to use part of the energy of ultraviolet radiation and other sources that had been

Figure 2.10
Development of a complex coacervate by the attraction of water molecules to a charged protein particle. A structure similar to this may have been a forerunner of the true cell.

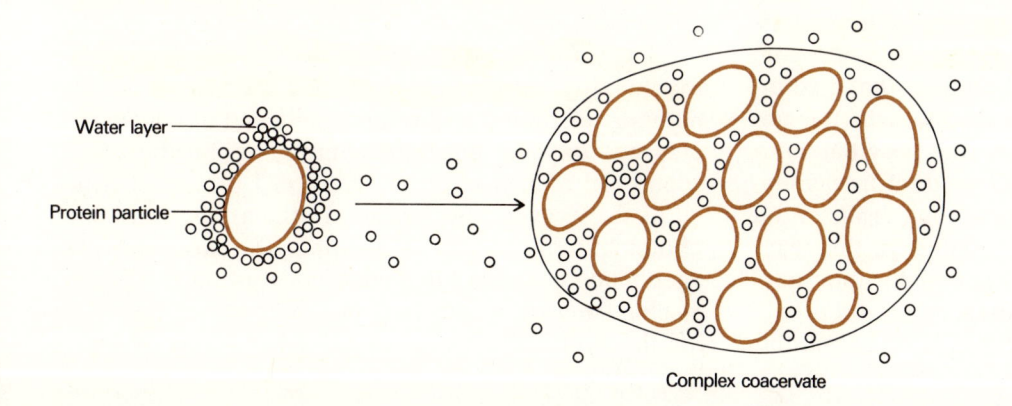

"locked up" in the organic molecules when they were synthesized. We generally associate respiration with oxygen, but since little or no free oxygen existed on the primal earth, it must be assumed that the first cells carried on *anaerobic* respiration, or respiration in the absence of oxygen. Anaerobic respiration is relatively inefficient because only a portion of the energy that is potentially available from an organic molecule is extracted. Even today, however, certain organisms such as yeast and some types of bacteria are anaerobic.

And so, after roughly two billion years of molecular genesis, true cells were drifting in the primordial seas, and life was at long last a reality.

The Grand Strategy

No sooner than it had begun, the emerging life was to face a new crisis. The newly evolved cells needed a constant supply of organic molecules for growth and energy, just as today we need a constant supply of organic food molecules. But the stores of sugars and amino acids were still dwindling in the watery environment because the ultraviolet radiation and other external energy sources had been reduced. At this point it must have appeared that life was doomed to extinction by starvation.

There was something missing at that point. That "something" was a mechanism that would allow living cells to use some reliable external energy source to directly synthesize organic molecules from inorganic substances. As a simple analogy, suppose that a country can no longer import a certain product which is necessary to that country's welfare. One possible solution would be to manufacture the product internally, provided of course that the necessary materials and a suitable source of energy were available. In this way, the country would no longer be dependent upon outside suppliers for a critically necessary product. In a similar fashion, if life was to survive, it was essential that the cell become independent of the diminishing environmental supplies of ready-made molecules.

So the new life that had been born in the ancient seas hung by a tenuous thread, waiting for the miracle (or lucky mutations, depending on your philosophy) that would provide the food necessary to guarantee its continued existence.

It is possible that some of the early forms developed a method of synthesizing organic molecules by using energy derived from the oxidation of inorganic materials such as sulfur and iron compounds, just as certain types of bacteria called *chemosynthesizers* do today. This method of food production, however, is too slow and inefficient to permit organisms to flourish on a large scale.

The answer came in the form of *chlorophyll,* the chemical in present-day green plants that allows them to use light as a source of energy for the manufacture of organic molecules from ever-plentiful supplies of carbon dioxide and water. Those cells that incorporated chlorophyll into their systems thus became the first *autotrophs* (self-feeders)—they were able to make food materials from the inorganic environment by using the energy of light from the sun. Meanwhile, other cells that lacked chlorophyll continued as *heterotrophs* (other feeders), and a balance between these two major nutritional groups was established that persists today. Thus it was that the grand strategy of life emerged, a pattern for existence that we will consider in greater detail in Chapter 3.

Excess oxygen is an important by-product of the foodmaking process called *photosynthesis.* As the numbers of photosynthesizing cells and organisms increased, oxygen began accumulating in the atmosphere. In fact, it is generally agreed that most if not all of the oxygen now present in the atmosphere is of photosynthetic origin. This supply of atmospheric oxygen made possible the development of more efficient mechanisms of respiration and helped set the stage for the eventual emergence of higher forms of life.

Spontaneous Generation

So far in this chapter, we have described the concept and origin of life in chemical terms, and we have suggested that life may be thought of as a highly complex organization of carefully controlled reactions. In other words, life is a natural process that obeys the laws of chemistry.

On the other hand, our description of life as originating from simple inorganic materials may at first glance appear to revive what was once a widely held belief, namely, spontaneous generation. This is the idea that living organisms are regularly generated from nonliving materials. For example, it was at one time commonly believed that rats were produced by garbage, maggots by decayed meat, and that snakes appeared in watering troughs. Even Shakespeare's Hamlet is heard to proclaim, "For if the sun breeds maggots in a dead dog..."

The theory of spontaneous generation had both supporters and opponents among scientists until a little over a century ago when Louis Pasteur performed experiments that supposedly disproved once and for all the concept that living organisms can spontaneously arise from nonliving substances. Pasteur boiled a nutrient mixture in flasks having long necks drawn out in an S-shape (Figure 2.11). As each mixture was boiled, steam condensed and collected in the lower portion of the neck of the flask. As the flask cooled, dust and microorganisms were removed from the air as it was forced by atmospheric pressure back through the water. Since most of the flasks so treated remained clear of bacterial growth until the neck was broken, Pasteur concluded that microorganisms were not generated by the nutrient mixture, but were brought into the mixture by dust particles. Thus for over a century, Pasteur's experiments have been routinely hailed as positive proof that spontaneous generation cannot and does not occur. Thus we have the time-honored biological principle which states that "life can come only from life."

In the interest of good scientific thinking, however, we should point out that contrary to widely held belief, Pasteur's experiments did *not* disprove the concept of spontaneous generation! All he proved was that spontaneous generation would not occur in a specific medium that was treated in a specific way. He then proceeded to generalize his results to include the millions or *billions* of possible material and treatment combinations that had not been tested. Pasteur, who incidentally was a giant among scientists of the nineteenth century and who made immeasurable con-

Figure 2.11
Pasteur's flasks. After the flasks were sterilized, the neck of the first flask (a) was broken and its contents exposed to the air. Later it contained living organisms. The contents of the second flask (b) was not exposed to the air and remained sterile.

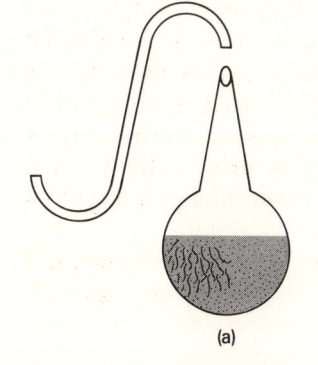

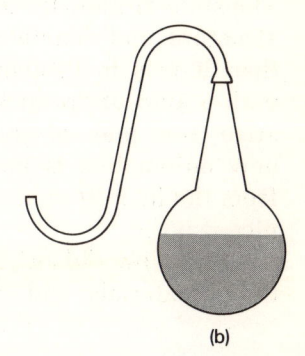

tributions to mankind, was therefore guilty of the common error of generalizing well beyond his experimental data.

Does this mean that the ghost of spontaneous generation has not really been laid to rest? After all, we have spent much of this chapter explaining how we think life *did* arise on earth from nonliving materials. At this point, it is important to make a careful distinction between Hamlet's "maggots arising in a dead dog" and the hypothesis that attempts to explain the origin of life. No modern biologist would entertain the notion that a living organism, no matter how simple, could spring into existence fully formed like the goddess Aphrodite emerging from the sea. On the other hand, to conceive of the origin of life as a step-by-step molecular evolutionary process over a period of two billion years or more is quite another matter.

In point of fact, it is probably impossible to "scientifically" disprove the *concept* of spontaneous generation by experimentation because one would have to perform millions of different experiments in order to account for all possible material-treatment combinations. Even then there could be still other possible combinations that have not even been discovered. Actually, we are probably safe in assuming that the modern biologist has no need of elaborate experimentation to convince him that organisms do not suddenly arise from garbage heaps or nutrient broths. The state of knowledge concerning such things as cell structure and function is much more sophisticated than it was in Pasteur's time, and this provides us with a kind of "prior knowledge" that worms do not arise from piles of grain, just as you "know" that a new automobile is not likely to arise fully formed from the floor of your garage, even though it might be nice if it did.

Still, life *did* originate at least once from nonliving materials, and you may therefore ask how we can be sure that it cannot happen again. Is it possible that new life is even now forming in today's oceans and tidepools? With few exceptions, biologists believe that those molecules that evolved in the primeval seas would be doomed to rapid extinction on today's earth. For one thing, they would be quickly gobbled up by the myriad of organisms that now hungrily roam the earth and its waters. Also, the present atmosphere has a high percentage of oxygen, and organic molecules tend to be unstable in the presence of oxygen. We can assume that the original forerunners of life were spared destruction by earth's organisms since no such organisms yet existed; you will also recall that the ancient atmosphere contained little or no oxygen.

Summary

In a very real sense, life did indeed originate from the "dust of the earth." Inorganic substances from the primeval atmosphere, from the primordial seas, and from the barren rocks were shaped over eons of time into living forms. Recalling the concept that life is in one sense a matter of complexity of chemical organization, we can see that our story of life's origins, filled though it is with gaps and uncertainties, is actually an account of a step-by-step organization of atoms and molecules through levels of increasing complexity.

It is estimated that the origin of simple life forms was completed somewhat over two billion years ago, and from that time to now the diverse forms of earth's life have been hammered out and shaped on the forge of evolution. And so, as we are told in the Book of Genesis, in the beginning the earth was indeed void, and darkness was on the face of the deep. And the waters brought forth abundant life—branching into the various groups and species that inhabit the earth today.

Questions

1. Compare the vitalistic theory of life with the mechanistic theory.

2. Suppose that a biologist happens to be deeply religious in the Judeo-Christian tradition. Could he as a biologist still adhere to a mechanistic view of life? How could he reconcile these apparently divergent views?

3. Was the origin of life on earth an unlikely accident or was it inevitable? Explain.

4. Could events similar to those that occurred on earth have happened, or be happening now, in other parts of the universe? Explain.

5. How do the experiments of Miller substantiate the hypothesis of Oparin?

6. We have said that man is unique in that he is the only animal that knows he must someday die. In what ways has this knowledge affected man's history, culture, and traditions?

7. The abortion controversy has stimulated arguments concerning the specific time at which a human life begins. Discuss this issue in terms of the concepts presented in this chapter.

8. In this chapter we have discussed the nature of life. Assuming that our assumptions concerning the nature of life are reasonably correct, how and in what terms would you describe death?

3

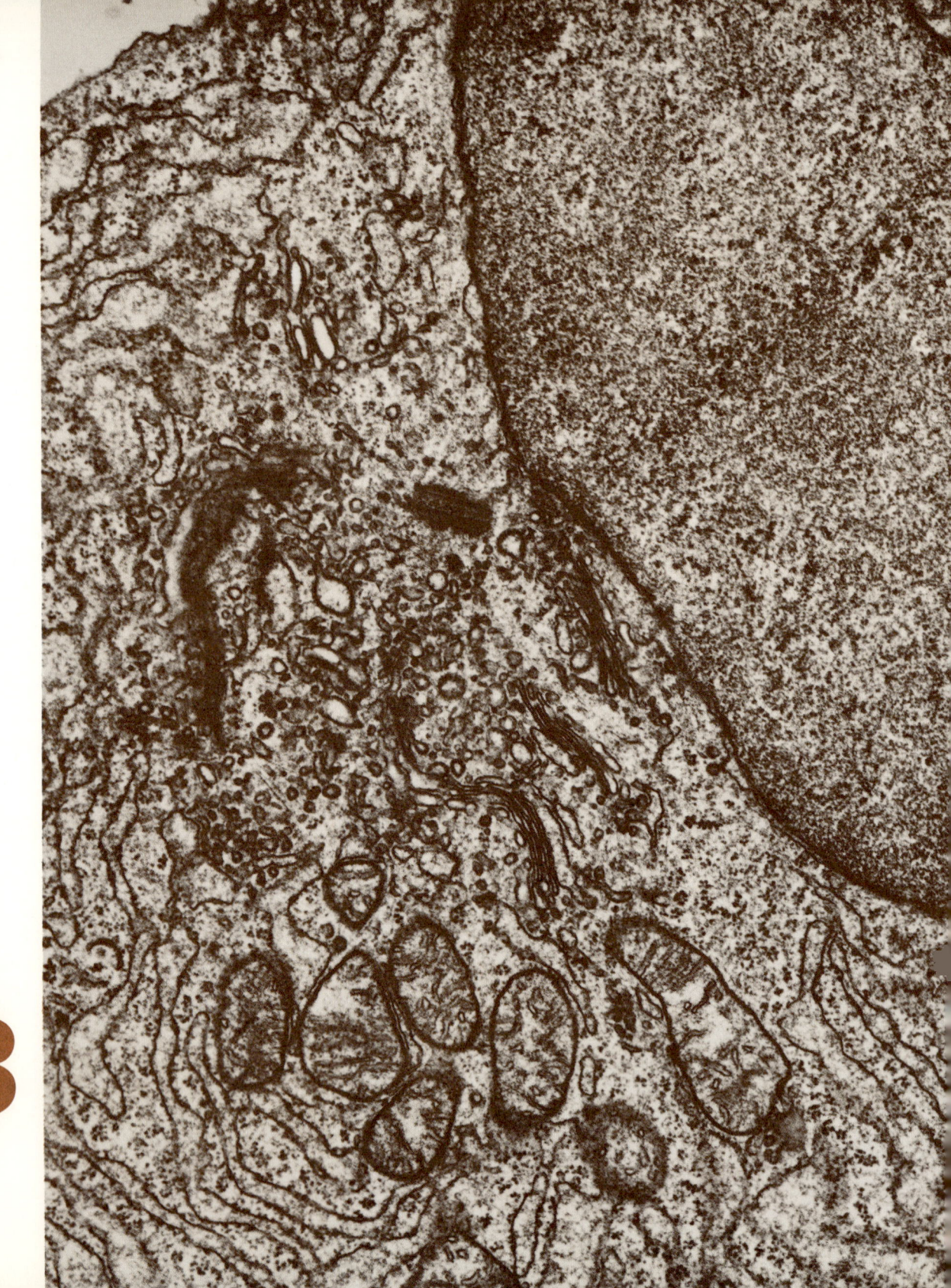

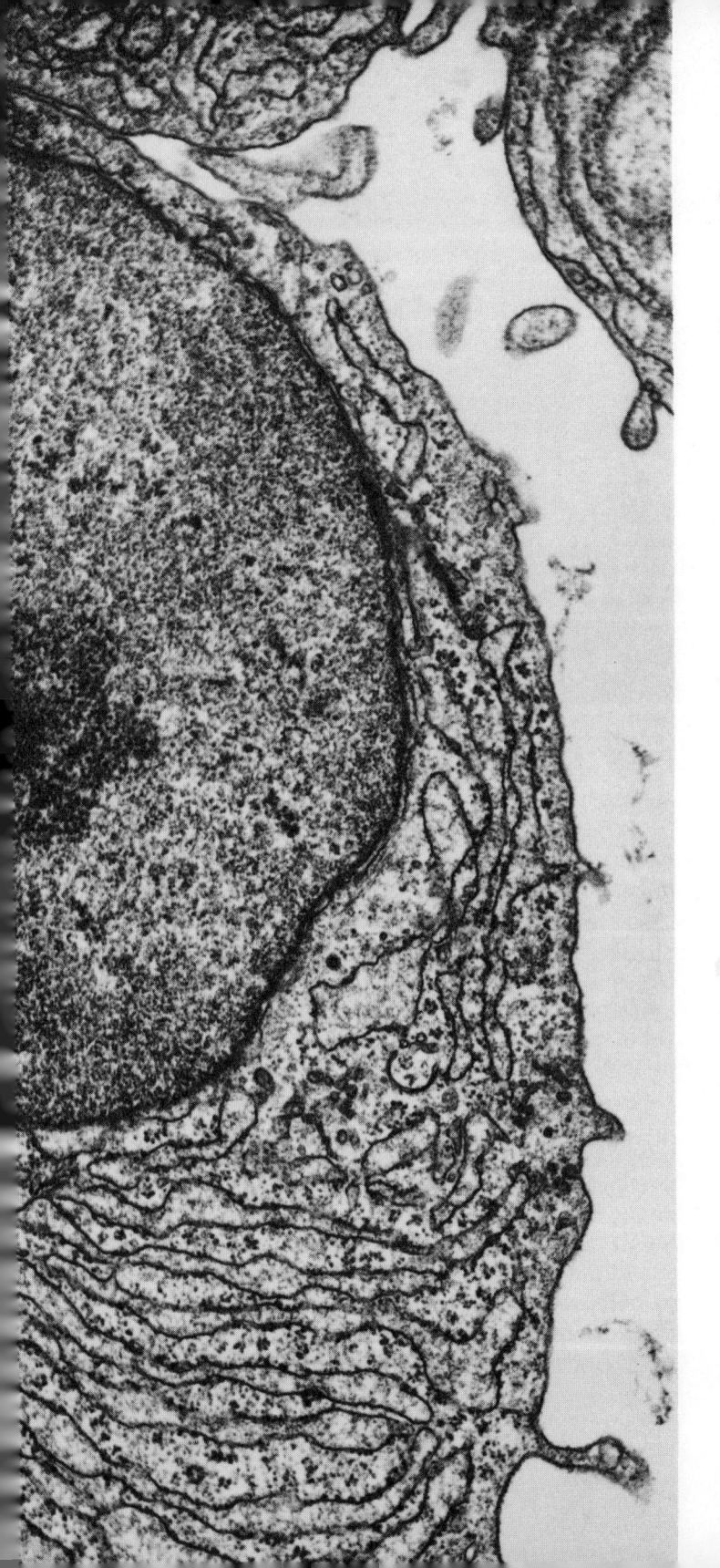

The Unit of Life

The Cell

We have seen that life was finally established on earth when it evolved to the level of the cell. These new cells were self-contained units of life, able to carry on complex activities that assured their continued existence. As time went on, cells gathered into colonies, and over millions of years of evolution they diversified and specialized, forming the tissues, organs, and organ systems characteristic of the multicellular organisms that came to dominate the earth.

The cell is sometimes described as the "building block" of organisms such as man. Unfortunately this concept implies that cells, like the bricks of a house, are merely structural units that give form and shape to the plant or animal of which they are a part. Actually the cell is much more than this; it carries on a variety of complex processes within its boundaries, and a living organism is itself a sum total of those

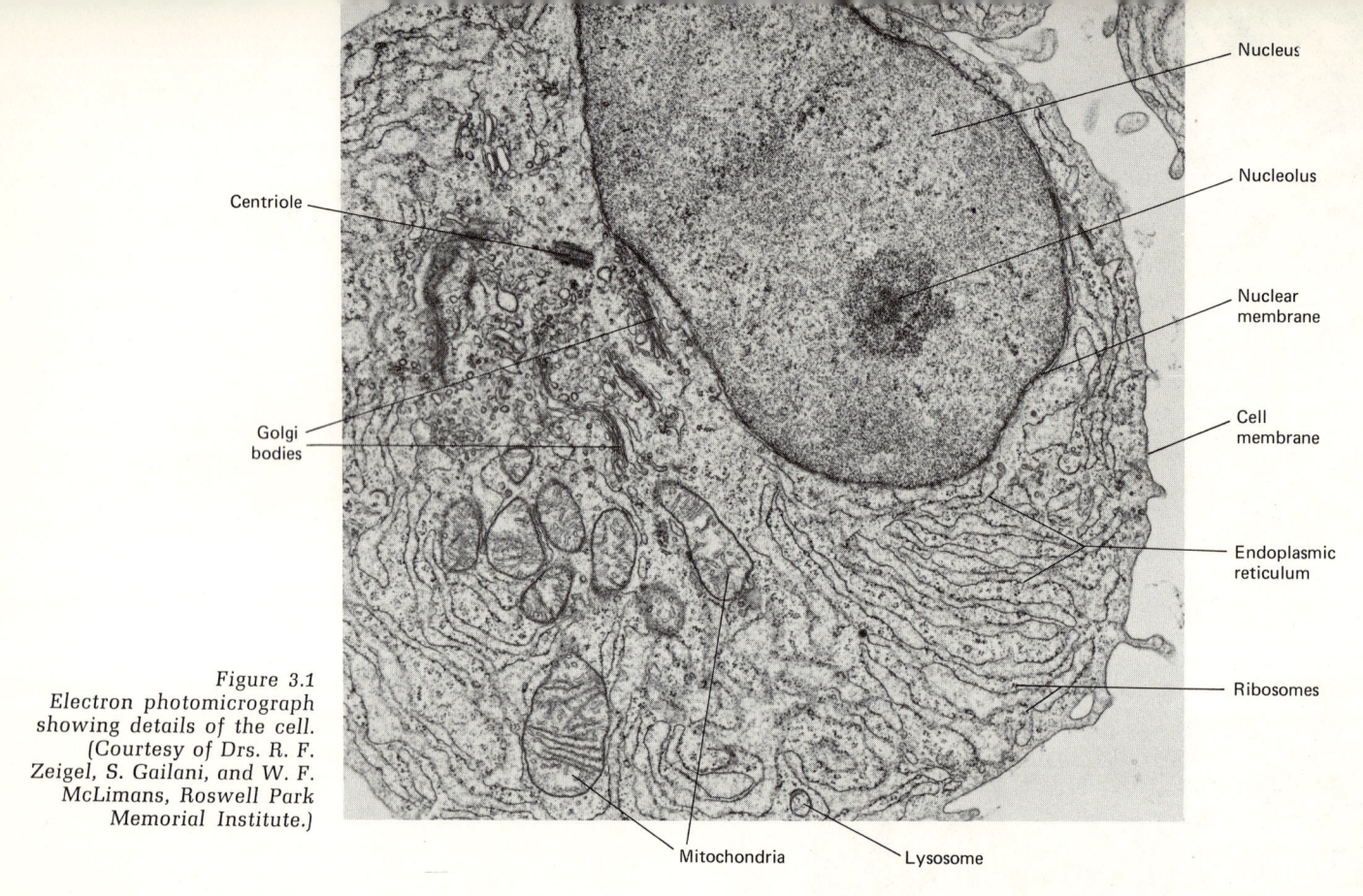

Figure 3.1
Electron photomicrograph showing details of the cell. (Courtesy of Drs. R. F. Zeigel, S. Gailani, and W. F. McLimans, Roswell Park Memorial Institute.)

activities. Complete death, for example, does not occur at the instant the heart stops or the brain ceases to function. We pass forever into that shadowy land only when cell activities cease and the complex chemical organization that we have strived all of our lives to maintain begins its inevitable and irreversible deterioration to the simpler materials of the earth.

In 1665, the British physicist Robert Hooke was looking through the newly invented microscope at a piece of thinly sliced cork. Hooke saw that the cork was like a honeycomb, consisting of small chambers which he called *cells,* from the Latin word for "small room." As time went on and better microscopes were developed, the study of cells intensified. As long ago as 1838, scientists were beginning to regard the cell as the fundamental unit of structure and function in living organisms.

Early microscopes revealed the cell as a blob of jellylike material containing a dense nucleus and held together by what was presumed to be a membrane. Then, as light microscopes became more powerful, biologists began to identify structures which they called organelles. However, even the best light microscope today can magnify a maximum of only about 1500 times, revealing only a part of the details of cell structure. The invention of the *electron microscope,* therefore, provided a welcome addition to the tools of the cell biologist. By using beams of electrons instead of light rays, the electron microscope is capable of magnifications up to 200,000 times! As might be expected, this opened a whole new era of inquiry into the mysteries of the cell (see Figure 3.1).

Merely to be able to look at the physical details of cell structures, however, was not enough to enable scientists to explain their functions. For one thing, only dead cells and tissues can be explored

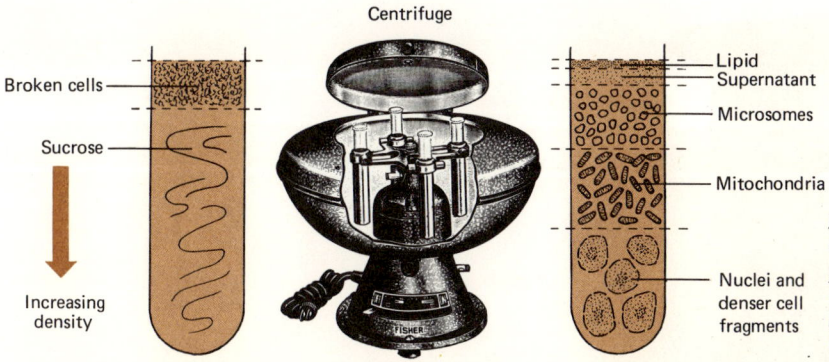

Figure 3.2
Centrifuge with tubes showing how cellular components are separated by centrifugal force according to their weights. (Centrifuge reproduced by permission of the Fisher Scientific Company, Chicago, Illinois.)

with the electron microscope because of the preparation procedures that are required, and dead cells of course do not function. Fortunately, biochemists had been developing methods by which living cells could be literally taken apart and separated into their various components so that their individual functions could be studied and analyzed. A common technique, for example, consists of mincing or grinding up tissue, mixing it with a solution, and spinning the mixture in a *centrifuge* at various rates of speed. Centrifugal force is thereby used to spin down various cell components and molecules according to their weight, thus separating them for collection in pure form for further biochemical study (Figure 3.2). In addition, techniques from radiation biology have also helped provide knowledge of cell processes and activities. Amino acids, for example, can be "tagged" with radioactive atoms and then followed through the various steps of the reactions that occur in the living cell.

These new instruments and techniques have provided considerable information concerning cell structure and function. As always, of course, there is much that remains to be learned.

Cell Structures

The diagram in Figure 3.3 represents a composite or "typical" cell. Actually there is really no such thing as a "typical" cell, and Figure 3.3 is therefore mainly a device to use as a basis for discussion. The living world, not to mention your own body, is filled with innumerable varieties of cell types, each type specialized to perform certain functions. How, then, can we compare a skin cell with a nerve cell flashing impulses to the brain or with a sperm cell whipping its way toward an egg cell and a glimpse of immortality? Actually, in spite of differences in shape, form, and specialization, all cells must carry on fundamental life processes and most of them have a number of basic structures in common.

The Cell Membrane

One of these structures is the *cell membrane*. Even nineteenth-century biologists assumed that the cell was held together by some kind of envelope, or membrane. This assumption appeared logical even though conventional light microscopes could not provide the details of membrane structure. Now, however, the biochemist and the electron microscope together have revealed the cell membrane as a complex structure that selectively regulates the flow of raw materials into the cell and the exit of waste products of cell activities as well as vital cell secretions.

Even now there is considerable doubt as to the specific details of membrane structure, but it apparently consists of lipids and proteins arranged in

34   The unit of life

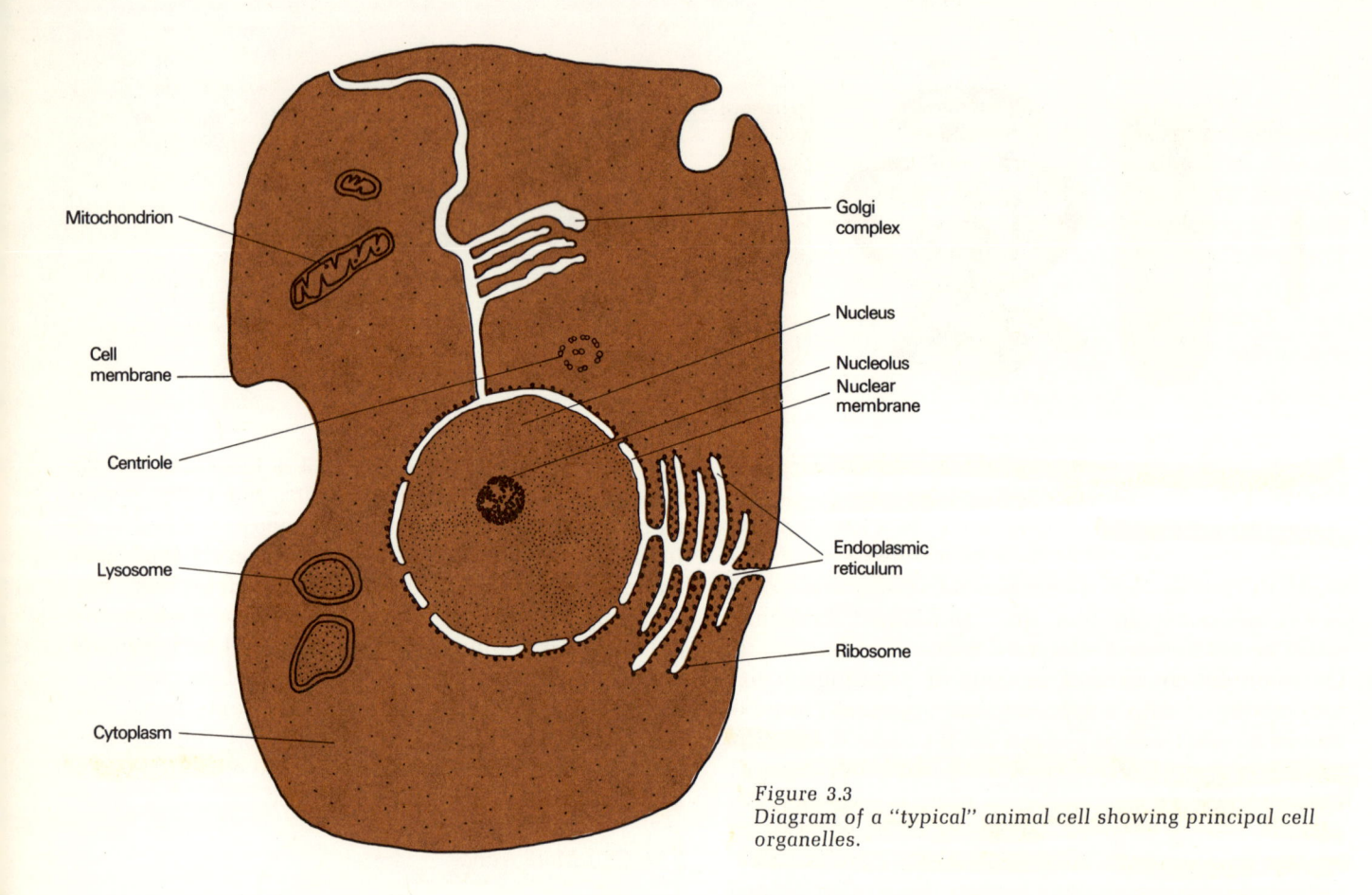

Figure 3.3
Diagram of a "typical" animal cell showing principal cell organelles.

the form of a mosaic. (You will recall that *lipids* is a general term for fatty substances.) Most importantly, the cell membrane is a *selectively permeable* structure; that is, it has the capability to "decide" the rate at which materials can move into or out of the cell at any specific time, as well as the ability to keep still other materials from moving across its barrier at all. To use a very rough analogy, the membrane might be compared to guards posted along the border between two countries. Depending upon the conditions, the guards may permit certain people to cross at one time but not at another. In addition, certain other people may be prohibited from crossing the border at any time.

It is this characteristic of selective permeability that underlies the principal function of the membrane, which is to regulate the flow of materials into and out of the cell. In another section of this chapter, we will take a look at some of the details of how materials are transported across cell membranes.

The Lysosome

Another kind of organelle, called the *lysosome,* contains within its membrane certain enzymes that are capable of breaking down carbohydrates, fats, and proteins. The lysosome is therefore potentially capable of digesting the very cell of which it is a part. This led Christian DeDuve (1963) to call lysosomes "suicide bags," since they apparently provide the cell with a means for self-destruction. Actually it has been suggested that this self-destruction, or *autolysis,* may be a normal part of certain life processes in which the lysosomal membrane may burst and release the destructive enzymes into the cell. Bone growth, for example, involves the systematic destruction of bone cells as well as the addition of new cells. Also, the regression of the tail of a tadpole as it changes to a mature frog may involve lysosomal activity. There is also evidence that lysosomes found in the white blood cells are responsible for the destruction of bacteria that are ingested by these cells. Quite obviously, lysosomal activities must be under very strict control in the healthy organism, but the nature of that control is yet to be understood. We might, on the other hand, speculate that the irreversibility of death may be due in part to a loss of this control, leading to a final self-destruction of the cells as the lysosomal membranes break down. Incidentally, although lysosomes are commonly found in animal cells, they have yet to be clearly observed in plant cells.

The Endoplasmic Reticulum and Ribosomes

The *endoplasmic reticulum* and *ribosomes* refer to a system of channels that extends throughout the *cytoplasm,* which is a term used to describe the general contents of the cell outside the nucleus. They appear to connect with the double membrane that surrounds the nucleus and in some instances appear to lead directly from the cell membrane into the cytoplasm. It has therefore been suggested that these endoplasmic reticulum channels may play a role in speeding the movements of materials into the cell and also from the cell nucleus to various parts of the cytoplasm. The endoplasmic reticulum is lined with small bodies called *ribosomes;* these are known to be the locations at which proteins are manufactured under the direction of the genetic material found in the nucleus. In a later chapter, we will discuss this process of protein synthesis and its relationship to heredity.

The Golgi Complex

The *Golgi complex* appears to be a specialized portion of the endoplasmic reticulum except that it does not contain ribosomes. The exact function of the Golgi complex is not known for certain, but there is evidence that it plays an important role in the secretion or "packaging" of substances by the cell. Gland cells and mucous-producing cells, for example, seem to be well supplied with Golgi bodies. Figure 3.4 shows how the Golgi complex participates in the release of

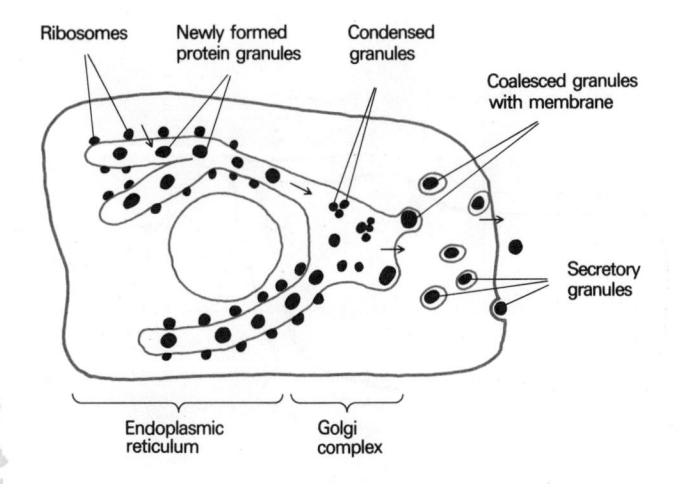

Figure 3.4
Golgi complex participating in secretion of digestive enzymes from the pancreas. (Adapted from A. C. Guyton, Function of the Human Body, 4th ed. Philadelphia: Saunders, 1974, p. 26.)

digestive enzymes by cells in the pancreas. Evidently the enzymes are synthesized at the ribosomes that line the endoplasmic reticulum. Enzyme granules form and then move into the Golgi complex where they balloon outward through the membrane of the Golgi complex into the cytoplasm to form *vesicles*. Each of these vesicles is therefore surrounded by a membrane obtained from the Golgi complex that isolates the enzymes from the rest of the cytoplasm. The vesicles then move to the surface of the cell where the vesicle membrane appears to fuse with the cell membrane and its content of enzymes is discharged from the cell. Thus, although the details of how the Golgi complex functions are not definitely known, the best evidence suggests that this organelle plays an important role in the secretory mechanisms of cells.

The Mitochondrion

Referring back to Figure 3.3, a cell organelle known as a *mitochondrion* is shown. Mitochondria have aptly been called the "powerhouses" of the cell. Each mitochondrion is bounded by a double membrane and contains shelflike folds that project into the interior. These folds are called *cristae*, and they contain enzymes necessary for the uninterrupted production of energy needed for the various activities of the organism and for the maintenance of life itself. It is within the mitochondria that the process of cellular respiration releases the energy that the green plant has taken from the sun and stored in the form of organic molecules such as sugar. The shape, size, and number of these mitochondria may vary in different cell types, but their essential role is the same. As might be expected, cells that are involved with large expenditures of energy are likely to contain correspondingly greater numbers of mitochondria.

In Figure 3.5, we see a plant cell showing the *chloroplasts* that are complementary to mitochondria in the general scheme of life. Chloroplasts are the green structures that contain *chlorophyll*, which is the chemical that enables green plants to "trap" the energy of the sun and store it in the chemical bonds of sugars, starches, and other basic food materials. It should be noted that chloroplasts are confined to plant cells, making all life on earth ultimately dependent upon green plants for its existence. We will further explore this energy scheme in the next section. Chloroplasts are actually one type of plant cell structures called *plastids*; some of these plastids store starch, some provide the color of flowers and fruits, and others give autumn foliage its coloration.

Another structure that is found only in plant cells is the cell wall, which is present in addition to the cell membrane. The *primary wall* is made of the carbohydrate *cellulose*; it is pliable and therefore capable of enlarging as the cell grows. Some plant cells may also secrete a thicker, stiffer wall called the *secondary wall*. The presence of a substance called *lignin* in the secondary wall gives it the kind of strength and rigidity that we associate with woody plants.

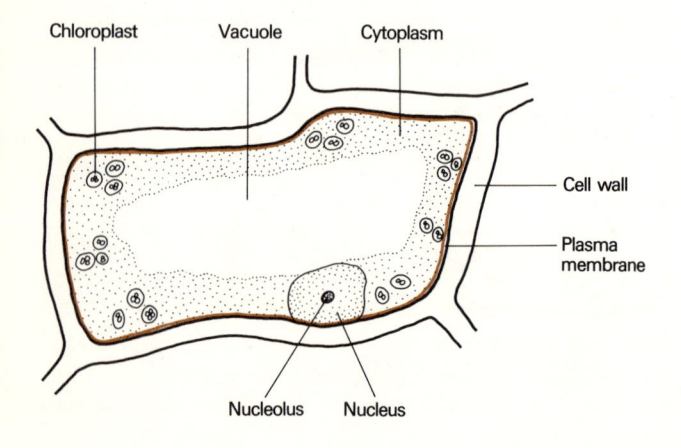

Figure 3.5
Diagram of a "typical" plant cell (a). Note the presence of chloroplasts and a cell wall (b) not found in animal cells.

Energy

Energy is an essential ingredient of life. To maintain life, the cell must carry on an incredible number of processes and reactions, the sum total of which is called *metabolism*. A constant supply of energy is necessary to drive those reactions, and if that energy is not available, the cell will die.

Energy is a concept that is difficult to define. We can perhaps describe it most easily as the *ability to do work*. Work, as the scientist defines it, is the movement of matter through a distance. Work is therefore done when a giant airliner is moved from New York to California. Work is also done when a tiny electron is moved from one part of an atom to another part.

Energy may be described as potential or kinetic. *Potential energy* is *stored energy* and can be represented by the untapped energy in a can of gasoline, in sugar molecules, or in an axe suspended over a block of wood. This potential energy may be converted to *kinetic energy, which is the energy of motion*. For example, as the gasoline or sugar burns, molecules move more rapidly. Potential energy is converted to kinetic energy. Some of this kinetic energy is available to do *work*, the work done by the controlled explosion of gasoline-air mixtures in an engine, or by an axe rushing downward to split a block of wood.

Molecules are specific associations of atoms which are linked together by *bonds*. These bonds result from interactions between the electrons of the linked atoms and contain potential energy. During a chemical reaction, kinetic energy is used to move electrons into different arrangements. When this happens, bonds are altered and different "associations of atoms" (molecules) result. More importantly, during this process there is a release of energy. This energy is referred to as *chemical energy* because it is released during a chemical reaction. Energy released in this fashion plays a major role in life processes. Other more familiar forms of energy include light, electricity, mechanical energy, and heat.

The First Law of Thermodynamics holds that energy can be converted from one form to another, and in the process of conversion the *total* amount of energy is unchanged. In other words, under ordinary conditions energy is neither created nor destroyed, but only changed in form. The conversion of electrical energy by a light bulb into light energy and heat energy is a case in point.

On the other hand, if we burn a quantity of gasoline in an automobile engine, is *all* of the energy that was locked up in the chemical bonds of the gasoline available to do work, to run the car? We know it is not because a cooling system is required to dissipate the large quantity of heat produced. The Second Law of Thermodynamics tells us that no matter how carefully we harness the energy released from the gasoline, some of that energy will not be available to perform work.

Every form of energy discussed so far has two components: one component is the energy available to do work, and the other component is the energy not available to do work. We can term these *useful* and *useless* energy. In any conversion of energy from one form to another (potential to kinetic; chemical to mechanical), both components are released. The useful energy determines the amount of work that can be done. This work may be the growth of a plant, the contraction of a muscle, or the acceleration of an automobile. In all energy conversions, the amount of useful energy is always reduced while the amount of useless energy increases. This increase of useless energy is spoken of as an increase in *entropy*. Entropy contributes to the disordered randomness of the universe. The fact that there is always a net increase of entropy occurring in the universe gives rise to the view that the universe is "running downhill" from an abundant supply of useful energy to a sparse supply. While this generalization holds for the universe "as a whole," there may be certain situations wherein a continuous input of energy (as from the sun) permits the evolution of more and more complex structures—the structures we know as *life*.

In simplest language, the Second Law says that no process of energy conversion is 100 percent efficient. In other words, 100 percent of the chemical energy available in gasoline does not go toward making the car run; part of it is lost as heat. An important biological application of this principle involves the energy relationships in an ecological system. We will discuss this further in a later section, but for now it is of interest to note that the total potential energy contained in the grasslands of a meadow is not available to the deer that graze upon them. By the same token, not all of the potential energy contained in the deer is available to the wolves and mountain lions that may prey upon those deer. Entropy takes its toll at every level.

The Energy Scheme

The appearance of life on earth would have been brief indeed if it had not been for the evolution of a strategy to provide a constant flow of energy with which to drive the countless metabolic reactions that take place in the living cell. This "energy scheme," which utilizes the practically limitless supply of energy from the sun, may be briefly summarized as follows.

In the process of *photosynthesis*, light energy from the sun is "captured" by the chlorophyll of green plants and stored as chemical energy in the molecular bonds of simple sugars such as glucose. Plants then use these simple sugars as building blocks and raw materials for the further synthesis of complex carbohydrates such as sucrose (table sugar), starches, and cellulose. In addition, the plant can convert simple sugars to glycerine and fatty acids, which are then combined to form fats, used primarily for energy storage. In still another type of synthesis, plants combine sugars with soil nutrients such as nitrates and phosphates to make amino acids. These amino acids are then linked together to form proteins, which are major structural and functional materials of the living cell.

The processes by which the plant creates complex molecules from smaller building blocks require energy, just as energy is required to build a wall from stones gathered in the fields. This energy is provided by a complex process called *biological oxidation*, or *respiration*. In this process, the energy that was stored in the molecular bonds of glucose is released. Thus, some of the simple sugar molecules are used to provide energy needed to synthesize the remaining sugar molecules into foods such as carbohydrates, fats, and proteins.

The green plant is therefore self-sufficient because it can provide its own food and energy needs from the two complementary processes of photosynthesis and respiration. This assumes that environmental factors such as light and soil nutrients are available. Animals, on the other hand, lack chlorophyll and therefore cannot carry on photosynthesis. They must ultimately depend upon green plants for the carbohydrates, fats, and proteins that are needed as building materials and energy sources. In one sense, therefore, "all flesh is grass"!

Animal digestive systems break down complex carbohydrates, fats, and proteins into simpler molecules such as glucose, fatty acids, glycerine, and amino acids. These are then used as structural materials and as sources of the energy that is released by respiration.

To summarize, photosynthesis captures light energy and converts it to chemical energy which is stored in the bonds of food molecules. Part of the food molecules that are synthesized during photosynthesis are broken down by the process of respiration, thus releasing the energy to drive metabolic reactions in the cell. The raw materials of photosynthesis are carbon dioxide (CO_2) and water (H_2O), and oxygen is given off as a by-product. Oxygen is used in the process of respiration, which in turn produces carbon dioxide and water. The carbon dioxide and water given off by respiration is thus available to the

Photosynthesis   39

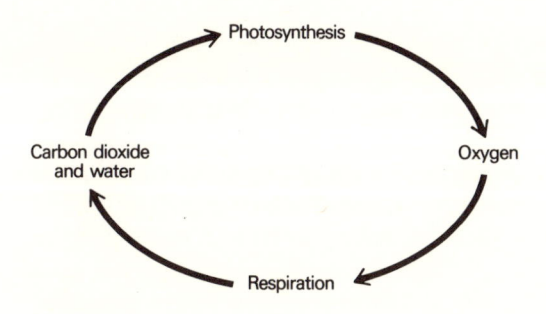

Figure 3.6
The photosynthesis-respiration cycle. (See text for details.)

plant for photosynthesis as part of an endless cycle (see Figure 3.6).

Photosynthesis

The importance of photosynthesis to the grand strategy of life cannot be overemphasized. The process by which the green plant uses solar energy to convert water and carbon dioxide into food is fundamental to the nutrition of all living organisms, with the exception of a few bacteria.

Photosynthesis means literally "to combine in the presence of light." In this process, the energy of light is used to drive a series of reactions that combine carbon dioxide (CO_2) and water (H_2O) into the sugar glucose, giving off oxygen as a vital by-product. This may be summarized by the following "overall" equation for photosynthesis:

$$6CO_2 + 6H_2O \rightarrow C_6H_{12}O_6 + 6O_2 \uparrow$$

The above equation shows that six molecules of carbon dioxide are combined with six molecules of water to produce one molecule of glucose and six molecules of oxygen gas. It is estimated that as much as 200 billion tons of carbon are cycled through this photosynthetic process annually, and it is believed that all of the oxygen now present in the atmosphere could be produced by photosynthesis in approximately 2000 years.

In addition to carbon dioxide and water, photosynthesis requires the presence of *chlorophyll*. Chlorophyll (*chloro* = green; *phyll* = leaf) is a green substance found in algae, mosses, and in the stems and leaves of higher forms of green plants. It is chlorophyll that gives the plant its color. Nongreen plants such as mushrooms and other fungus plants do not contain chlorophyll and therefore do not carry on photosynthesis. In most green plants, the chlorophyll is contained in small structures called *chloroplasts* (Figure 3.7), which tend to be most abundant in cells near the upper surface of the leaf where there is maximum sunlight. It is within these chloroplasts that the chemical reactions of photosynthesis take place.

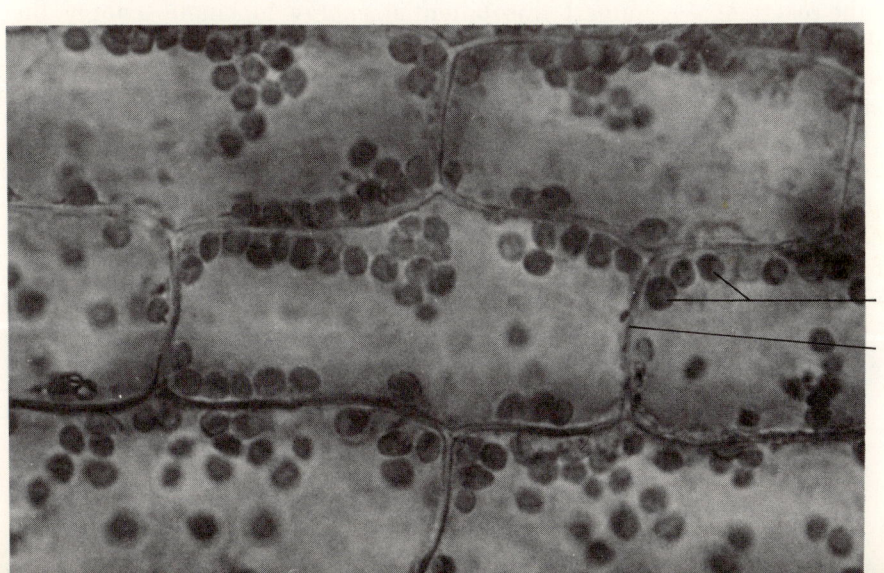

Figure 3.7
Photomicrograph of a plant cell showing chloroplasts.

40   The unit of life

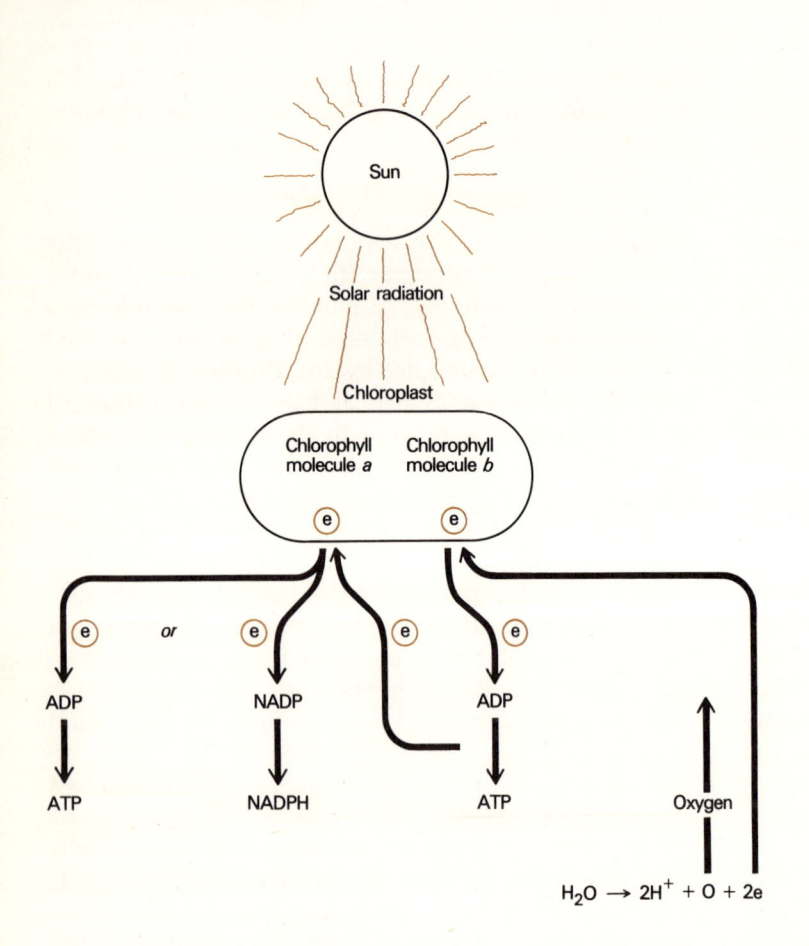

Figure 3.8
Diagram summarizing the light-phase reactions of photosynthesis.

Photosynthesis involves a series of complex reactions that even now are only partially understood. What follows is therefore a summary of the reactions; it must be emphasized that many details are left out.

The reactions take place in two stages: (1) the *light phase* and (2) the *dark phase*. The light phase (Figure 3.8) is based upon the fact that light energy can cause an electron to be displaced from its normal position in a chlorophyll molecule. The light energy from solar radiation causes an electron to move through a distance, thus performing work and imparting energy to the electron. This may be compared to performing work by lifting a book a few feet above a table. The book, like the chlorophyll electron, then contains potential energy relative to the table top. To convert this potential energy to kinetic energy that can do work, we need only to drop the book back to its original position or "ground state" on the table top. In a similar though much more complex fashion, the potential energy of the displaced electron is converted to *chemical* energy as it returns to its ground state in the chlorophyll molecule. This chemical energy is then used to drive further reactions that result in the manufacture of glucose.

Two varieties of chlorophyll, *a* and *b*, are believed to participate in the light phase. When light energy

Photosynthesis

"excites" an electron in chlorophyll molecule a, the electron may be transferred to one of two systems. One possibility is that it may be transferred to an acceptor molecule called *nicotinamide adenine dinucleotide phosphate,* which fortunately is abbreviated NADP. This produces NADPH, which will be used in later reactions in the photosynthetic process. Another possibility is the transfer of the chlorophyll a electron to a system in the cell which returns it in a step-by-step fashion to the ground state. In this process, the energy that the electron gained from the sunlight is converted to chemical energy in the form of *adenosine triphosphate,* or ATP. ATP subsequently provides energy to drive glucose-making reactions in the dark phase. We will say more about ATP in the next section.

If the electron from chlorophyll molecule a combines with NADP, it does not return to the ground state in chlorophyll a. This is evidently where chlorophyll molecule b comes in. When an "excited" electron from chlorophyll b enters the ATP-producing system, it returns to chlorophyll a, thus replacing the electron that chlorophyll a contributed to NADP.

Meanwhile, an electron is supplied to chlorophyll b by a molecule of water (H_2O). In this latter step a molecule of water is split, releasing oxygen, but the details of this water-splitting reaction are still far from clear. At any rate, it may be seen that oxygen is produced during the light phase, and it should be noted that any excess oxygen which is not used by the plant in respiration is released through openings in the leaf called *stomata* (Figure 3.9).

The second phase, called the *dark reactions* (Figure 3.10), involves the combination of carbon dioxide with the hydrogen that was produced during the light phase. The term "dark reactions" does not imply that this phase must proceed in the dark; it is used sim-

Figure 3.10
Diagram summarizing the dark-phase reactions of photosynthesis.

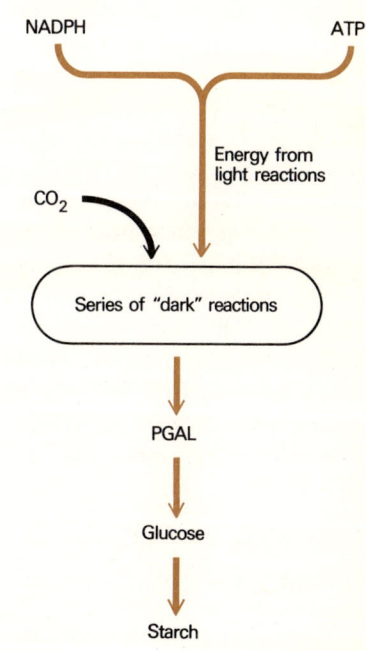

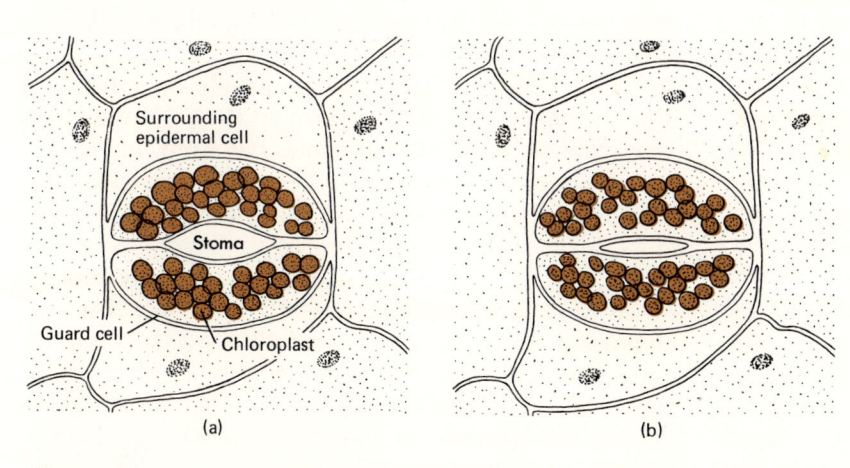

Figure 3.9
Stomata: (a) open and (b) closed. Stomata open when pressure builds within the guard cells that surround the opening.

ply to signify that light does not play a direct role. This second phase involves a complex series of reactions that involve enzymes as well as the NADPH and ATP that were produced in the light phase. We will merely summarize it as follows:

$$CO_2 + 2H_2 \rightarrow CH_2O + H_2O$$

where CH_2O is the basic carbohydrate unit.

At the conclusion of the photosynthetic process, the chloroplast has produced oxygen and a simple form of carbohydrate called *phosphoglyceraldehyde* (PGAL). Both PGAL and oxygen are used directly by the plant cell for its own nutritional requirements and for building or rebuilding the cell's structural components.

On the other hand, the chloroplast normally produces more PGAL and oxygen than the cell needs. The excess oxygen is lost by diffusion through the stomata, and the PGAL is "packaged" into other materials for export to other parts of the plant and for storage in the roots and stems. PGAL is apparently too chemically reactive to be transported as such, and so it is converted to sugars such as glucose, fructose, or sucrose. Excess carbohydrates are stored by the plant in the form of starch, or in some cases as fats such as olive oil or castor oil.

Soil nutrients provide the plant with raw materials necessary for the synthesis of proteins, nucleic acids, and other molecules which are essential to the structure and function of the living organism. These soil nutrients include nitrogen, phosphorus, potassium, and sulfur, as well as traces of metallic elements such as magnesium, which is at the heart of the chlorophyll molecule.

Harvesting the Sun's Energy

We have seen that photosynthesis "traps" solar energy and converts it to the chemical energy locked up in the molecular bonds of carbohydrates and the other food materials available from the green plant. This "locked up" energy is of no use to either plant or animal, however, until it is released and made available to drive the complex chemical processes that maintain life. One way to release the energy contained in a quantity of glucose is to burn it, just as we would burn a piece of wood in a fireplace. When the wood burns, the solar energy that was stored in that wood by photosynthesis is released in the forms of light and heat. Oxygen must be present if the wood is to burn, because the process will not go forward unless oxygen is available to "accept" the hydrogen that is part of the wood cellulose molecule. It is quickly evident, however, that energy derived in this fashion cannot be used by the living cell, because such energy is produced at excessively high temperatures and is quickly dissipated.

Instead, living organisms derive chemical energy from foods such as carbohydrates and fats through a complex series of reactions that constitute the process of *cellular respiration*. Although slower, cellular respiration is similar to combustion in that hydrogen atoms are removed from a carbohydrate and combined with oxygen to yield water (H_2O). In this process, energy is released from the molecular bonds slowly, a few at a time, and a significant portion of the food energy is captured as usable chemical energy in the form of ATP (Figure 3.11). ATP, or adenosine triphosphate, is sometimes referred to as the "energy money" of the body, because *almost all* reactions and processes of the cell *that require energy* must derive that energy from ATP.

In cellular respiration, glucose is partially broken down in a complex sequence of reactions known as *glycolysis*. In the process, some energy is extracted and incorporated into the high-energy bonds of ATP molecules. The glucose fragments that result from this partial breakdown are then carried into the mitochondria, where another series of reactions completes the final breakdown of the glucose to carbon dioxide and water. In the process, ADP (*adenosine*

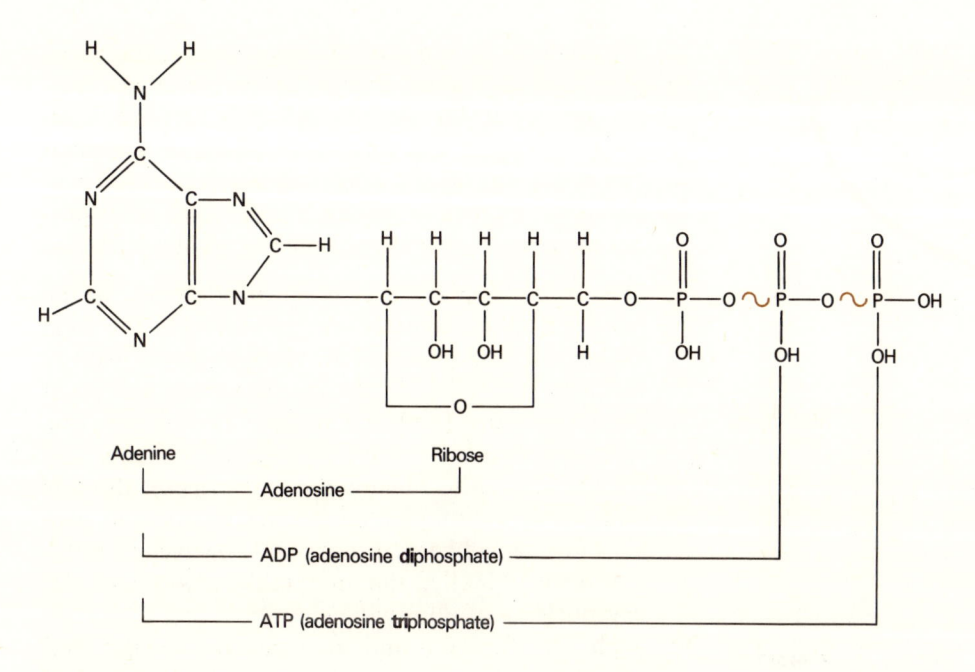

Figure 3.11
Structure of adenosine triphosphate (ATP). Note that three phosphate groups are attached to the adenosine molecule. The third phosphate group is attached with a high-energy bond (∽).

diphosphate) is converted to ATP by the addition of an energy-rich phosphate bond. The energy thus stored in the high-energy bond of ATP is used to drive reactions (Figure 3.12). It should be emphasized that not all of the energy stored in food molecules is converted to the usable energy of ATP. The Second Law of Thermodynamics is not to be denied, and part of the potential energy of food is inevitably lost as heat. This heat is not entirely useless, however, since it helps to maintain the organism's body temperature. To summarize,

1. Energy from food + Phosphate + ADP → ATP + Heat
2. ATP → ADP + Phosphate + Energy for reactions

With the exception of lower organisms such as yeast cells and certain forms of bacteria, oxygen is an indispensable part of respiration. It is used as the final "acceptor" of the hydrogen that is present in the food molecule; without oxygen, life in all higher organisms would soon cease to be.

As we look back at photosynthesis and respiration, we can see that the interaction of these two processes form the basis of the "grand strategy" of life on earth. Essentially, photosynthesis pushes the car up the energy hill (Figure 3.13), and respiration takes it down the hill as it makes energy available for the maintenance of life. We might once again briefly summarize this grand strategy as follows.

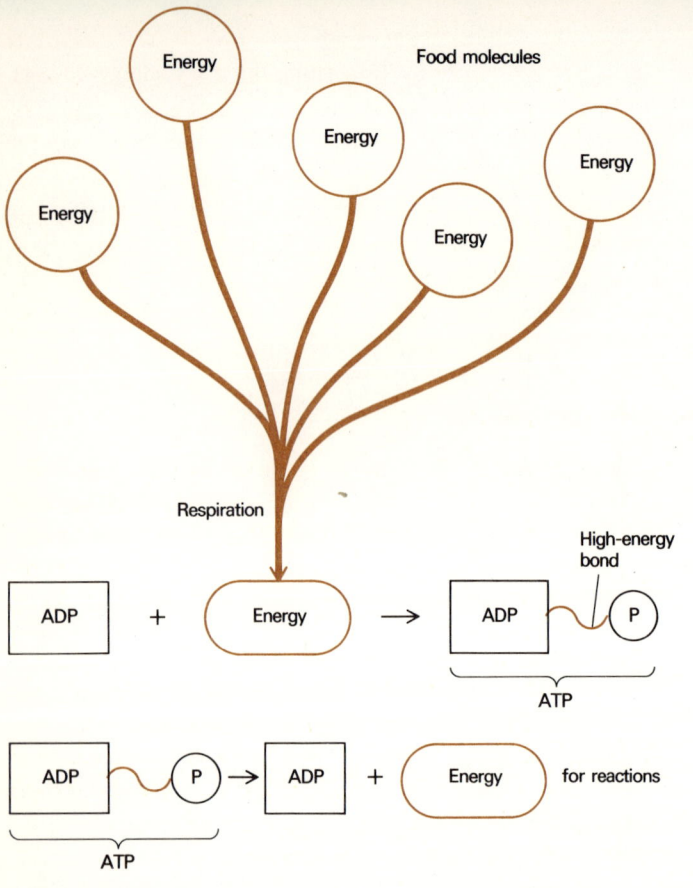

1. Through the process of photosynthesis, the green plant captures solar energy and stores it as chemical energy in the molecular bonds of food materials such as carbohydrates. In addition, the plant synthesizes proteins, nucleic acids, and other fundamental molecules of life. Finally, the plant produces oxygen as a by-product of photosynthesis.

2. In cellular respiration, oxygen participates in the release of the chemical energy stored in the food materials. The green plant thereby gains energy to carry on its life functions; however, the plant produces more food and oxygen than it can use, and this excess is used to support animals and nongreen plants. Meanwhile, the carbon dioxide and water vapor yielded by the respiration process are cycled back to the plant for use in photosynthesis. Finally, the inevitable death and decomposition of all organisms return raw materials such as nitrogen and phosphorus to the soil where they can be reused by the plant as part of a never-ending cycle.

Figure 3.12
The process of respiration extracts energy from the bonds of food molecules and incorporates it into the high-energy phosphate bond of ATP. When ATP is converted to ADP, the energy released is used to drive reactions.

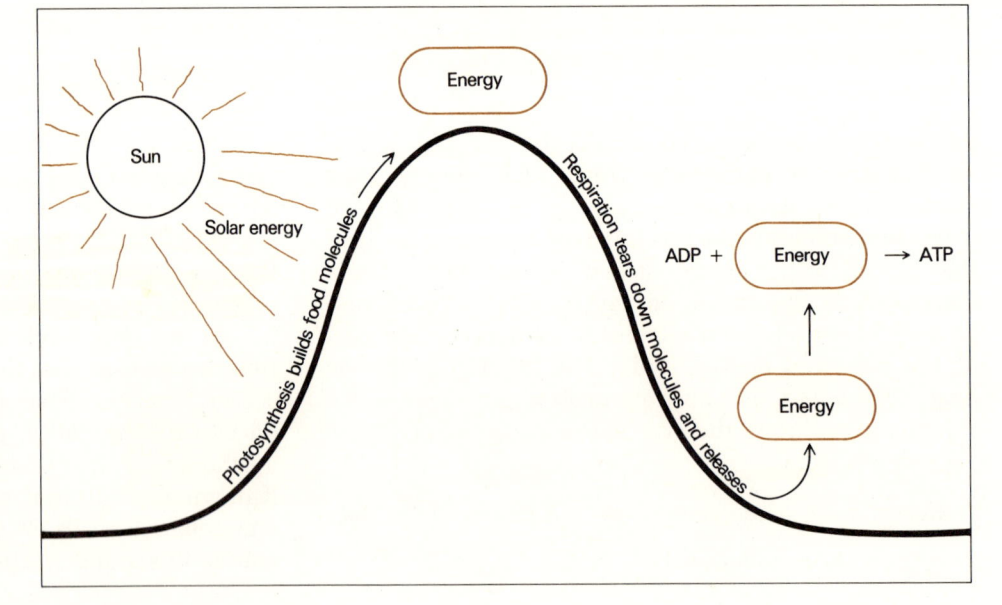

Figure 3.13
Diagram summarizing the energy relationship between photosynthesis and respiration. Photosynthesis builds food molecules and stores within them energy from the sun. Respiration releases that energy and incorporates it into ATP.

The Flow of Cellular Materials

In an earlier section, we described the cell membrane as a protein-lipid structure that selectively regulates the flow of materials into and out of the cell. What controls this movement of various molecules across the membrane barrier? What forces or processes are involved in making certain that the cell has an adequate supply of oxygen and nutrients, and that waste products are efficiently removed?

For one thing, it is considered likely that substances cross the membrane in accordance with their solubility in lipids (fats). Thus oxygen, carbon dioxide, alcohol, and fatty acids are fat-soluble and are able to cross the lipid barrier with no difficulty. On the other hand, substances which are not soluble in fat, such as sugar, do not pass through the lipid portion and may have to be shepherded across by protein molecules acting as carriers. Thus, it is probably the case that both the protein and lipid portions of the membrane play important roles in regulating the entry and exit of materials.

It has also been suggested that water and certain other substances enter the cell through pores present in the membrane. These "pores" should not be thought of simply as holes in the membrane, but rather as channels lined with charged protein molecules that conduct materials into the cell. It is also possible that these pores or channels conduct raw materials directly to various parts of the cell in response to signals that "inform" the membrane that some component is needed.

Diffusion

Molecules are constantly in motion. In fact, the concepts of heat and temperature are regarded by the physicist as expressions of molecular motion. A quantity of *heat,* for example, is measured by the number of molecules in motion, whereas *temperature* is a function of the average velocity with which molecules move. All molecular motion, therefore, would theoretically cease at the point of "absolute zero," or 273 degrees below zero on the Celsius scale.

This constant, random motion of molecules is the basis of a kind of molecular mass-movement called *diffusion.* A time-honored illustration of this principle makes use of a beaker of water with a small amount of concentrated dye placed at the bottom of the beaker. Figure 3.14 shows that over a period of time, the dye is dispersed throughout the water as the dye molecules move *toward* those areas where they are

Figure 3.14
Dispersal of dye molecules throughout a beaker of water.

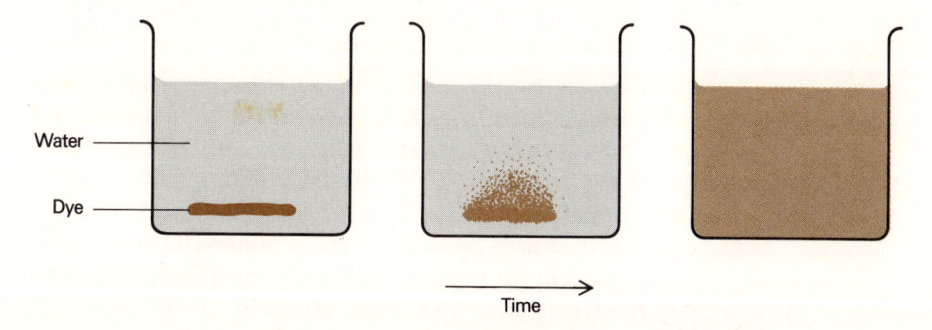

least concentrated. Eventually, as the dye molecules become evenly dispersed throughout the water, the *net* movement of the mass of molecules stops, although the individual molecules continue to move about in a random fashion.

In Figure 3.14 there is therefore a movement from an area of high concentration of dye molecules to an area of comparatively low concentration of dye molecules. This illustrates a concept called the *concentration gradient* or *diffusion gradient*. The ability of a mass of molecules to move from point *A* to point *B* will therefore depend upon the concentration gradient between points *A* and *B*. Figure 3.15 shows three possible situations involving diffusion gradients.

Figure 3.15(a) shows a gradient existing between a high concentration at point *A* to a comparatively low concentration at point *B*. Like a car coasting downhill, the net movement of molecules in this case will be from *A* to *B*, with the result that the concentration at *B* will increase. Figure 3.15(b) shows a different situation—in this case the molecular concentration is greater at *B* than at *A*, and therefore the concentration or diffusion gradient leading from *A* to *B* is uphill. Taking our analogy further, we know that a car will not run uphill *unless energy is expended to make it do so*. In the situation represented by Figure 3.15(b), therefore, there will be no net movement of molecules from *A* to *B* unless external energy is applied. Figure 3.15(c) shows a situation in which the molecular concentrations at *A* and *B* are the same. While there will be equal random movement of molecules from *A* to *B* and from *B* to *A*, there will be no *net* movement of molecules between *A* and *B* because the diffusion gradient is zero.

We can now relate the principle of diffusion to the movement of materials across a cell membrane. As long as the concentration of a specific substance is greater on the outside of a cell than on the inside, there will be a net movement *into* the cell, provided that the substance in question can pass through the membrane barrier. In the same way, materials such as waste products or cell secretions move out of the cell because of their relatively higher concentration within the cell.

As we said earlier, some substances such as certain sugars apparently cannot pass through the lipid barrier because of their insolubility in fats. It is now thought that these substances pass through the membrane by *facilitated diffusion*. In this process, molecules such as those of sugar are transported across the membrane by protein molecules, called *carriers*, much like the way passengers are shuttled across a river by a ferry boat. It should be emphasized, however, that in facilitated diffusion, the substance in question is still moving *down* a concentration gradient, or from an area of high concentration to an area of relatively lower concentration.

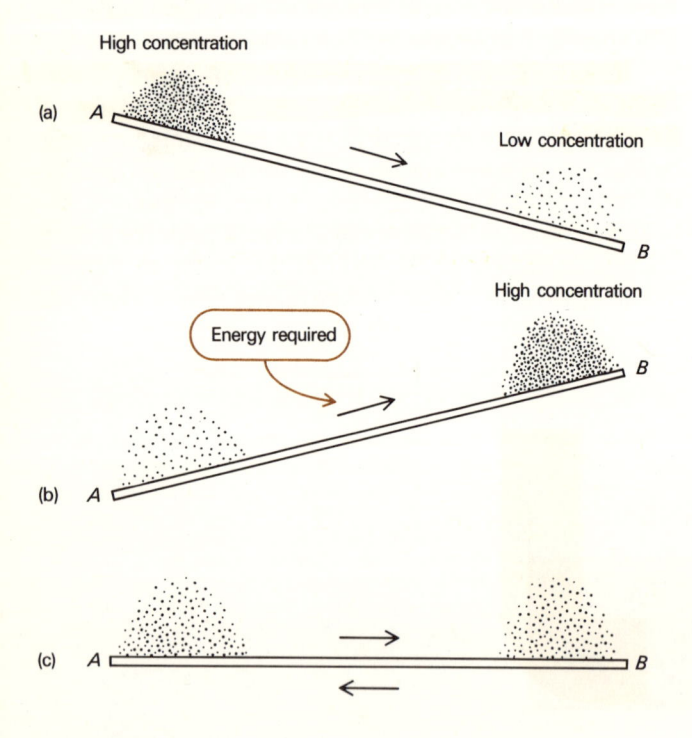

Figure 3.15
The concentration gradient. (See text for details.)

Osmosis

Osmosis may be defined as the *movement of water through a semipermeable membrane.* Again, the direction of net movement of water molecules will involve a concentration gradient, and the net movement will be from that side of the membrane where water molecules exist in high concentration to the side where the concentration of water molecules is comparatively low.

Figure 3.16 illustrates the principle of osmosis by using a U–tube which is partitioned by a membrane that is permeable only to water molecules. In Figure 3.16(a), you will note that the left side of the partition contains distilled water and the right side holds a sugar solution. Since the distilled water theoretically contains only water molecules, we can assume that the concentration of water molecules, that is, *the number of molecules per unit volume,* is very high. On the other hand, since the solution in the right arm of the U–tube consists of water molecules *plus* sugar molecules, we can therefore assume that the number of water molecules per unit volume is relatively lower than that on the left. A concentration gradient therefore exists between the left side of the membrane and the right side. The net movement of water molecules will consequently be through the membrane from left to right, or toward the side where the water has been mixed with the sugar. This results in a rise in the level in the right arm of the U-tube, as shown in Figure 3.16(b). This increase in level on the right-hand side will continue until the pressure due to the height of the water on the right side equalizes the *osmotic* pressure that is exerted by the left side. At that point, there will continue to be random movement of water molecules back and forth through the membrane, but the *net* movement will be zero.

Now consider the three sugar solutions in Figure 3.17. Assuming that the total volumes of all three solutions are the same, it follows that solution (a) has more water molecules per unit volume than either solutions (b) or (c). Following the principle of the concentration gradient, there could therefore be a net movement of water molecules from solution (a) to solution (b) or (c), but since the concentration of water molecules in solutions (b) and (c) are equal, there would be no net movement of water molecules between (b) and (c).

We therefore say that solution (a) is *hypotonic* to the other two solutions, and they in turn are *hyper-*

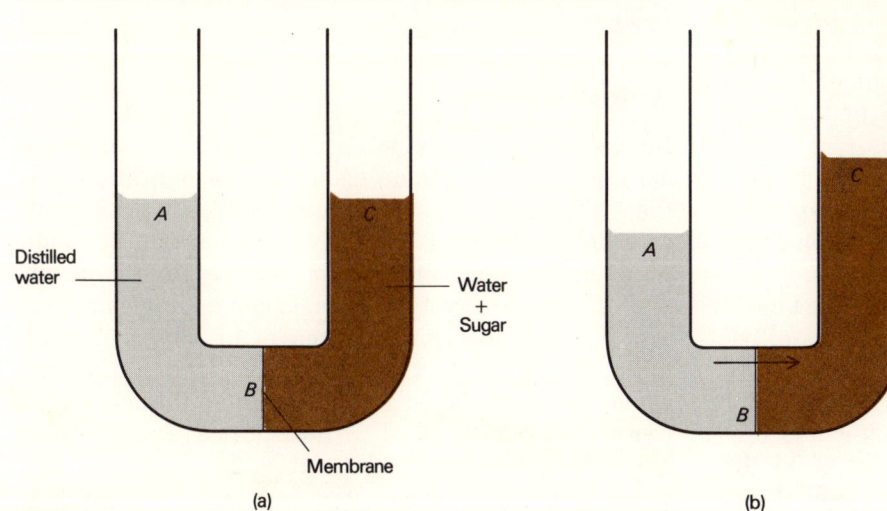

Figure 3.16
Osmosis demonstrated by movement of water molecules through a permeable membrane. (See text for details.)

48   The unit of life

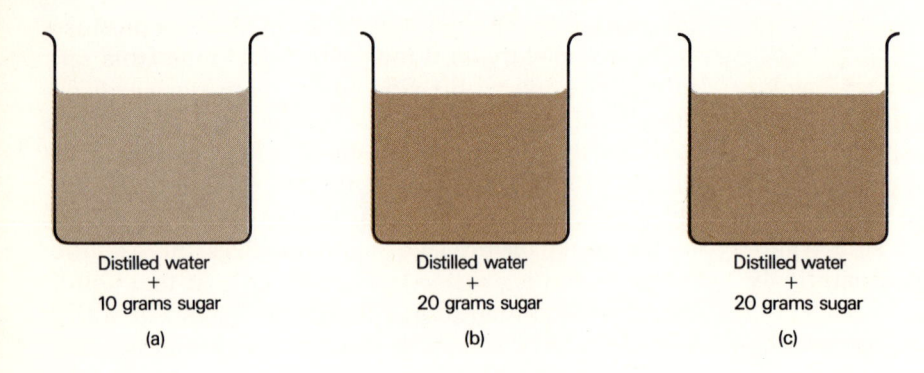

Fig. 3.17
Three sugar solutions. Solution (a) is hypotonic to (b) and (c), and they are both hypertonic to (a). Solutions (b) and (c) are of the same concentration, so they are isotonic to each other.

tonic to solution (a). Solutions (b) and (c) have the same water-molecule concentrations, so they are *isotonic* to each other. Therefore, osmotic movement of water through a membrane goes *from a hypotonic solution toward a hypertonic solution.*

This principle has some interesting applications. You may recall that the Ancient Mariner complained that there was "water, water, everywhere, nor any drop to drink!" If the shipwrecked sailor were to drink the seawater that abounds around him, the concentration of salts in that seawater would be *hypertonic* to the solutions in his tissues and circulatory system. Water would therefore move out of his tissues and he would die of dehydration. In the same way, a fish that is strictly adapted to fresh water would die of dehydration if it were placed in seawater. Marine fish, on the other hand, are adapted to salt water and consequently cannot survive in fresh water.

When injecting substances directly into the bloodstream, the physician must consider the effects of osmotic principles on the blood cells and other tissues. For example, if a quantity of distilled water were introduced into the circulatory system, this would create an environment that would be *hypotonic* to the contents of the red blood cells. Water would therefore move into the cells and they would expand to the bursting point. On the other hand, an injection of a concentrated solution would have the opposite effect —the environment would then be *hypertonic* to the red cell contents and water would move out of the cells, leaving them in a shrunken or *crenated* state. For this reason, substances that are injected directly into the bloodstream are given in a *physiological saline* solution, which is equivalent to a 0.9 percent salt solution. This solution is *isotonic* to cells and tissues.

Active Transport

In some cases it may be necessary to move materials across a cell membrane *against* a concentration gradient. This, as we said before, is like expecting an automobile to coast uphill. Now, a car *can* be made to go uphill if we provide an external source of energy. In the same way, materials can be moved across a cell membrane from an area of low concentration to an area of high concentration, provided a source of energy is available. As in facilitated diffusion, carrier molecules are used to transport molecules from one side of the membrane to the other. Unlike facilitated diffusion, however, the body must provide energy in order to force the molecules to move "uphill" against the diffusion gradient. In this kind of situation, the molecules are said to move through the membrane by *active transport* (Figure 3.18).

One example of where an active transport system is needed involves the absorption of substances such as glucose and amino acids from the intestinal tract. The concentration of glucose and amino acids in the mass of digested food material moving through the intestine is relatively low. The active transport mechanism is therefore necessary to the movement of these extremely important nutrients into the circulatory system so they may be taken to the cells.

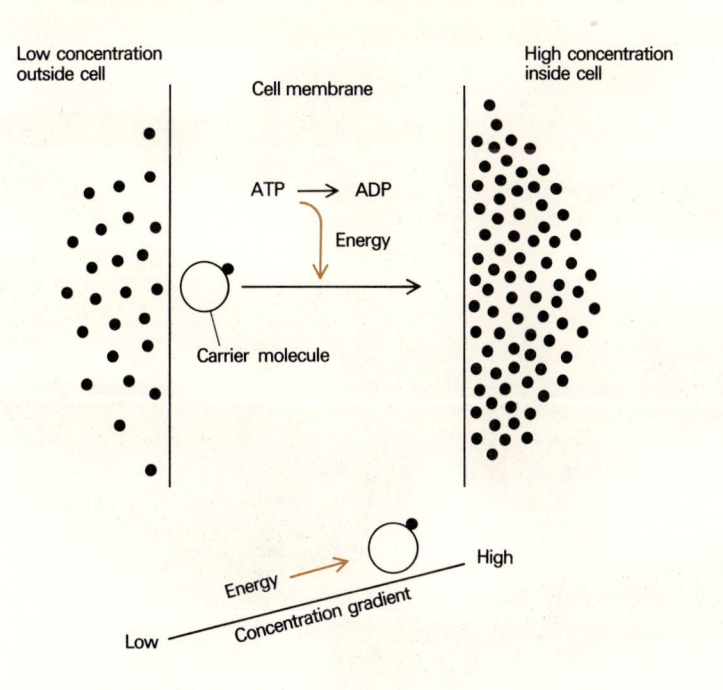

Figure 3.18
Diagram illustrating active transport. One possible mechanism is that molecules are carried from an area of low concentration to one of high concentration by a "carrier molecule," which is probably a protein. ATP supplies the energy to go "uphill."

Questions

1. Discuss the following statement: The cell is the minimum unit of structure and function of all life.

2. Compare plant cells and animal cells in terms of basic similarities and differences.

3. The "cell theory" is said to be a unifying principle for all life. Discuss and explain this statement.

4. Describe the principal function of each of the following:
   a) Cell membrane
   b) Lysosomes
   c) Mitochondria
   d) Chloroplasts
   e) Golgi bodies

5. The victim in a murder mystery was killed by a massive injection of distilled water directly into a vein. Explain the biological principle underlying this perfect crime.

6. Discuss the "energy scheme" and its relationship to cell function.

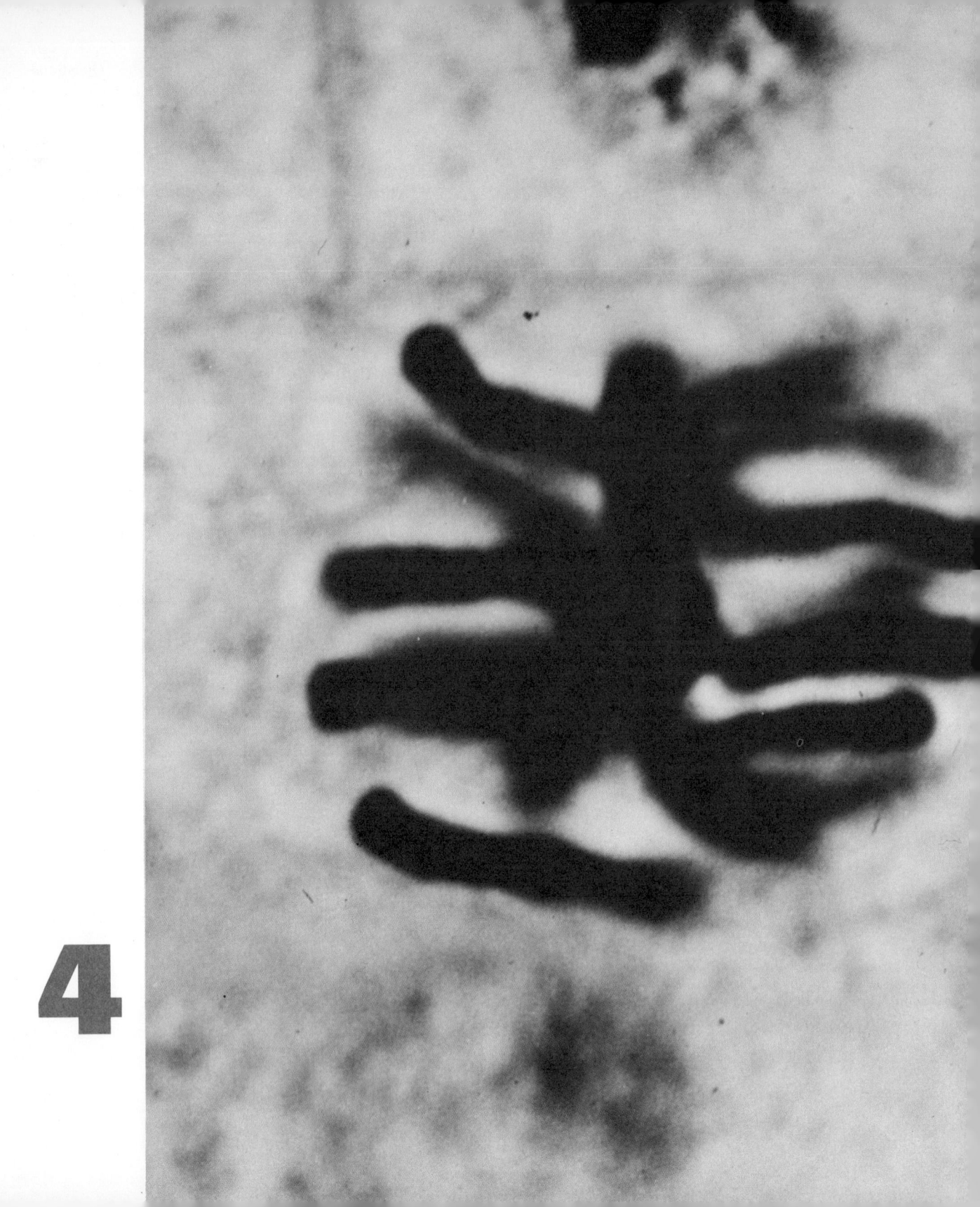

4

The Cell Nucleus

Chromosomes

The dense structure shown in Figure 4.1 is the cell *nucleus*. Note that the nucleus is enclosed by a double membrane that appears to be connected to the *endoplasmic reticulum*. Also, within the nucleus you will find a small, even denser area called the *nucleolus*. The nucleolus appears to be rich in a nucleic acid called *ribonucleic acid* (RNA). In a later chapter, we will discuss RNA and the important role that it plays in heredity.

During recent years, the nucleus and its function have been the subject of intensive research activities. This interest in the nucleus is not surprising, since it contains the hereditary material given to us by our parents, and much of what we are as individuals is predestined by the way this nuclear material functions. The cell nucleus, therefore, has far-reaching sig-

The cell nucleus

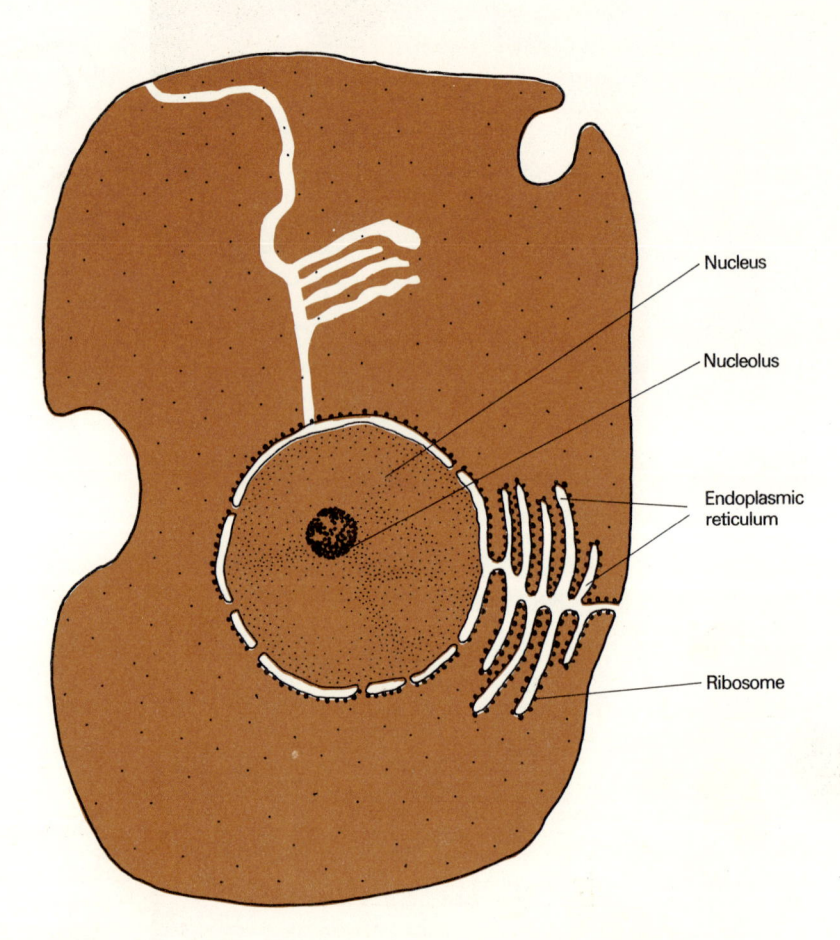

Figure 4.1
Simplified diagram of a typical cell showing only a few of its organelles.

nificance for human society and for the general human condition; we will discuss the details of nuclear function in a later portion of this book.

The nuclear material is organized into structures called *chromosomes*. The nucleus of each of the billions of body cells, or *somatic* cells, that make up the human organism normally contains 46 chromosomes, half of which were contributed by each parent. The best way to observe chromosomes is by a method that "catches" a dividing cell at a point when the chromosomes are well formed and most visible. This is done by taking a sample of blood and growing the white cells in culture. The cells are then chemically treated at the midpoint of division when the chromosomes are most evident. The result as seen through a microscope is shown in Figure 4.2.

The chromosomes, as shown in Figure 4.2, can be photographed and the individual chromosomes are cut from the photograph and arranged according to size and special features. This deceptively simple procedure yields a very useful device called a *karyotype*. The karyotype is a kind of species "fingerprint," since different species have chromosomes that differ in number, shape, and size. Figure 4.3 illustrates two normal human karyotypes, one obtained from a male and the other from a female.

Now look carefully at Figure 4.3. First of all, note that the chromosomes exist in pairs. They are called *homologous* pairs, meaning that the chromosomes in each pair are partners in terms of their effects on the

Figure 4.2
Photomicrograph of a human chromosome smear. Note that the chromosomes appear as "doublets" because they were caught in the metaphase stage of mitosis. (Courtesy of the Carolina Biological Supply Company.)

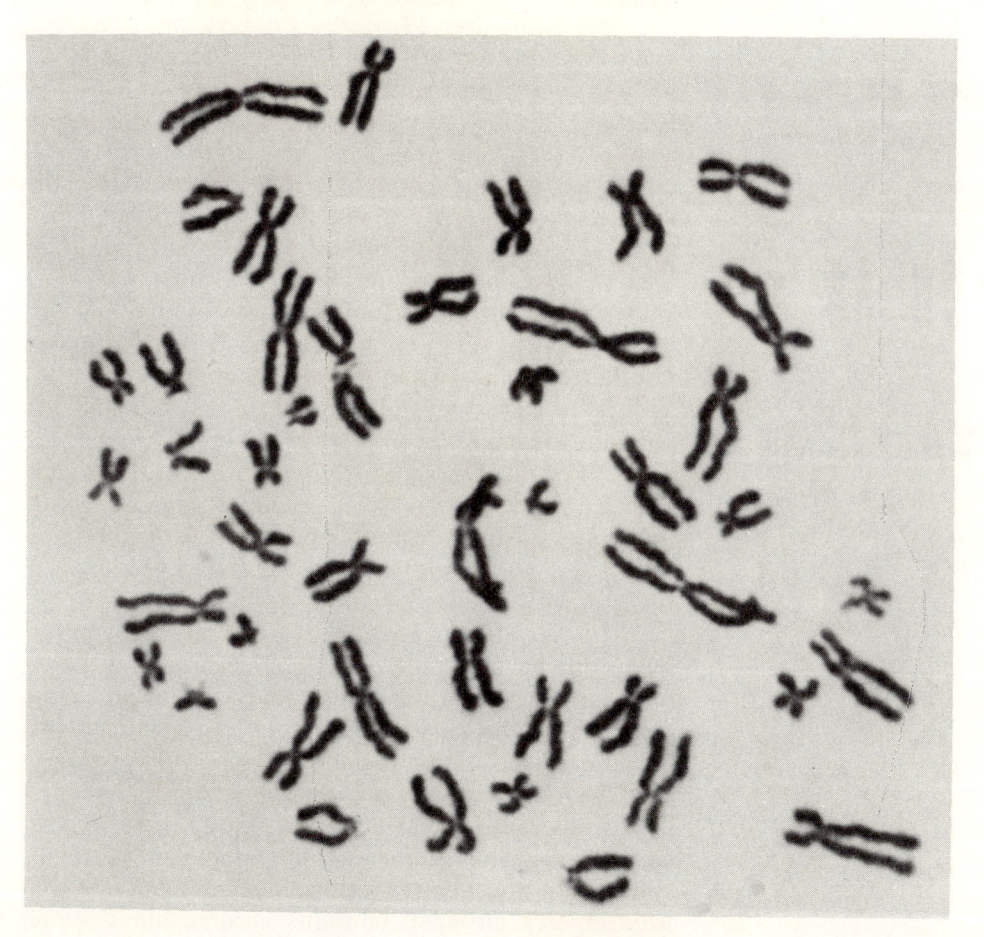

54   The cell nucleus

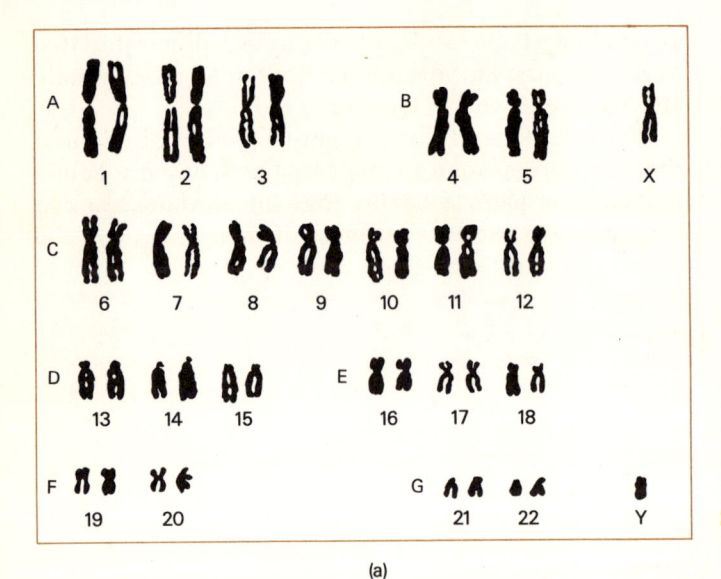

Figure 4.3
A normal human male karyotype (a), and a normal human female karyotype (b). (See text for details.) (From "Guide to Human Chromosome Defects," National Foundation–March of Dimes, 1968.)

inherited traits of the individual. For example, if one member of a homologous pair carries the genetic material that determines the shape of your earlobes, then the other member will also carry an earlobe shape determiner at the same location. They may both call for the same type of earlobe, or one may call for a certain type and the other may call for a different type. In the latter situation, the kind of earlobes that you have depends upon which of the two different earlobe factors is *dominant*. A factor is dominant if it tends to overcome or mask the effects of its partner, which in turn is regarded as *recessive*.

Also note that 22 of the total 23 pairs are the same for both sexes. These 22 pairs that are common to both male and female karyotypes are called *autosomes*. The 23rd pair, however, differs according to sex. In the female, the 23rd pair consists of two X chromosomes, but the male karyotype shows that the sex chromosomes of the male consist of one X chromosome and one Y chromosome. We will see later that the presence of the Y chromosome has been shown to determine maleness in the human species.

Looking again at the karyotypes in Figure 4.3, you will notice that each chromosome is actually doubled. This is because they had doubled prior to cell division in order to ensure a complete set of chromosomes for each new cell. The two members of a doublet, called *chromatids*, are temporarily joined at a point called the *centromere*, and they carry identical genetic material. We will come back to this when we explore the process of cell division in the next section.

The karyotypes show that the chromosomes divided into seven groups: A, B, C, D, E, F, and G. Looking closely, you will see that they are grouped according to size and position of the centromere. In group A, for example, the chromosomes are of a large variety and the centromere is *metacentric*; that is, it is located in the exact middle of the chromatid pair. Group G, on the other hand, contains small chromosomes and the centromere is located near one end (*telocentric*). Group C contains chromosomes which are of an intermediate size and in which the centromere is *submetacentric,* or located slightly off center. In addition to being grouped according to A, B, C, etc., the chromosome pairs are also numbered. How, then, do we distinguish between, say, chromosome pairs 6 and 7? Actually it is not easy to do, and to some extent this numbering system has been arbitrary. However, new and more sophisticated techniques that make use of the fact that different chromosomes show specific banding patterns are now helping scientists identify individual chromosomes with greater precision.

The human karyotype and the development of procedures by which it may be obtained represent an important scientific discovery, and we will follow it up in detail in a later chapter when we consider some of the problems associated with abnormal chromosome patterns.

Cell Division

Mitosis

We all begin life as a single cell, our destinies forged at the moment that an egg cell combines with a sperm cell. At that point we are a bit like the first cells struggling in the primeval seas, because we depend upon a sequence of events that will guarantee our development and survival.

The single cell that was originally "you" began a series of divisions, forming a mass of cells that continued to grow, develop, and differentiate into the specialized tissues, organs, and organ systems that now make up your body. This growth and development by cell division continued after you were born, and even today cell division goes on in certain tissues such as the intestinal tract, the skin, and the linings of the kidney tubules. The bone marrow contains cells which continuously divide, producing the blood cells that are basic to the proper function of the blood. The cells of certain other tissues, on the other hand, divide infrequently, if at all, once development of the organism is completed. Division of cells in the brain, for example, practically stops after the first few months of life.

The kind of cell division that maintains both the number and the specific chromosome content in the daughter cells is called *mitosis.* There is a good deal about mitosis that is not fully understood, but the descriptive details are fairly simple. It is also interesting to note that the details of mitotic cell division are essentially the same for all organisms except bacteria and blue-green algae. This has led to speculation that this basic process was "perfected" and became universal very early in the evolution of life.

Figure 4.4 shows the major stages that a cell goes through when dividing by mitosis. You should understand that these "stages" are arbitrary and useful for purposes of illustration, but a timelapse movie of an actual cell undergoing division would show it to be a smooth, continuous process.

Referring to Figure 4.4, we will examine each stage as follows.

1. *Prophase.* In the first stage of mitosis (a), the chromosome material begins to gather as tangled threads called *chromatin* in the nucleus. Prior to this, during *interphase,* the chromosomes were not apparent. Located near the nucleus in an animal cell is a small structure called the *centriole.* Centrioles have not been observed in higher plant cells, which could mean either that they are not present in plant cells or that they have not yet been detected. In prophase, the centriole has divided and the two daughter centrioles are migrating to the poles of the cell. The *aster* consists of filamentous rays in a starlike cluster around each centriole. The *mitotic spindle* is also forming between the two centrioles. This spindle consists of a series of parallel, gel-like filaments.

2. *Metaphase.* By metaphase (b and c), the nuclear membranes have disappeared. Note that the chromosomes have lined up on the equatorial plane of the

56   The cell nucleus

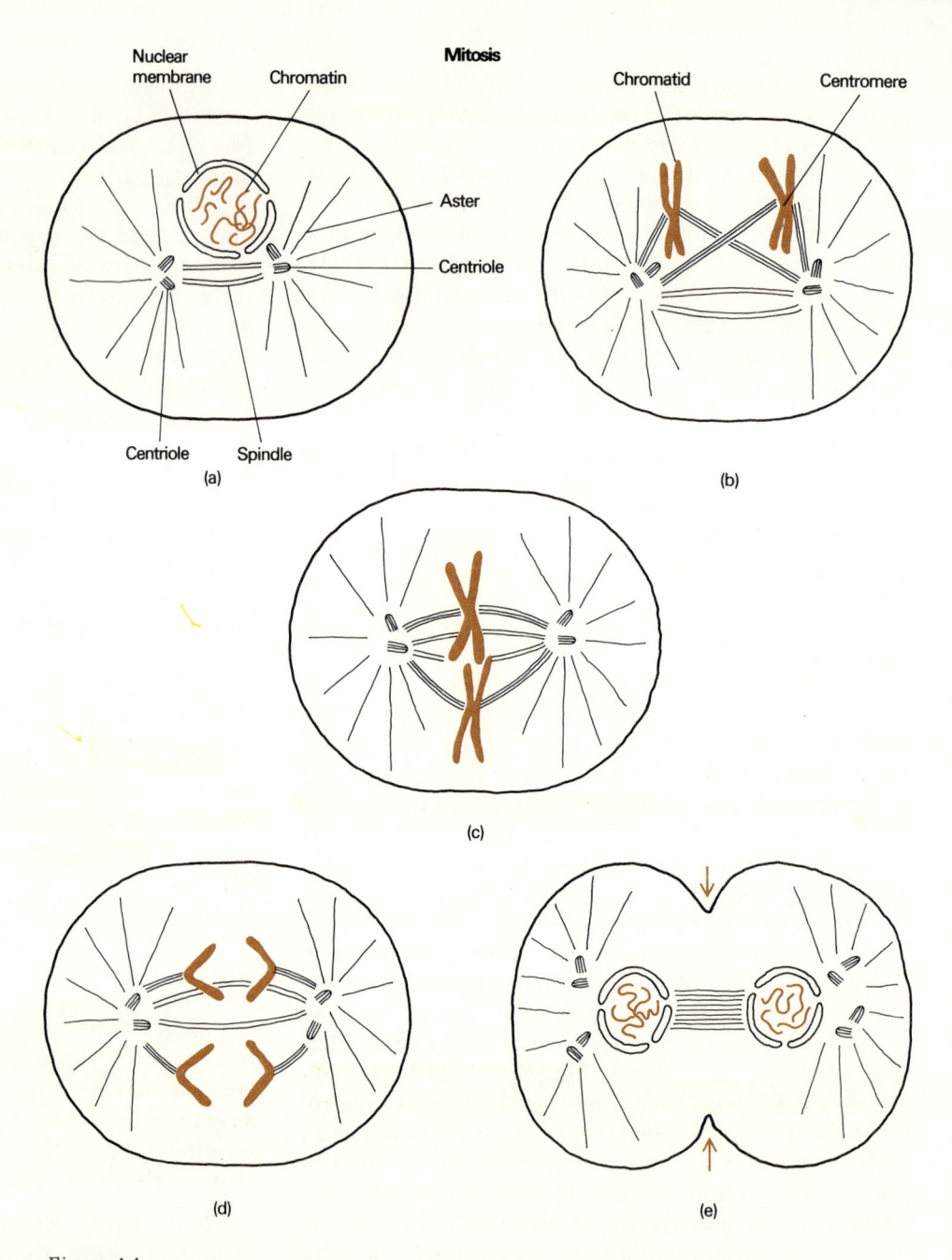

Figure 4.4
Series of diagrams illustrating the process of mitosis in an animal cell. (See text for details.)

cell on a line perpendicular to the spindle. Also be sure to note that the chromosomes are doubled as they were in the karyotypes shown in Figure 4.3. (You will recall that karyotypes are derived from cells that are in the metaphase stage.) This doubling, which constitutes an exact duplication, took place while the cell was in *interphase,* which is the period between divisions. The duplicates or *chromatids* are joined at the centromere. The chromatids will now separate as the centromere divides and the chromosome duplicates start to move toward the poles at opposite ends of the cell. This movement marks the beginning of *anaphase.*

3. *Anaphase.* This stage (d) shows the duplicated chromosomes moving toward the poles. Note that from the V-shape of the chromosomes it appears that they are being dragged through the cytoplasm. The actual force that moves the separated duplicates toward the poles is not known. One theory suggests that it may involve structural changes in certain proteins.

4. *Telophase.* After reaching the poles (e), the chromosomes elongate and again become a jumbled mass. The nuclear membrane reappears and the cytoplasm divides to complete the formation of two cells that theoretically have the exact same genetic material as the original cell.

Mitosis occurs with different frequencies in different tissues, and the amount of time required for a cell to completely divide varies according to a number of internal and external factors.

Mitosis is essentially the same process in both animal and plant cells, except that centrioles and asters have not been observed in plant cells. Also in plant cells, the final step in division differs somewhat from telophase in animal cells. Due to the presence of the cell wall in the plant cell, a *cell plate* forms between the two daughter cells during telophase, and the cell wall reforms at that point. Figures 4.5 and 4.6 are photographs which compare various stages of mitosis in plant and animal cells, respectively.

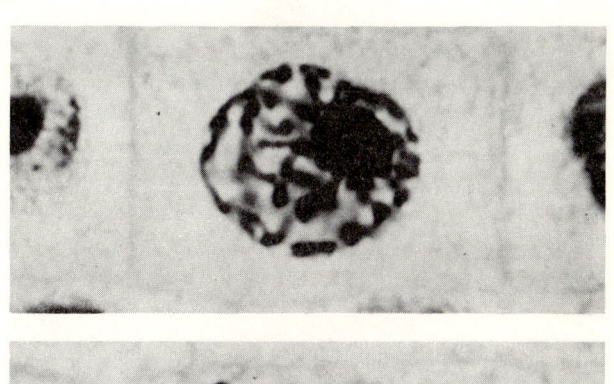

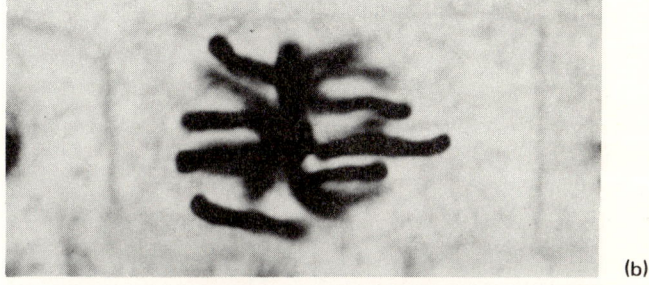

Figure 4.5
Photomicrographs showing stages of mitosis in a plant cell: (a) prophase, (b) metaphase, (c) anaphase, and (d) telophase. (Courtesy of the Carolina Biological Supply Company.)

(a)
(b)
(c)
(d)

Figure 4.6
Photomicrographs showing stages of mitosis in an animal cell: (a) prophase, (b) metaphase, (c) anaphase, and (d) telophase. (Courtesy of the Carolina Biological Supply Company.)

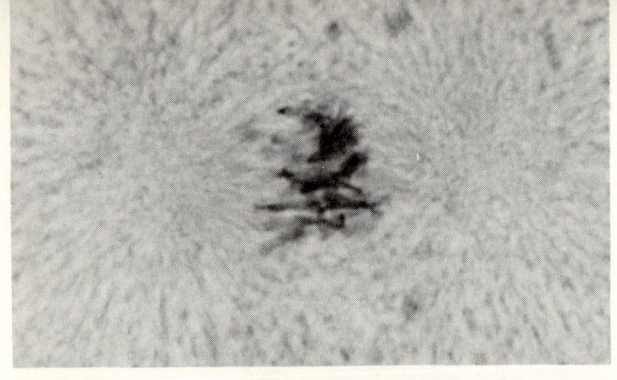

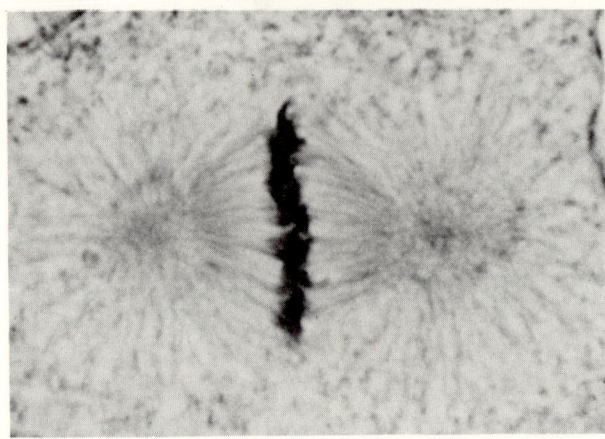

Meiosis

Mitosis is the process of cell division which is important to embryological development as well as to the continued division and replacement of body cells throughout life. We might summarize mitosis and its relationship to chromosome number as follows, using the human chromosome number as an example.

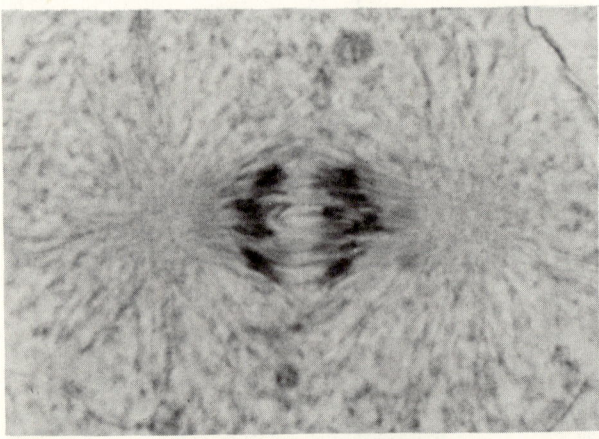

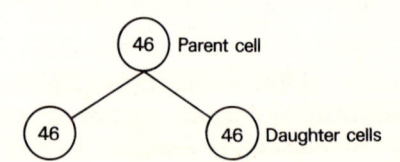

```
        ( 46 ) Parent cell
        /      \
   ( 46 )    ( 46 ) Daughter cells
```

On the other hand, sexual reproduction involves the combination of a male sperm cell with a female egg cell to produce a zygote, which is the single-cell beginning stage of the new individual. Therefore, if the sperm and egg cells each contained 46 chromosomes, the zygote would carry 92 chromosomes, which would be a radical departure from the species number of 46 that has been established through millions of years of evolution. Furthermore, you can see that the chromosome number would continue to increase through several generations from 92, to 184, to 368, and so on, resulting in an obviously impossible situation.

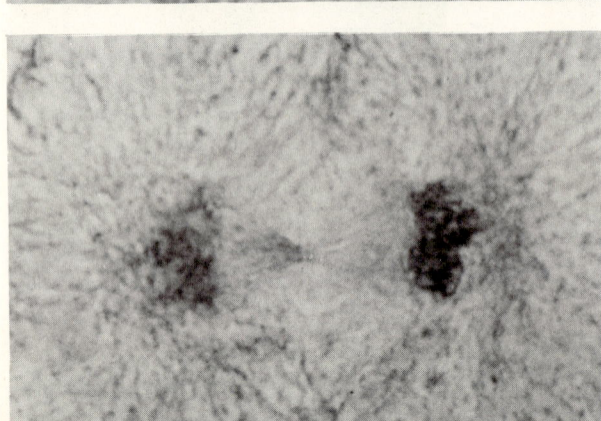

What we need, therefore, is a process whereby the chromosome number in the sex cells can be cut in half, the full species number being restored when the sperm and egg combine. This situation can be illustrated as follows:

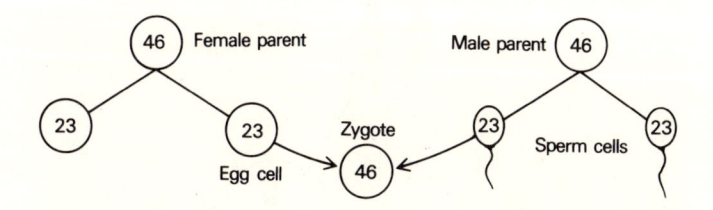

```
         46  Female parent         Male parent  46
        /  \                                   /  \
      23    23      Zygote        23          23
     Egg cell       46          Sperm cells
```

Once the zygote has been produced with the correct species number of chromosomes, *mitotic* division will then ensure the development of an individual with the appropriate species number of chromosomes in each body cell.

The kind of cell division that reduces the chromosome number by half is called *meiosis,* or reduction division. As you can see, it is critically important to sexually reproducing organisms. Meiotic cell division produces sperm and egg cells; it therefore takes place in the *testes* of the male and in the *ovaries* of the female. We will consider the details of these organs and their functions in the next chapter.

Meiosis is a relatively complex process. For our purposes we will do our best to simplify it while retaining the basic idea and significance. Although Figure 4.7 shows the basic process by which sperm cells are produced, it must be emphasized that this presentation leaves out considerable terminology and also ignores certain complications.

In Figure 4.7 we are considering, for the sake of simplicity, a hypothetical male organism that has two pairs of chromosomes; however, the same principles shown here can be extended to the human male with his 23 chromosome pairs. Refer to this figure as you go through the following steps of *spermatogenesis.*

1. The *spermatogonium* is a specialized cell found in the male sex glands, or *testes.* Since our example possesses a species chromosome number of four, the

Spermatogenesis

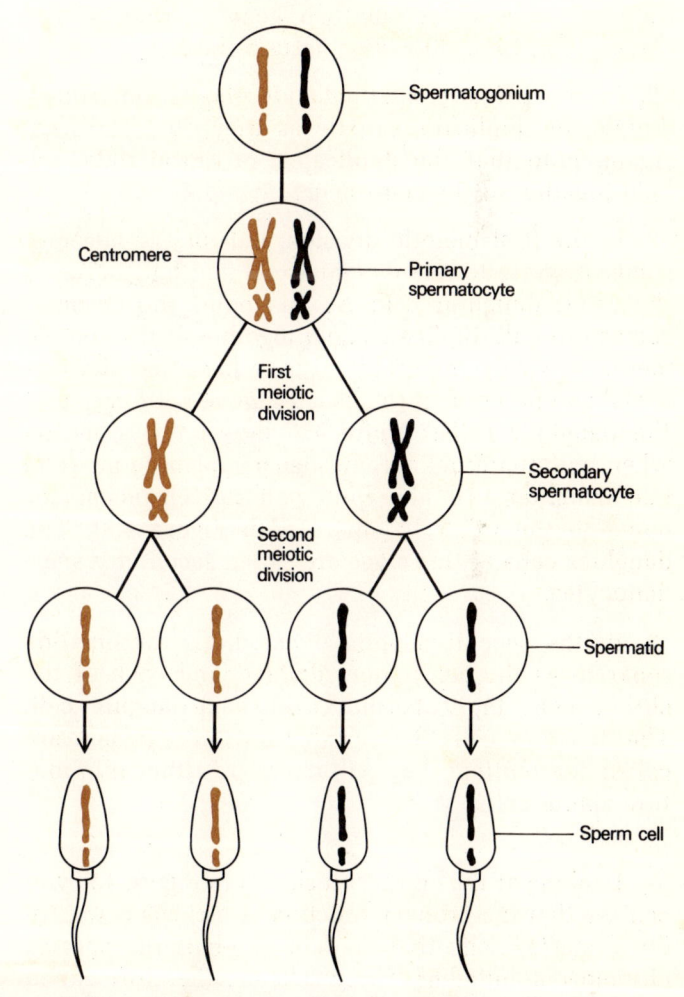

*Figure 4.7
Series of diagrams illustrating the process of meiosis in a hypothetical male organism (spermatogenesis).*

spermatogonium contains two pairs, or four chromosomes. Since it contains *pairs* of each of the homologous chromosomes, it is called a *diploid* cell. Note that the colored chromosomes are maternal chromosomes; that is, they were obtained originally from our example's mother. The black chromosomes, on the other hand, came from the father and are therefore labeled paternal chromosomes. It is very important that you keep in mind that although the two chromosome pairs are homologous, they will undoubtedly differ to some extent in their genetic material and therefore in their effects on certain traits.

2. In the *primary spermatocyte* the chromosomes double, or duplicate, exactly as they do in mitosis. Again, note that the duplicates, or chromatids, are held together at the centromere.

3. In the first meiotic division, half of the chromosomes migrate to one daughter cell and half move to the other daughter cell. At this point the chromosomes are still duplicates held together at the centromere. It is very important to note here that the maternal and paternal chromosomes move *randomly* into the daughter cells. Figure 4.7 shows the situation when both maternal chromosomes happen to move to one daughter cell and both paternal chromosomes move to the other (compare with Figure 4.8). The daughter cells at this stage are called *secondary spermatocytes*.

4. In the second meiotic division, the chromatids separate as the centromere divides, and each of the chromosome duplicates moves into each daughter cell. The daughter cells that result from this division are called *spermatids;* they will develop further into mature sperm cells.

Looking at the end result shown in Figure 4.7, you can see that the primary function of meiosis is to produce special cells that contain one-half the species chromosome number. These cells are therefore called

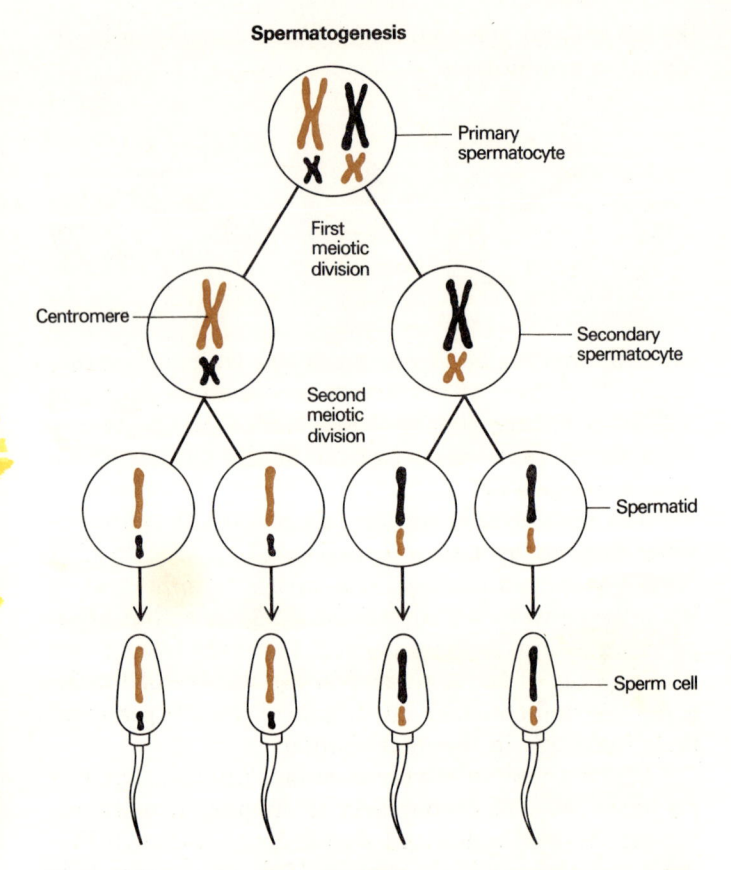

Figure 4.8
The results of meiosis when paternal and maternal chromosomes rotate positions in the primary spermatocyte.

haploid. Taking another approach, we can say that one member of each chromosome pair went into each sex cell.

You can also see that *two* kinds of sperm cells were produced in terms of genetic content. In other words, two of the four sperm cells carry only paternal chromosomes and the other two contain only maternal chromosomes.

Figure 4.8 shows the primary spermatocyte stage with a small but significant difference. This time the paternal and maternal chromosomes have rotated positions. Following through the subsequent meiotic divisions, we can see that the chromosome content of the resulting sperm cells is different from that carried by the sperm cells shown in Figure 4.7. Now, assuming that individual chromosomes may rotate their positions during meiosis, and also assuming that this rotation is random or accidental and occurs about 50 percent of the time, then it is theoretically possible for our hypothetical male organism to produce *four different kinds* of sperm cells. When we say different, we mean different in terms of the genetic content or message that they carry. The four types produced from our example are shown in Figure 4.9.

The egg cell is produced in a similar fashion, except that where four sperm cells are produced by one spermatogonium, only one egg cell is formed by the corresponding *oögonium*. The other cells are called *polar bodies* and do not have a direct function in reproduction. As in sperm production, however, one oögonium can produce *four possible kinds* of egg cells in terms of chromosome content and therefore in terms of the specific genetic message that they carry. Figure 4.10 shows the general process of egg cell production, or *oögenesis*.

The variety of chromosome combinations found in the sex cells of a given individual is the basis of an important principle. For example, if we were to follow the meiotic process in an organism having *three* pairs of chromosomes, we would obtain *eight* different sperm cell types or *eight* different egg cell types. An organism with *four* chromosome pairs would yield

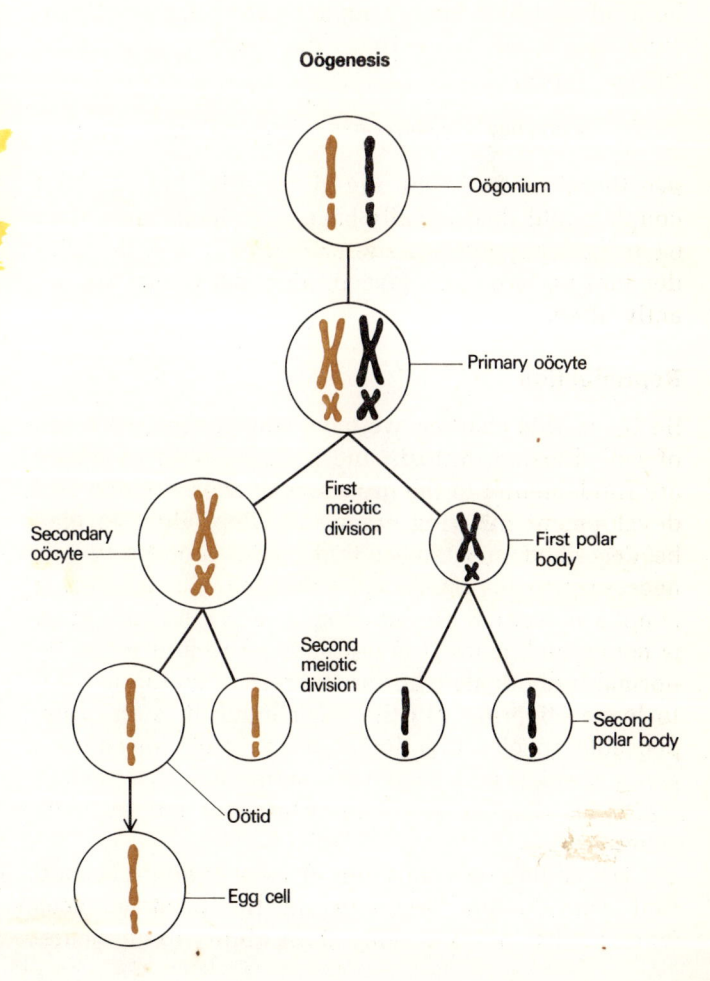

Oögenesis

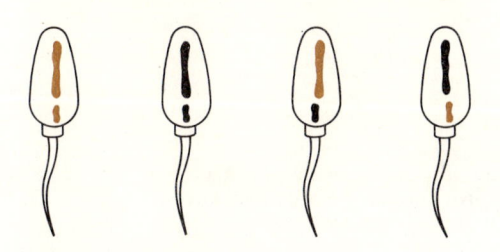

Figure 4.9
Four possible kinds of sperm cells that could be produced by meiosis in a hypothetical male organism with two pairs of chromosomes. Note that the paternal and maternal chromosome combinations are different in each; therefore the genetic message is different in each case.

Figure 4.10
Diagrams illustrating the process of oögenesis in a hypothetical female organism. Note that the process is the same as in spermatogenesis, except that instead of four eggs, only one egg and three nonfunctional polar bodies are produced.

sixteen sex cell types, and so on. We can therefore generalize and state that the number of genetically different sperm or egg cells produced by a sexually reproducing organism that is not self-fertilizing is 2^n, where n signifies the number of chromosome pairs.

Now, since man carries 23 chromosome pairs, the number of possible kinds of sperm cells that a human male can produce is 2^{23}, which is well over eight million! Since the same may be said of the egg cells produced by a human female, we can compute the number of different genetic types of offspring that could be produced by a given couple by the number of possible sperm cell types times the number of possible egg cell types, or

$$8{,}000{,}000 \times 8{,}000{,}000 = 64{,}000{,}000{,}000$$

We therefore find that any given child born to that couple could theoretically have any one of more than 64 trillion possible genetic makeups. It is little wonder that no two of us (except identical twins) are exactly alike.

Reproduction

So far in this chapter, we have considered two types of cell division, mitosis and meiosis, both of which are fundamental to the processes of reproduction and development in living organisms. Reproduction may be described in simplest terms as a process that is necessary to the continued existence of a species. It is not an essential life function as far as the individual is concerned, since it is not unusual to find perfectly normal individuals who for one reason or another fail to leave offspring. On the other hand, it is an indispensable species function, since it is obvious that a given species will soon face extinction if a certain minimum number of offspring are not left by each generation.

Depending on your point of view, it might be said that reproduction originated in the primeval seas when nucleic acids developed the ability to duplicate.

You will recall that these ancient forerunners of life are thought to have accomplished this by utilizing organic molecules from their environment as the necessary building blocks.

Asexual Reproduction

As life developed and as various types of simple organisms evolved, it is probable that *asexual reproduction* became the predominant method by which the continuation of life was assured. Asexual reproduction is found among many life forms today; it is especially common in the simple organisms that comprise the group known as the protozoa. The *amoeba*, for example, is a protozoan that produces by *fission*, which is perhaps the simplest form of asexual reproduction (Figure 4.11). In fission, the organism simply splits by mitotic division into two more-or-less equal parts. Figure 4.12 shows how another protozoan, the

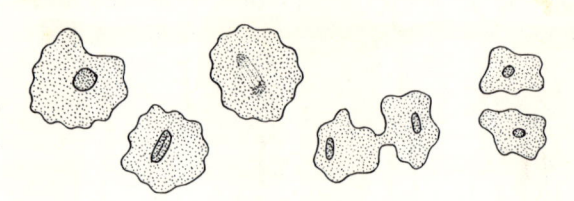

Figure 4.11
Diagram illustrating fission in amoeba. In this form of asexual reproduction, the organism simply splits in two by mitotic division.

paramecium, may also reproduce asexually by simple fission. An alga known as *pleurococcus* (Figure 4.13), which is famous among Boy Scouts as the "moss" that grows on the north side of a tree, provides an illustration of fission from the plant world.

In another form of asexual reproduction called *budding*, a small part of the parent cell splits off and develops into a new individual. Yeast cells reproduce in this way (Figure 4.14), and so may a small aquatic animal called *hydra*.

Asexual reproduction 63

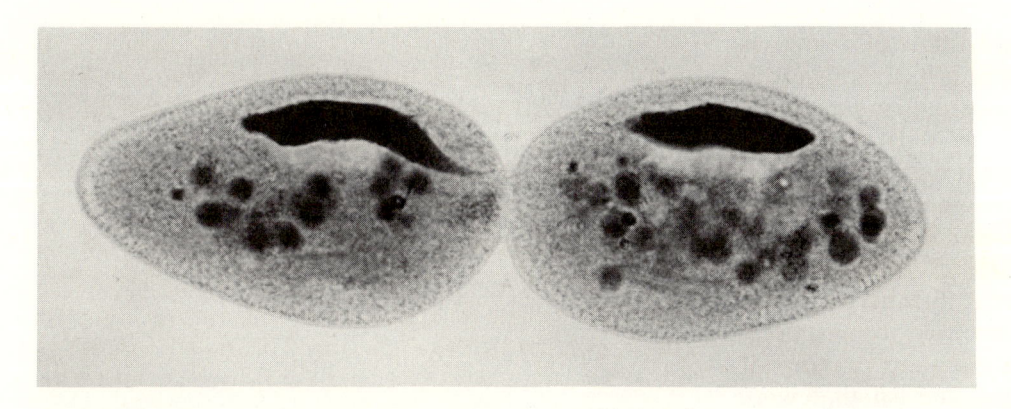

Figure 4.12
Diagram illustrating paramecium reproducing asexually by fission. (Courtesy of the Carolina Biological Supply Company.)

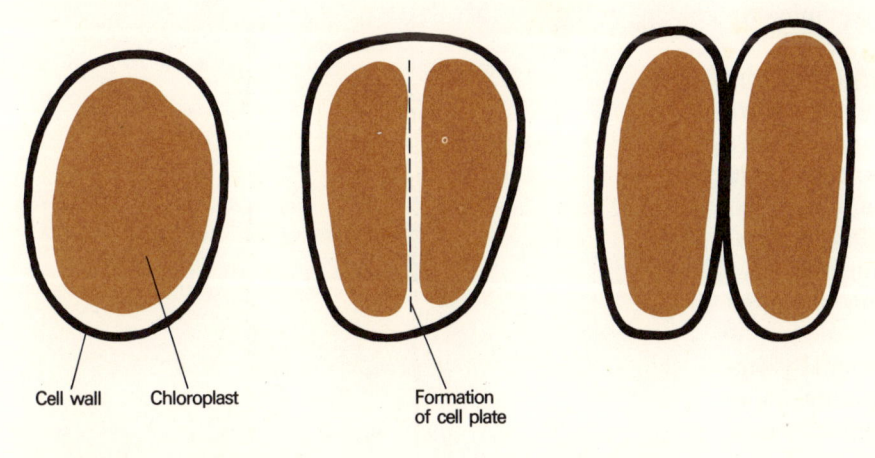

Cell wall   Chloroplast   Formation of cell plate

Figure 4.13
Diagram illustrating fission in pleurococcus. This form of asexual reproduction is similar to division in amoeba, except that a cell plate forms which develops into a cell wall.

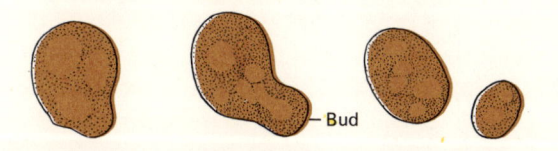

Bud

Figure 4.14
Diagram illustrating asexual reproduction by budding in a yeast cell. Note that a small part of the parent organism splits off and develops into a new cell.

64   The cell nucleus

Some animals can regenerate a new part such as a leg or tail if the original one is lost. Salamanders, for example, can grow a new leg or tail, and a certain type of lizard may deliberately sever its own tail in order to distract a predator while it makes its escape, following which it grows another tail. In special cases, this ability to regenerate parts may provide another variety of asexual reproduction called *fragmentation*. For example, a starfish that is cut in two becomes two starfish, because each half will proceed to regenerate the half that is lost. There is a well-known story of some fishermen who tried to kill starfish that were preying on their oyster beds by catching the starfish, chopping them in two, and tossing them back overboard. This actually doubled the number of starfish they wanted to eliminate!

Plants, too, may be asexually propagated in a variety of ways. Strawberry plants, for example, increase in number by sending out *stolons*, which are horizontal stems growing along the surface of the ground (Figure 4.15). Potatoes may be and usually are propagated by planting *tubers* (the potato itself is a tuber). Farmers do not plant potato *seeds*; they plant seed *potatoes*, or tubers. A rose bush or a pussy willow may be asexually reproduced by planting *cuttings*, in which a branch is simply cut from the main plant, placed in water until roots develop, and then planted.

Grafting is an interesting and economically important form of asexual reproduction in plants, especially fruit trees. Some years ago, for example, a new variety of apple known as the Cortland was developed. The Cortland has many of the eating qualities of the MacIntosh apple but has the added advantage

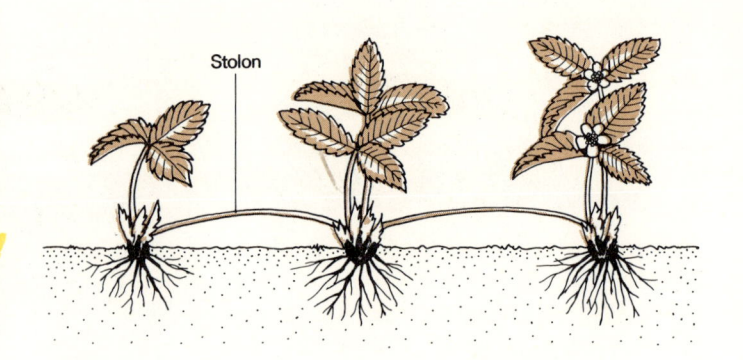

Figure 4.15
Vegetative or asexual reproduction in the strawberry plant. Stolons or "runners" grow above ground and take root, forming new plants.

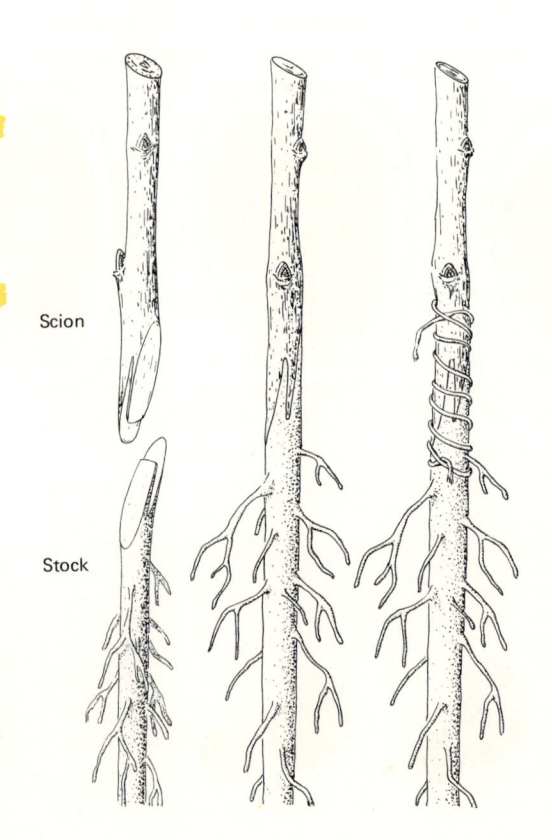

Figure 4.16
Apple trees propagated asexually by grafting. A stem (scion) from the desired variety is inserted into a notch cut in the stump of any variety (stock). The apples that are produced will be determined by the scion, not the stock. In this way, it is possible to have several varieties of apples growing on the same tree.

of remaining firm and unspoiled for longer periods during winter storage. How can we propagate a new variety such as the Cortland and still maintain its unique characteristics? If we simply planted the seeds, we would be disappointed because seeds result from *sexual* reproduction, and therefore contain the variations that we pointed out when discussing meiosis. In other words, the Cortland must be propagated *asexually* if we are to maintain the same qualities that make the fruit economically valuable. This is done by *grafting,* a process in which branches are removed from a Cortland tree and grafted onto the stock of any apple tree (Figure 4.16). In this way, thousands of descendants of the original Cortland tree have been produced. This same method is used with other types of fruit trees. How, for example, would you propagate a *seedless* orange?

Sexual Reproduction

Asexual reproduction has the advantage of relative simplicity because only one parent is required. Only one amoeba, for example, is necessary in order to produce more of its own kind. On the other hand, it is important to note that the "daughter" amoebae are, in most cases, exactly like the parent cell; therefore, unless significant mutations occur, there will be a remarkable lack of variation throughout the species. So, as we suggested earlier, asexual reproduction does have the advantage of simplicity, and the life of the lowly amoeba is certainly untroubled by the pitfalls and problems of romance. In addition, certain forms of asexual reproduction such as grafting, cuttings, stolons, etc. are important to man because they permit us to maintain better qualities in certain food crops and flowers.

On the other hand, lack of variation among the members of a species can be a distinct disadvantage, especially if there should be a significant change in the environmental conditions to which that species is adapted. For example, if all the members of a given species were adapted to survive only within a narrow temperature range, extinction would very likely follow if shifting climatic conditions should produce a change in the mean temperature of the environment. On the other hand, if there were sufficient *variation* in certain physiological characteristics that allowed at least some members of the species to tolerate a broader temperature range, then they might survive and possibly pass on this variation to the next generation. In a later chapter, we will see that this very mechanism is an important driving force behind the process of evolution.

Therefore, while sexual reproduction may render life more complicated as well as more interesting, the wide variation which results from the enormous number of possible combinations of two parent cells is a distinct advantage to a species that is struggling to adapt to changing environmental conditions. You will recall that in an earlier section, we pointed out that meiotic division produces at least eight million different kinds of human sperm cells and an equal number of kinds of egg cells. Quite obviously, with sexual reproduction there is an almost limitless variety of individuals within a sexually reproducing species. In the next chapter, we will consider sexual reproduction as it occurs in the human organism.

Questions

1. Describe the process of mitosis and explain its significance.

2. Describe the process of meiosis and explain its significance. Compare meiosis with mitosis.

3. What is the significance of the random movement of paternal and maternal chromosomes toward the poles in the type of cell division known as meiosis?

4. Distinguish between sexual and asexual forms of reproduction.

5. What are some advantages and disadvantages of asexual reproduction? of sexual reproduction?

5

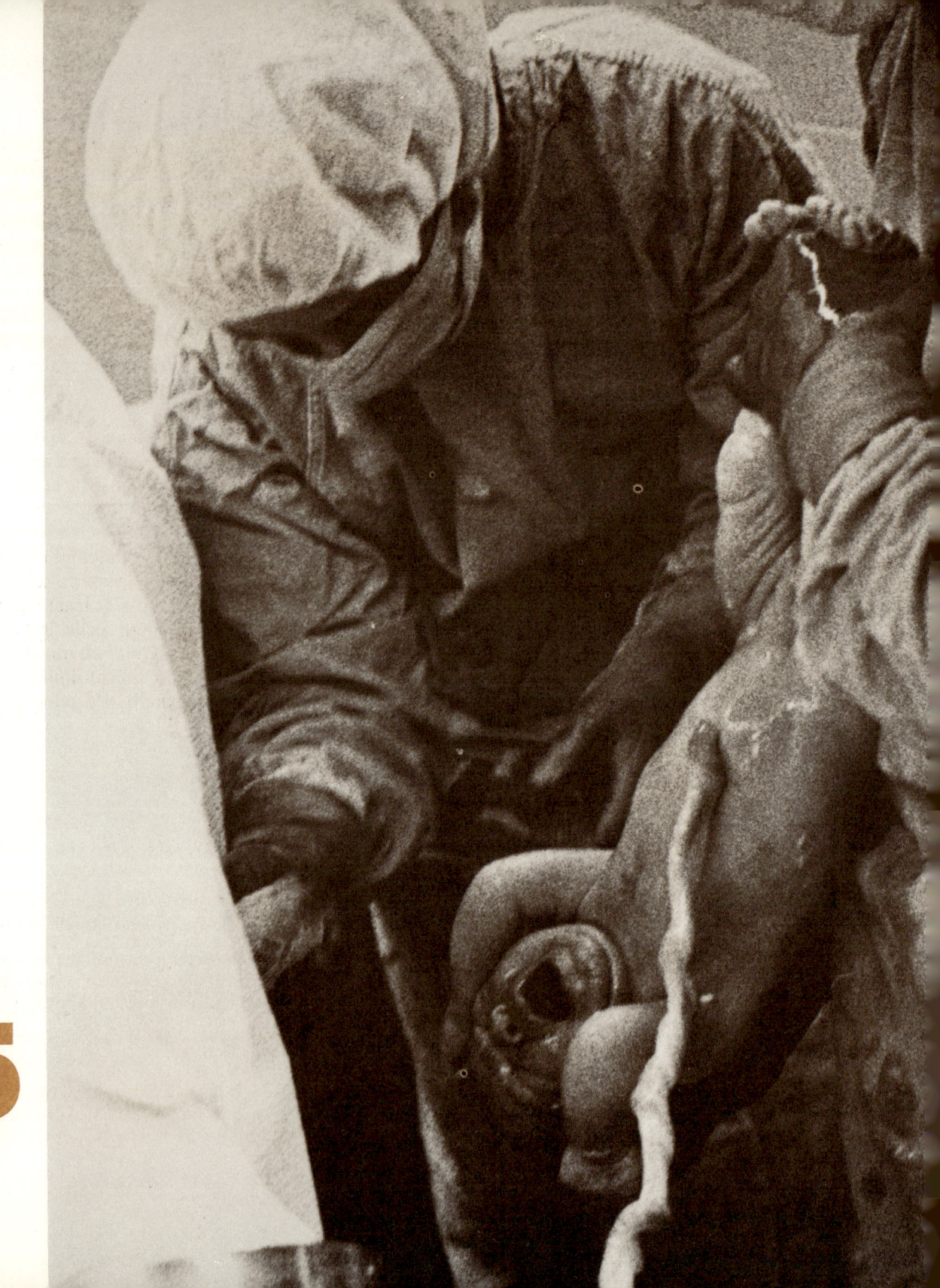

"Be Fruitful, and Multiply..."

Human Sexuality

The role of reproduction in the human species is unique in the world of life. Whether they come from primitive societies or from modern suburbia, human beings engage in sexual activities *not* necessarily connected with reproduction. In modern societies, sexual activities are as often as not carried on with the reproductive aspect deliberately removed by sterilization or some form of contraception. There is probably no other human activity that has been more ritualized and romanticized. Its emphasis surrounds us everywhere, emanating from the movie marquee, the magazine racks, and from the incessant television commercials that offer modern love potions guaranteed to produce everlasting beauty and sexual attractiveness.

Human *sexuality* has had a turbulent history. The more frank and liberal attitudes characteristic of the eighteenth century gave way to what might be called the sexual dark ages, which consisted of over 150 years of taboos, many of which persist today. Sex has been regarded by various groups at various times as ugly, sinful, and shameful—even within marriage—to be tolerated only because no one could come up with a better way to produce children. Now, in the last third of the present century, we have seen many of the old taboos fall away. The new attitude, and undoubtedly the far more healthy one, holds that sex is a normal, natural, and beautiful expression of love and affection between two human beings.

It is interesting to note that anthropologists have suggested that the evolution of certain aspects of human sexuality has contributed significantly to the survival of our species. They point to the fact that the human female, unlike those of other species, is receptive to the male at practically all times, whether she is fertile or not. Therefore, it has been suggested that this unique *nonreproductive* aspect of human sexuality constitutes an important bond between the parents in a family unit. This family unit, in turn, has guaranteed adequate care of offspring during the long formative years of the human child.

In this chapter, we will consider some of the biological principles associated with human reproduction. Then in the next chapter, we will apply these principles to the various methods of contraception which permit human beings to engage in sexual activities without pregnancy as a consequence.

The Master Controls

Human reproduction involves a complex series of hormonal relationships. *Hormones* are chemicals secreted by *endocrine glands;* endocrine glands that play major roles in reproduction include the *pituitary gland,* the male *testis,* and the female *ovary.* The testis and ovary are known as the *gonads,* or sex glands, and are also the source of reproductive cells.

Endocrine glands are ductless; that is, the hormones that they secrete are picked up directly by the bloodstream and carried to various "target" organs throughout the body. The mechanism by which hormones function is not completely understood, but there is evidence that in some cases they alter the permeability of cell membranes, and in other instances they may speed the synthesis of proteins by cells. At any rate, hormones are powerful chemical agents, and a small amount of a given hormone will produce significant effects.

The "master controls" of reproduction are located in the area of the brain. They consist of (1) the *hypothalamus,* a lower brain center located at the base of the brain, and (2) the *pituitary gland,* an endocrine gland located beneath and adjacent to the hypothalamus.

Figure 5.1 shows the anatomical relationship of the hypothalamus to the pituitary gland. You should note that the pituitary actually consists of two separate structures or *lobes*. These are called the anterior (front) and posterior (back) lobes.

The *anterior lobe* is a gland that has cells capable of producing hormones. The *posterior lobe,* on the other hand, appears to be merely a storage place for hormones manufactured by nerve cells in the hypothalamus.

The relationship between the hypothalamus and the pituitary is of great significance to reproduction. Nerve tissue in the hypothalamus secretes protein substances known as GRF, or *gonadotropic releasing factors*. These factors are then transported by blood vessels to the anterior lobe of the pituitary, as shown in Figure 5.1. At the anterior lobe, they stimulate the glandular cells to produce *gonadotropic hormones,* which are then carried by the bloodstream to the gonads. *Tropic* means "responding to a specified stimulus"; gonadotropic hormones therefore signal the testis or ovary to secrete sex hormones.

This situation involves a *negative feedback* system which is designed to control the level of sex hormones in the blood. Figure 5.2 uses a common home-heating system to illustrate the principle of negative feedback. When the temperature in a house rises to a

The master controls 69

certain point, the thermostat signals the furnace to turn off and stop producing heat. As the temperature falls, the thermostat signals the furnace to renew the production of heat until the temperature returns to the level at which the thermostat signals the furnace to turn off again, and so on.

This example may be roughly compared to the relationships involving the hypothalamus, pituitary, and gonads. Looking once again at Figure 5.2, you can see that when the sex hormones increase to a certain level in the blood, the hypothalamus (thermostat) is caused to reduce its production of GRF. This, in turn, inhibits the secretion of gonadotropic hormones from

Figure 5.1
Diagram illustrating the anatomical relationship between the hypothalamus and the pituitary gland. Nerve cells in the hypothalamus secrete gonadotropic releasing factors which are carried by the portal blood vessel to the anterior lobe of the pituitary. This stimulates the production of gonadotropic hormones by the glandular cells of the anterior lobe.

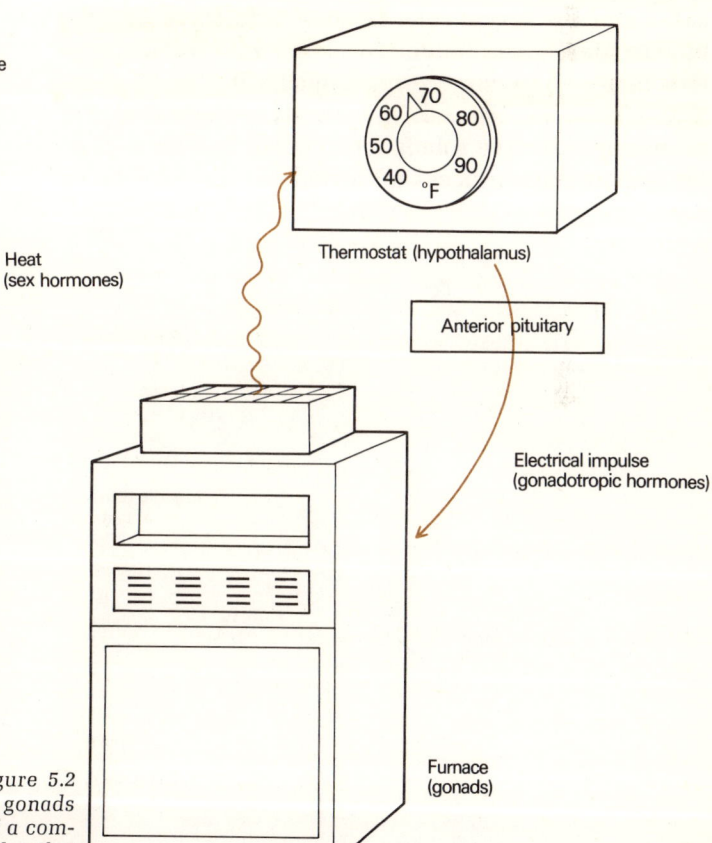

Figure 5.2
The negative feedback system between the gonads and the pituitary gland, using the analogy of a common home-heating system. (See text for details.)

the pituitary, and therefore the production of sex hormones (heat) by the gonads (furnace) is reduced. Conversely, a low level of sex hormones (heat) in the bloodstream will "turn on" the hypothalamus (thermostat), which will then call for increased secretion by the gonads (furnace).

A simple illustration of *positive feedback* may help to clarify the concept of negative feedback. Suppose that the more heat a furnace produces, the more the thermostat calls for heat. This would represent a positive feedback situation.

The Male Reproductive System

If we leave aside the myriad social and legal complications of his sexual role, the primary function of the human male is to deliver 23 of his 46 chromosomes to a point where they can be joined with the 23 chromosomes of the egg cell. The strictly biological aspect of his reproductive role is then ended, and the rest of the babymaking process is carried out by the female.

In this section, we will look at the anatomy and physiology of this male "delivery system" along with some of the problems involved in its operation.

Normally, the male child is reproductively sterile until sometime between the ages of eleven and thirteen years. At this time, the phenomenon known as *puberty* signals the beginning of adolescence, which leads to sexual maturity. At puberty the hypothalamus steps up production of the gonadotropic releasing factors, and the anterior lobe of the pituitary responds by secreting gonadotropic hormones. These, in turn, are carried to the testes where they initiate sperm cell production and also stimulate the testes to secrete male hormones, the most important of which is *testosterone*. Under the influence of testosterone,

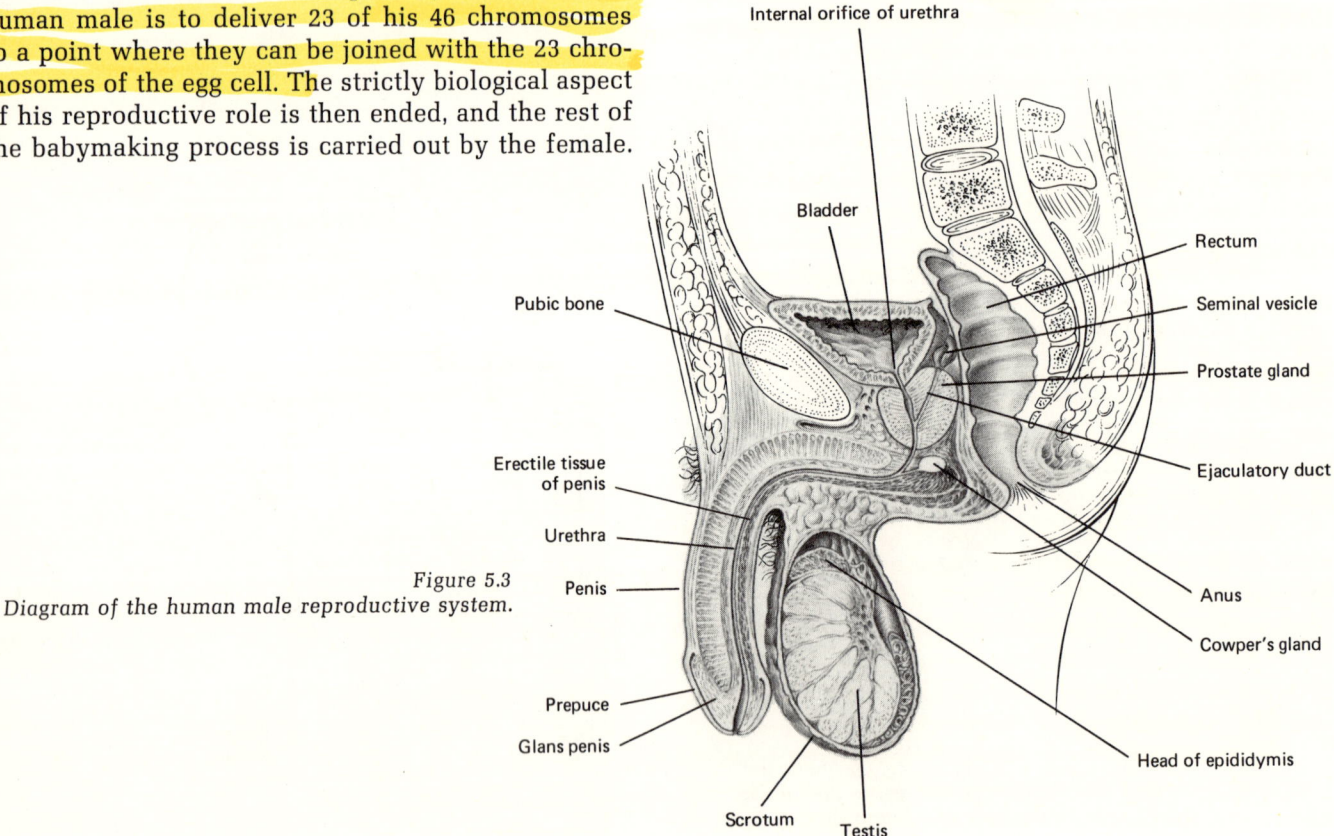

Figure 5.3
Diagram of the human male reproductive system.

the male secondary sex characteristics develop, which include the beard, the deeper voice, broad shoulders, and narrow pelvis. Testosterone also plays an important role in the development of the male sex drive. Although the level of this hormone may dwindle as middle age approaches, most men continue to produce testosterone at some level until well into old age. The same may be said of sperm cell production; cases of men fathering children when in their seventies or even eighties are not unknown.

The Testes

The *testes* produce both the sperm cells and the male hormone testosterone. Looking at Figure 5.3, you will note that the testes are suspended outside the abdominal cavity in a saclike structure called the *scrotum*. The testes do not function properly at body temperature; their position in the scrotum places them in an environment that is about seven degrees Fahrenheit cooler. The testes actually develop inside the abdominal cavity, but they normally descend into the scrotal sac shortly before birth. Not infrequently, the testes fail to descend at birth, a condition known as *cryptorchidism*, or "hidden testes." In most of these cases, they descend into the scrotum during the first few years of life, and in almost all cases they will descend by puberty. If not, the physician may elect to place them in the scrotum by surgical means, since the individual will almost certainly be sterile if they remain exposed to the higher temperature of the abdominal cavity.

The testes may also be attacked by infections such as that caused by the mumps virus. Mumps can be especially serious in the mature male because of a tendency for the virus to invade the testes, producing an inflammation called *orchiditis*. This may render the testes unable to produce sperm cells and the individual will consequently be permanently sterile.

Figure 5.4 shows a cross section of a human testis. You can see that the testis is divided into compartments and that each of these compartments is packed with a long, coiled tubule called the *seminiferous tubule*. The seminiferous tubules are lined with special cells called *spermatogonia*, which go through a series of meiotic divisions that result in the formation of sperm cells (Figure 5.5). At this point, you may wish to return to Chapter 3 and review this process. Figure 5.6 summarizes the events that take place in the seminiferous tubule. As you look at this diagram, you should recall how meiosis reduces the chromosome number by half, and also that each sperm represents one of approximately 8,000,000 possibilities in terms of specific genetic content.

Figure 5.4
Cross section of the human testis. Note the relationship of the seminiferous tubules, epididymis, and vas deferens.

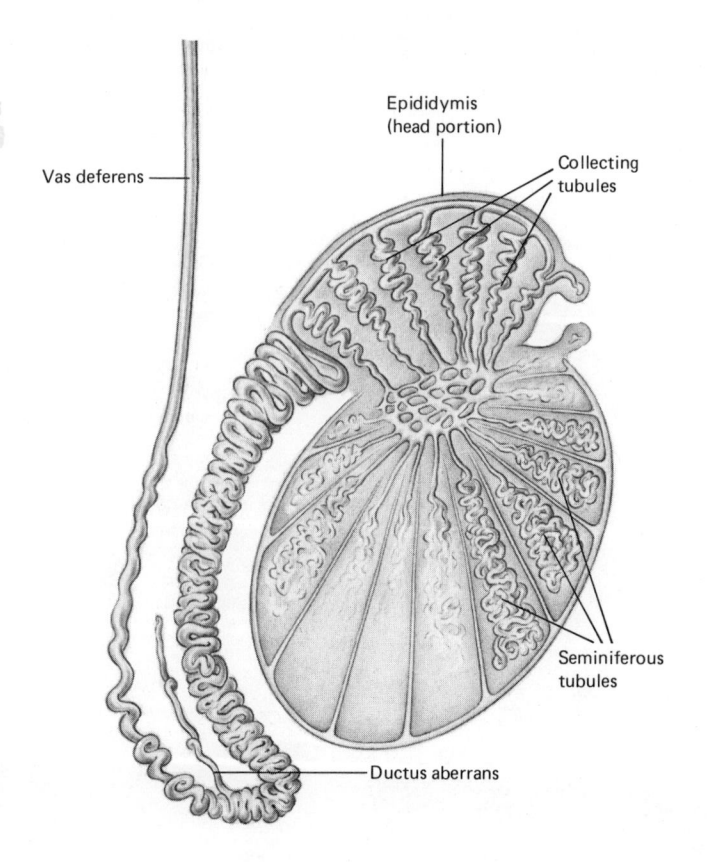

72   "Be fruitful, and multiply..."

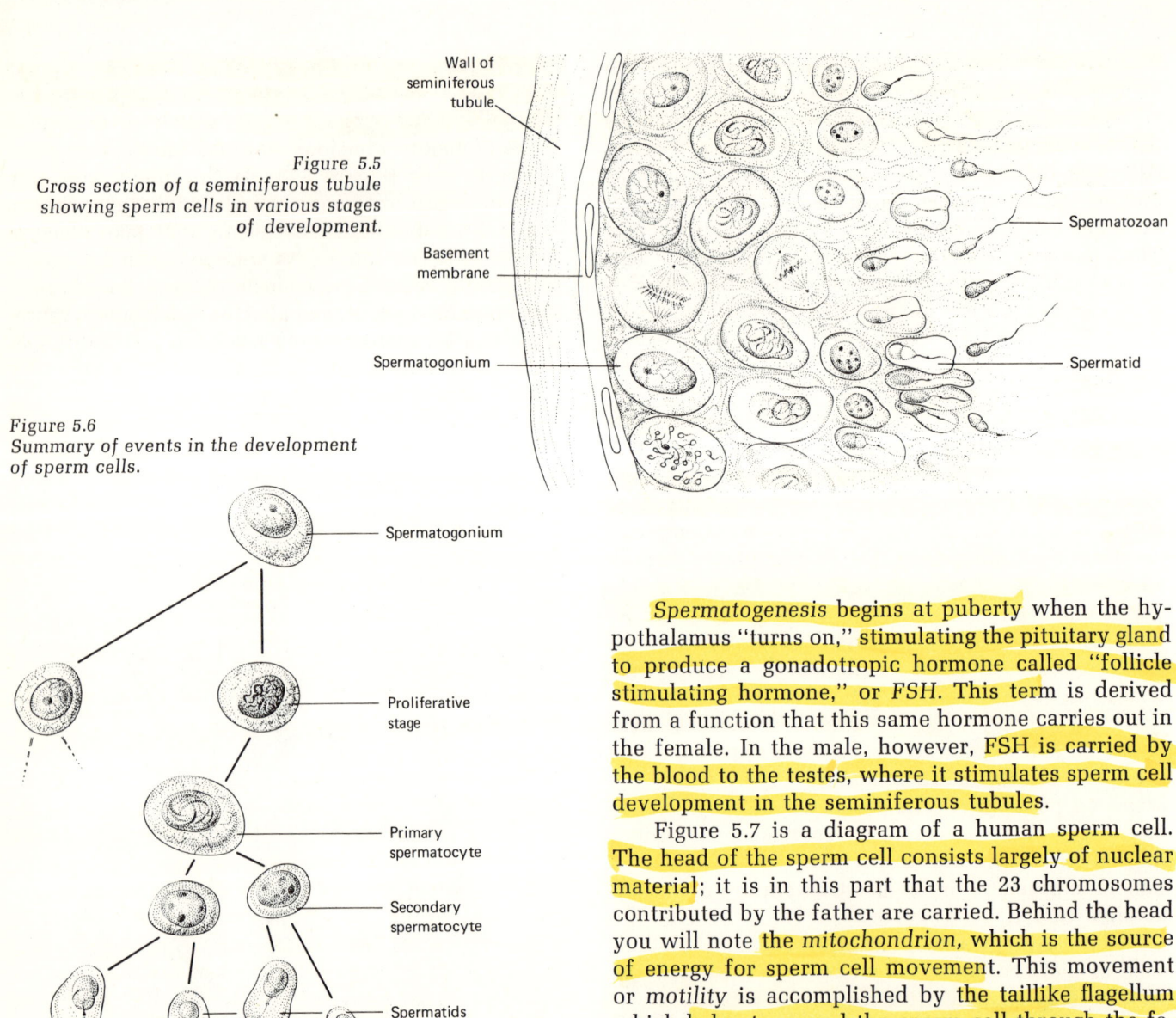

Figure 5.5
Cross section of a seminiferous tubule showing sperm cells in various stages of development.

Wall of seminiferous tubule
Basement membrane
Spermatogonium
Spermatozoan
Spermatid

Figure 5.6
Summary of events in the development of sperm cells.

Spermatogonium
Proliferative stage
Primary spermatocyte
Secondary spermatocyte
Spermatids
Sperm

Spermatogenesis begins at puberty when the hypothalamus "turns on," stimulating the pituitary gland to produce a gonadotropic hormone called "follicle stimulating hormone," or *FSH*. This term is derived from a function that this same hormone carries out in the female. In the male, however, FSH is carried by the blood to the testes, where it stimulates sperm cell development in the seminiferous tubules.

Figure 5.7 is a diagram of a human sperm cell. The head of the sperm cell consists largely of nuclear material; it is in this part that the 23 chromosomes contributed by the father are carried. Behind the head you will note the *mitochondrion*, which is the source of energy for sperm cell movement. This movement or *motility* is accomplished by the taillike flagellum which helps to propel the sperm cell through the female reproductive tract. At the tip of the sperm cell, there is a structure called the *acrosome*; it is believed that this structure secretes an enzyme called *hyaluronidase* which aids in the process of fertilization.

In addition to FSH, the pituitary gland releases a gonadotropic hormone called "luteinizing hormone,"

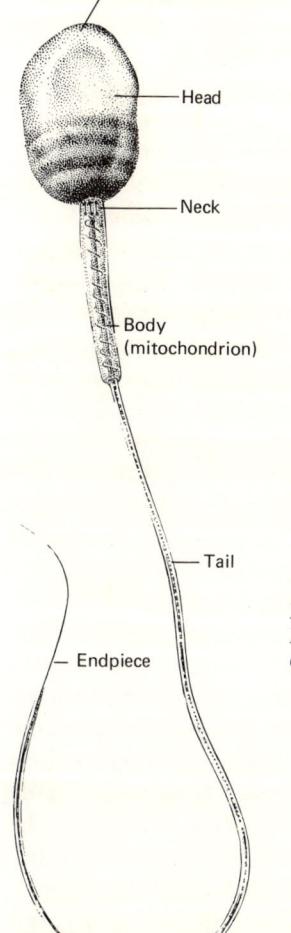

Figure 5.7
Diagram of a mature human sperm cell.

or *LH*. Again, this hormone is identical to a pituitary hormone which has a very different function in the female, so in an attempt to avoid confusion, LH in the male is sometimes called "interstitial cell stimulating hormone," or *ICSH*. LH, or ICSH if you prefer, stimulates the *cells of Leydig* to produce testosterone, which has already been described as the major male sex hormone and the one most responsible for the development and maintenance of masculine traits. The cells of Leydig are located in the spaces between the seminiferous tubules, as shown in Figure 5.8.

Removal of the testes is called *castration*; the surgical procedure is called an *orchidectomy*. In the prepubertal male, castration results in the development of a *eunuch*, an individual who is usually of large stature and lacking in normal sex drive. In the past, castration was deliberately performed on individuals in order to produce harem guards, since their lack of normal sex drive made them especially trustworthy for this purpose. Incredibly, the practice of castration was at one time also sanctioned for the purpose of producing males with special voices for church choirs! Castration of the mature male with previous sexual experience will probably reduce normal sex drive, but it may not be lost entirely. This suggests that once established, sex drive in the human male may be partly psychological in origin.

Path of the Sperm Cells

We will now follow the path of the sperm cells as they leave the seminiferous tubules where they are formed (refer to Figures 5.5 and 5.9). Note that a series of collecting tubules connect the main body of the testis with a coiled tube which rests on its surface. This coiled tube is called the *epididymis*. From the epididymis, we can follow the path of the sperm cells to another tube called the *vas deferens*, which can be traced up into the abdominal cavity where it eventually enters the *urethra*, or the opening to the outside. Although it is still not completely clear, it is considered probable that inactive sperm cells are stored in the epididymis and to some extent in the vas deferens. While thus stored they are not motile; that is, they do not move by whipping their taillike flagella as they do when they are released from the male reproductive tract. In this way they conserve the relatively small amount of energy that they possess.

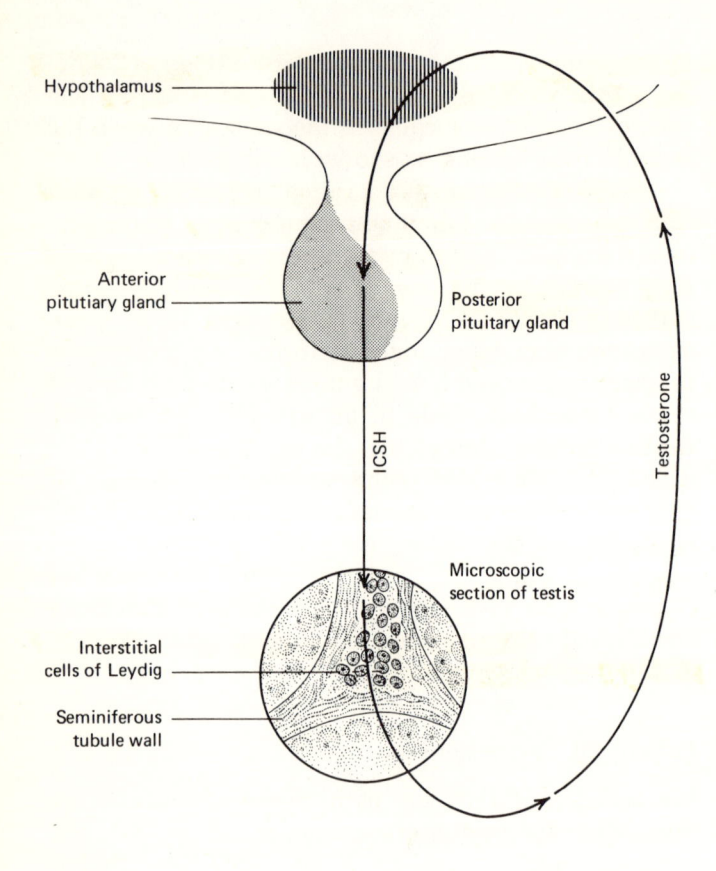

Figure 5.8
Interaction of ICSH (LH) and testosterone in the human male.

Accessory Glands

Looking again at Figure 5.9 and also back to Figure 5.3, we find the *prostate gland*. This structure consists of glandular and muscular tissue and surrounds the opening of the urinary bladder. When sperm cells are released by the male, an event called *ejaculation*, rhythmic contractions of the prostate contribute a viscous, milky substance that makes up most of the seminal fluid, in which the sperm cells are transported into the vagina of the female. The *seminal vesicles* also contribute to the seminal fluid. The prostate and seminal vesicles are important because the fluid that they add to the sperm cells is alkaline. This alkalinity (1) stimulates motility in the sperm cells, and (2) helps to neutralize the acidity of the female vagina. The vagina is actually hostile to sperm cells because it is normally acidic; sperm cells will not long survive in an acidic environment. The alkaline nature of the seminal fluid thereby increases the probability that a significant number of sperm cells will survive and participate in the fertilization process.

The prostate gland is a frequent source of difficulty because of its unfortunate position around the opening of the bladder. A significant number of men, especially those beyond middle age, suffer from *prostatitis,* or inflammation of the prostate. In this condition, the prostate enlarges and constricts the neck of the bladder, making urination difficult or even impossible. The retention of urine is a serious matter and may require surgical intervention. Even more serious is cancer of the prostate, as it accounts for as much as three percent of all cancer cases in human males.

Also note the structure labeled *Cowper's gland*, located at the beginning of the urethra. These glands supply mucous to the urethra. During sexual excitement prior to intercourse, Cowper's glands secrete increased amounts of mucous, providing a lubricant which facilitates the insertion of the penis into the vagina.

The Penis

The penis is the male copulatory organ and it functions to introduce sperm cells into the female reproductive tract. The *urethra,* or tubular opening through the penis, permits passage of urine from the bladder to the outside. It also carries the seminal fluid with millions of sperm cells into the female vagina, where they

The testes   75

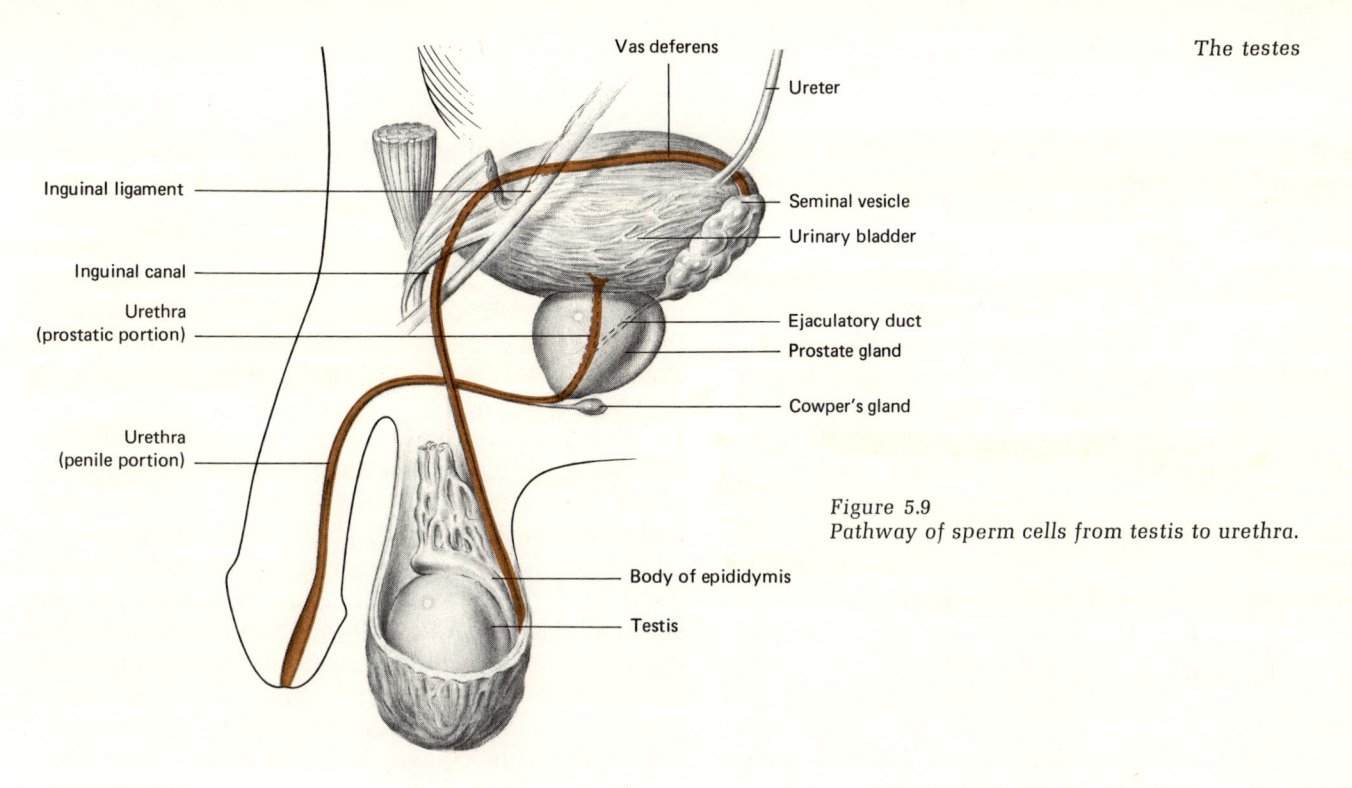

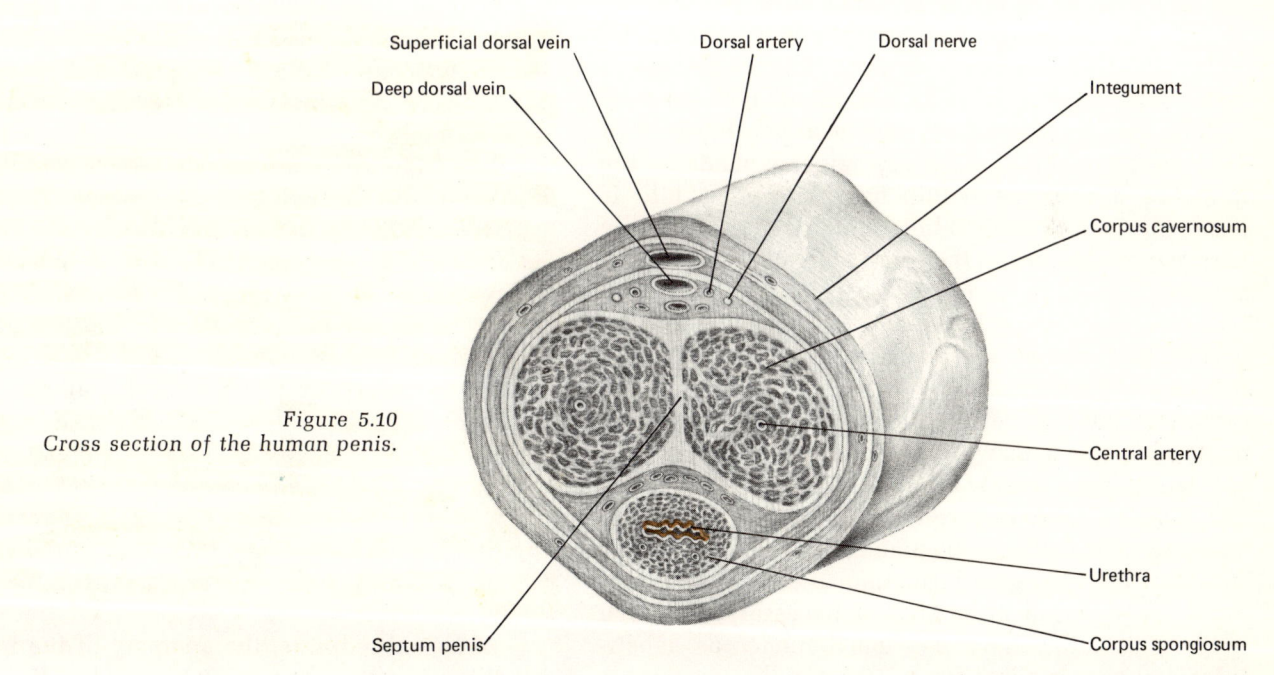

Figure 5.9
Pathway of sperm cells from testis to urethra.

Figure 5.10
Cross section of the human penis.

are deposited as close to the opening of the uterus as possible. The penis contains three cylindrical bodies of spongy tissue which become engorged with blood during periods of sexual excitement (Figure 5.10). This causes the organ to become enlarged and rigid, a phenomenon known as *erection*. In its erect state, it can be inserted into the vagina during the act of intercourse. At the climax of sexual intercourse, the entire male reproductive tract contracts rhythmically, beginning with the epididymis and extending through the vas deferens and the urethra. Fluid from the prostate and seminal vesicles is added to the sperm cells, and a total of three cubic centimeters or more of seminal fluid containing around 400 million sperm cells is deposited into the vagina.

The Female Reproductive System

The reproductive function of the female is undeniably more complex than that of the male. Like the male, she furnishes 23 of the 46 chromosomes that will determine the offspring's genetic traits, but whereas the male's biological function ends at that point, hers is just beginning. For the next nine months, or approximately 266 days, her body will bear the burdens of extra nutritional and excretory demands made by the new life developing within her. After the child is born, the major responsibility for its care is usually hers, especially during the early years of growth. Considering this great discrepancy between the reproductive responsibilities of men and women, it is not surprising that women are finally beginning to question the logical basis for the fact that both secular and sectarian laws governing human reproduction are made almost exclusively by men!

The human female is well equipped to carry on her complex reproductive function, to harbor the developing life from its beginning as a single cell to full-term baby, to give birth, and to provide nourishment for a period after birth if necessary, although modern societies have developed numerous substitutes and supplements for mother's milk.

Figure 5.11 illustrates the basic anatomy of the female reproductive system. The external portion is called the *vulva* and it consists of two sets of folds. The outermost, larger folds are called the *labia majora* and the inner, smaller set is called the *labia minora*. The labia minora surrounds the entrance to the *vagina*, which is a muscular tube about three or four inches in length and is regarded as the female copulatory organ.

In young girls, a membrane called the *hymen* usually partially blocks the entrance to the vagina. An intact hymen is mistakenly regarded by uninformed males as a sure sign of virginity, while its absence is considered to be equal proof of nonvirginal status. In point of fact, hymens come in various shapes and sizes; a large percentage of young girls tear the hymen during normal childhood physical activities and, conversely, intact hymens may be present in those whose virginity is long beyond recall.

Located just above the labia minora is a small mass of erectile tissue known as the *clitoris*. The clitoris is homologous to the male penis, having similar embryological origins. During periods of sexual excitement, it becomes engorged with blood and highly sensitive.

The vagina extends inward to the *cervix*, which is the opening to the *uterus*, or womb. The uterus is a pear-shaped, muscular organ that is well adapted to its function of housing the developing embryo. Its lining, called the *endometrium*, changes in thickness and in other ways during the monthly cycle as it is prepared anew to receive and nourish the embryo.

From the upper portion of the uterus, the *oviducts*, or *fallopian tubes*, lead to the *ovaries*, which are the female gonads and are thus comparable to the male testes. Each fallopian tube terminates in a finger-like structure called the *ostium*. The ostium rests on the surface of the ovary and picks up the egg cell when it is released by the ovary during the monthly cycle.

As described here, the anatomy of the female reproductive system is relatively simple. We shall see,

The female reproductive system 77

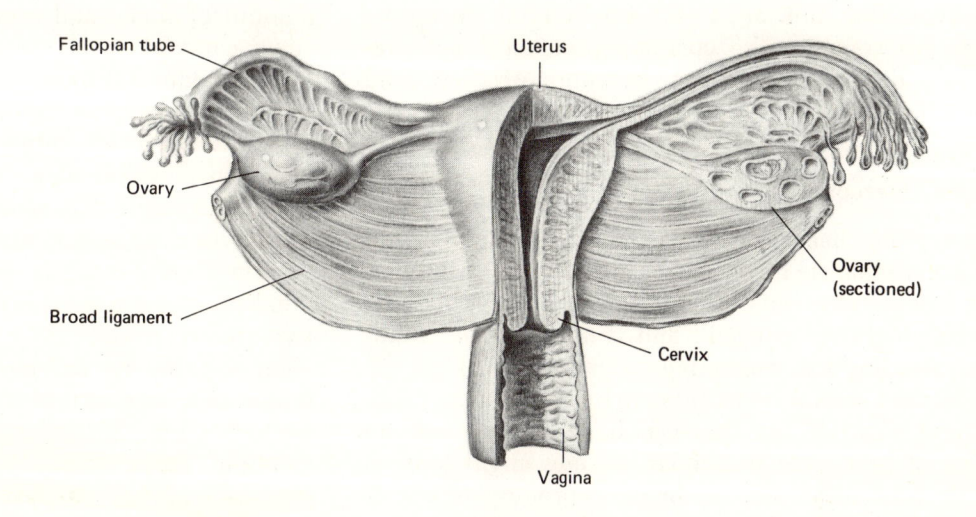

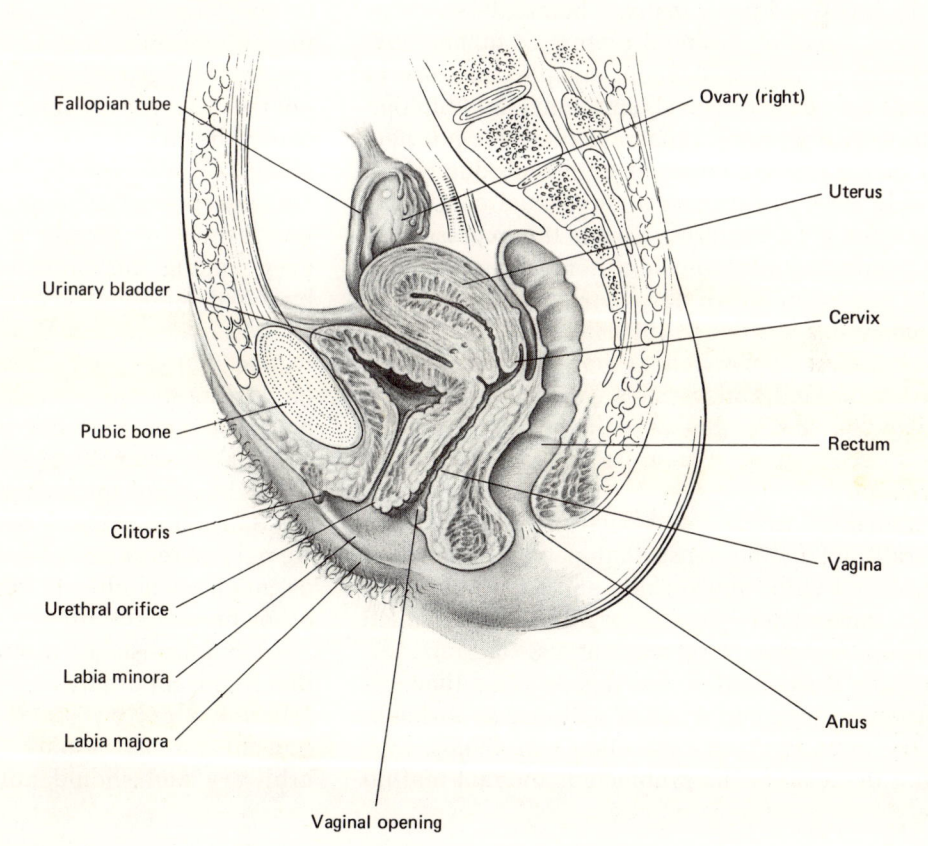

Figure 5.11
Diagram of the human female reproductive system.

however, that this apparent simplicity is deceptive when we consider the complex processes that prepare the uterine lining for the reception of a fertilized egg cell.

Sexual Intercourse

A complete discussion of sexual intercourse would require a separate book. An excellent and highly readable treatment of this topic has been written by E. T. Pengelley (1974); you will find it listed in the Annotated Reading List beginning on page 361.

Human sexual intercourse is by no means a "simple and natural" act that can be described only in terms of biological principles, as one might look at the division of an amoeba or the reproductive activities of the birds and bees. Instead, human intercourse is only one aspect of the total picture of *human sexuality*, which is an unbelievably complex mixture of social, cultural, psychological, emotional, and biological factors. Furthermore, human sexuality is *not* necessarily or explicitly connected with reproduction. It is probably rare, in fact, for heterosexual couples to enter into the act of intercourse for the express purpose of producing offspring, and homosexual couples certainly never do. There now appears to be a growing tendency to accept or to at least tolerate such variations in sexual expression as homosexuality. This may well be a tacit and perhaps overdue recognition of the significant role that sexuality plays in human relationships, quite apart from its *potential* reproductive function.

As stated before, human sexual intercourse is not a "natural" act in the sense that people come to it fully equipped with inborn skills and "instincts" that ensure an immediately satisfactory and well-adjusted relationship. A couple must work at a sexual relationship to *make* it successful, and this requires time, effort, patience, and, above all, a willingness to listen to and try to understand each other's problems. Certainly not the least of the problems is the fact that no amount of social and economic equality between the sexes can make their obvious physical, physiological, and behavioral differences disappear. If we now add the influence of the rigorous sex-role conditioning to which we are all subjected from the day we are wrapped in either blue or pink receiving blankets, it is not surprising that males and females tend to differ greatly in their sexual attitudes and responses.

Until quite recently, sexual intercourse and its related problems were hidden under a cloak of secrecy. Even today, despite the emphasis—and overemphasis—on sex that we see in today's movies, novels, and magazines, our attitudes still bear the mark of the Victorian era. Sex education in public schools, for example, remains a controversial subject in many places and may bring violent reactions from parents or special groups that are self-appointed guardians of public morality.

In the puritanical atmosphere that characterized our society for so long, human sexuality and especially the act of intercourse were not considered fit subjects for research. As a consequence, ignorance and misconceptions concerning this critically important aspect of human life continued to be the rule even among the educated and, indeed, even among biologists and physicians!

W. Masters and V. Johnson (1966) were the first to use human subjects to conduct extensive studies on the nature of sexual intercourse. Their research has produced a wealth of new information and has debunked a number of ideas that have long been dear to the hearts of the writers of marriage manuals. Even so, there is still much to be learned; we must be extremely careful not to generalize, since people bring as much variability to sexual activities as they do to other human activities.

For purposes of discussion, Masters and Johnson divide the act of physical intercourse into four phases: (1) excitement phase, (2) plateau phase, (3) orgasmic phase, and (4) resolution phase. These phases are arbitrary and should not be thought of as separate and well defined, since one phase grades imperceptibly into the next.

The *excitement phase* usually starts as a result of sensory stimuli such as touch, sight, sound, or smell. In some cases, an active imagination helps to build sexual excitement. In general, the excitement phase prepares both sexes, physically and probably emotionally, for the actual act of intercourse. In the male, this preparation stage is relatively simple and consists of the establishment of an erection as blood fills the spongy cylindrical spaces in the penis. The erect penis is capable of entering the vagina of the female; without such erection, insertion or *intromission* is not possible.

This preparation phase is extremely important to the female because her body is essentially preparing itself for penetration. As Pengelley (1974) wisely points out, it is much different to be penetrated than to penetrate. It follows that if the female is to gain enjoyment from intercourse, she must be thoroughly prepared, largely by the attentions of the male. This preintercourse preparation, sometimes called *foreplay*, may vary considerably according to the needs of different women, but it is essential that it go on as long as necessary if intercourse is to be enjoyable and satisfying for *both* partners. Masters and Johnson (1966) found that the female typically undergoes a number of physical changes during this excitement phase. First of all, the vaginal barrel is lubricated by glands located in the vaginal wall, thus helping prepare it for intromission. Also, the inner two-thirds of the vagina lengthens and enlarges, and the cervix, which normally extends down into the vaginal tract, is pulled upward, helping to clear the way for intromission. The labia minora increase in size as they become engorged with blood; the labia majora tend to separate. The clitoris, which you will recall is homologous to the male penis, increases in sensitivity to touch. The breasts tend to increase in size and the nipples become erect and sensitive.

The *plateau phase* begins when the penis is inserted into the vagina and active intercourse starts. During this phase the testes enlarge and are elevated toward the body wall. In the female, the clitoris retracts beneath a fold of tissue called the *clitoral hood*, and the opening of the vagina narrows. The outer one-third of the vagina and the labia minora develop a tension response. These areas make up what is called the *orgasmic platform*, because the female climax reaction, or *orgasm*, appears to originate from there. Meanwhile, in both partners there are increases in heart rate, respiration, and blood pressure.

The *orgasmic phase* is the climax of intercourse. The male orgasm consists of rhythmic muscular contractions of the epididymis, prostate, seminal vesicles, and penis, all of which result in the *ejaculation* of seminal fluid and sperm cells. This orgasmic phase lasts for a short time—actually only a few seconds—and it is accompanied by an intense pleasurable sensation that begins in the pelvic region and spreads throughout the body. The elevation of the testes that occurred during the plateau phase appears to be associated with the force of ejaculation. In older men, for example, this elevation is reduced, and thus the force with which the seminal fluid is discharged is also reduced.

In the female, the orgasmic response appears to begin with rhythmic contractions of the orgasmic platform which then spread through the vagina to the uterus and to the anal sphincter muscle. These vaginal contractions vary in number from three to fifteen and are accompanied, as in the male, by an intense sensation of pleasure that begins at the clitoris and spreads through the pelvic area to the rest of the body. Some females, incidentally, may reach orgasm more than once during the same act of intercourse. This again varies in different women and depends upon a variety of factors, not the least of which is the skill and patience of her partner. The male, on the other hand, can achieve orgasm only once and must wait for a period that varies from a few minutes to several hours before he is once again capable of carrying on intercourse. It is therefore important that the male exercise the patience and control necessary to ensure that the female has maximum opportunity to achieve orgasm. This, by the way, is much more important than both partners reaching climax at the same time, which is a marriage-manual ideal but which most people find is hardly worth the effort.

The *resolution phase* follows orgasm in both the male and female. In the male, the penis returns to its nonerect state and the testes lower and return to normal size. In the female, the vagina and labia return to normal and the cervix of the uterus lowers back into the upper portion of the vaginal canal. Both partners usually experience a release of tension, a feeling of relaxation, and a desire to sleep.

The preceding is a coldly clinical description of human sexual intercourse and represents the "ideal" situation in which all physical responses occur as they are supposed to. Unfortunately, things do not always go so well. For example, problems related to female orgasm continue to be the subject of innumerable books and articles; indeed, in some circles it seems that female orgasm has become the entire focus of human sexuality. It is true that some women achieve climax only rarely or not at all, even though hormone levels and physical structures appear to be normal. A recent study (Bell) of a large group of American wives showed that of those studied, only 17 percent reported that they "always" achieved orgasm. Another 42 percent said that they reached climax "most of the time," 32 percent said "some of the time," and 8 percent reported that they never experienced orgasm. Similar results have been reported by other studies. It is especially interesting to note that these figures are not significantly different from those published by A. C. Kinsey well over twenty years ago.

Sex drive and the ability to achieve orgasm in the female is probably influenced originally by the presence of female hormones, but psychological and emotional factors undoubtedly play a significant role as well. Older women, for example, often experience an increased sex drive after the ovaries have stopped producing high levels of hormones. It has been suggested that this increase in sexual activity may be due in part to the elimination of the fear of pregnancy that haunts many women for most of their reproductive years.

Actually, it is entirely normal for the female to occasionally fail to reach climax. In fact, unfortunately for some wives and apparently for some husbands as well, the current emphasis on orgasm has created unnecessary feelings of sexual inadequacy. Indeed, a substantial number of women object to climax as an unfailing measure of total sexual satisfaction. On the other hand, if a female never or almost never achieves orgasm, it may strongly suggest psychological influences, although physical problems cannot be ruled out entirely in all cases.

In some cases, failure to respond may reflect a married woman's inability to suddenly think of sex as a normal and wholesome relationship when she has been conditioned for most of her life to regard it as a trap to be avoided at all costs. After all, it is not that long ago when women spoke of "giving themselves" or "submitting" to the male rather than approaching intercourse as a mutually enjoyable relationship in which neither partner is necessarily dominant. It seems highly likely that these same attitudes still strongly influence today's supposedly sexually liberated society.

In addition, many females do not achieve orgasm due to failure of the male partner to recognize that adequate preintercourse preparation of the female is necessary, and that sex drive in the female is complicated by emotional factors to a far greater degree than in the male. Finally, sex drive in the female may vary considerably over her monthly cycle. She is not equally responsive at all times, whereas the male's sex drive is more consistent.

The Menstrual Cycle

The "hypothalamic clock" that governs the human female's reproductive life begins to function at around eight or nine years of age. This leads to regular menstrual cycles, usually beginning somewhere between the ages of eleven and fifteen. This onset of menstruation in the young girl is one of a number of changes related to puberty that correspond to those that take place in the adolescent male.

The beginning of female adolescence is accompanied by the release of female sex hormones from the ovaries. The principal female hormones are *estradiol, estrone,* and *progesterone.* Estradiol and estrone are known collectively as *estrogens;* we will simplify matters somewhat by referring to them as *estrogen.* Estrogen stimulates the full development of the female sex organs and also participates in the growth spurt which characterizes girls during early adolescence. It also appears to bring about the eventual cessation of growth by halting growth of the long bones. In this latter respect, estrogen seems to have a stronger influence than testosterone; this may be part of the reason why the average height of females is less than that of males. Estrogen also influences the development of the secondary sex characteristics of the female. These include a distribution of fat in the hips, the development of the mature female breast, growth of hair in the groin and armpit areas, and a voice that is typically higher pitched than that of the male because of a difference in the size of the larynx, or "voice box."

The menstrual cycle is a phenomenon that occurs only in the human female, in apes, and in Old-World monkeys. In other mammals, the female goes through a period called *estrus,* or "heat," during which she is fertile and is also receptive to the male. Willingness on the part of the female to enter into intercourse therefore coincides with maximum fertility. In the human species, however, the female may be receptive to the male at any time during her cycle, including those times when she is not likely to be fertile.

Menstruation itself is sometimes referred to by women as the "curse," not only because it is obviously an inconvenience, but also because it is often accompanied by symptoms such as abdominal cramps, headache, nausea, and general irritability. Still, although the menstrual cycle may be inconvenient and at times even unpleasant, it is nevertheless essential to the reproductive process. If the egg cell is fertilized, the hormones released during the cycle ensure that the uterine lining is thoroughly prepared to receive the embryo and provide nourishment during its early days of development. Secondly, the complex hormonal relationships of the cycle ensure that no more egg cells are produced until after development and birth of the offspring. Otherwise there could possibly be several embryos at various stages of development in the uterus at the same time.

The length of the menstrual cycle varies considerably among women, and in some cases it may vary in the same woman. Cycle irregularity is especially common among adolescent girls and among older women approaching *menopause,* or "change of life," which is that time of life when the menstrual cycles stop. Normal cycles may vary in length from about 20 to 40 or more days, but the "average" cycle in the "average" woman lasts about 28 days. For our discussion, we will therefore use the 28-day cycle as our model, keeping in mind as we do so that many perfectly normal women have cycles that differ considerably from the "textbook average."

What follows below is a step-by-step summary of the events that take place during the normal menstrual cycle. This will require careful study. As you go along, it will help if you refer often and carefully to Figures 5.12, 5.13, and 5.14.

1. The menstrual cycle begins on the first day of the menstrual *period.* This period, also called *menstruation,* lasts from three to six days, during which there is a vaginal discharge that contains blood and mucous from the breakdown of the uterine lining. At this time, the pituitary is stimulated by the hypothalamus to secrete FSH, or *follicle stimulating hormone.* You will recall that this is the same hormone that stimulates the production of sperm cells in the male. In the female,

82   "*Be fruitful, and multiply...*"

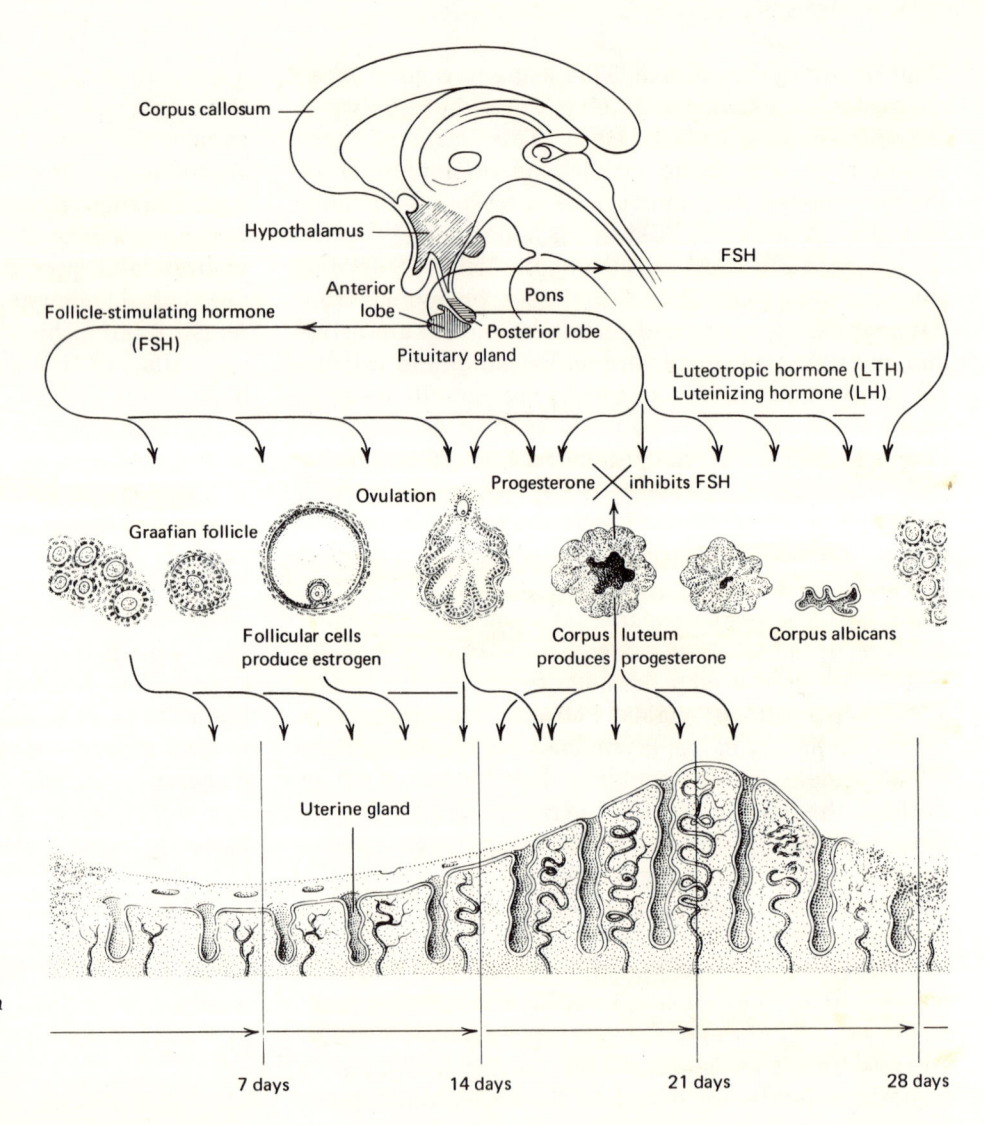

Figure 5.12
Diagram illustrating the relationships between the menstrual and the ovarian cycles. (See text for details.)

however, **FSH is carried by the bloodstream to the ovary where it initiates the development of a** *follicle* (Figure 5.14). **There are several thousand of these primordial, or immature, follicles in the ovaries of a female, and each one consists of an egg cell surrounded by a single layer of cells.** Only about 400 follicles will actually mature during a female's average reproductive life of 30 to 40 years.

2. **As the follicle develops under the direction of FSH, its cells secrete the hormone** *estrogen* **and a smaller quantity of another hormone,** *progesterone*. **Estrogen brings about a thickening of the**

The menstrual cycle   83

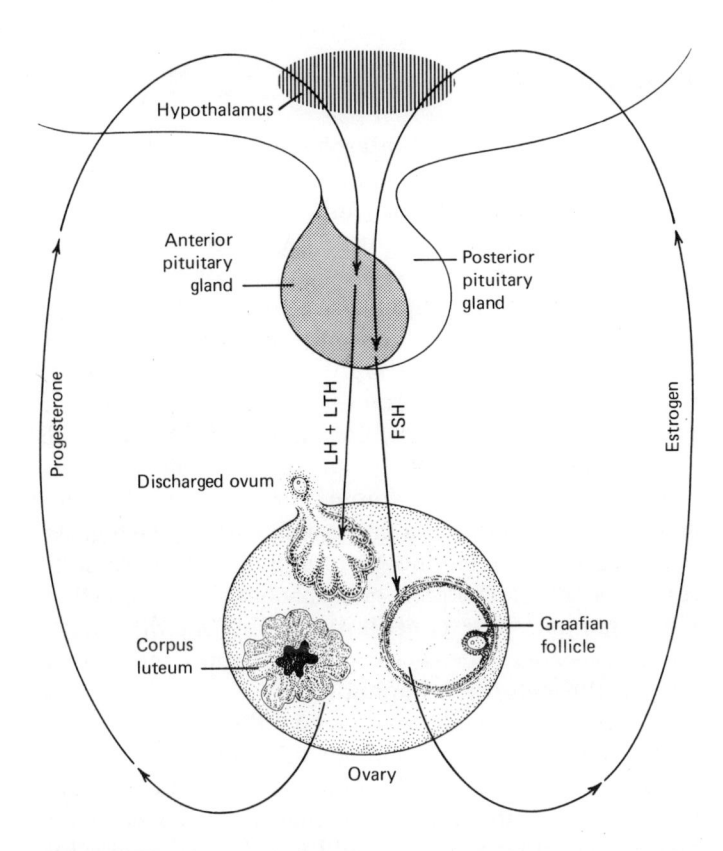

Figure 5.13
Interaction of the various female hormones. (See text for details.)

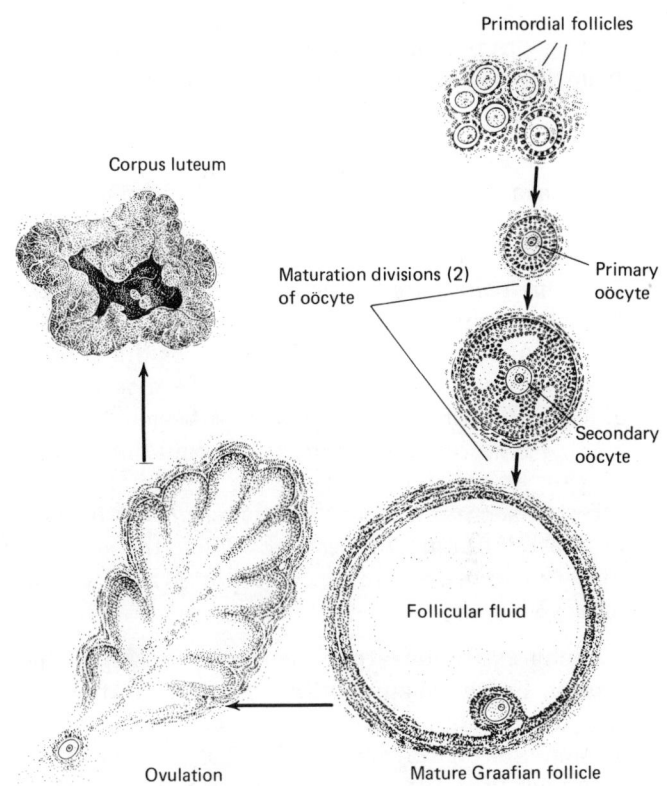

Figure 5.14
Diagram showing various stages in the development of the follicle and the corpus luteum.

uterine lining which includes the distribution of blood vessels and glandular structures (Figure 5.12). This is the beginning of the process whereby the uterine lining is prepared to receive the embryo.

3. As the level of estrogen in the blood increases, the negative feedback system calls on the pituitary to secrete lesser amounts of FSH. Meanwhile, the hypothalamus is stimulating the pituitary to secrete the *luteinizing hormone,* or LH (Figure 5.13). You will recall that LH is the hormone that brings about the production of testosterone in the male.

4. By the midpoint, or approximately day 14 of the 28-day cycle, the follicle and egg cell have developed to maturity. At this time the "hypothalamic clock" calls for a surge of LH to be released from the pituitary. This sudden increase in the level of LH acts together with FSH to break the egg cell out of the follicle (Figures 5.12, 5.13, 5.14), after which the egg enters the fallopian tube (Figure 5.11). This release of the egg cell from the ovary is a critical point in the menstrual cycle and is called *ovulation.* As you can see, it is essential that the "hypothalamic clock" turns on at exactly the right time if a mature egg is to be available for fertilization. The exact nature of the

female "hypothalamic clock" and why it differs from the male hypothalamus is one of the mysteries of human reproduction.

5. Ovulation is accompanied by a change in the follicle to a structure called the *corpus luteum,* or "yellow body." This conversion of the follicle to the corpus luteum is influenced by the pituitary hormone called the "luteinizing hormone," or LH. LH, along with another pituitary hormone called the *luteotropic hormone* (LTH), causes the corpus luteum to produce large amounts of *progesterone* along with smaller amounts of estrogen (Figure 5.13). Progesterone continues the buildup of the uterine lining and stimulates its glandular structures to secrete nutrient substances. This is part of the continuing process by which the uterine lining is prepared for the possible arrival of an embryo.

6. As long as the level of progesterone is high, the uterine lining remains intact. The preparation of the uterine lining reaches its maximum point around the eighth day following ovulation. This coincides with the approximate time at which the embryo is due to arrive, provided that an embryo has been produced. By the tenth to the twelfth day after ovulation, the rising level of progesterone inhibits the pituitary's production of LH and LTH. As a result, the corpus luteum deteriorates and the level of progesterone drops, permitting the breakdown of the uterine lining and renewed menstrual flow. At this point, the levels of estrogen and progesterone are low; thus through the negative feedback system, the hypothalamus once again calls on the pituitary for FSH, thereby beginning a new cycle (Figure 5.13).

The foregoing summary represents the events of the menstrual cycle as they occur when the egg cell is *not* fertilized. In the following sections, we will consider the process of fertilization and what happens when an egg cell is fertilized.

Fertilization

At the climax of sexual intercourse, the fertile human male releases seminal fluid containing somewhere around 400 million sperm cells into the vagina close to the opening of the uterus. A sperm cell swims through the vaginal fluids at a rate of about two-tenths of an inch per minute, which is not fast enough to account entirely for the fact that sperms are found well up into the uterus only 90 seconds following their ejaculation by the male. It is believed that the presence of the seminal fluid causes contractions of the female tract which help speed the sperm cells into the uterus. It has been found, for example, that the seminal fluid contains a substance called *prostaglandin,* which has been shown to stimulate uterine contractions. Fluid currents and the beating of hairlike cilia also possibly contribute to the rapidity with which the sperm cells are propelled into the uterus and up through the fallopian tube.

Those sperm cells that survive the journey to the upper reaches of the fallopian tube may or may not meet an egg cell, depending on how close to the time of ovulation intercourse took place. If they do meet an egg cell, however, they will find it surrounded by a barrier of follicle cells that adhered to its surface when it broke out of the ovary. If the egg is to be fertilized, therefore, this barrier must first be broken down. The sperm cells do this by secreting an enzyme called *hyaluronidase,* which digests away a portion of the cell barrier that surrounds the egg. Since a single sperm does not secrete enough hyaluronidase to effectively break down the barrier, it is believed that this must be done by a large number of sperm cells. This may be one reason why fertility depends in part upon the release of many millions of sperm cells by the male.

Normally, a single sperm makes contact with the egg cell membrane. The egg cell responds by engulfing the head of the sperm, and the male nucleus combines with the female nucleus, thus bringing together the 23 chromosomes carried by the sperm and the

23 chromosomes contributed by the egg cell. The result is the single-cell stage of a new human being that may or may not successfully develop and enter the world nine months later.

It should be emphasized that although many sperm cells are necessary to accomplish fertilization, one and only one is normally permitted to enter the egg cell. The mechanism that permits one sperm to enter and to exclude the rest is not definitely known, but its importance is obvious if we consider the necessity to restore *exactly* the human chromosome number of 46.

It is generally estimated that sperm cells can live no more than 24 to 48 hours outside of the male reproductive tract, but some experts maintain that 72 hours is closer to the upper limit. It is also estimated that the egg cell will live no more than 24 hours after it leaves the ovary. Now, if we combine the preceding estimates with the fact that only one egg cell is normally released during each menstrual cycle, we might conclude that the female fertile period is limited to a span of two or three days each side of ovulation. In the next chapter, we will look at a method of contraception that is based on this apparently limited period of fertility; we will also see why this method is woefully unreliable because of the difficulty in predicting the exact time of ovulation.

Fertility

It is generally estimated that around 10 percent of marriages in the United States are involuntarily sterile. As an educated guess, it is probable that something over half of these sterile marriages are due to a lack of fertility on the part of the female.

In discussing female fertility, it is necessary to distinguish between the ability to *conceive* and the ability to carry the baby to full term. It is possible that a so-called sterile woman may regularly conceive, but because the uterine lining is not thoroughly prepared, the embryo fails to implant in the uterine wall and is therefore lost with the menstrual flow.

Such women are therefore actually "pregnant" for short periods without knowing it. Failure to conceive is another matter, and this problem is sometimes caused by a failure to ovulate. This may be due to unexplained failure of the "hypothalamic clock" to fire at the appropriate time and cause the surge of LH, which seems to be necessary to successful ovulation. This cause of sterility may be corrected by hormones given at the appropriate time during the menstrual cycle. In other cases, an abnormally thick capsule of tissue may surround the ovary, making it impossible for the egg cell to break out even though hormone levels are normal. There is also a small percentage of cases in which the ovaries are underdeveloped and do not produce egg cells.

If the egg cell is to be fertilized, it must enter the fallopian tube, and at the same time sperm cells must be able to make their way up the fallopian tube. A common cause of female sterility, therefore, involves a blockage of the fallopian tubes by scar tissue caused by an inflammation. Inflammation of a fallopian tube is called *salpingitis,* and one of the leading causes of salpingitis is *gonorrhea.* Gonorrhea is one of the two most common venereal diseases in the United States, and as the term *venereal* implies (Venus is the goddess of love), it is spread by sexual contact. Unfortunately, gonorrhea often goes undetected in the female and an inflammatory process may go on for a considerable time before she receives treatment.

Male infertility is by no means a small factor in sterile marriages. Lack of fertility in the male is commonly caused by (1) inability to produce any sperm cells, (2) inability to produce a sufficient number of sperm cells, or (3) production of an abnormally high proportion of defective sperm cells.

We have already mentioned that the fertile male releases 400 million or more mature sperms into the vagina during intercourse. If this number falls below 100 million, the male is likely to be infertile, even though it would be unwise to assume complete sterility. Also, while it is probable that most men produce at least some defective sperm cells, it is only when the proportion of defective sperms becomes too high that infertility results.

There is evidence that some marriages are sterile even though both partners possess normal fertility. It has been suggested that in these cases, the female tract rejects the sperm produced by her husband because of a specific kind of antibody reaction. Such a couple is therefore "sterile" with each other but would normally be fertile with other partners.

In situations where male infertility is a problem, pregnancy can often be achieved by *artificial insemination*. This is a procedure by which sperm from an anonymous donor or from a "sperm bank," where sperm cells have been kept in a frozen state, are artificially introduced into the vagina as closely as possible to the time of ovulation. Sometimes the husband's seminal fluid is mixed with that of the donor in an attempt to avoid certain psychological problems, since by this method the couple could never be absolutely certain that the resulting child was *not* fathered by the husband.

When artificial insemination was first introduced, there was an initial flurry of objections raised on both legal and moral grounds. There were even those who condemned artificial insemination as a form of adultery. With changing attitudes toward sexual matters, however, this moral objection now appears to most people as silly as arguing about how many angels can dance on the head of a pin. The legal problems, however, are potentially more serious. Who, for example, is the legal father of the child? Must the husband go through formal adoption procedures to ensure his legal rights as the child's father? Several states have already passed legislation that minimizes such problems, and most others give tacit approval by assuming that these problems need not arise.

Yet, whether or not to use artificial insemination remains a very personal decision which can only be made by the couple involved. It is obvious that both must agree and that neither the husband nor the wife should have any unspoken reservations that could later develop into severe psychological barriers. With such agreement, artificial insemination can provide a childless couple with the experience of pregnancy and birth, plus a child that is genetically at least half theirs.

While sterility may be a major problem for an individual couple that desires children, it is by no means the major reproductive problem that faces mankind today. To the contrary, excessive human fertility is probably one of the major sources of mankind's woes, and far more attention is being given today to methods of *preventing* conception than to promoting it.

Pregnancy

You will recall from our discussion of the menstrual cycle that if the egg cell is not fertilized, the corpus luteum will stop secreting progesterone and the endometrium will break down into the menstrual flow. (Menstruation has therefore been called, poetically if not always realistically, the "weeping of the disappointed womb.") In this section, we will consider the complex chain of events that is set in motion when the egg cell *is* fertilized.

After the male nucleus combines with the female nucleus in the egg cell, the resulting *zygote* goes through a series of mitotic divisions called *cleavage*. The first of these divisions produces two identical cells. A few hours later each of these two new cells again divides, producing four cells, and then these four cells each divide to yield eight, and so on. As the cells divide, they are held together by a membrane called the *zona pellucida,* as shown in Figure 5.15. In a small number of cases, the cells resulting from cleavage do not stay together and develop separately into two embryos. Since the process of mitosis supposedly guarantees that their chromosome contents are identical, the two embryos develop into *identical twins. Fraternal twins,* on the other hand, are formed when two different egg cells are fertilized by two different sperm cells; they are therefore no different

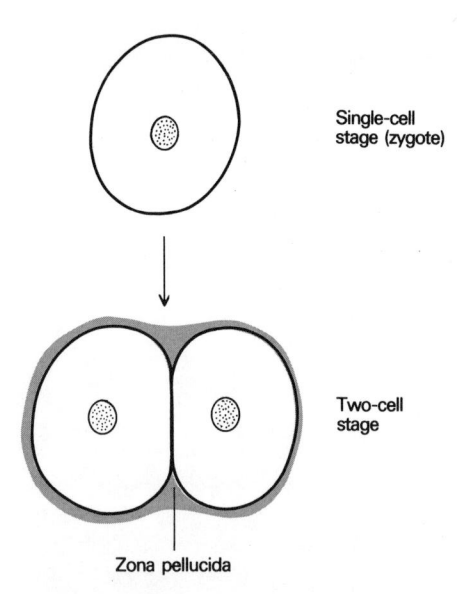

Figure 5.15
Division of the fertilized egg to form the two-cell stage, which is held together by the zona pellucida.

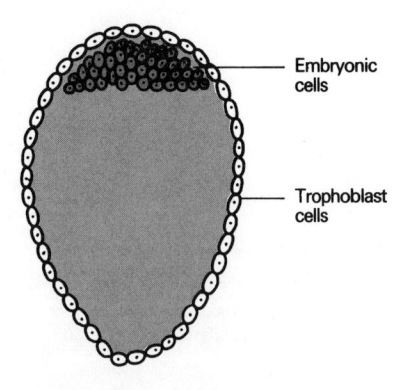

Figure 5.16
Blastocyst, showing trophoblast cells and embryonic cells that will develop into the embryo.

from regular brothers and sisters except that they have the same birthday.

The number of embryonic cells increases rapidly as cleavage progresses. With each division the cells become smaller, so the total size of the cell mass does not increase much beyond that of the original egg cell during the time that the cell mass is in the fallopian tube. Meanwhile, the mass of embryonic cells is propelled through the tube by wavelike contractions of the tube itself. Although the exact time is largely an educated guess, the journey through the fallopian tube probably takes three or four days.

Implantation

The next step, a very critical one, is called *implantation*. After the ball of embryonic cells has reached the uterus, further division results in a hollow structure called a *blastocyst*. Figure 5.16 shows that one side of the blastocyst consists of *trophoblast cells* and that the other side gives rise to the embryo itself.

The trophoblast cells secrete a substance that enables the blastocyst to burrow into the uterine lining. This happens on approximately the eighth day following ovulation, which you will recall is the point in the menstrual cycle when the uterine lining is at its peak of preparation. By around the twelfth day following ovulation, the blastocyst is safely embedded in the uterine lining, where it is bathed with nutrients from the mother's blood and from glandular secretions. Implantation is now completed (Figure 5.17).

As the embryo implants into the lining of the uterus, the trophoblast cells secrete a hormone called *chorionic gonadotropin*. Chorionic gonadotropin maintains the corpus luteum, which thereby continues production of progesterone and estrogen. Since this prevents the onset of the menstrual period, a "missed period" is a primary symptom of pregnancy. (We

88   "Be fruitful, and multiply..."

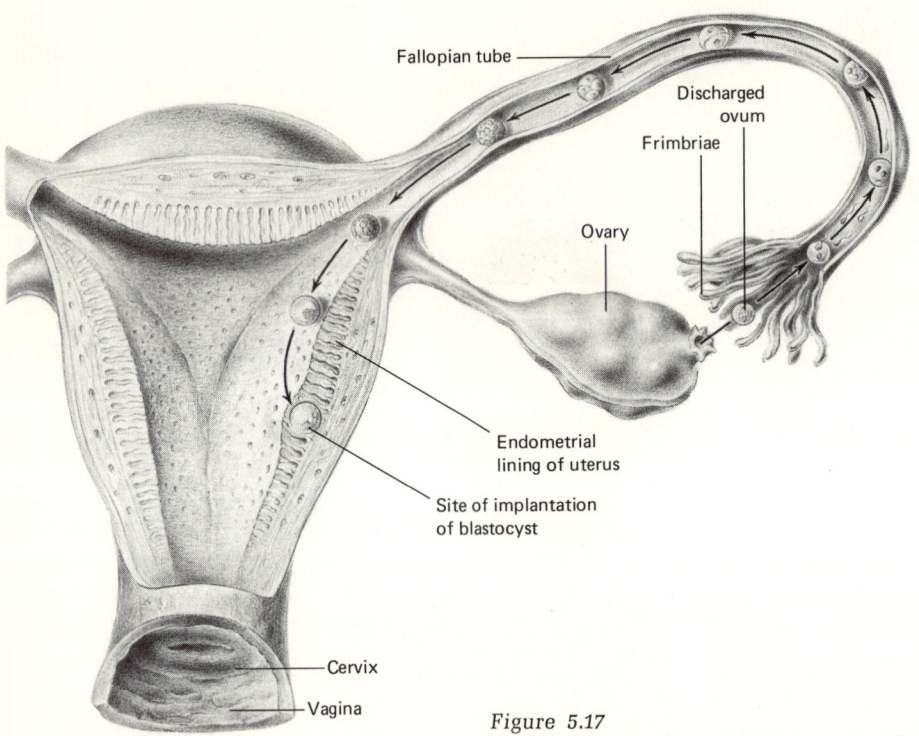

Figure 5.17
Implantation in the endometrium. Fertilization has occurred high up in the fallopian tube.

should hasten to add, however, that other conditions such as anemia or even worry and emotional upset can cause missed or late periods.) In addition, the continued high levels of progesterone and estrogen keep the pituitary from secreting FSH, thus preventing the development of new follicles. The uterus is thereby reserved for the exclusive occupancy of the new embryo for the next nine months.

Actually it is called an embryo only for the first eight weeks or so, during which the basic body structures develop from the original formless mass of cells. From that point up until birth, the developing and growing new individual is called a fetus. It is only after birth, when it takes its first breaths and ends its parasitic dependence on its mother's body, that it is dignified with the term "baby"—or John or Susan or whatever else that may be appropriate.

The Placenta

After implantation, the embryo receives oxygen and nutrients directly from the mother's blood and from uterine secretions. This is only a temporary arrangement, however, as a more complicated system is necessary to meet the increasing demands of the embryo as it develops over the next nine months.

As embryonic growth continues, projections reach out from the trophoblast and penetrate the uterine tis-

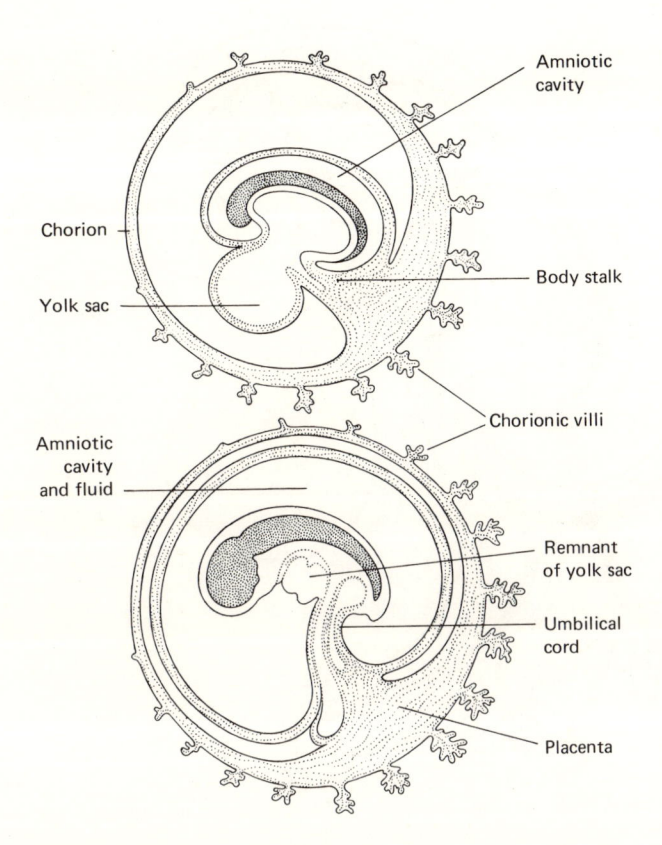

Figure 5.18
Relationship between the embryo and extra embryonic structures, including the placenta and amnion.

sue. These embryonic tissues together with maternal tissues develop into the *placenta*. The placenta is a disc-shaped structure about nine inches in diameter, and it serves as the point where transfer of nutrient and waste materials between mother and fetus will take place through most of the pregnancy.

In addition to its function as a transfer organ, the placenta produces large quantities of progesterone and estrogen. You will recall that chorionic gonadotropin causes the corpus luteum to remain intact during the early part of pregnancy. However, after approximately three months, the corpus luteum degenerates and the progesterone level starts to drop. By this time, however, the placenta is taking over by secreting large amounts of progesterone and estrogen. Under the control of these hormones, the uterine environment is maintained for the remainder of pregnancy. In some women, a tendency toward *miscarriage*, or spontaneously terminated pregnancy, may be traced to the fact that the placenta does not take over hormone production from the corpus luteum in time.

Figure 5.18 shows the relationship between the embryo and the placenta, and Figure 5.19 provides a close-up diagram of placental structure. Note that the embryo is attached to the placenta by the *umbilical cord*. This structure contains the umbilical artery and the umbilical veins; these are blood vessels which are extensions of the embryonic circulatory

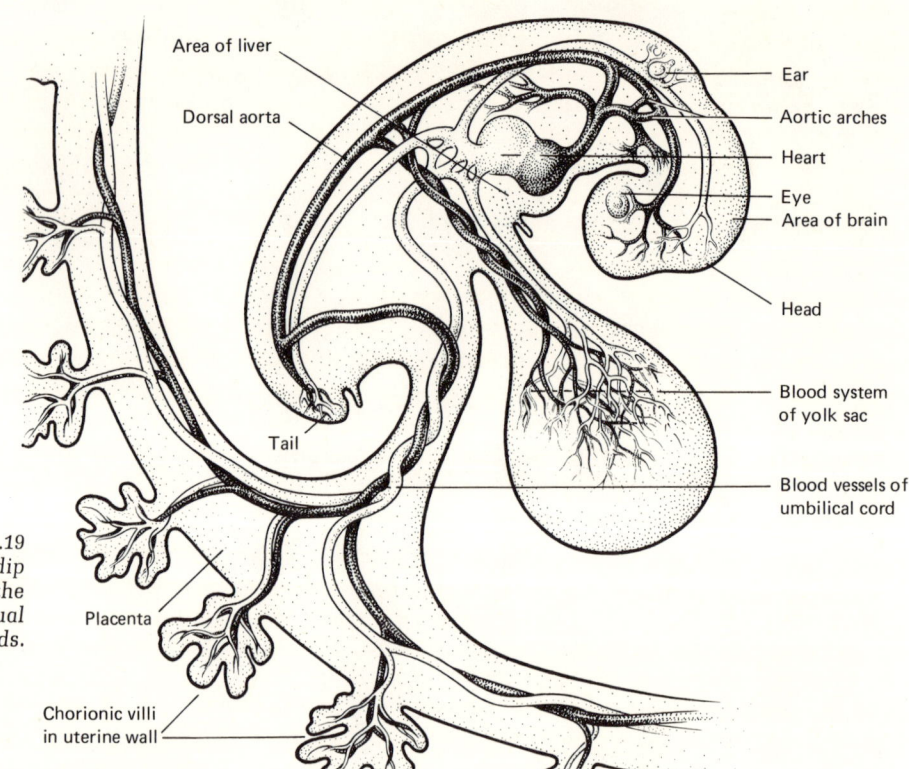

Figure 5.19
Details of placental structure. The villi dip into spaces which are filled with the mother's blood. Note that there is no actual mixing of maternal and fetal bloods.

system, so that the embryonic heart pumps blood through the placenta as well as through the embryo. The blood goes through the umbilical artery to the placenta, where it breaks down into capillaries that extend into the fingerlike villi. Figure 5.19 shows how these villi are bathed by the maternal blood as it passes through the sinuses (spaces) in the uterine portion of the placenta. Substances such as oxygen, glucose, and amino acids pass by diffusion from the maternal blood to the fetal blood. At the same time, waste materials diffuse through the villi from the fetal blood to the maternal blood. These wastes are then excreted by the mother through her own excretory organs. Note that the fetal blood, which is now richer in oxygen and nutrients, is returned to the fetal circulatory system through the umbilical veins. It may be seen that under normal circumstances, the maternal blood is entirely separate from the fetal blood, and mixing does not usually occur except in instances where there is a breakdown of the placental barrier.

Unfortunately, the placenta is not necessarily a barrier to substances that may be harmful to the developing embryo. For example, the virus that causes "three day measles," or *rubella*, passes through the placenta easily. If this should occur in the early stages of pregnancy, a variety of serious birth defects may result. Alcohol also passes through easily, and pregnant women should be aware that alcohol may be harmful to the fetus, especially in its later stages of development. Antibodies of all kinds pass through the placenta; some of these provide the fetus with temporary immunity to common infectious diseases. Others, however, such as those associated with blood types, may be harmful. We will consider some of the problems related to blood types in a later chapter.

Medical science has only recently become aware of the effects, known and unknown, that a wide variety of drugs can have on fetal development. The dangers during pregnancy were emphasized in a particularly shocking fashion during the early 1960's, when thousands of deformed children were born in West Germany, England, and Canada as a result of their mothers' taking *thalidomide,* a supposedly harmless sedative. As a result of this tragedy, *knowledgeable* physicians are now extremely careful about prescribing drugs for pregnant women and caution their patients to avoid all drugs insofar as possible. The fact is that no one is *certain* that even such things as aspirin may not have some detrimental effect on the development of the fetus.

Looking once again at Figure 5.18, you will note that the embryo develops in a fluid bag called the *amniotic sac.* The membrane forming the sac is called the *amnion,* and the fluid in which the embryo is immersed is called the *amniotic fluid.* This fluid acts as a pneumatic shock absorber to protect the embryo against mechanical injury, and the high specific heat of the watery fluid keeps temperature changes to a minimum. Since the lungs of the fetus do not function until immediately after birth, there is no danger that it may drown. This amniotic fluid has provided scientists with a means of detecting certain genetic defects in the fetus long before birth. We will discuss this process, called *amniocentesis,* in a later chapter.

Development

A detailed account of human embryological development is well beyond the scope of this book; in this section, therefore, we will briefly summarize the stages that the fetus goes through as it is prepared for the momentous and traumatic experience of birth.

The nine months of human pregnancy are divided into trimesters, or three periods of approximately three months each. During the first trimester, there is rapid development of major organ systems. The heart, for example, begins to beat at approximately four weeks, accelerating quickly to the normal fetal rate of 180 beats per minute. By the end of the second month, the fetus is a little more than one inch in length. At this point its arms, legs, ears, nose, and mouth have recognizable beginnings.

During the second trimester, the reproductive organs begin to form and the skeleton starts to change from cartilage to bone. By the fourth month, the fetus begins to move in the uterus, kicking its legs and stretching its arms. These first movements are felt by the mother as a fluttering sensation, known in the past as the time of "quickening." At one time it was assumed that quickening signaled the actual beginning of life.

By the end of the second trimester, the fetus is covered by a cheesy substance which apparently acts as a protective skin covering. The fetus is now about twelve inches long and developed to the point that it might survive outside the mother's body if provided with artificial life support. However, the probability of survival at this stage is very low.

During the last trimester, the fetus matures and "fills out" in preparation for birth. From the seventh month on, the chances that a premature infant will survive increases significantly. During this last period, the fetus makes increasing demands on the mother's body and takes significant percentages of the nutrients in her diet. This is especially true of calcium, iron, and nitrogen. The mother's heart must work harder to keep blood flowing through the placenta, and her breathing may become difficult due to the pressure exerted on her diaphragm by the greatly enlarged uterus. It is little wonder, therefore, that by the end of nine months, the average mother-to-be is eagerly looking forward to ending the joys of pregnancy.

Increasing attention is now being given to the

problems of nutrition that relate to pregnancy, particularly to the effects of malnutrition on both mother and fetus. It is generally agreed that both maternal health and normal fetal development depend in part on a diet that is relatively rich in proteins, minerals, and vitamins. Experiments with rats, for example, have shown that embryos never develop when the mother is fed a diet that is completely lacking in protein.

It is especially important that the mother's diet be supplemented during the last three months of pregnancy when rapid fetal growth is taking place. Iron deficiency anemia, for example, is found in the latter stages of pregnancy almost universally among women in countries where poverty is the rule, and it is relatively common even among more affluent Americans.

The old saying "a tooth for every child" refers to the tendency to transfer calcium from the teeth and bones of the mother to meet the needs of the fetus in situations where there is a calcium deficiency. In fact, it appears to be a general rule that the fetus will appropriate scarce nutrients for its own development,

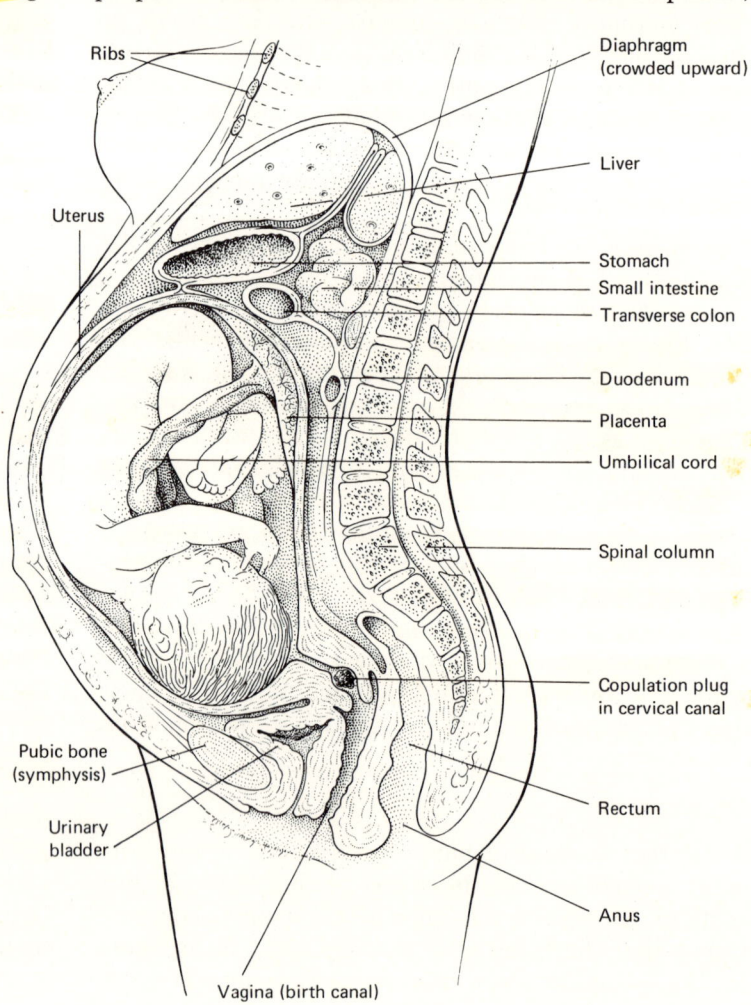

*Figure 5.20
Position of the fetus in the uterus about one month before birth.*

thus affecting the health of the mother in cases where the diet is inadequate to meet the needs of both.

There is considerable controversy over the effects of malnutrition on nervous-system development and the level of intelligence of persons whose mothers received inadequate diets during pregnancy. Studies that involved people born during a short-term famine in the Netherlands during World War II and on those born in concentration camps during the same period do not seem to support the idea that deficient diets always cause irreparable brain damage in the fetus. This evidence is not conclusive, however, and the possibility that fetal brain damage is a common side effect of poverty cannot be discounted. We will discuss this problem further in a later chapter.

Birth

The process of birth is generally divided into three stages. The first stage, called *labor,* involves a series of uterine contractions which occur with increasing frequency and last until the opening of the uterus (cervix) is dilated. During this stage, the amniotic sac usually breaks and the fluid is expelled through the vagina.

After dilation of the cervix is complete, the next stage involves the expulsion of the baby through the birth canal. This is usually accomplished in a few minutes by powerful contractions of the uterus (see Figures 5.20 and 5.21).

Anywhere from ten minutes to an hour after the baby is delivered, the uterus again contracts, shearing the placenta from the uterine wall. The placenta and amnion are then expelled as the *afterbirth.* Considerable bleeding occurs from the uterine wall at the point where the placenta was attached, but as the uterus rapidly contracts to its normal or near-normal size, the blood vessels are constricted and bleeding is thereby controlled.

The factors actually initiating the process of birth are not definitely known, but the posterior lobe of the pituitary apparently contributes to the process by se-

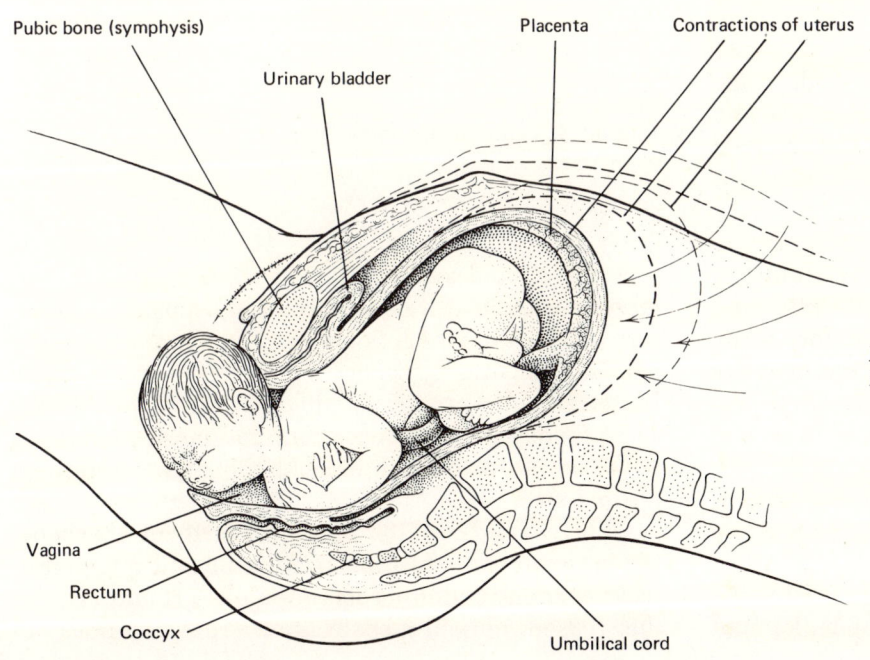

Figure 5.21
Expulsion of the fetus through the birth canal.

creting *oxytocin,* which is a hormone that is known to cause uterine contractions.

The fetus has at last become a baby, and it is now recognized by the courts as a "legal person." It has left the warm, dark, moist, and secure environment of the uterus to enter a world where night alternates with day. From now on it must breathe on its own, and its lungs may soon be tinged with the soot of air pollution. It must also eat on its own in a world where hunger is a way of life for too many. The organism that began nine months before as a microscopic cell has joined the sometimes joyful and sometimes sorrowful world of humanity.

Lactation

In the human female, breast development begins at puberty (actually somewhat before) and is apparently under the control of estrogen.

During pregnancy, the high levels of progesterone and estrogen secreted by the placenta initiate changes in the breast that will prepare it for milk production after the baby is born. Figure 5.22 shows the internal structure of the female breast. Note the *alveoli,* which are complex saclike structures lined with milk-producing cells. Under the influence of progesterone, the ducts in the breast increase in size and number, and the breasts enlarge considerably over whatever size they were in the nonpregnant state.

The secretion of milk by the cells lining the alveoli is initiated by *prolactin,* which is yet another hormone secreted by the anterior lobe of the pituitary. Prolactin means "to produce lactation." Before birth, prolactin secretion is inhibited through the negative feedback system by the high levels of placental progesterone and estrogen. After the baby is born and the placenta is expelled, the levels of progesterone and estrogen drop rapidly and the production of prolactin is no longer inhibited. This leads to the production of milk a few days following birth.

Even then, the suckling action provided by the baby is an important last step because the milk is not

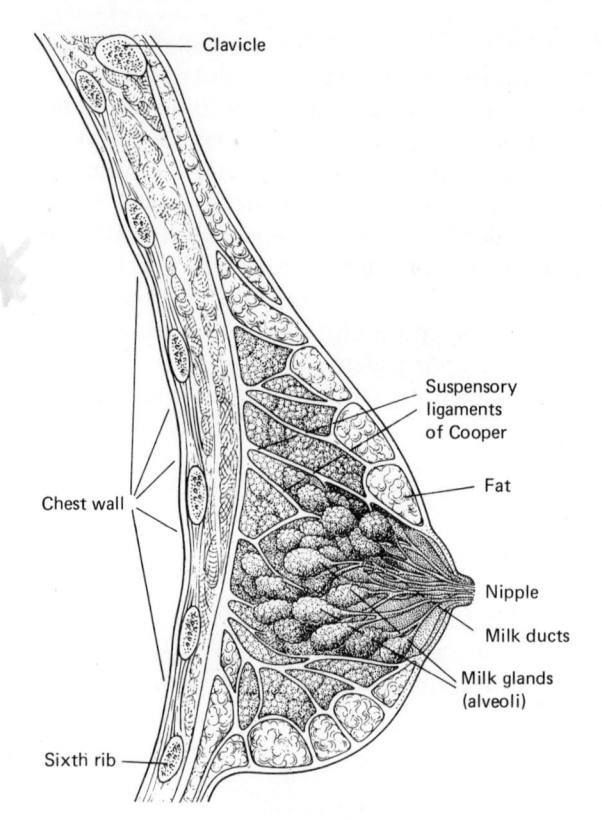

Figure 5.22
Internal structure of the female breast.

"let down" and available at the nipple until the hormone *oxytocin* is released from the posterior pituitary. Oxytocin, which is released by the "suckling response," apparently causes the alveoli to contract and force the milk through the ducts to the nipple. At the same time, oxytocin causes contractions of the uterus; it is therefore thought to help the uterus return to normal size.

For several months after the birth of the baby, the large amounts of prolactin produced by the pituitary apparently inhibits secretion of FSH and LH. For this reason, normal menstruation ordinarily does not resume for several months after the baby is born. This has led to an unfortunate and widespread misconception that a nursing mother cannot become pregnant. Actually she does not know when her normal menstrual cycles will begin, and more importantly, she has no way of knowing when she will ovulate. Many a nursing mother has found that pregnancy was indeed possible at a time when it was least welcome.

Summary

Looking back on the material of this chapter, we can see that human reproduction is a very complicated affair, and if we consider the many potential roadblocks that threaten to prevent the "miracle of birth," then it may indeed seem miraculous that any of us are here at all.

The fact remains, of course, that we *are* here and that we are part of a human population that approaches four billion souls and which threatens to double by the time today's average college student reaches the age of 50. In spite of the inherent difficulties, the consequences of excessive human fertility is one of the fundamental issues of our time. As such, it concerns the individual family, it concerns whole nations, and it is of grave concern to the future of the human species.

In subsequent chapters, we will apply the principles considered in this chapter to a discussion of human fertility as a human problem. In Chapter 6 we will consider various ways of avoiding the consequences of fertility, and in Chapter 11 we will discuss the problem of uncontrolled population growth.

Questions

1. Discuss the application of the negative feedback principle to hormonal changes in the human female.

2. Artificial insemination using an anonymous donor is a possible solution to male sterility in a marriage. What are some of the advantages of this procedure? What are some of the possible problems and pitfalls?

3. Briefly compare the reproductive role of the female with that of the male. What influences have these differences had on the establishment of the traditional social roles of men and women?

4. The nineteenth century saw the establishment of widespread and rigid taboos related to human sexual behavior. In what ways have those taboos and their lingering effects been injurious to human society?

5. State the primary functions of the menstrual cycle. Discuss the menstrual cycle in terms of how those functions are performed.

6. What are the possible implications of the fact that a fetus is not considered by the courts to be a "legal person"?

7. What widespread and long-term effects might poor nutrition during pregnancy have on populations located in poverty areas?

8. Since only one sperm cell normally enters the egg cell during fertilization, why is it that a minimum of 100 million sperm cells must be released by the male if fertilization is likely to be successful?

9. The human being has been described as "the only animal that can make love all year round." What is the relationship of this statement to the evolution of the human family as we know it today?

6

Birth Control

The Need for Birth Control

Everybody loves a baby. It is difficult for many of us to believe that there could be such a thing as an unwanted child. It is traditional to look upon the birth of a healthy baby as a blessed event complete with baby showers, doting grandparents, birth announcements, and cigars from the proud father. This is the tradition of the home that welcomes a new baby as a human being to be fed and clothed, to be loved and cared for, and to be taught to live a productive life as a member of society.

But what about the homes of that other half of humanity where the spectre of starvation is a constant visitor? Here the traditional blessed event may be just another stomach to fill from a food supply that is already too meagre to meet the family's needs. Furthermore, poverty and hunger are not confined to underdeveloped countries located in far corners of the earth; poverty and hunger are found in the most affluent countries, the United States for one.

There are also couples who may have the necessary financial resources but lack the *emotional* resources to provide the love and attention that children need in order to grow into psychologically stable adults. And then there is the woman whose body is literally worn out from the traumas of repeated pregnancies and the responsibilities of mothering a large family. Does she welcome another baby, and another, and yet another, as "blessed events"?

Women have always been interested in birth control; this is understandable because women become pregnant, women go through childbirth, and women care for the children during their early years. Most women apparently welcome this experience, but too much of any good thing is more than enough.

Men, on the other hand, have not always shared the interest that women have in birth control; this is equally understandable because men do *not* go through nine months of pregnancy, men do *not* go through childbirth, and even the most understanding and helpful husband may share only a small part of his wife's responsibility for dirty diapers, 2:00 A.M. feedings, and hours spent at the bedside of a sick child. Yet in the past, and to a large degree even now, men have had almost complete control over laws governing reproductive functions, functions which are largely the business of women. National and state lawmaking bodies are controlled by men, and the hierarchies of Judeo-Christian religious institutions are controlled by men.

In 1873, Congress passed the notorious Comstock Law, which forbade interstate commerce involving contraceptives or birth control information. This was done as a result of agitation by Anthony Comstock, who headed the "Society for the Suppression of Vice" and who apparently was determined to stamp out sex for all time. Similar, even more restrictive statutes were passed by state legislatures. Thus in the United States, the stage was set for a contraceptive "crime wave."

St. Augustine (another man) said, "Sexual intercourse, even with one's lawful wife is unlawful and sinful if offspring are prevented." But the title of Champion Male Chauvinist must be reserved for the Reverend William John Knox Little, who proclaimed in 1880,

> Wifehood is the crowning glory of a woman. To her husband she owes the duty of unqualified obedience. If he be a bad or wicked man, she may gently remonstrate with him, but refuse him never. I am the father of many children, and there are those who have ventured to pity me. Keep your pity for yourself, I have replied. *They have never cost me a single pang!*

Considering all this, it is not surprising that the term "birth control" was invented by a woman. Prior to World War I, Margaret Sanger, an American nurse, was shocked by what she saw of the human misery that came with uncontrolled fertility. Sights such as that of a young woman already worn out by repeated childbearing and dying of a clumsy, self-induced abortion led Mrs. Sanger to campaign for freedom to provide birth control information to those who needed and wanted it. Mrs. Sanger, herself a mother of three, was jailed several times for her efforts, but her dedication and determination helped awaken the public (especially other women) to the need for more enlightened attitudes.

Since World War II, most of the rigid legal and moral restrictions on birth control have either fallen by the wayside or are largely ignored. This is probably due to an increased number of individuals with a higher level of education, to changing attitudes toward sexual matters in general, and to pressure exerted by increasingly undeniable worldwide population problems. There are still many people, especially those of the Catholic faith, to whom "artificial" contraceptive devices are contrary to deep-seated religious beliefs. At this writing, the church has not changed its firm position on this matter. Abortion is still decidedly controversial, and objections come from other groups as well as from Catholics.

Surveys indicate that approximately 85 percent of all married women in the United States use some form of birth control at least part of the time. A percentage this large must necessarily contain women of all faiths and ethnic groups. Unfortunately, as we will see later in this chapter, contraceptive methods vary in effectiveness, and many women may be using methods that are only slightly better than nothing at all.

Methods of Birth Control

What to do about excessive fertility is an ancient human concern. Old writings and the annals of folklore contain a wealth of interesting—if ineffective—recipes and incantations designed to prevent conception. They range from teas and potions made from various herbs to having a woman sew pebbles in her garments or step across a grave three times by the light of a full moon. When it was discovered that babies result from intercourse and that male seminal fluid is a factor in conception, the logical approach was to prevent the deposit of semen into the vagina. The most primitive method of this type, known as *coitus interruptus*, is described in the Old Testament and is still in use to some extent today.

Infanticide, or the deliberate killing of newborn babies, is a method of fertility control that is unacceptable to modern civilized societies, although it is said to have been extensively practiced in eighteenth century Europe. Infanticide is still found among some primitive peoples, and it is said to be committed as an act of desperation by mothers in certain Latin American countries where contraception is forbidden.

The single most prevalent approach to birth control on a worldwide basis is abortion, which also happens to be the most controversial of modern methods. It is estimated that twenty to thirty million legal and illegal abortions are performed in the world annually, and, as might be expected, the incidence is highest where methods of *preventing* conception are forbidden by legal or moral sanctions.

In general, methods of birth control can be divided into two major types: (1) contraception, and (2) abortion. *Contraception,* or "contra-conception," involves methods whereby the sperm cell is kept from uniting with the egg cell. *Abortion,* on the other hand, may be roughly compared to the act of shutting the barn door after the horse has escaped. Conception has already occurred, and the function of abortion is therefore to remove the developing embryo or fetus before it goes to full term.

In general, contraceptive methods consist of (1) those that seek to prevent union of the sperm cell with the egg cell, and (2) those that make use of the principles underlying the menstrual cycle. The following are brief descriptions of some common methods of contraception.

Coitus Interruptus

Coitus interruptus is actually one of the oldest known forms of contraception. It involves withdrawal of the male just before ejaculation; the seminal fluid is therefore not deposited into the female tract. It can be fairly effective, but there is always a significant danger that sperm cells will escape prior to ejaculation. Coitus interruptus requires an unusual degree of control by the male and is generally not conducive to a satisfactory sexual relationship. It is therefore not especially popular, particularly in the United States and similar nations where more satisfactory forms of contraception are available.

Post-Coital Douche

The method known as *post-coital douche* consists of washing seminal fluid from the vaginal tract with a syringe immediately following intercourse. Everything from plain water to vinegar solution to Coca Cola has been used for this purpose, the latter two because of their acidity and tendency to kill sperm cells. This approach to contraception is mentioned here mainly to point out that it is highly ineffective; in fact, many experts regard it as little better than nothing at all. You will recall that seminal fluid and sperm cells are taken into the cervix very quickly. In fact, active sperm have actually been found in the cervix within ninety seconds following ejaculation by the male.

Condom

The *condom* is a rubber sheath worn by the male to prevent entrance of sperm cells into the vagina (Figure 6.1). As near as can be determined, the condom originated in eighteenth century England for use as protection against venereal disease. Until recent years, it was commonly used as a contraceptive in more-or-less casual extramarital relationships, but since oral contraceptives have come into the picture, the use of the condom has been significantly reduced. It has been suggested that this may be partially responsible for a general increase in cases of venereal disease. As a contraceptive, the condom is highly effective, and with proper use its failure rate is probably only around 2 percent. Its chief disadvantages include the necessity to interrupt sexual activities at a critical time and a reduction in sexual satisfaction for both partners.

Diaphragm

The *diaphragm* is a flexible rubber cup that fits over the cervix, forming a mechanical barrier against the entrance of sperm cells into the uterus (Figure 6.2). In order to be effective, it must be used in conjunction with a spermicidal jelly which is placed around the rim of the cup. In addition, the diaphragm must be fitted by a physician and refitted following changes resulting from pregnancy or weight change.

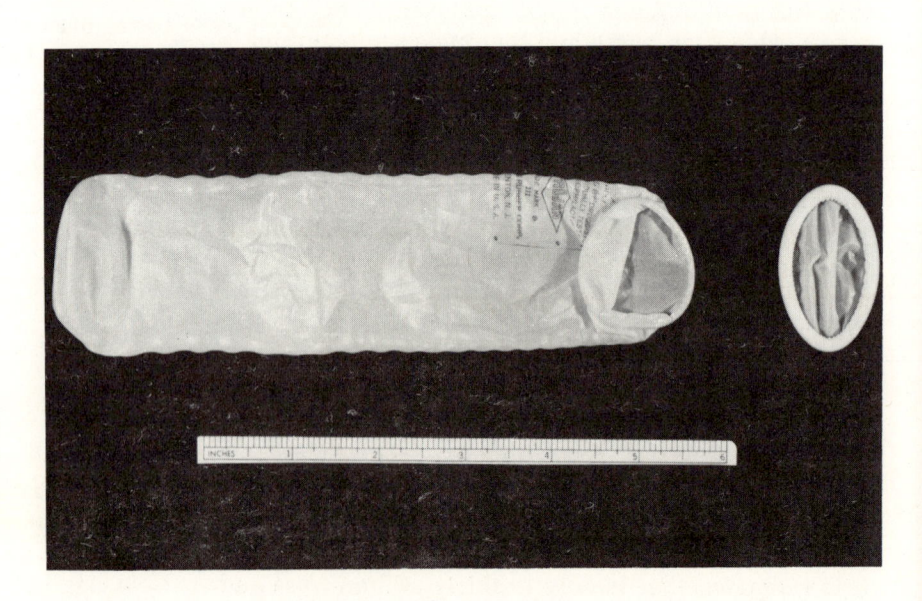

*Figure 6.1
Condoms, unrolled and rolled.*

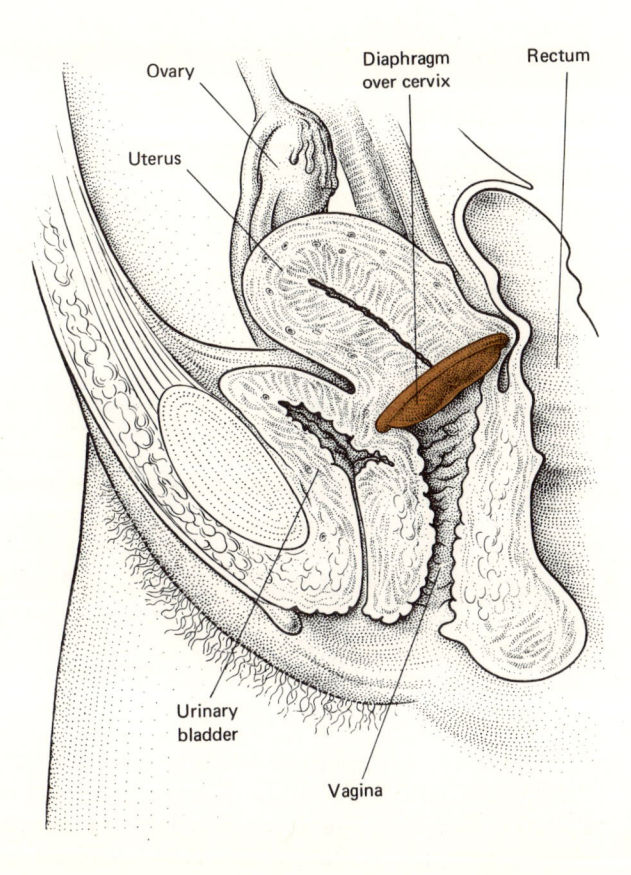

Figure 6.2
Cross section of the pelvis showing the diaphragm in place.

The failure rate of the diaphragm is about 5 percent. A portion of this failure rate can be accounted for by the possibility of displacement during intercourse, but most of it is probably due to failure to use the diaphragm consistently. One disadvantage of the diaphragm is that its use requires a high degree of motivation to avoid conception. The fact that it is "intercourse-connected," that it must be inserted prior to intercourse, is considered by some women to be another disadvantage. Yet when used properly, it is a highly effective method of contraception, and it is useful to women who for some reason are unable to use oral contraceptives.

Contraceptive Foams

Contraceptive foams are spermicidal agents that are spread over the cervix with an applicator. They are simple to use and can be purchased directly over the drug counter in most places. They are claimed to be effective, but they are probably much less reliable than the diaphragm. Whether a foam is used or not should depend on just how important it is to avoid pregnancy. For the highly motivated couple, a very safe nonprescription method is a combination of the condom and foam.

Rhythm

There are two means of fertility control that are currently approved by the Roman Catholic Church. One involves abstinence from sexual intercourse, which is 100 percent foolproof but is not likely to achieve great popularity. The other is the so-called *rhythm method,* which is based on principles underlying the menstrual cycle.

If we consider that ovulation normally occurs only once during a menstrual cycle, and if we further consider that the sperm and egg cells each have a theoretical average maximum life of about 48 to 72 hours, it then follows that abstinence from intercourse during the period covering about three days

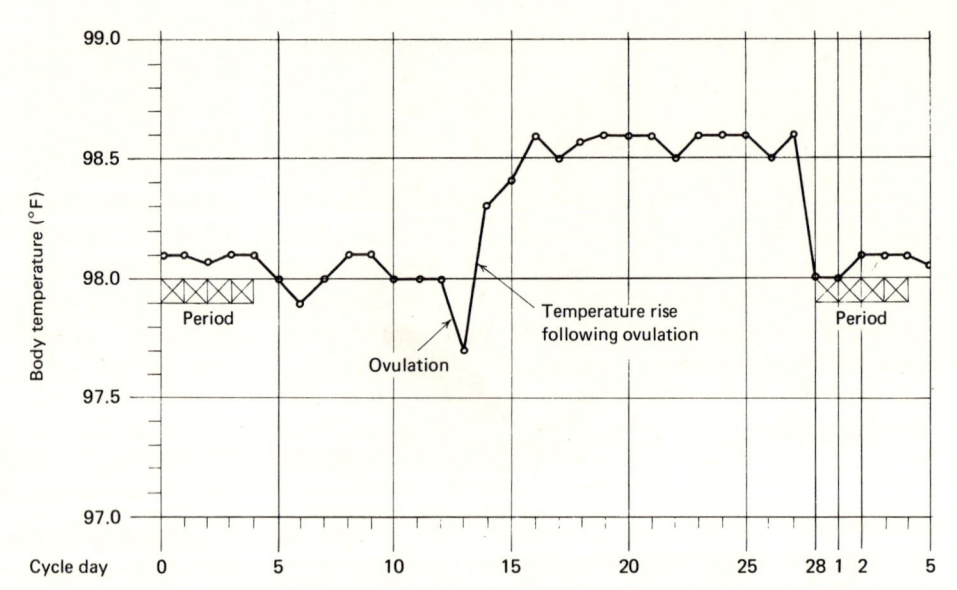

Figure 6.3
Temperature chart used in the rhythm method. Note the sudden dip in body temperature at the time of ovulation.

before and three days following ovulation will prevent conception. This is the main theoretical principle upon which the rhythm method is based.

It is obvious that if this system is to be successful, we must be able to pin down the exact time of ovulation. When ovulation does occur, there is a dip and then a sharp rise in the woman's body temperature. Therefore, if she takes and records her temperature every morning before arising, she can safely proceed with intercourse after her temperature has been elevated again for three days in a row. This leaves about ten "safe" days until her period begins again (Figure 6.3).

What about the time prior to ovulation? Unfortunately, the temperature method can only indicate that ovulation *has* occurred; there is no foolproof way of predicting when ovulation *will* occur. Even if we could assume that ovulation usually occurs around the 14th day in a woman with a regular 28-day cycle, there is no guarantee that in any given month ovulation will not shift to an earlier time. Indeed, it is not impossible for ovulation to occur as early as the end of the menstrual period itself. We are therefore faced with the fact that the only safe time in our hypothetical example is during the last ten days of the cycle, and that intercourse prior to ovulation must therefore be regarded as a variety of "Russian roulette."

Women with irregular menstrual cycles present even greater problems. To make matters worse, irregularity is common in very young women, when pregnancy may well be unfortunate for both mother and child. Also, irregularity and even skipped periods is practically the rule in women approaching menopause; many of these women are grandmothers and few grandmothers look forward with great anticipation to becoming mothers! Then too, as we shall see in a later chapter, the risk of having severely defective babies, such as "mongoloids," increases significantly with the age of the mother. This in itself is an excellent reason why older women should avoid pregnancy.

Even in otherwise regular women, there are temporary influences such as stress, worry, and even long trips involving time-zone changes that can affect the regularity of ovulation and increase the "Russian roulette" aspect of the rhythm method.

You will also recall from Chapter 5 that nursing mothers usually do not menstruate for several months following childbirth. Ovulation is therefore clearly unpredictable, and there is no sensible theoretical basis on which the rhythm method can be applied to the nursing mother. This is therefore yet another obviously critical situation in which the rhythm method fails to provide adequate protection against unwanted pregnancy.

From all of the foregoing, we would predict that the rhythm method is likely to be a highly ineffective means of contraception. This prediction is indeed confirmed by the failure rate, which is estimated to be as high as 38 percent for all women using it.

On the other hand, rhythm does have one advantage that cannot be ignored. If used properly, it can provide an alternative to repeated pregnancies for many women who may reject more effective methods because of conscience. The "failure rate" associated with using no contraceptive device or method is about 90 percent. Compared to this, a failure rate of 38 percent is at least better than nothing.

Oral Contraceptives

Oral contraceptives are a partial realization of the ancient dream of finding a magic potion that can be taken by a woman to prevent conception. They are known collectively as the "Pill," but in fact, there are at least twenty varieties, each containing specific quantities and combinations of hormones.

Oral contraceptives were first developed by John Rock and Gregory Pincus back in the early 1950's and were officially approved for public use in 1960. Since then, their use has become widespread; as long ago as 1969 it was estimated that over eight million American women relied on oral contraceptives as their primary method of contraception. Despite occasional scare headlines in newspapers and women's magazines, the Pill now appears to be firmly established as a leading force in the battle against excessive or unwanted fertility. Although in 1968 Pope Paul VI reaffirmed the opposition of the Church to all forms of "artificial" contraception including the Pill, this decision created an unusual division of opinion within the Church and among Catholic laymen. Studies and surveys strongly suggest that an increasing majority of devout Catholics contend that use of the Pill is every bit as compatible with conscience as is the "natural" method of rhythm. As a matter of fact, it might be said that oral contraceptives make use of nature's own method of contraception. You will again recall from Chapter 5 that the principles underlying the menstrual cycle ensure that only one egg cell will be released during an individual cycle, and if fertilization occurs, that *no more* egg cells will be produced for the remainder of pregnancy and for some time thereafter.

Ovulation is dependent upon adequate secretions of FSH and LH by the anterior pituitary gland. During pregnancy, these hormones are suppressed because of the consistent levels of estrogen and progesterone maintained first by the corpus luteum and then by the placenta. It therefore follows that if FSH and LH could be suppressed by deliberately maintaining a minimum level of estrogen and progesterone, ovulation would be prevented and conception would be impossible. This, in brief, is the main principle of oral contraceptive function. Essentially, the Pill contains the hormones of pregnancy and the ovary is "fooled" into acting as it would during the pregnant state.

There are essentially two kinds of oral contraceptives. One, the so-called *combination type,* contains chemicals that are similar in structure to estrogen and progesterone and have the same suppressant effect on the pituitary—or actually, on the hypothalamus. These combination pills are generally taken for 20 or 21 consecutive days beginning on day 5 of the cycle. Day one, incidentally, is considered to be the first day of menstruation. At the end of 20 or 21 days, the pills are stopped and a menstrual flow will follow as a kind of "withdrawal symptom." Since the endometrium has not built up to the extent that it does normally, the menstrual flow tends to be lessened and problems with painful menstruation tend to be reduced. For this reason, the Pill may also be used as a treatment for menstrual difficulties. In addition to suppressing ovulation, the combination Pill apparently changes the nature of the cervical tissue, making it more difficult for sperm to enter the uterus. Also, the reduced endometrial buildup possibly reduces the chances of successful implantation even if ovulation and fertilization should occur.

The *sequential type* of oral contraceptive more closely follows the hormonal changes that take place during the regular cycle. Pills containing estrogen are given for the first 14 to 16 days, and progestin, a chemical imitation of natural progesterone, is given for the next 5 to 6 days.

As a contraceptive measure, the Pill is highly effective. The failure rate is estimated to be less than 1 percent, but most clinicians maintain that even this low failure rate is not due to the Pill itself, but rather to failure to take the Pill as prescribed. Physicians generally feel that a woman who carefully follows the prescribed regimen is afforded 100 percent protection against unwanted pregnancy, although this might be difficult to prove scientifically.

It must be pointed out, however, that a number of objections to oral contraceptives have been raised. For example, it has been suggested that a link may exist between the Pill and cancer of the breast and cervix. No evidence is available from well-designed studies to support this claim, although there is a possibility that oral contraceptives may *aggravate* such cancers if they already exist. It has also been pointed out that a woman who is taking the Pill under the careful supervision of a competent physician is actually afforded a special degree of protection from cancer, since he will likely insist on regular breast examinations as well as "pap" smears. The pap smear is a technique used to discover early cancer of the cervix. This early detection is especially important because both cervical and breast cancers respond well to early treatment.

Since oral contraceptives contain hormones that mimic those of pregnancy, it is not surprising that some women exhibit the characteristic symptoms of early pregnancy when they first begin taking the Pill. Dizziness, nausea, and sore breasts are common side effects, but they usually disappear with continued use of the Pill. Also, some physicians hesitate to prescribe oral contraceptives for women having a history of migraine, because there is some evidence that this particularly miserable type of headache may be aggravated by the Pill. A history of liver disease may also be reason for caution, since the hormones present in the Pill are ultimately broken down by the liver. Diabetes, or potential diabetes, may also be considered as a possible reason to avoid oral contraceptives, although some physicians point out that pregnancy could be a more serious complication for the diabetic than use of the Pill.

The most publicized problem that has been associated with oral contraceptives is *thromboembolism,* which is the formation of blood clots within blood vessels. Deaths have occurred among Pill users as a result of *pulmonary embolism,* which results when a blood clot reaches the lungs. Again, while the studies that have been done are not as conclusive as they could be, the evidence suggests that *three* deaths from thromboembolism can be expected to occur among 100,000 Pill users compared to *one* death from this cause among 100,000 nonusers of the Pill. It is important, however, to compare this statistic with the number of deaths that can be expected from complications associated with pregnancy, which is 28 per

100,000 in the United States. Therefore it would appear to logically follow that it is much safer to take the Pill than to become pregnant. It may help to put both of these statistics into perspective if we consider that the risk of death from either the Pill or from pregnancy is no greater than the risk taken by riding in an automobile.

In addition, it is important to emphasize that our discussion of possible dangers associated with oral contraceptives has dealt with populations and not with individuals. Whether oral contraceptives should or should not be prescribed for an individual woman is a matter between her and her physician. An eminent gynecologist, Dr. Robert W. Kistner, expressed this distinction especially well when he wrote,

> In the *science* of medicine certain generalizations may be made by statistical data; but in the *practice* of medicine, each patient is an individual and generalizations are not always applicable. No two patients are exactly alike and in the interest of optimal medical care, every patient must be observed as an individual and appropriate therapy prescribed.

To summarize, oral contraceptives are apparently here to stay, at least until a better method comes along. They are certainly not troublefree and they are not for every woman, but their high degree of reliability coupled with the advantage that they are psychologically and physically separated from the act of intercourse have made oral contraceptives a popular and useful means of fertility control.

The Intrauterine Device (IUD)

It is said that camel drivers used to place pebbles in the uterus of a female camel before beginning a long trek across the desert. Apparently they had somehow discovered that the presence of a foreign object in the uterus would prevent pregnancy and the inconvenience that would go along with a pregnant camel. It appears that they also knew that when the pebbles were removed, fertility would be restored.

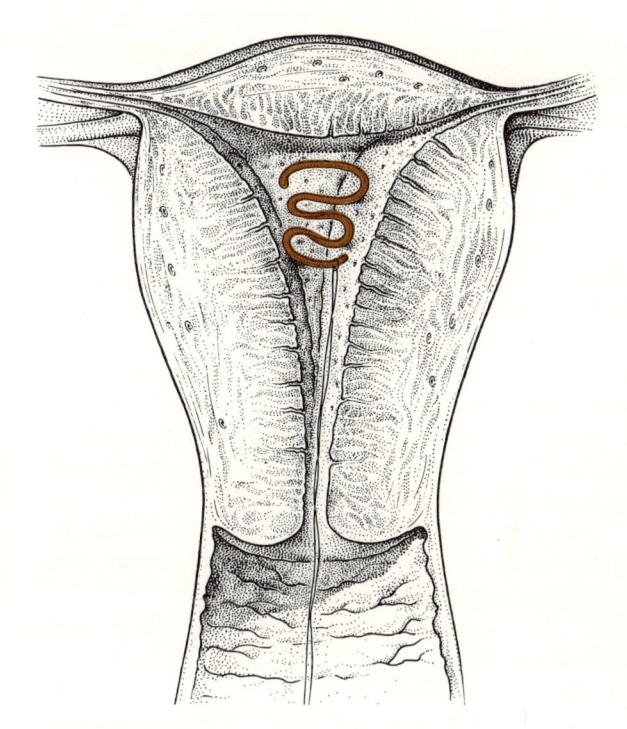

Figure 6.4
Lippe's loop in place inside the uterus.

How this device prevents pregnancy is not much better understood than in the camel driver's time. One theory suggests that the presence of the foreign body in the uterus speeds the egg cell down the fallopian tube and prevents fertilization. In that sense, the *intrauterine device*, or *IUD*, is a contraceptive. A more widely accepted theory suggests, however, that the IUD prevents implantation, which in the view of some people makes it an *abortifacient*, which is a device or drug that causes abortion.

The IUD is a small plastic device that must be inserted into the uterus by a physician or other qualified person. IUD's come in a variety of shapes and sizes, but one of the most popular is Lippe's loop. Figure 6.4 shows Lippe's loop in place inside the uterus. A fairly recent development is the addition of copper to the IUD; it has been found that copper increases reliability, but the reason for this is unknown. Figure 6.4 also shows the threads hanging from the IUD and extending into the vagina. These threads permit the patient to check periodically to make certain that the IUD is still in place; this is important because undetected expulsion of the IUD is one of the principal causes of failure.

In general, the IUD is considered to be very effective, having a failure rate estimated somewhere around 3 percent. Its advantages include the fact that once it has been inserted into the uterus, it can be forgotten except for routine checks to make sure that it has not been expelled. It eliminates the need for oral medications with their possible side effects, and it is especially useful in situations where a low literacy rate makes it difficult to use contraceptive methods that involve elaborate directions.

On the other hand, not all women can tolerate the IUD. In some cases it is repeatedly expelled, and in others it is associated with cramps and profuse bleeding at menstruation. Still, it is estimated that at least one million American women are successfully using IUD's.

Sterilization

If it were not for the fact that sterilization is usually irreversible, it might be considered a near perfect form of contraception. It involves no fuss, no bother, no equipment, and no medication; it is as reliable as any form of birth control can be. Unfortunately, sterilization carries an implication of finality that many people are unwilling to accept. Even so, it is an increasingly popular method of birth control, and as of this writing, well over two million Americans have availed themselves of this procedure.

The simplest sterilization procedure is the *vasectomy,* which is performed on the male. In this operation, the physician makes a small incision on each side of the scrotum, takes out a small section of the vas deferens, and ties the remaining ends. Figure 6.5 illustrates this procedure. A vasectomy is an extremely simple procedure, it is relatively inexpensive, and it can be performed in a physician's office with a local anesthetic.

Although in some cases it may be possible to surgically connect the cut ends of the vas deferens and thereby restore the sperm pathway, this restoration procedure is successful less than 50 percent of the time. It is therefore imperative that the individual consider very carefully the fact that a vasectomy will probably result in a permanent end to his ability to have children. It must be emphasized, however, that a vasectomy has no effect on the quantity, quality, or duration of the male's sex life. In fact, studies show that it may well be enhanced, probably because of the resulting freedom from fear of producing pregnancy. Still, the suspicion that a vasectomy will reduce male sex prowess tends to persist, and this misconception, although completely baseless, undoubtedly keeps many men from seeking sterilization.

In women the operation is called *tubal ligation,* and it is comparable to tying off the sperm duct in the male (see Figure 6.6). It is most easily performed immediately after childbirth because the fallopian tubes are easily accessible at that time. At other times the operation is more extensive, but in either case an abdominal incision and a hospital stay are required.

New techniques have recently been developed that simplify the sterilization procedure in women. An instrument called a *laparoscope* is inserted through a small incision, and by use of this instrument the surgeon can fuse the fallopian tubes closed by the use of heat. Even with this simplified procedure, however, general anesthesia and usually an overnight stay in the hospital are required. The male vasectomy is still the simpler procedure.

Vasectomy

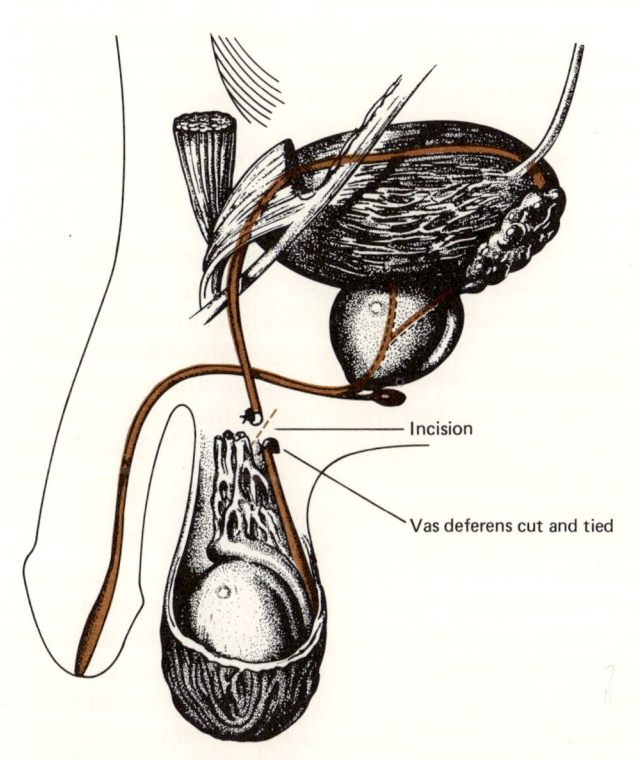

Figure 6.5
Vasectomy. Note that the sperm pathway has been permanently interrupted.

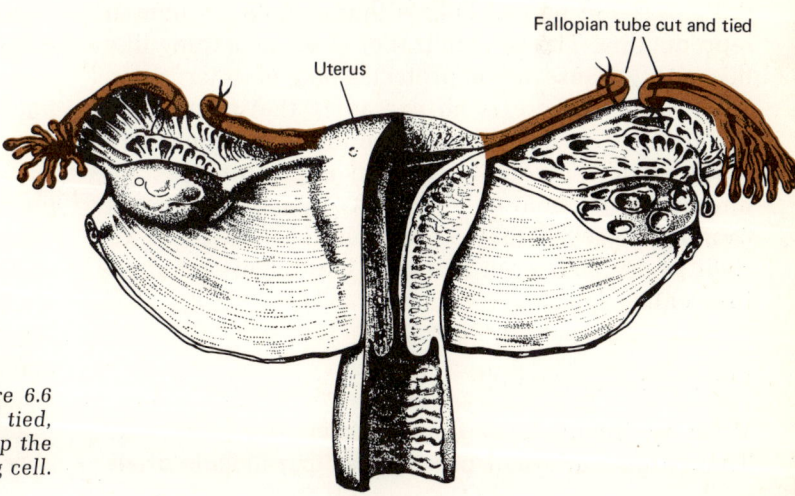

Figure 6.6
Tubal ligation. Both fallopian tubes are cut and tied, thus preventing the sperm from traveling up the tube to meet the egg cell.

It comes as a surprise to many people that sterilization is not always 100 percent effective, although as might be expected, it is very close to it. The failure rate is estimated to be around 0.03 percent. Failure can occur in a vasectomy, for example, when unprotected intercourse occurs too soon after the operation. This is due to the fact that viable sperm may be stored in the vas deferens. Or, failure can also result if the fallopian tube is not tied off properly and the sperm and egg cells find their way through the disturbed pathway.

Morning-After Pill

At this writing, a suitable "morning-after" pill has not been perfected. Such a drug would be useful, since it would only be taken following intercourse. It would therefore eliminate not only mechanical contrivances, but also the necessity to be on a daily regimen of pill-taking.

A drug called *DES* (*diethylstilbestrol*) is sometimes used in cases where an "afterthought" preventive measure is especially important. A victim of rape is a good case in point. It has also been used as a

Tubal ligation

"panic-button" drug for the lady who changes her mind following an unplanned, unprotected act of intercourse. It appears to effectively prevent pregnancy if dosage is begun within 72 hours after intercourse, or sooner if possible.

DES is not intended, however, for use as a regular "morning-after" pill. For one thing, the woman who takes it may suffer from nausea and vomiting, and secondly, there is suspicion that regular use of DES might either produce or seriously aggravate cancer of the breast or cervix. It must therefore be used only as an occasional emergency measure. There is also evidence which strongly suggests that if a woman uses DES *while pregnant,* there will be a tendency for female offspring to later develop cancer of the vagina. A number of recently reported cases of this kind will most certainly cause physicians to be more cautious in prescribing this drug.

It is not definitely known if DES functions as a contraceptive or as an abortifacient, but it appears probable that it interferes with implantation and is therefore an abortifacient. Thus there are potential moral and legal implications connected with its use, making this an excellent example of the degree to which man can rationalize his actions. It is also a demonstration of the frustrating inconsistencies in the moral and ethical beliefs that complicate human reproduction. The rationalization goes something like this. (1) Because an unprotected act of intercourse has occurred, there is no guarantee that conception has resulted. In fact, the chance that conception will result from a *randomly selected* unprotected act of intercourse is about one in twenty-five. (2) Because it cannot be proved that conception did occur, it is alright to use DES even though it is probably an abortifacient.

Male Contraceptive Pill

What are the prospects for a male contraceptive pill? There are a variety of possibilities that include drugs that would alter the seminal fluid so that it would kill the sperm, surgical procedures that would make vasectomies easily reversible, and periodic injections of hormones that would suppress the secretion of FSH by the pituitary and thus inhibit the formation of sperm cells.

On the other hand, even if a "male pill" should be developed, we still must face the fact that the average man is not as interested as the average woman in preventing pregnancy. The most obvious reason is the fact that it is not *his* pregnancy that is being prevented! From a practical point of view, therefore, it is probable that the most successful contraceptive methods will continue to be those that are under the control of the female, since she stands the most to lose if the method should fail.

Abortion

Both older and more modern methods of contraception have one thing in common—they all have failure rates. Now, if it is true that contraceptive means are employed to prevent unwanted pregnancies, and if it is also true that no contraceptive measure is 100 percent effective, it must therefore follow that unwanted pregnancies will occur even in a society in which contraception is an accepted practice. It was on this basis that Garrett Hardin calculated a few years ago that at least 250,000 unwanted babies were born annually in the United States. Furthermore, studies have suggested that between 36 and 49 percent of the babies born to the American middle class are unplanned and therefore, to some degree, are unwanted. On a worldwide basis, it has been estimated that 30 million unwanted births occur each year.

For the reluctantly pregnant woman, abortion is the only alternative to having a baby for which she may be unable to provide adequate physical or emotional care. Now we have the ingredients of a classic dilemma, because of all methods of birth control, abortion is the one most fraught with legal, moral, religious, and emotional complications. Considering this, it usually comes as a surprise to learn that abortion has always been and continues to be the single most widely used method of birth control in the world! Between 20 and 30 million abortions are performed annually, and in those countries where contraception is considered to be illegal or immoral, the number of abortions may meet or exceed the number of live births per year. A large percentage of these abortions are illegal, and before relaxation of abortion laws in the United States, it was estimated that over 1000 women died annually from complications resulting from back-alley or self-induced abortions.

Opponents of abortion point out that it involves the destruction of a human life. Furthermore, they emphasize that the terminated life is that of a fetus, which is the most helpless of human beings and which has no control over its own destruction. The more intense opponents argue that destruction of human life at any stage should be regarded as murder.

On the other hand, those who have argued for liberalized abortion laws maintain that a human embryo is not a human being; it merely has the potential for becoming a human being. What is now a classic argument in favor of abortion involves a comparison of the embryo to the plans for a house. If one accepts this comparison, then it follows that the destruction of an embryo is not the same as killing a human being any more than burning the plans for a house is equivalent to burning the house itself. Whether this is a valid argument depends largely on your own fundamental position. An opponent of abortion, for example, might well answer by saying that the plans for a house represent a critical and irreplaceable stage in the life of that house. Furthermore, he might well argue that in the case of an embryo, the house is already under construction. Since, as we discovered earlier in this book, the "plans" for each human being is unique, it follows that these plans are indeed irreplaceable, although the architects (the parents) could replace them with another, but different, set of plans.

Closely related to the abortion controversy is the question of when a human life begins. Does it begin at the moment of conception, when the "plans" for the new individual are first laid? Does it begin when implantation is completed? Does it begin when the placenta is firmly established? Does it begin at the moment of "quickening," when the fetus begins to move within the uterus? Or does it begin only when the baby enters the outside world and an official birth certificate is issued? Many words have been written on this subject, and each "beginning time" has its advocates. Again, the position that a given individual takes is largely determined by his or her basic position and philosophical convictions concerning abortion. Now, if we were to carefully review the principles that we have considered so far in this book, we would very likely conclude that arguments over the point at which a human life begins are basically nonsense. The "new" life developing in the uterus is merely a continuation and a branch of the life that began billions of years ago in the primeval seas.

And so it goes, pro and con, with most of the arguments on both sides of the issue generating more heat than light. Perhaps the most logical arguments for liberal abortion laws include the following points. First, an abortion is often less undesirable than the birth of an unwanted baby that may suffer physical or emotional starvation, or both. Second, an abortion is less undesirable than is a world in which overpopulation becomes a sickness that pervades all of humanity. And third, termination of a pregnancy by a qualified physician in a hospital is a far better alternative than the pain, remorse, and deaths that are the fruits of backroom abortions performed by quacks, or of abortions clumsily self-induced by desperate women.

Abortion is a simple and safe procedure when performed by a qualified physician under medically suitable conditions. In cases of early pregnancy, usually twelve weeks or less, the procedure involves either a *D and C* or a newer method called *vacuum aspiration.* "D and C" is short for dilation and curettage; it is done by opening the cervix using a series of dilating instruments and then scraping the uterine lining with a device called a curette. Vacuum aspiration is even simpler and is now being used extensively. This involves removal of the uterine contents with a device that works on the principle of a vacuum cleaner. As pregnancy continues the situation naturally becomes more complicated; in later stages of pregnancy, a saline solution or prostaglandins are usually injected into the uterus and abortion follows spontaneously in a few days.

The legal aspects of abortion in the United States are changing rapidly, probably as a result of widespread changes in public opinion. Until recently, the laws of most states permitted abortion "only when necessary to save the life of the mother," a concept that has relatively little practical meaning in these days of modern medical care. In a few states, abortion has been allowed in special circumstances such as those involving rape, incest, or when there is a high probability of a seriously defective fetus. The first totally liberal abortion law was passed in New York state, and even this law came under repeated attacks by antiabortion forces seeking its repeal.

Then in January, 1973, The Supreme Court struck down all antiabortion laws in the United States by ruling that American women have a right to have abortions during the first six months of pregnancy. The court further ruled that individual states may interfere with this "right of privacy" only during the last three months of pregnancy, when the fetus has a chance of maintaining life by itself outside the uterus. This landmark decision by the Court apparently has the effect of making all abortion laws in the nation essentially similar to the New York law.

The New York state abortion law went into effect in 1970, and during the first year of the law's existence, an estimated 215,500 legal abortions were performed throughout the state. Many of these cases involved out-of-state residents who came from every state in the union in search of escape from unwanted pregnancies. During subsequent years, the annual number of legal abortions has increased despite concerted efforts by "right-to-life" groups to have the laws repealed.

The abortion experience of New York state has some important lessons for the nation and for the world at large. For one thing, it demonstrated that abortions can be performed with a maximum degree of safety. In the early 1970's, the death rate due to complications following *legal* abortion was 3.7 per 100,000 women. If we now compare this figure to that of 28 deaths per 100,000 women resulting from complications due to pregnancy, we might draw the conclusion that it is far safer to obtain a legal abortion than it is to remain pregnant.

It was also shown that a liberal abortion law provides a long-overdue service to the poor—to those who need it most. Rigid abortion laws have always discriminated against the poor, driving them in desperation to the back-alley quack. The wealthy have always been able to purchase reasonably safe abortions, either locally or, in the case of the very affluent, in other countries such as Japan. With the legalization of abortion and the subsequent establishment of clinics that charge their fees on a sliding scale, safe and competently performed abortions became available to the poor as well as to the wealthy.

In spite of all this, it is doubtful that abortion is the best solution to the problem of birth control. As we pointed out earlier, abortion might be compared to the act of closing the barn door after the horse has escaped. It would therefore be infinitely better to improve our methods of keeping the barn door shut. In other words, there is a need to constantly improve contraceptive methods; there is a need for widespread education concerning contraception; there is a need for changed attitudes toward the moral and ethical aspects of so-called artificial methods of contraception. While abortion may continue to be a necessary backup procedure for contraceptive failure, it is hardly a suitable or even intelligent *substitute* for contraception, artificial or otherwise.

In this chapter, we have considered some of the major individual needs for birth control and the various ways by which those needs can be met. In a later chapter, we will consider in detail what might be called a *global* need for birth control resulting from the continuing explosion of humanity.

Questions

1. In our discussion of the rhythm method, we pointed out some of the reasons why it has a high failure rate. Thinking in terms of marriage and family relationships, what other disadvantages of the rhythm method can you think of?

2. The rhythm method has been characterized as a "natural" method of contraception and oral contraceptives are considered to be "artificial." What arguments might you offer in favor of classifying oral contraceptives as a "natural" means of birth control?

3. How have the social and sexual roles of men and women influenced their respective attitudes toward birth control? In what ways have these factors affected progress toward solving the problems of excessive human fertility?

4. In terms of reality, what are some important consequences of opposition to effective methods of contraception? Why might opposition to effective artificial contraceptives present a moral dilemma to the individual who regards abortion as criminal destruction of human life?

5. What effect might increased use of oral contraceptives have on the prevalence of venereal disease? Why?

6. Suppose that you are in charge of a research team that is given the task of developing the ideal contraceptive. List the characteristics that you would look for in the ideal contraceptive that you would hope to develop.

7. There has been considerable controversy over the possible effects of oral contraceptives on the general health of women. Discuss the concept of "risk" as it applies to oral contraceptives, pregnancy, other drugs, driving automobiles, and the general activities associated with normal living.

8. How might the effectiveness of oral contraceptives and IUD's compare in areas of the world where education is lacking and the illiteracy rate is high?

9. A man and his wife are 30 years old and 28 years old, respectively. They have two children and they agree that they want to limit their family to its present size. On this basis, they decide that the husband should have a vasectomy. What factors and possibilities should they consider before the final step is taken?

10. Abortion is the single most prevalent form of birth control that is used on a worldwide basis. Still, a large number of people condemn abortion, regarding it as murder of an unborn child. Take a position on this issue and give arguments to defend your point of view.

7

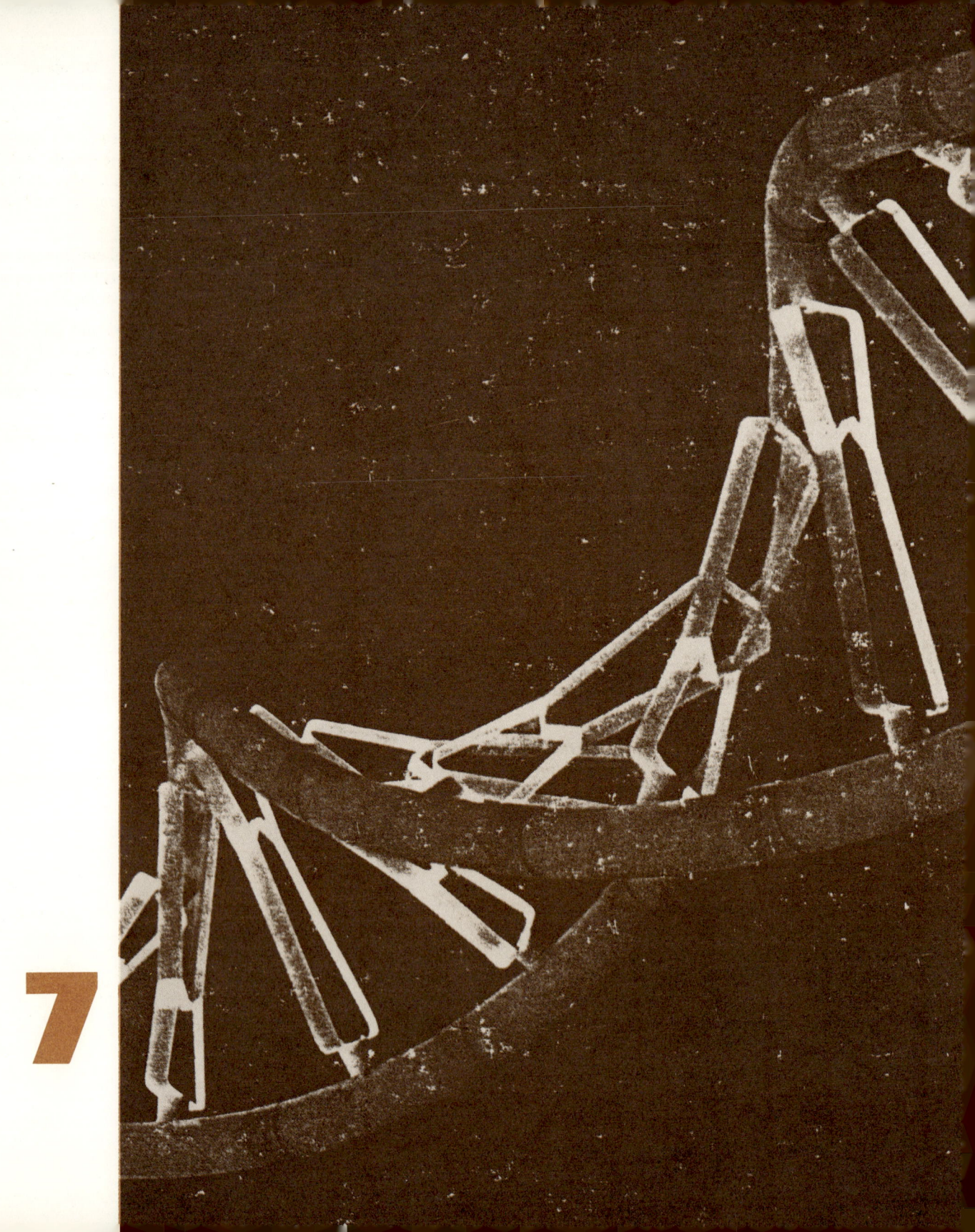

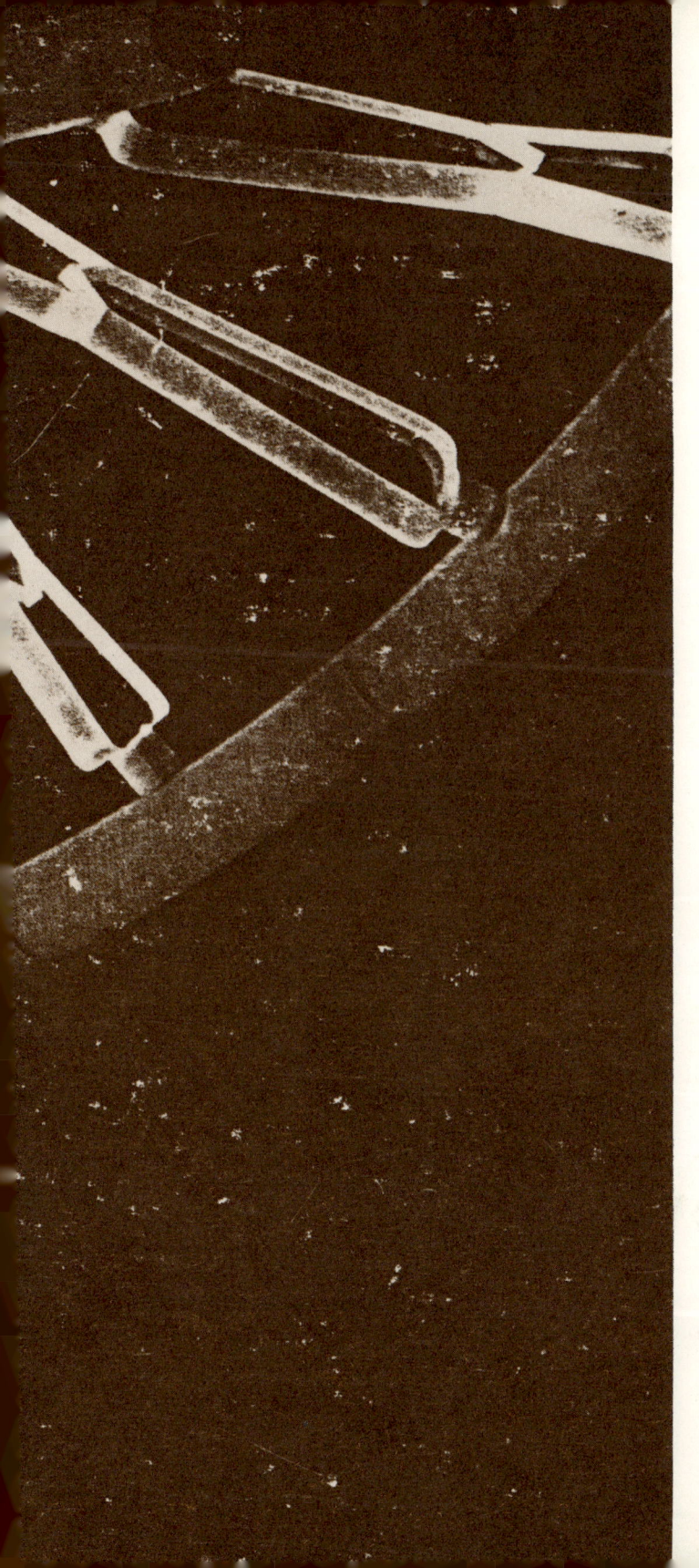

The Legacy – DNA

The Importance of the Gene

It is said that we bring nothing into this world and take nothing from it when we leave. Actually, whether we are born wealthy or poor by material standards, we all enter the world with an inborn legacy from our parents and from long-forgotten ancestors reaching back into the dim past. The nature of this endowment was decided by a single—and irrevocable—roll of the meiotic dice, and the specifics of the will were written at the moment of fertilization when one of over 64 trillion possible genetic combinations were produced from your parents alone. You are therefore a unique individual, created by a complex interaction between your genetic inheritance and the environmental forces that began even before birth to shape your individuality. It is probable that no one exactly like you has ever been on earth before, and it is highly unlikely that anyone exactly like you will appear again.

One thing is certain: We have no voice in the choosing of our ancestors, and we have no alternative but to accept the legacy that the genetic dice have decreed shall be ours. Just how critical is this genetic endowment to the development of the individual? Are we forever limited by the relative quality of our genes, or can environmental forces significantly modify our individual genetic potential? Therein lies the basis of the "nature-nurture" controversy. Like most issues important enough to argue about, this "nature-nurture" conflict has had extreme views expressed by both sides. Some have argued that we are fashioned mainly by genetic factors, while others have insisted that environment plays the dominant role. Better knowledge of genetics, however, has shown that both of these views oversimplify what is undoubtedly a very complex relationship between heredity and environment.

For example, there are many traits that are fixed by heredity and which are subject to little or no change by environmental influences. Such traits include sex, blood type, fingerprints, skin color, facial features, and a host of others obvious and not so obvious.

On the other hand, a child born with the inherited condition known as *phenylketonuria* (PKU) may be saved from severe mental retardation and may develop normally if his diet is suitably altered immediately after birth. Many people with inherited diabetes are also living normal lives due to the discovery of insulin. Also, a person with an unacceptable nose or a pronounced overbite has available the skills of a plastic surgeon or an orthodontist. The foregoing are just a few examples that illustrate the fact that we are not necessarily always prisoners of our genes. The expression of our genetic legacy can be, and often is, modified by environmental factors.

This is not to say, however, that environment can always come to the rescue of the individual regardless of his genetic endowment. For example, a child with *Down's syndrome* (mongolism) will be severely retarded mentally and physically. Although proper training may enable him to function on a somewhat higher level, no amount of exposure to the best of environmental conditions will enable that child to become a normally functioning member of society.

An especially good example of the "nature-nurture" aspect of development is that peculiarly human trait known as intelligence. The problem is exceedingly complex because intelligence is difficult to define and even more difficult to measure. One thing seems certain: Intelligence *is* a function of the brain and the *potential* for intelligence *is* therefore fixed by genetic control over brain development. On the other hand, the brain needs certain nutrients in order to develop properly, and poor nutrition during pregnancy may cause irreparable harm to the developing fetal brain. Even if this does not occur and the brain develops to its full capacity as decreed by the genes, the realization of complete genetic potential for intelligence still involves complex environmental forces that begin operating shortly after birth. Intelligence is therefore a trait that effectively supports the old saying, "Heredity loads the gun, but environment pulls the trigger."

Yet despite the unquestionable influence of environmental factors on the total development of the individual, the fact remains that the nature and quality of the genetic legacy play a critically important and fundamental role. In order to fully appreciate the importance of our genetic heritage, we must first understand some of the more fundamental aspects of the structure and function of the genetic material. In this chapter, we will attempt to build our understanding of genetic function in a step-by-step fashion, beginning with a review of some simple but necessary chemical principles.

The Energy of Chemical Reactions

Suppose we were to mix quantities of oxygen and hydrogen in a container and then seal the container so that none of the mixture could escape. What would happen? Actually nothing would happen, at least nothing that we would be able to detect. In fact, we could leave the sealed container sitting around for months on end without anything exciting happening or any significant change in the contents.

On the other hand, suppose we were to break the seal and hold the mouth of the container to an open flame. If you have ever seen this classic demonstration, you know that we would be rewarded with a loud explosion. The hydrogen and oxygen gases combine chemically to produce water according to the following well-known reaction:

$$2H_2 + O_2 \rightarrow 2H_2O + \text{Energy}$$

You can see that the above reaction does not take place unless it is "encouraged" to do so by energy from an outside source, which in this case is the open flame. Once initiated, however, it proceeds very definitely on its own until water formation is complete.

A more common example of this type of reaction is the burning of a piece of paper. Suppose that you decide to burn this book—a decision that by now you may regard as having great merit. First of all, you would probably tear out the pages in order to increase the surface area, and then touch a match to the pile. You know from experience that the match would provide a small amount of "starter" energy, after which you could gleefully watch the reaction proceed on its own.

Why do these reactions need energy to get them started? And why, once started, do they proceed on their own without a further supply of energy from an external source?

In considering the first question, we need to recall the importance of molecular velocity as a basis of chemical reactions. At room temperature, the average velocity of the hydrogen and oxygen molecules is simply not great enough to bring about a sufficient number of collisions per unit of time. Also, the hydrogen molecule (H_2) and the oxygen molecule (O_2) are each composed of two atoms held together by a covalent bond. Before they can be rearranged in the form of water (H_2O), they must first be pulled apart and their covalent bonds broken. This requires energy.

This "starter energy" which is needed to initiate reactions that then take off by themselves is called *activation energy*. In the hydrogen-oxygen reaction, the open flame furnished the activation energy, while in the more satisfying book-burning incident it was supplied by the match.

Now to consider the second question, we should recall that both reactions obviously produce large amounts of energy in the forms of heat and light. At this point it is important to recall that a molecule contains a certain amount of chemical energy due to the arrangement of the electrons that are associated with it. A molecule of hydrogen contains a certain amount of chemical energy, and so does a molecule of oxygen. On the other hand, a molecule of water possesses *less* energy than the *sum* of the separate energies of the hydrogen and oxygen molecules. Therefore, when hydrogen and oxygen combine to form water, something must happen to that excess chemical energy. At this point the First Law of Thermodynamics enters the picture. This very basic physical law tells us that energy is neither created nor destroyed, but is merely changed to some other form of energy. Thus in the hydrogen-oxygen reaction, the excess chemical energy appears explosively in the forms of heat and light.

Now let us consider the question of what would happen if we attempted to reverse the reaction. We would then have

$$\text{Energy} + 2H_2O \rightarrow 2H_2 + O_2$$

which is a reversal of the reaction by which water was formed. In this case, water is broken down into hydrogen and oxygen; therefore, the same amount of energy that was lost when the water was formed must be restored. Consequently, instead of a reaction that proceeds on its own after a little encouragement, we now have a reaction that requires a constant supply of energy. Water can be broken down by supplying electrical energy, but this energy must be supplied constantly. If it is turned off at any point, the reaction will stop.

Reactions of the first kind, that is, those that go forward on their own after a little push, are called *exergonic* (energy-liberating) reactions. The second type, illustrated by the decomposition of water, is called an *endergonic* (energy-absorbing) reaction.

At this point we are ready to consider the "energy hill" diagram shown in Figure 7.1. Looking at this diagram, suppose that we are riding in a car and have the misfortune to run out of gas at point *A*. We can see that it would be easy to reach point *C* provided we all got out and pushed it over the hump represented by point *B*. We could all climb aboard at that point and happily coast down to point *C*. The hump that is represented by point *B* is the *activation-energy barrier;* this is the barrier we overcame when we held the hydrogen-oxygen mixture to the open flame. The reaction, like the car, then coasted down the hill to point *C*.

But now suppose that when we reach point *C* we are told that the bridge ahead is washed out and we must return to point *B*! We must then push the car back up the hill, exerting a constant and unfailing push as we go. This is what happens when we apply a constant source of electrical energy in order to decompose water.

Let us return for a moment to the problem of the activation-energy barrier. In the case of the hydrogen-oxygen reaction, we had no real problem in pushing the reaction over the hump at point *B* because we had a hot, open flame available to supply the necessary energy. On the other hand, the human body normally operates at a temperature of around 98.6°F, and this is hardly enough to push chemical reactions over activation-energy barriers.

Fortunately there is a solution. Since we lack sufficient energy to push the car over the hill at point *B*, we will send for a bulldozer to lower the height of the hill, thus making it possible to push the car over the activation-energy barrier with even the small amount of energy that we can muster. This is essentially the procedure used by living systems, and it is accomplished by the use of *catalysts*. A catalyst is a chemical that accelerates a reaction without being changed in any way by its participation in the reaction. At this point we should emphasize the fact that a catalyst does *not* cause a reaction to happen unless it would occur anyway, even though very slowly. A catalyst, then, increases the *rate* of a chemical reaction.

Enzymes

Chemical reactions that take place in the body do so because of the presence of special catalysts known

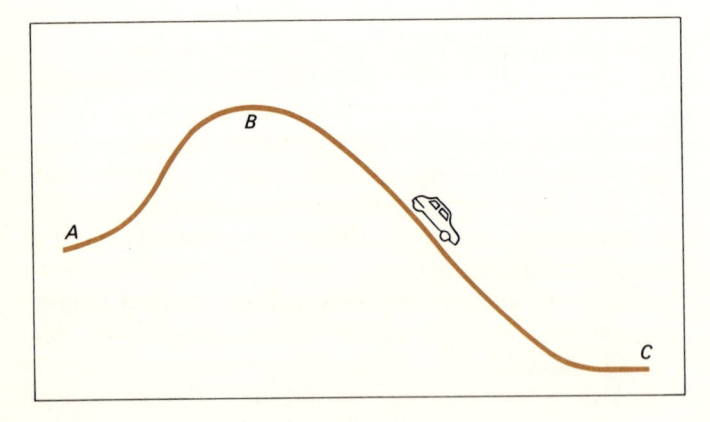

Figure 7.1
The "energy hill." (See text for details.)

The energy of chemical reactions 117

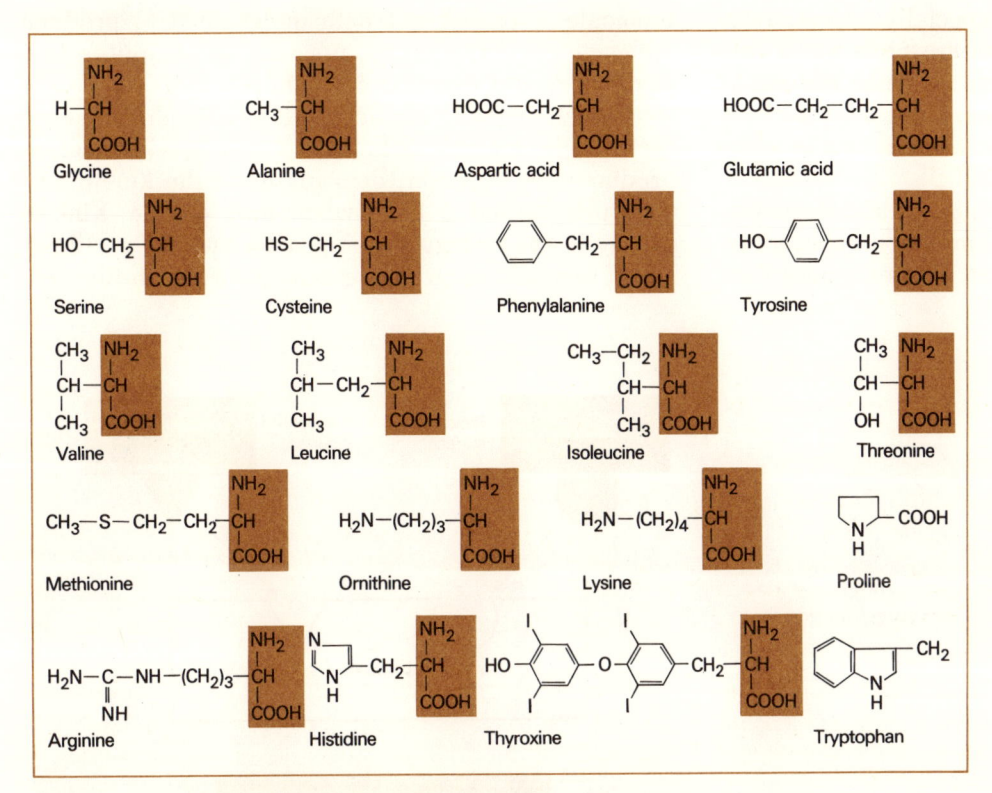

Box 7.1
Structures of the twenty amino acids that are important building blocks of proteins. Note how they differ according to their side chains.

as *enzymes.* Enzymes are therefore necessary to the normal operation of living systems and are essential to the maintenance of life itself. Unfortunately, enzymes do not always function as they are supposed to. Some of the consequences of the failure of enzymes to properly catalyze reactions will be discussed in this and later chapters.

All enzymes are protein molecules, although some may consist of a large protein molecule combined with a smaller nonprotein molecule called a *coenzyme.* Vitamins are sometimes important to the structure of coenzymes, which partially explains the importance of vitamins to normal body function.

You will recall from Chapter 2 that proteins are tremendously complex molecules made up from a practically limitless number and variety of combinations of the twenty amino acids that are used as their structural "building blocks." Box 7.1 contains the structural formulas of the twenty basic amino acids used in living systems. Be sure to note how these amino acids differ in structure according to the shape of their side chains. This large variety of possible shapes of protein molecules is especially important to enzyme function. Since enzymes are proteins, it follows that the patterns on which enzymes are made have almost limitless possibilities.

Enzyme Function

The theory of how enzymes catalyze reactions is explained by the "lock-and-key" model illustrated in Figure 7.2. Referring to this figure, let us suppose that molecule A is to react with molecule B to produce molecule AB. Without the presence of an appropriate enzyme, A and B will occasionally collide and sometimes their reacting surfaces will be brought into contact, producing AB. Whether A and B will react to produce AB, however, depends upon the *kinetic energy* that A and B have when they collide. Kinetic energy is the energy developed by moving particles, and it depends upon the velocity with which they are

Figure 7.2
The "lock-and-key" theory of enzyme function.

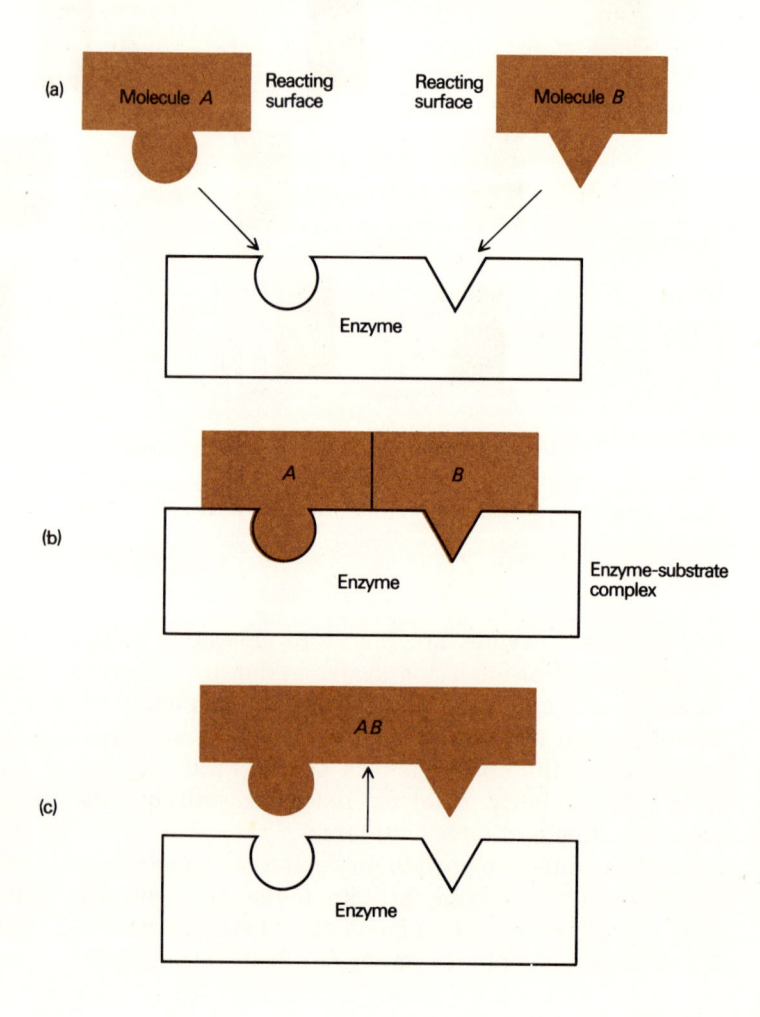

moving. At body temperature, there is not enough activation energy available to provide sufficient kinetic energy to molecules A and B so that a reaction will take place.

Molecules A and B have definite and specific shapes, represented diagrammatically in Figure 7.2. Note that the enzyme molecule also has a specific shape and that it closely fits molecules A and B. Step (b) of the figure shows how A and B lock onto the enzyme with their reacting surfaces held in contact in order to produce AB. In the third step, the finished product is released by the enzyme, which is now free to catalyze more reactions between other A and B molecules.

This lock-and-key model, with certain complications, is the generally accepted explanation of enzyme function. From this model it is apparent that an enzyme must have a specific structure in order to catalyze a specific reaction; if the enzyme structure should vary, its ability to catalyze the necessary reaction may be drastically reduced. This fact is of critical importance, and we will discuss its significance in greater detail later in this chapter.

The chemical (or chemicals) upon which an enzyme acts makes up the *substrate,* and the enyzme-substrate complex is generally but not always *specific.* For example, sucrose (cane sugar) is broken down in the digestive process to glucose and fructose by the digestive enzyme sucrase:

$$\text{Sucrose} \atop (\text{Glucose} + \text{Fructose}) \xrightleftharpoons{\text{Sucrase}} \text{Glucose} + \text{Fructose}$$

Suc*rase* specifically attacks suc*rose* and no other molecule. Some enzymes, on the other hand, will attack several different substrates. For example, certain digestive enzymes attack different types of protein molecules because their specific function is to break the *peptide bonds* between specific pairs of amino acids. Since such amino-acid pairs exist in a variety of protein molecules, the enzyme is not choosy about which protein molecules it attacks. In general, however, *specificity* is a good word to apply to enzyme function. Enzymes work mainly on only certain substrates, and an enzyme must have a specific structure to do the job.

It is not unusual to find a key that will almost, but not quite, fit a certain lock. Perhaps you have had the experience of inserting such a key into a lock and then finding it difficult or even impossible to pull it out again! In this way, enzymes can sometimes be locked up and their function destroyed by chemicals that are similar enough to the genuine substrate to effectively "fool" the enzyme. Certain poisons such as cyanide work on this principle. Cyanide, for example, renders an important respiratory enzyme useless by permanently combining with it, thereby removing it from the field of action.

As we have seen so far, enzymes are essential to the maintenance of life itself. Furthermore, the structures of enzymes are critically important to the smooth operation of the living system. A poorly manufactured enzyme can be disastrous, and so enzyme synthesis, or more generally, *protein synthesis,* is therefore one of the most fundamental processes carried on by the cell. We will begin our discussion of this important and fascinating process by examining the structure of our genetic legacy—DNA.

DNA

In 1953, James D. Watson and Francis E. Crick first described the basic structure of the material that makes up our genetic heritage. This discovery was an important milestone in biology, opening up an area of research that has since seen man's knowledge of heredity increase at a rate that is probably unequaled in any other field. In this and subsequent chapters, we will see some of the implications of the new genetics for human society and for the individual. First, however, we need to look at the basic structure and function of the genetic material.

The active hereditary material of chromosomes is *deoxyribonucleic acid,* or DNA. This somewhat lengthy name is derived from the fact that DNA is a *nucleic acid* and contains a five-carbon sugar called *deoxyribose.* Box 7.2 shows the chemical structures of the components of DNA; with each chemical formula is the symbol we will use in our simplified approach to DNA structure.

Referring to Box 7.2, we can see that the basic building blocks of DNA consist of (1) a phosphate group, (2) deoxyribose sugar, and (3) four *bases* consisting of *adenine, thymine, guanine,* and *cytosine.* Thymine and cytosine belong to a class of chemical compounds called *pyrimidines,* while adenine and guanine are classified as *purines* (see Chapter 2). Note that adenine and guanine are double-ring compounds, while thymine and cytosine each consist of a single ring. In the DNA molecule, *adenine normally combines only with thymine and guanine normally combines only with cytosine.*

Box 7.2
Chemical structures of the components of DNA. An appropriate symbol is used in place of the actual molecular structure.

Let us now proceed to build a DNA molecule in a step-by-step fashion using the symbols of its chemical components as illustrated in Box 7.2. As a first step, we will arrange the deoxyribose sugar molecules as follows.

Next, we will add the phosphate groups which are bonded to the sugars as follows.

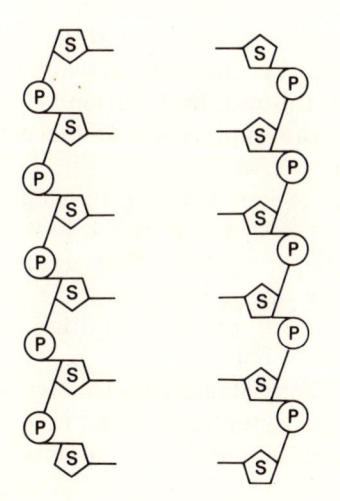

At this point, we can compare the construction of the DNA molecule to that of a ladder. We may think of the sugar-phosphate chains as the sides of the ladder; the next step is to add the rungs. The "rungs" are composed of various combinations of the bases as shown below.

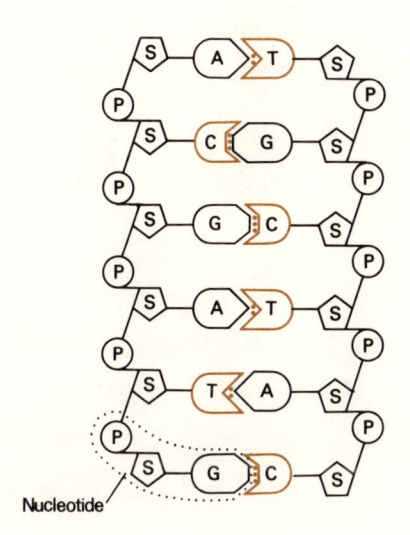

Now that we have completed the "ladder" as shown above, we give it a twist to produce a double helix; the result is the Watson-Crick model of DNA as shown in Figure 7.3.

Before we go further, let us refer back to our "ladder" version of DNA. Note the two or three dots that appear between each of the base pairs. These dots symbolize *hydrogen bonds,* which are relatively weak chemical bonds and permit the DNA molecule to "unzip" down the center with relative ease. Also note the group which is encircled by a dotted line. This group consists of a base, a sugar, and a phosphate, and is called a *nucleotide.* Each complementary chain making up the DNA molecule is therefore a string of nucleotides; it is therefore called a *polynucleotide chain.*

Now we need to consider a very important aspect of the DNA molecule. Looking at one of the polynucleotide chains in the above diagram, you will note that there are six base positions, each of which is to be filled by one of the four possible bases—A, T, C, or G. This means that there are four possibilities for position one, four for position two, four for position three, and so on. If we now multiply $4 \times 4 \times 4 \times 4 \times 4 \times 4$, we obtain a total of 4096 different possible sequences of the bases in a polynucleotide chain consisting of only six positions! Now, if we consider that our illustration is by no means realistic and that a real DNA molecule contains thousands or possibly millions of base positions in the chain, it follows that the number of possible base sequences is practically limitless. This is an extremely important aspect of DNA in terms of its role in the synthesis of enzymes and other proteins, because it can provide an almost unlimited variety of patterns for an equal number of enzymes and other proteins.

Replication of DNA

You will recall from Chapter 4 that the chromosomes duplicate themselves in both mitotic and meiotic cell division. Since chromosomes are made up of protein and DNA, it follows that chromosome duplication must include the duplication of the genetic message carried by the DNA.

DNA replication is a step-by-step process (refer back to Figure 2.9a, b, c). In the first step, the DNA molecule "unzips" at the points represented by the weak hydrogen bonds between the base pairs. In the second step, each of the resulting polynucleotide chains randomly collides with the bases, phosphate groups, and deoxyribose sugar molecules which are available in the cytoplasm of the cell as nucleotides. With these raw materials, a complementary chain is

Figure 7.3
The DNA model in this photograph is a simplified illustration of the molecule's helical shape. Note how C combines with G and A with T. (Courtesy of Science Related Materials, Inc., Janesville, Wisconsin.)

constructed (step 3). The last step shows two DNA molecules that are identical to each other and to the original molecule from which they were formed.

Ribonucleic Acid

Ribonucleic acid (RNA) differs from DNA in three respects. First, RNA does not contain the base thymine, but contains a different base, *uracil,* in place of thymine. Therefore in RNA, uracil is substituted for thymine at any point at which thymine would be found in the base sequence of DNA. Second, RNA contains the sugar *ribose* in place of deoxyribose, which explains the name *ribo*nucleic acid. Finally, RNA is usually single-stranded and not in the form of a helix. The basic differences in molecular structure between the building blocks of DNA and RNA are shown in Figure 7.4.

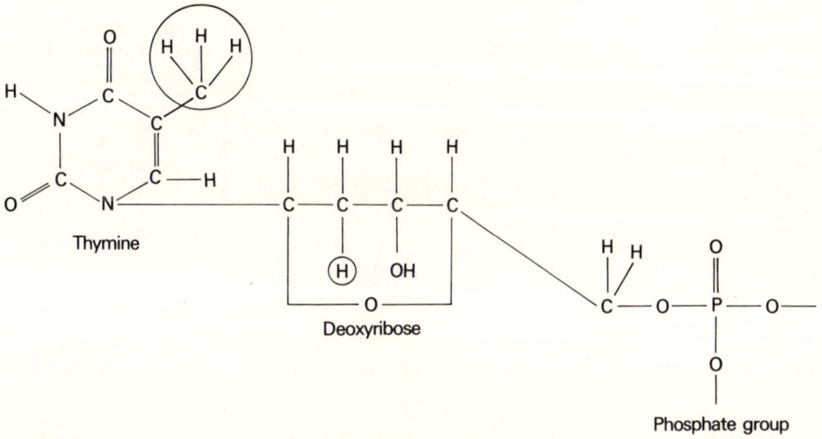

(a) **DNA nucleotide**

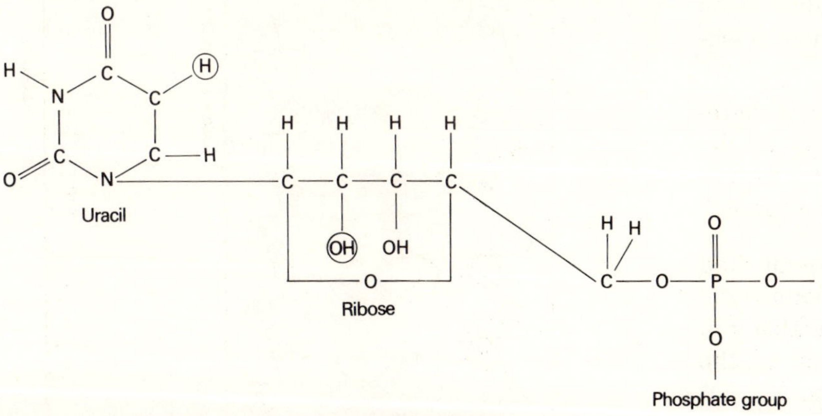

(b) **RNA nucleotide**

Figure 7.4
Basic differences in molecular structure between DNA and RNA nucleotide components. These differences are indicated by circles.

Protein Synthesis

The manufacture of proteins, which include the all-important enzymes, usually takes place on that part of the cell called the *endoplasmic reticulum* (see Figure 4.1). Specifically, the sites of protein synthesis are the *ribosomes* located along the tortuous pathways of the endoplasmic reticulum.

We will begin our discussion of protein synthesis with an overview of the process. Then we will proceed to the details of how the genetic code interprets the genetic message of DNA into the structure of the protein itself. The overview of protein synthesis is as follows.

1. DNA is the active part of the inherited genetic material, and it acts as the "master blueprint" from which the protein is ultimately constructed.
2. DNA is first translated into ribonucleic acid (RNA).
3. This RNA, called *messenger RNA* (mRNA), carries the genetic message to the ribosome where it provides a blueprint for the construction of a specific protein.
4. Another type of RNA, called *transfer RNA* (tRNA), brings specific amino acids from the cell cytoplasm to the ribosome. These amino acids are then arranged in a definite sequence in accordance with the mRNA blueprint.
5. The basic structure of the finished protein is therefore directly dependent on the structure of the messenger RNA, which in turn is directly dependent upon the structure of the DNA, or "gene."

Synthesis of proteins by the cell is accomplished according to a system that involves the *genetic code*. Box 7.3 presents a series of illustrations to which you should refer as we examine the mechanism of the genetic code in the following series of steps.

1. A polynucleotide chain of DNA in the nucleus acts as a pattern for the formation of a complementary chain of RNA. Note how the thymine of DNA "calls for" adenine, cytosine calls for guanine, guanine for cytosine, and adenine for thymine. Note also that in RNA, uracil substitutes for thymine and calls for adenine.

2. The chain of RNA produced in step (1) is called *messenger RNA*, or mRNA. In the second step, you can see that mRNA has carried the genetic

Box 7.3
The steps in protein synthesis using a DNA strand as a fundamental template. (See text for details.)

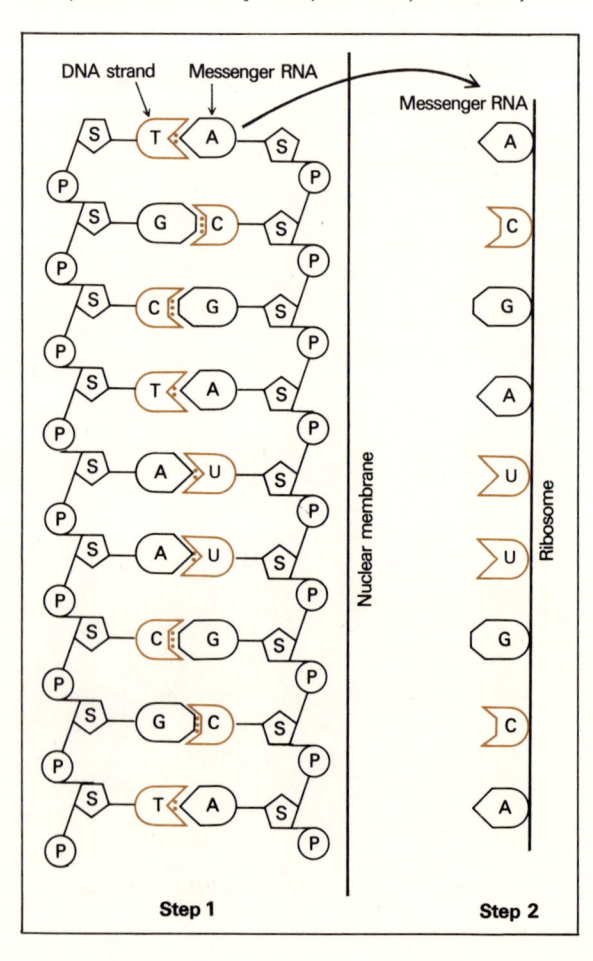

message of the gene (DNA) to the ribosome, where it has been established as a kind of pattern "workbench" upon which a *polypeptide chain* will be built. You will recall from Chapter 2 that a polypeptide chain is a string of amino acids linked together.

3. You can now see a molecule of *transfer RNA*, or tRNA, combining with a specific amino acid from the cell's cytoplasm. Note that the tRNA has the three bases C-G-U as part of its molecule. This specific sequence of bases will determine the location on the mRNA "workbench" to which the amino acid will be brought by the tRNA. You can readily see that the C-G-U of the tRNA will "code in" on that part of the mRNA which has

Box 7.3 (continued)

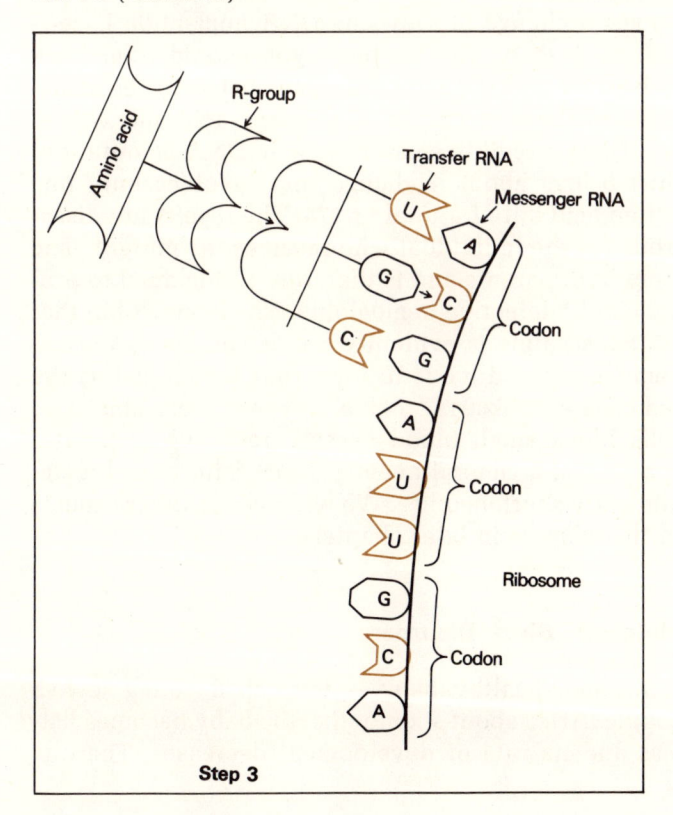

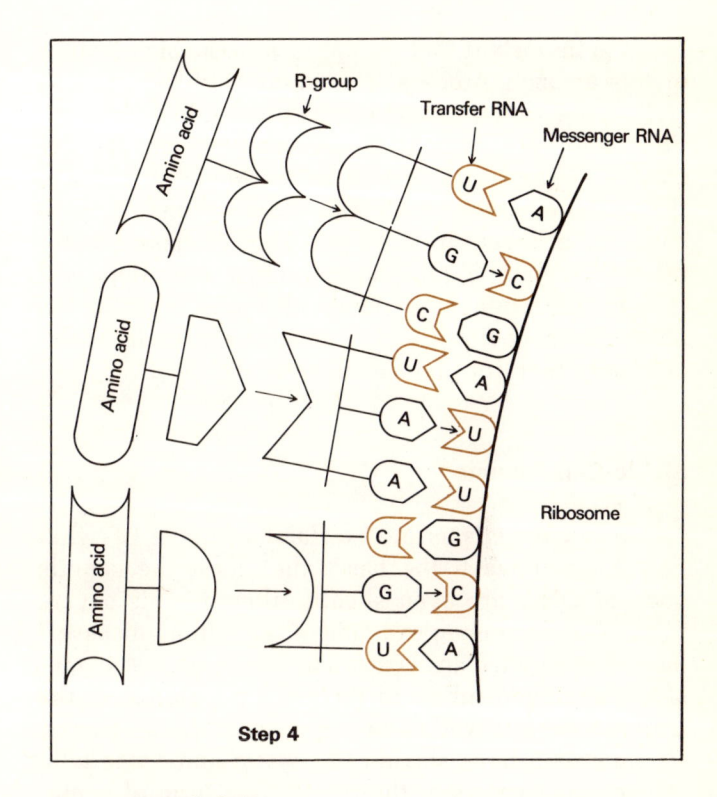

Step 3

Step 4

Box 7.3 (concluded)

the sequence G-C-A. C-G-U "fits" G-C-A but does not, for example, fit the next sequence of three bases on the mRNA, which is U-U-A. It should also be apparent by now that each of the *codons* which carry one of the twenty available amino acids consists of a combination of *three* bases.

4. Next, you can see that other tRNA molecules have "coded in" at appropriate locations on the mRNA. Each of these tRNA molecules carries a certain amino acid; we therefore have three specific amino acids arranged in a definite sequence.

5. Finally, bonds form between the amino acids and the polypeptide chain is complete. This polypeptide chain now forms part of the primary structure of a protein molecule.

It is important that we appreciate the significance of this amazing process. The amino-acid *sequence* as well as the *kinds* of amino acids present in a polypeptide chain are ultimately dependent upon the *sequence* of the bases as well as the *kinds* of bases in the inherited DNA. A shift in the positions of these bases or a substitution of one kind of base for another is called a *mutation*. Such a mutation will change the sequence or kinds of amino acids in the polypeptide chain, and the resulting protein molecule may therefore be significantly different.

Sickle-Cell Anemia

Sickle-cell anemia is an excellent example of what may happen when the bases that form the genetic code of DNA are even slightly altered. This severe form of anemia is an example of an "ethnic disease," having originated in Africa and found almost exclusively among members of the black population of the United States as well as among certain African blacks. Persons who have sickle-cell anemia and those who are genetic carriers of the trait have a form of hemoglobin that differs slightly but significantly from the normal type. You will recall that hemoglobin is the substance in red blood cells that combines with and carries oxygen.

Hemoglobin is made up of an iron compound combined with four polypeptide chains, two of which are called *alpha* chains and the other two of which are called *beta* chains. Each pair of alpha and beta chains contains 287 amino acids; the hemoglobin molecule therefore consists of a total of 574 amino acids. Now, if we consider a short segment of one of these polypeptide chains, we find that it consists of the following sequence of amino acids:

Proline—**Glutamic acid**—Glutamic acid—Lysine

On the other hand, the form of hemoglobin associated with sickle-cell anemia differs in this same segment as follows:

Proline—**Valine**—Glutamic acid—Lysine

From this it is apparent that the sickle-cell form of hemoglobin differs from the normal variety only in that *valine* has been substituted at one point for *glutamic acid*. Now, it is established that the messenger RNA triplets G-A-A and G-U-A call for glutamic acid and valine, respectively. Furthermore, these messenger-RNA triplets would be created by the *DNA* triplets C-T-T and C-A-T, respectively. Following this through carefully and recalling the principles of protein synthesis, you will see that a change from C-T-T to C-A-T in one segment of DNA would call for a substitution of valine for glutamic acid in the polypeptide chain. This, therefore, involves a change in only one DNA nucleotide base—that of T to A! At this point you should refer back to Box 7.1 and note the difference between the structures of the side chains of glutamic acid and valine.

This incredibly small change in DNA (a mutation) thus brings about a change that involves only one amino acid out of a total of 574. The results are disastrous for the individual who receives the mutant gene from both parents and is therefore condemned to produce only defective hemoglobin. The hemoglobin that contains valine is evidently not as soluble as normal hemoglobin and tends to form rods which distort the red cells into bizarre "sickle" shapes. These abnormal cells block small blood vessels, resulting in painful symptoms, a general anemia, heart failure, and a significantly shortened life. We will discuss other aspects of this disease in later chapters.

Metabolic Block Disorders

A normal, healthy baby is born to a young Jewish couple. After about six months, the baby becomes listless and its rate of development decreases. There is increasing evidence of nervous-system involvement. The diagnosis indicates *Tay-Sachs disease.* Like sickle-cell anemia, this is also an "ethnic" disease and is found almost exclusively among Jews. The Tay-Sachs child does not produce an active form of the enzyme *hexosaminidase-A,* and as a result accumulates a fatty substance in the brain which leads to brain deterioration. This diagnosis guarantees that the child's condition will slowly deteriorate while the parents stand by helplessly, until at the age of two or three, the child's death finally puts an end to their anguish and to the mounting medical bills.

Another couple has a child with fair skin and blond hair, apparently normal at birth. They begin to notice that he is lagging behind the timetable of normal development. Leg development seems especially slow; there are jerky body movements and a musty body odor. Tests show evidence of mental retardation and an abnormally high level of a certain amino acid in the blood. The child's condition is diagnosed as *phenylketonuria* (PKU). If this condition is not diagnosed very early, the child will always be mentally retarded; as he grows to maturity, he will more than likely live out his life in an institution (see Figure 7.5).

Still another couple has a child, also normal at birth, that begins to develop symptoms in early infancy. The liver becomes enlarged and the child begins to develop cataracts. If an early diagnosis is not made, the child will become blind, mentally retarded, and die. The diagnosis indicates *galactosemia*—the baby is being poisoned by its mother's milk!

These three conditions are examples of a long list of human afflictions known as *metabolic block disorders.* The theoretical basis of this type of disorder may be summarized in terms of what we have learned so far:

1. Chemical reactions in the body are catalyzed by enzymes, and these reactions will not proceed effectively without appropriate enzymes.

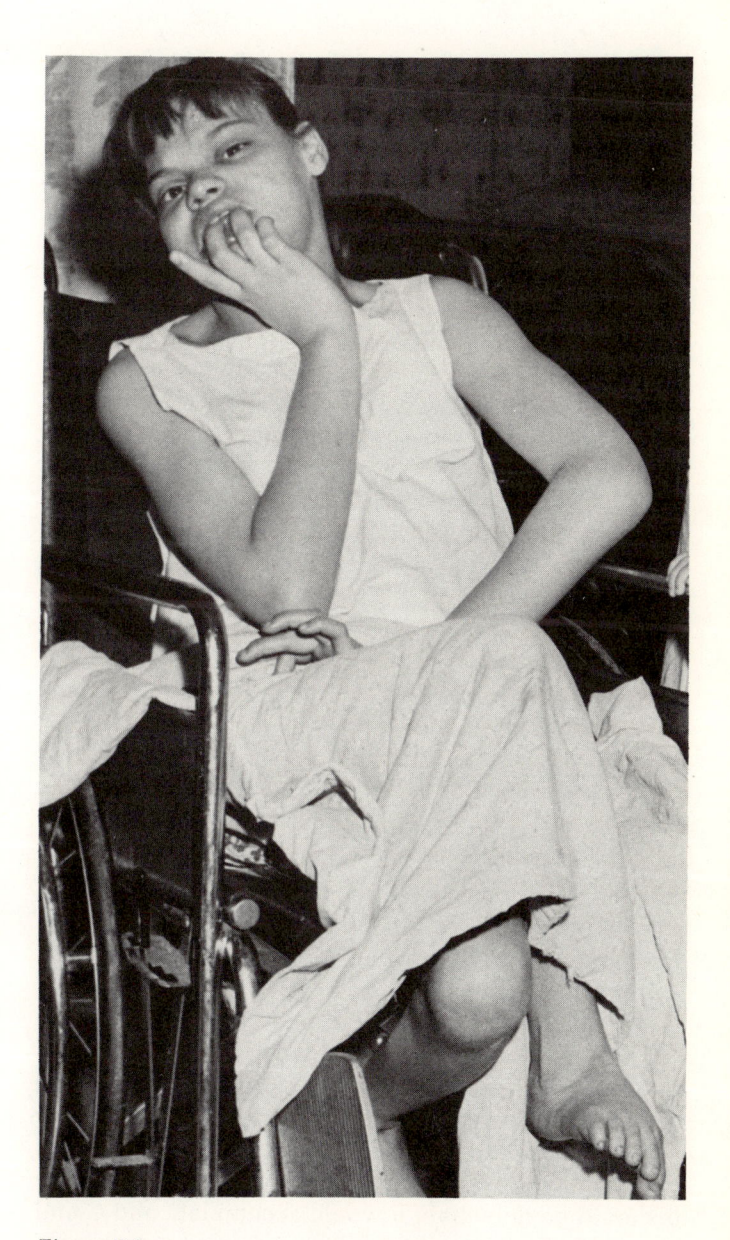

Figure 7.5
A patient afflicted with phenylketonuria (PKU). She is severely retarded and normal leg development has been impaired. (Courtesy of the National Foundation–March of Dimes.)

2. Enzymes work on the "lock-and-key" principle; it is therefore essential that the enzyme be shaped to fit the substrate on which it acts.

3. The structures of proteins, *including* enzymes, are determined in part by specific DNA patterns carried by the chromosomes.

Many reactions in the living system occur in a chainlike fashion, as represented by the following model:

$$A \xrightarrow{a} B \xrightarrow{b} C \xrightarrow{c} D$$

Such a series of reactions may be referred to as a *metabolic pathway,* and in our model this pathway converts molecule A to molecule D. However, this is done in a series of steps which involve the conversion of A to B, then B to C, and finally C to D. Above the arrow in each step is a symbol for the appropriate enzyme that is required to catalyze that specific step. Thus, enzyme a is required to convert A to B, and so on.

Now let us consider some of the things that could go wrong in this pathway. Suppose, for example, that enzyme a is constructed from a faulty DNA pattern and as a result does not fit the substrate, just as the wrong key will not open a lock. As a result, molecule A is not converted to B, and therefore production of B, C, and D is blocked at that point. Furthermore, we can see that substance A will begin to accumulate in the system. We have, in other words, a block in the metabolic pathway. The lack of enzyme a is only one possible source of difficulty; if you look back at our metabolic pathway model, you can see that if enzyme b should be defective, B would accumulate and C and D would not be produced. This principle can be applied throughout the pathway. It is a bit like the situation in which a bucket brigade is formed to put out a fire. If any one person is removed, the water buckets pile up at that point and no water arrives at the fire.

Still another possibility for failure is illustrated by the following model:

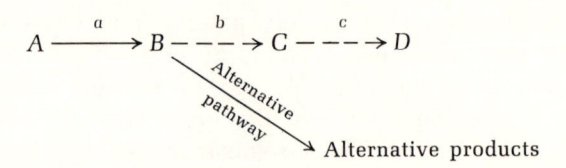

$$A \xrightarrow{a} B \dashrightarrow{b} C \dashrightarrow{c} D$$
$$B \searrow \text{Alternative products (Alternative pathway)}$$

In this case, note that B has accumulated due to the failure of enzyme b to catalyze the reaction involving the conversion of B to C. As the level of concentration of B increases, an alternative reaction pathway takes over and B is broken down into various other products which may be toxic to the living system.

Phenylketonuria

Now we can apply the above principles to phenylketonuria, which was our second example of a metabolic block disorder. Phenylketonuria, or PKU, results from the body's inability to convert the amino acid *phenylalanine* to another amino acid, *tyrosine,* as follows:

$$\text{Phenylalanine} \xrightarrow{\text{Phenylalanine hydroxylase}} \text{Tyrosine}$$

Note that the above reaction is catalyzed by a specific enzyme, *phenylalanine hydroxylase.*

In order to explain the mechanism of PKU, we must resort to another of our imperfect analogies. Suppose that we have two keys, both of which fit the lock on a particular door. Since both keys fit the lock, we may go through the door one-hundred times with little or no difficulty. If, on the other hand, we were to file off one of the small ridges on one of the keys, then that key will no longer fit the lock. Now, if we go through the door one-hundred times, we will still get through, but more slowly due to the occasional fumbling that results from the fact that one of the two keys does not fit the lock. If we now carry our analogy further and file off the same ridge on the sec-

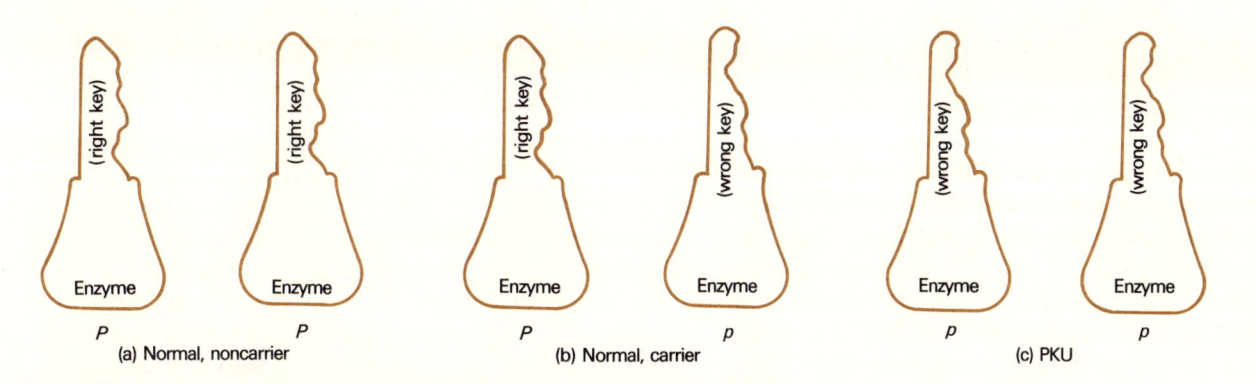

Figure 7.6
Comparison of an enzyme to a key. Note that the PKU child has two "wrong keys." He therefore cannot "open the lock," that is, convert phenylalanine to tyrosine.

ond key, we will not go through the door at all because now neither key fits the lock.

In applying our analogy to PKU, we must first recall that our chromosomes exist in pairs, one member of each pair having come from each parent. We therefore have two DNA patterns that code for the enzyme phenylalanine hydroxylase. If both patterns code for the correct enzyme structure, then "both keys fit the lock" and we convert phenylalanine to tyrosine with no difficulty (see Figure 7.6). If, on the other hand, we have inherited a defective DNA pattern from one parent and a correct pattern from the other, we are then in a position similar to that of the individual with one right and one wrong key. We will still be able to convert phenylalanine to tyrosine, but more slowly. It follows, therefore, that the PKU child is in the unfortunate position of having inherited a defective DNA pattern from *each* parent. He therefore has two "wrong keys," or defective enzymes, and is unable to convert phenylalanine to tyrosine, producing the following situation:

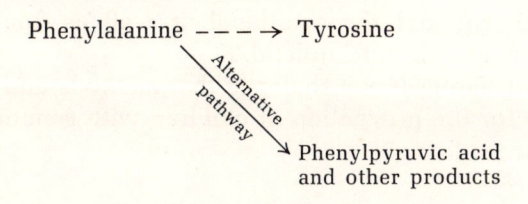

Phenylalanine $\dashrightarrow$ Tyrosine
$\searrow$ *Alternative pathway*
Phenylpyruvic acid and other products

Therefore, phenylalanine builds up to abnormally high levels. Since it is not converted to tyrosine, the level of tyrosine tends to be low, consisting of only the tyrosine that is brought into the body in the diet. The phenylpyruvic acid and other breakdown products and possibly even the phenylalanine itself attack the brain cells of the infant, producing severe mental retardation. It should also be noted at this point that tyrosine is necessary to the production of *melanin*, which is the brown pigment that provides skin, hair, and eye coloration. Since the PKU child may carry a relatively low concentration of tyrosine, this explains the tendency for such children to have fair skin, blonde hair, and blue eyes, though this varies with individuals.

Since the basic problem is the accumulation of abnormally high levels of phenylalanine, it follows that brain damage may be avoided to a greater or lesser degree by giving the child a diet that is low in phenylalanine. Such a low-phenylalanine diet has been shown to be successful in the management of PKU children and is especially successful if begun immediately after birth. Unfortunately the damage is irreversible, and dietary treatment will have no effect in cases where the child has been on a regular diet for a period of time. This treatment does not always prevent retardation, and more research will be required to fully assess its value. For example, it is not known for certain at this writing if the PKU patient can assume a regular diet after reaching maturity or if he must remain on the special diet for the rest of his life. Unfortunately, the diet is unpalatable and requires close supervision by the parents, who live in constant fear that the child may be given improper food by well-meaning neighbors or by school cafeterias.

If the dietary treatment is to be successful, it is essential that it be started immediately after birth. This requires that a method of very early identification be available. The *Guthrie test* for PKU is now practically routine in most hospitals throughout the United States. This test requires a few drops of the infant's blood, which is then tested for phenylalanine level by checking the blood's effects on bacterial growth. The Guthrie test has come under criticism from some quarters because of the possibility that a high phenylalanine level may result from certain other conditions which are only temporary and less serious than PKU, with the consequence that the child is needlessly put on a difficult and unpalatable diet. At this writing, however, it appears that most experts would advise against taking a chance. In other words, if the Guthrie test is positive, it would appear preferable to initiate the low-phenylalanine diet rather than risk severe mental retardation.

Galactosemia

Galactosemia is another example of a metabolic block disorder, and it results from the inability to convert the sugar *galactose* to the sugar *glucose*.

As you will recall, milk sugar is called *lactose*, and it is broken down by the digestive process to glucose and galactose as follows:

$$\text{Lactose} \xrightarrow{\text{Lactase}} \text{Galactose} + \text{Glucose}$$

The galactose thus produced must then be converted to glucose. This is done by a series of stepwise reactions which are mediated by enzymes, one of which is *galactose-1-phosphouridyl transferase*. The child with galactosemia does not produce an effective form of this enzyme, and the level of galactose therefore builds in the blood with the disastrous consequences described earlier in this chapter. Fortunately, in this situation the problem can easily be solved by using a milk substitute made from soybean products which do not contain lactose. In some cases, adults seem to develop an intolerance to milk sugar, but this is apparently due to problems with a different enzyme.

Tay-Sachs Disease

From our discussions of PKU and galactosemia, it is apparent that the hereditary defect causing these conditions may be partially or completely overcome by appropriate manipulation of the environmental factor of diet. Unfortunately, at this writing no such hope can be offered in the case of Tay-Sachs disease. The child with Tay-Sachs disease lacks the enzyme *hexosaminidase-A,* which is required to process fats. As a result, fatty materials build up in the nervous system and gradually destroy nerve cells. Death is inevitable, usually by the age of three or four years.

Carrier Identification

Our examples of inherited metabolic block disorders are merely a few in a long list of such diseases that plague mankind. We have seen that in some cases, the effects of defective genes can be reduced or controlled by manipulating environmental factors, but there are those who argue that it is far better to prevent the birth of such children if possible. This is an obviously logical argument when conditions such as Tay-Sachs disease are involved; where treatment is nonexistent and the emotional as well as financial burdens are all but intolerable.

In Chapter 9, we shall see that methods and programs for the prevention of children with genetic defects depend in part on the ability to identify those normal parents who are carriers of the defective gene. With PKU, for example, this may be done by testing the individual's ability to convert phenylalanine to tyrosine by "loading" the blood with phenylalanine and then measuring the time it takes for the phenylalanine level to be reduced. Because of the fact that carriers have one "right key" and one "wrong key," this reduction time is usually significantly greater in the PKU carrier than in the non-carrier, although a rather large variation in reduction time causes some degree of overlap between the carrier and the noncarrier, making the reliability of the test somewhat less than perfect.

The Tay-Sachs carrier may also be identified with a simple blood test that reflects a decreased production of hexosaminidase-A. A similar test can identify a carrier of galactosemia. As we shall see later, knowing that prospective parents are or are not carriers of a specific genetic defect can help predict the chances that they will produce afflicted children.

Questions

1. List several human traits that you think are (1) controlled exclusively by heredity, (2) determined entirely by environmental forces, and (3) determined by a combination of heredity and environment.

2. Discuss the basic mechanism by which enzymes promote chemical reactions. Why are enzymes especially important to body functions?

3. Discuss the critical nature of DNA structure in terms of enzyme function.

4. Review the basic principles associated with metabolic block disorders. What are some of the ways in which a metabolic block disorder differs from typhoid fever?

5. Discuss the following statement: In some respects we are forever limited by the nature of our genes, but in some cases we are not.

6. Why is it useful to be able to identify carriers of a gene that produces a genetic defect such as Tay-Sachs disease?

7. Suppose that you have the means to eliminate galactosemia, Tay-Sachs disease, and phenylketonuria, but you can attack only one disease at a time with a considerable lapse of time before you can eliminate the next one. In what order would you eliminate them? Why?

8

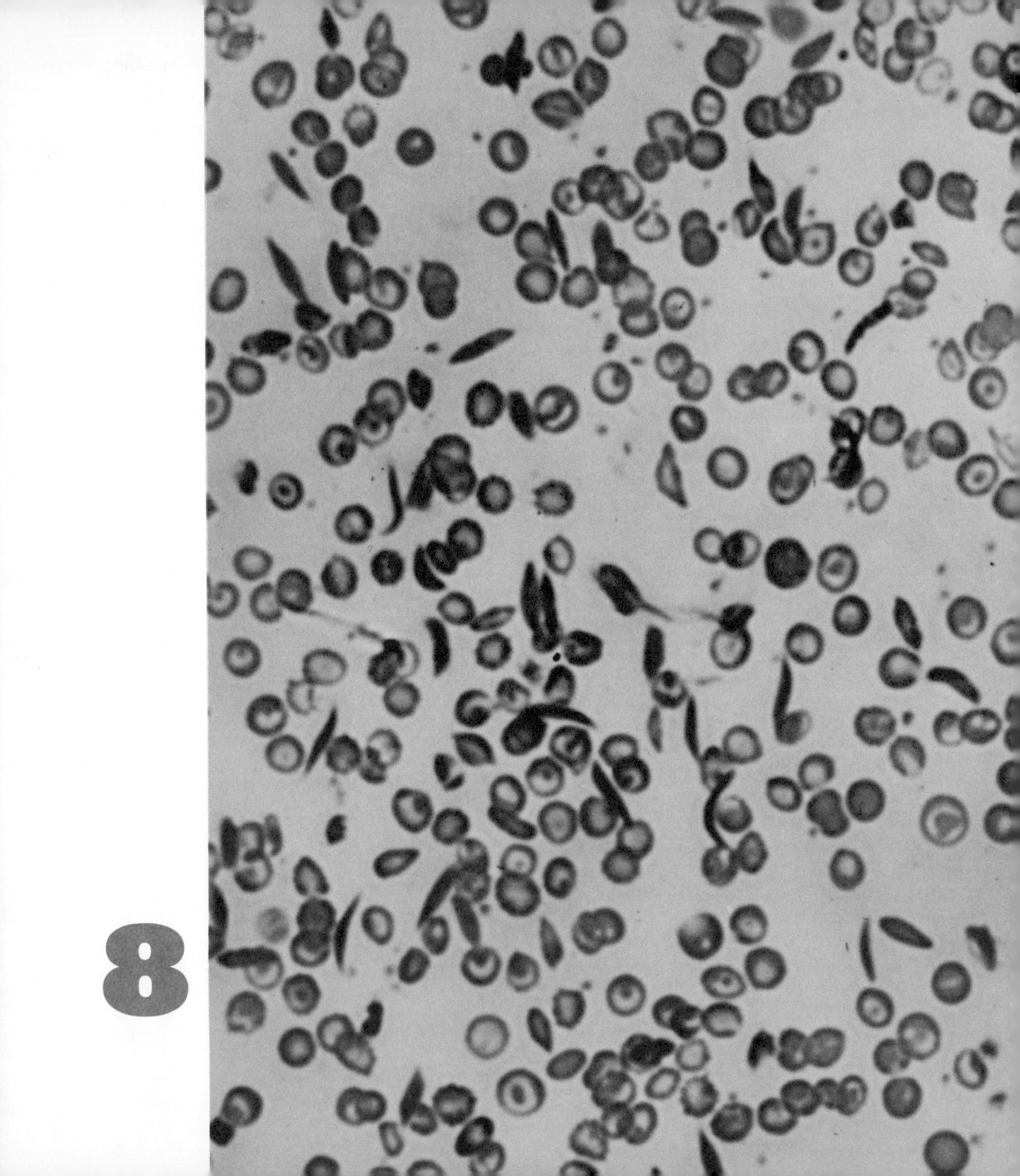

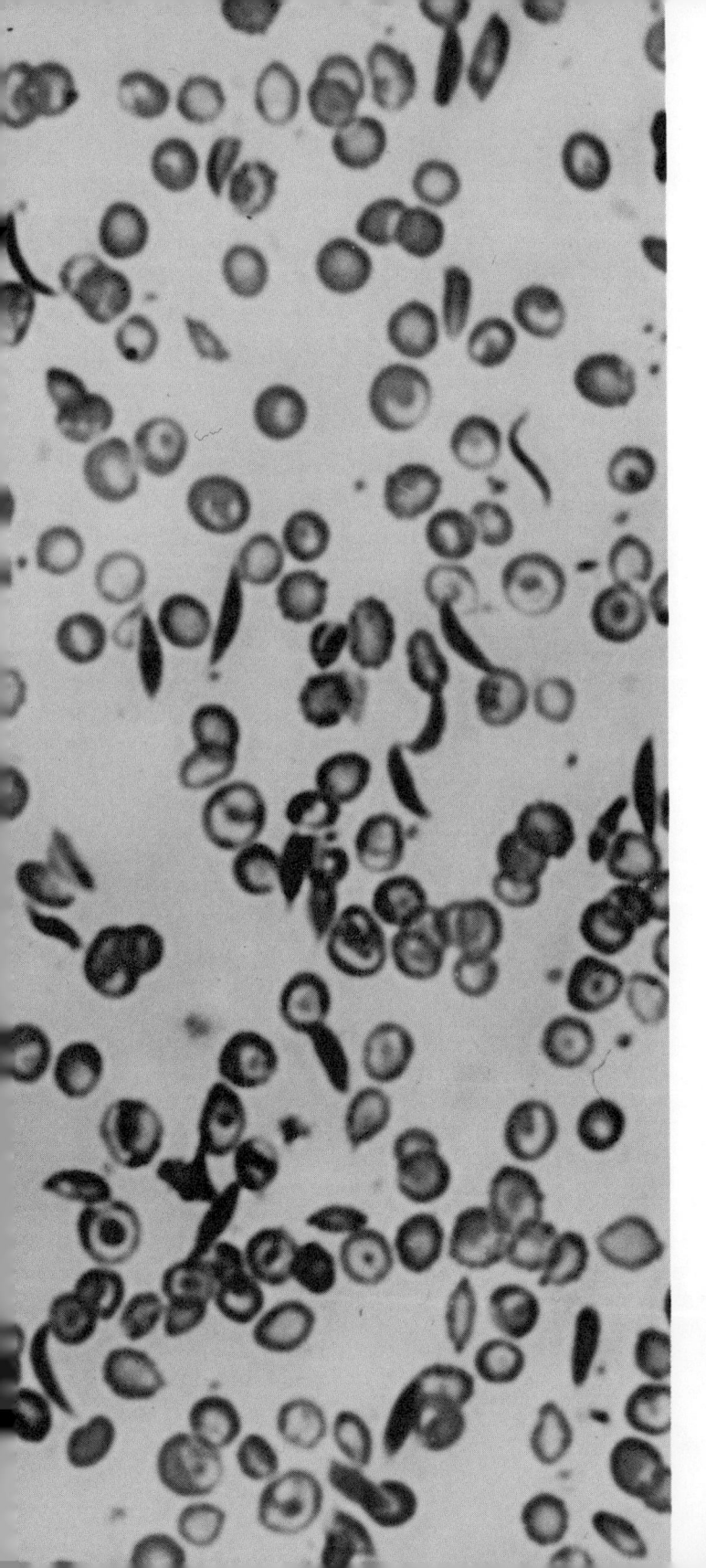

The Genetic Dice

The Rules of the Game

We come into existence because of a chance meeting between a sperm cell and an egg cell, each of which carries a set of genetic instructions that is one of approximately 8 million possible variations of the parental legacy. Meiosis shakes the genetic dice, and fertilization rolls one of roughly 64 trillion possible combinations. This is indeed life's supreme gamble. As with all games of chance, there are winners, there are those who break even, and there are some like PKU or Tay-Sachs children that are the losers.

In this chapter, we will consider some of the principles that determine how parental genes combine to produce the unique genetic makeup of the offspring. First, however, we will need to know the rules of the game, or more specifically, the rules of *probability*.

A probability statement may predict all the way from certainty that an event will *not* occur to certainty that it *will* occur. If there is no chance that it will occur, then the probability is zero. Probability statements therefore run from 0 to 1 and are expressed as fractions or decimals.

We will begin by using a coin to illustrate the basic rules of probability. Suppose that we have an honest coin with heads on one side and tails on the other. If our coin is a very thin dime, we can assume that if it is tossed into the air, it must land with either heads or tails facing up, since it would be impossible for it to land and balance on edge. Since there is no reason why one side should be favored over the other, it follows that the probability of obtaining a head is 1/2 and the probability of obtaining a tail is also 1/2. Since one or the other *must* occur, it follows logically that the probability that either a head *or* a tail will come up in any single toss must be certainty, or 1.

Now suppose that we toss two dimes at the same time. What is the probability that they will both come up heads at the same time? Since the probability of a head in each case is 1/2, we have the following situation:

Dime no. 1	Dime no. 2
H (1/2)	H (1/2)

Looking at this situation from an intuitive point of view, you might be willing to bet a considerable sum of money that a single coin will come up heads, but how would you feel about betting your life's savings that a thousand coins tossed into the air will *all* come up heads? You know instinctively that this is not a good bet; you can see that as the number of coins *increases*, the probability that they will *all* come up heads *decreases*. Since we express probability in terms of fractions, it follows that we must *multiply* the probabilities of separate events in order to obtain the *smaller* fraction which represents the *smaller* probability that those separate events will occur at the same time. We have therefore illustrated the *product rule* for calculating probabilities, and we can state that rule as follows:

> Product Rule: *The probability that two or more independent events will occur, either together or in a row, is found by multiplying the probabilities of the separate events.*

Applying this product rule to the case of the two dimes, we therefore multiply the probability of the first dime coming up heads (1/2) by the probability of the second dime coming up heads (1/2), obtaining 1/4 as the probability that they will both turn up heads at the same time. Be sure to note that the product rule aslo applies to two or more events occurring in a row. In other words, the probability of tossing the same dime twice and obtaining two heads would also be 1/2 × 1/2, or 1/4.

What is the probability of obtaining one head and one tail when the two coins are tossed? Consider the following:

	Dime no. 1	Dime no. 2
1.	H (1/2)	T (1/2)
2.	T (1/2)	H (1/2)

In this case we can win our bet in *two ways,* or by two different combinations. Either combination shown above satisfies the condition that specifies one head *and* one tail. Calculating the probabilities of the separate combinations we have:

$$\left. \begin{array}{l} 1.\ H\ (1/2) \times T\ (1/2) = 1/4 \\ 2.\ T\ (1/2) \times H\ (1/2) = 1/4 \end{array} \right\}\ 1/2$$

Since we can obtain one head and one tail by either combination (1) or combination (2), it follows that we have two ways by which one head and one tail can be produced. We therefore *add* the probabilities of the two ways, or 1/4 + 1/4 = 1/2. This illustrates the *addition rule,* which can be expressed as follows:

> Addition Rule: *When an event can occur in two or more ways, the probability of that event is calculated by adding the probabilities of the ways by which it could occur.*

It follows that the probability of two tails is the same as that of two heads, or 1/2 × 1/2 = 1/4. We can thus summarize the possibilities associated with two coins as follows:

Dime no. 1		Dime no. 2		
H (1/2)	×	H (1/2)	=	1/4
H (1/2)	×	T (1/2)	=	1/4
T (1/2)	×	H (1/2)	=	1/4
T (1/2)	×	T (1/2)	=	1/4

Note that when we add the probabilities of all the events that could occur when we toss two coins, they add up to 1, or certainty. This is logical since it is obvious that two heads, *or* two tails, *or* a head/tail combination *must* occur when two coins are tossed.

A coin provides us with what is called an *a priori* probability model. *A priori* means "established or determined before the fact." In other words, we know from the way a coin is constructed and from the laws of gravity what is likely to happen even before we actually try tossing it. Suppose, however, that we were to toss a given coin 100 times and obtain 90 heads. This result differs significantly from the average result of 50 heads that we would expect if we were to repeatedly toss an honest coin 100 times over a long period. Although we must admit that it would be *possible* to obtain 90 heads or more from 100 tosses of an honest coin, this significant difference from our expected average of 50 heads leads us to suspect that our coin is *not* honest and the *a priori* probability model of 1/2, 1/2 does not apply. Now, if we were to repeatedly toss this same coin 100 times and obtain an average of around 90 heads, we would then have an empirical probability model, which is a model based on experience or experimentation. As you can see, as far as this specific coin is concerned, the *empirical* model differs from the *a priori* model. We will see more pertinent biological examples of differences between *a priori* and empirical models as we go through this chapter and the next.

The Determination of Sex

Sex determination is a good example of the discrepancy that can occur between *a priori* and empirical probability models. You will recall from Chapter 4 that we normally possess 46 chromosomes, half of which come from each parent. You will also recall that these 46 chromosomes consist of 22 pairs of *autosomes* and one pair of *sex chromosomes,* the latter represented by the XX pair in the female and the XY pair in the male. Figure 8.1 shows two normal chromosome arrangements or *karyotypes,* one obtained from a human male and the other from a human female.

During sex-cell production, meiotic division normally ensures that one member of each pair of parental chromosomes will find its way into the sperm or egg cell, except in those cases when chromosomes go astray as we shall see in the next chapter. Therefore, as far as the sex chromosomes are concerned, we have the following situation:

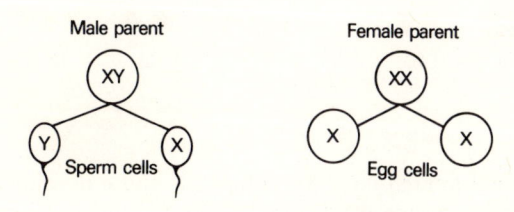

Male parent
XY
Y   X
Sperm cells

Female parent
XX
X   X
Egg cells

From the above, it is apparent that one-half of the sperm cells produced by the male parent will carry the X chromosome and one-half will carry the Y chromosome. All egg cells furnished by the female parent will normally contain one X chromosome. At fertilization, the possible combinations are therefore:

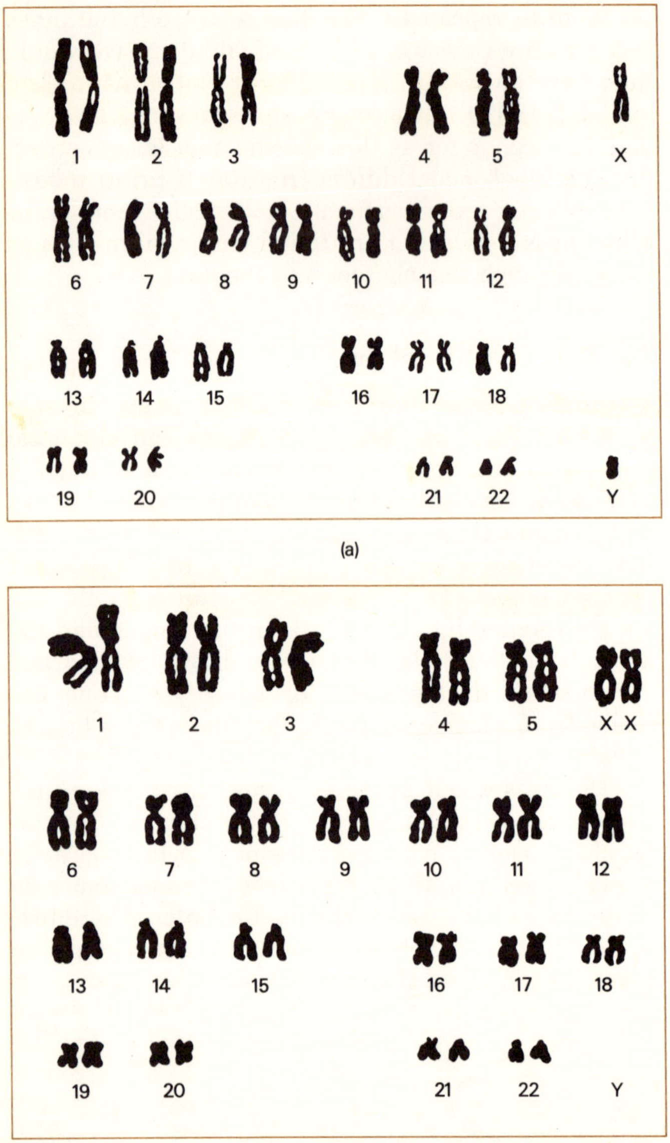

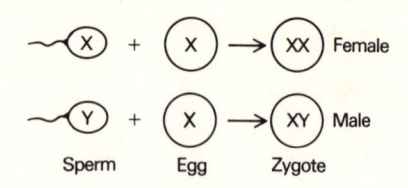

$$\text{Sperm} \ X + \text{Egg} \ X \rightarrow XX \ \text{Female}$$
$$\text{Sperm} \ Y + \text{Egg} \ X \rightarrow XY \ \text{Male (Zygote)}$$

This results in a theoretical 50:50 ratio of boys to girls. Our *a priori* probability model associated with any random birth is therefore:

$$\text{Boy} = 1/2$$
$$\text{Girl} = 1/2$$

However, this is not borne out by long-term observations and records on the *actual* sex ratio at birth! Among U.S. whites, the ratio of boys to girls is about 106 to 100; in the black population of the United States the ratio is closer to 103 to 100. The empirical probability that a boy will result from a random birth among U.S. whites is therefore close to 0.515, rather than 0.50 as our *a priori* model would imply. Various explanations for this discrepancy have been offered. One hypothesis considers the significant difference in size and weight between the X and Y chromosomes. According to this hypothesis, the Y-bearing sperm is like a racehorse that carries a lighter jockey. All other things being equal, he will win more often than a competitor carrying a heavier jockey. Detractors of this theory point out that sperm cells are moved toward the egg cell by uterine contractions and by cilia, and the winner of the race is therefore not solely dependent upon the speed of its own locomotion. At this point, it is probably safer to admit that the reason for the higher proportion of boys to girls at birth is simply not known.

Studies of early abortuses suggest that the ratio of boys to girls may be even higher at conception, per-

Figure 8.1
A normal human male karyotype (a) and a normal human female karyotype (b). (From "Guide to Human Chromosome Defects," National Foundation–March of Dimes, 1968.)

haps as high as 160 to 100. This ratio tends to decrease toward the time of birth, apparently because male embryos and fetuses have a smaller chance of completing intrauterine development than females. Thus the ratio of males to females decreases from conception on, and for various environmental as well as physiological reasons reaches its lowest point when we arrive at old age, making the characterization of women as the "weaker sex" one of the great misconceptions of all time.

Simple Gene Combinations

A number of human hereditary traits are of the type that are controlled by a single pair of genes carried on a specific chromosome pair. The members of such a gene pair each have either identical or different effects on the development of a given trait, depending on whether the specific base sequences in their DNA structures are the same or slightly different.

The Hapsburg Lip

An example of a genetic trait that involves a single pair of genes is the "Hapsburg lip," so called because it has been traced through the royal Hapsburg family from the sixteenth century down to modern times. The Hapsburg lip consists of a prominent underslung jaw and a protruding lower lip and is controlled by an autosomal dominant gene. It is *autosomal* because it is carried on one of the autosome pairs, not on the sex chromosomes. It is *dominant* because the trait will appear in an individual even if only one member of the chromosome pair carries the gene for the trait. This may be illustrated by letting capital H represent the gene for the Hapsburg lip and h represent the gene for normal features:

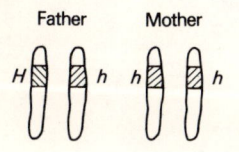

In the case illustrated above, the father has the gene for Hapsburg lip on one member of the chromosome pair and the gene for normal features on that chromosome's partner. Since the gene for the trait happens to be dominant, he will have the Hapsburg lip even though the gene for normal features is also present. The mother, meanwhile, carries both genes for normal features and therefore does not have the Hapsburg lip.

Now let us see what kinds of offspring can result from the individuals described above. Thinking in terms of the Hapsburg-lip gene only, the father can produce two types of sperm cells, H and h, presumably in equal numbers. The mother, on the other hand, can produce only one kind of egg cell, h. We thus have the following situation:

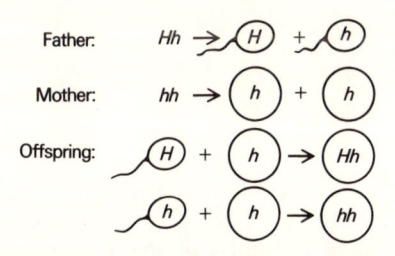

Looking at the various combination possibilities, we can see that half the offspring will have the gene, and therefore the trait, while the other half will not. At this point, however, it is important that we stress emphatically that the foregoing combination possibilities do *not* mean that if this couple has four children, two of the children will be affected by the trait and two will not! The offspring possibilities from a genetic cross provide us with an *a priori* probability model, and the model applies equally to each individual child. What happens at the first birth does not affect what happens at the second birth, and so on, any more than obtaining a head on the first toss of a coin changes the probability of obtaining a head on the second toss.

For example, it is perfectly possible for this couple to have two children in a row, both of which exhibit the Hapsburg lip. This would be like tossing a coin twice and having heads come up twice in a row. Going back to the probability model from our genetic cross, we can see that *with each individual birth, the probability of an afflicted child is 1/2.* Therefore, applying the product rule, the *probability* that this couple would have two afflicted children in a row is $1/2 \times 1/2$, or 1/4.

What is the probability that they could have four children, all of whom have the Hapsburg lip? This would be $1/2 \times 1/2 \times 1/2 \times 1/2$, or 1/16. You will note, however, that the more children they have, the less is the probability that they will *all* be afflicted with the condition. On the other hand, the probability that all the children will be normal is also less.

Simple dominant inheritance also occurs with many other traits, including a particularly vicious type of headache called migraine and a degenerative disease known as Huntington's Chorea.

Brachydactyly

A condition called *brachydactyly* involves the presence of abnormally short fingers due to a missing middle joint (see Figure 8.2). This is also an example of simple dominance. It has been noted, however, that two brachydactylous persons occasionally produce a child that dies in early infancy due to serious

Figure 8.2
The short fingers on the right hand are typical of brachydactyly. This is an inherited condition controlled by a dominant gene. (Courtesy of the American Genetic Association.)

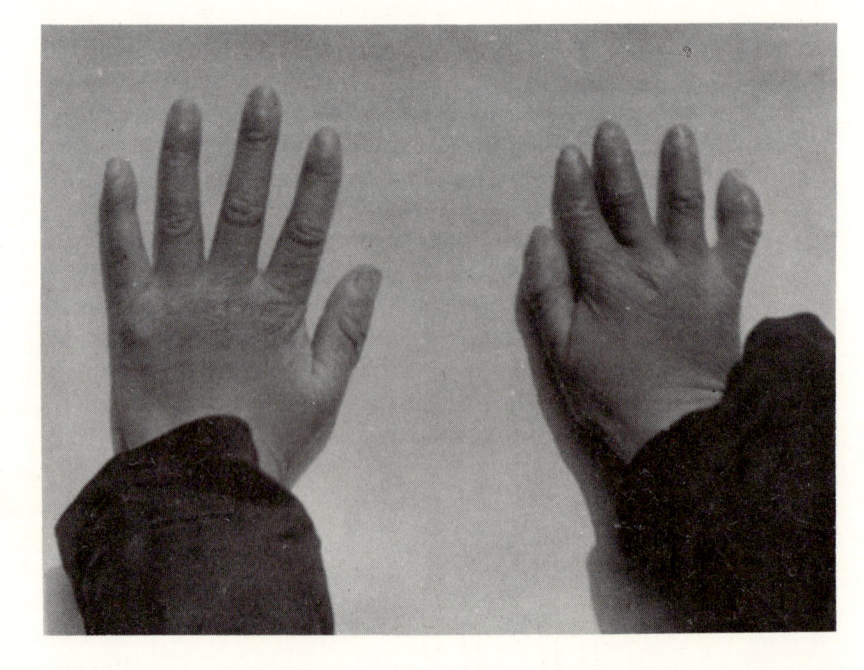

skeletal defects. This situation involves the following cross and offspring combinations:

>Brachydactylous parents: $Bb \times Bb$
>Offspring: BB, Bb, Bb, bb

Note that one of the offspring possibilities shows a "double dose" of the gene for brachydactyly (BB). The effects of this double dose apparently go far beyond a simple skeletal defect of the fingers; it produces skeletal defects so general and serious as to be incompatible with life. We thus have an example of a *lethal gene,* or a gene that causes death when it exists on both members of a chromosome pair.

Looking at the probability model associated with a brachydactylous couple, we can see that with each child, the probability the child will be brachydactylous (Bb) is 1/2, the probability that it will be normal (bb) is 1/4, and the probability that it will die in early infancy (BB) is 1/4. Once again, keep in mind that our probability model applies to each birth *independently*. It is not true, as many people still assume, that if the couple's first child dies of skeletal defects, they need not worry about the next three. Neither is it true that if the first child dies from this condition, so will all the others. It is *possible,* however, that they could have three children, all of whom die from the lethal combination of the gene. But the probability of this happening would be $1/4 \times 1/4 \times 1/4$, or 1/64, which is not very likely but certainly within the realm of possibility.

Albinism

Albinism is a classic example of a *recessive* gene. As we shall see, a recessive gene must exist either alone or in a double dose if its effects are to be expressed. Albinism (see Figure 8.3) is a metabolic block disorder of the type discussed in Chapter 7 and is due to a lack of the enzyme needed for the conversion of

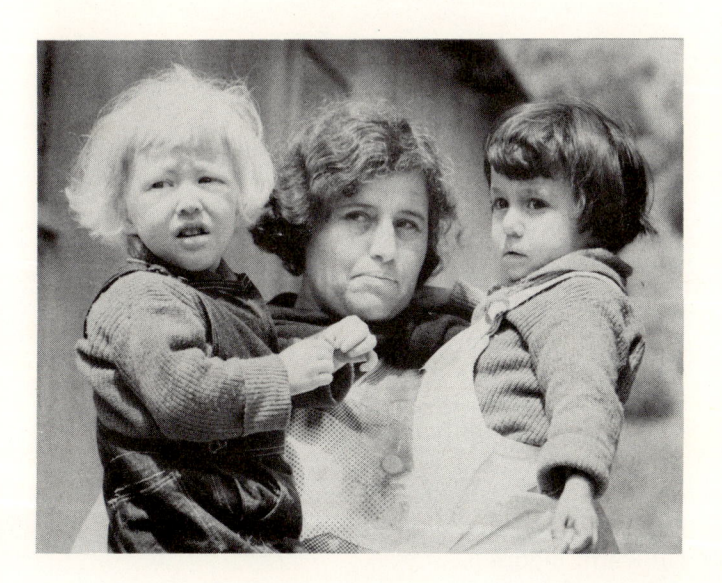

Figure 8.3
The boy on the left is an albino; he lacks normal pigmentation because he is unable to convert tyrosine to melanin. His brother (right) has normal pigmentation. (Courtesy of the American Genetic Association.)

the amino acid tyrosine to *melanin*. Melanin is the brown pigment that is normally found to a greater or lesser degree in the skin, in the hair, and in the iris of the eye.

There are three possible gene combinations related to albinism. The combination *AA* is a normally pigmented noncarrier. The combination *Aa* is also normally pigmented but carries the recessive gene for albinism (*a*). The combination *aa* is an albino because both genes, or DNA segments, code for an enzyme that does not "fit the lock." Consequently, this individual cannot convert tyrosine to melanin and therefore lacks normal pigmentation. At this point we should understand that there are other genes that control the degree to which melanin is deposited, and that this accounts for the gradations of skin color that exist between and among the various races of man. But since the individual with the gene combination (*aa*) has little or no melanin available, he or she will be an albino no matter what the other genes call for.

To produce an albino, both parents may be albinos themselves, resulting in the following offspring:

Parents: *aa* × *aa*
Offspring: All *aa*

Or, one parent could be an albino and the other a normally pigmented carrier, as follows:

Parents: *aa* × *Aa*
Offspring: *Aa, aa*

What is probably the most frequent situation involves two people that are both normally pigmented but who are both carriers of the recessive gene:

Parents: *Aa* × *Aa*
Offspring: *AA, Aa, Aa, aa*

Note that we have generated three probability models. In the first case, the probability that any given birth will result in an albino is 1, or certainty. In the sec-

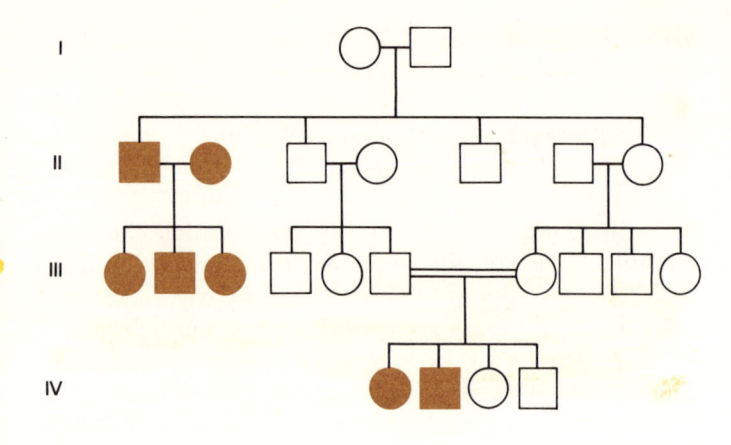

Figure 8.4
Example of a family pedigree of albinism, a recessive gene. Circles represent females and squares represent males. Circles and squares that are shaded represent afflicted persons. Note that two albinos produce all albino children (left). The double line represents a cousin marriage.

ond case, the probability of an albino is 1/2, and in the third and more usual situation, the probability is 1/4.

If only one parent is a carrier, then we have:

Parents: *AA* × *Aa*
Offspring: *AA, AA, Aa, Aa*

Note that in the above case, the probability of producing an albino child is zero, but it is interesting to note that the probability of producing a carrier is 1/2. From this we can see that a recessive gene may tend to persist in a family even though it may achieve full expression very infrequently. Figure 8.4 shows a family pedigree of albinism. Compare this to the family history of the dominant gene for migraine headache shown in Figure 8.5.

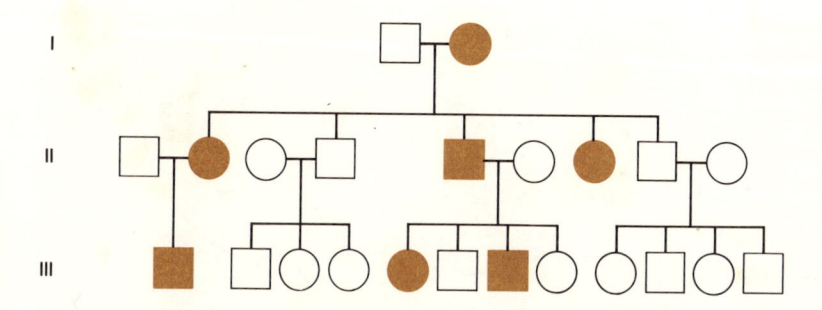

Figure 8.5
Family pedigree of migraine headache, a dominant gene. Note that if one parent is afflicted, approximately one-half of the children are afflicted.

Gene Frequencies

From the preceding discussion of albinism, we can see that there exist two forms of the gene that is involved in the conversion of tyrosine to melanin. One form (A) is the "correct" pattern and the other form (a) is the "wrong" pattern. Now, if we could gather all the A and a genes that are found in the population of the United States into a *gene pool,* and if we could count all of the A and a genes, we could then determine the relative *frequencies* with which A and a occur in the population. In other words, we would know how often each is found in comparison to the other. Quite obviously, we cannot actually gather all the genes together and count them, but we will see that there is a way of at least estimating such gene frequencies. One estimate, for example, suggests that out of every 100 A and a genes in the U.S. population, there is one a gene for every 99 A genes. The frequency of the gene for albinism is therefore estimated at 1/100, or 0.01.

The concept of gene frequencies has some very important applications. For instance, a knowledge of gene frequencies helps us to predict the occurrence of genetic defects of various kinds. It is therefore an important tool in the hands of the genetics counselor when he advises parents who have reason to be concerned about having defective children. As we go along, we will develop the concept of gene frequencies and apply it to specific cases that involve genetic defects. Then in Chapter 16, we will see that this same concept is an important part of the process of evolution.

Returning to our example of a gene pair consisting of A and a, it can be seen that, at least as far as this specific pair is concerned, a given population

142 The genetic dice

would consist of three different gene combinations —*AA*, *Aa*, and *aa*. Furthermore, as meiosis takes place in that population, two "kinds" of sperm and egg cells would be produced—those carrying the *A* gene and those carrying the *a* gene. We are, of course, ignoring the thousands of other genes that are carried by those making up the population and focusing on a single, specific pair.

We can therefore think of a population as consisting of (1) a group of individuals, (2) a collection of gene combinations, (3) a collection of sperm and egg cells, or (4) a population or "pool" of genes. Figure 8.6 shows a hypothetical population of sperm and egg cells. Note that some of the sex cells carry the *A* gene while others carry its *a* partner. We are now looking at the individuals making up the populations simply as vehicles which carry the sex cells that will form the next generation.

Following meiosis, the next step in reproduction is fertilization. As you see this illustrated below, take special note of the fact that the more general symbols p and q are used to represent the frequencies of the members of a given pair of genes. In the following, p represents the frequency of *A* and q the frequency of *a*.

[Diagram: Sex cells × Sex cells → Zygotes]

A (p) × A (p) → AA (p^2)
A (p) × a (q) → Aa (pq)
a (q) × A (p) → Aa (pq)
a (q) × a (q) → aa (q^2)

The gene combinations that result from fertilization are *AA*, *Aa*, and *aa*, or in terms of our more general symbols,

$$p^2 + 2pq + q^2$$
$$(AA)\ \ (Aa)\ \ (aa)$$

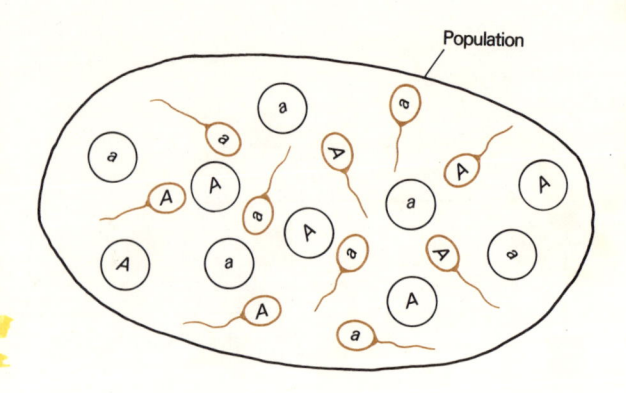

Figure 8.6
Hypothetical population of sperm and egg cells.

Now carefully consider this very obvious point. If we have a basket filled with apples, peaches, and pears and we add together all the apples, peaches, and pears that are in the basket, we will have *all* (100 percent) of the apples, peaches, and pears present in the basket. If we now substitute the gene combinations *AA*, *Aa*, and *aa* for apples, peaches, and pears and at the same time think of the basket as representing a population, it follows that:

$$p^2 + 2pq + q^2 = 1 \quad (8.1)$$
$$(AA)\ \ (Aa)\ \ (aa)$$

Keep in mind that 1 is another way of saying 100 percent. Furthermore, if we total all the *A* and *a* genes in a given population, it must follow that we have 100 percent of the *A* and *a* genes found in that population, or

$$p + q = 1 \quad (8.2)$$
$$(A)\ \ (a)$$

Statements (8.1) and (8.2) are the basis of the *Hardy-Weinberg Law,* which is the starting point for a complex and very important area of biology known as *population genetics*. In Chapter 16, we will see how this basic law helps us explain the process of evolution. In this section, however, we are mainly interested in how these concepts can be applied to individual genetic problems.

Tay-Sachs Disease

We can now consider applications of statements (8.1) and (8.2) to Tay-Sachs disease, which you will recall from Chapter 7 is a devastatingly tragic metabolic block disorder (Figure 8.7).

Tay-Sachs disease is inherited in the same way as albinism, but because the afflicted child dies long before reaching reproductive age, Tay-Sachs children can be produced only by two normal carriers. Thus we have

Parents: Tt × Tt
Offspring: TT, Tt, Tt, tt

yielding a probability of 1/4 that a child from any given birth will be afflicted.

Tay-Sachs disease is found almost exclusively among Ashkenazi Jews, who are members of the Jewish population whose ancestors came from Eastern Europe. For practical purposes, we can consider that our *population of interest* consists of American Jews that are of Ashkenazi origin; this includes most of the Jewish population in the United States. It has been estimated that the incidence of Tay-Sachs births within this population is approximately one in every 8100 births. It is well to point out here that the frequency of births which involve genetic defects such as Tay-Sachs disease, PKU, and similar conditions are at best only approximations and are subject to argument and constant revision.

Now, if we consider the statement

$$p^2 + 2pq + q^2 = 1$$
(TT) (Tt) (tt)

Figure 8.7
A child afflicted with Tay-Sachs disease. Beginning at about six months of age, the nervous system degenerates and the child inevitably dies by age three or four. (From B. W. Volk (ed.), Tay-Sachs Disease. New York: Grune & Stratton, 1964. By permission.)

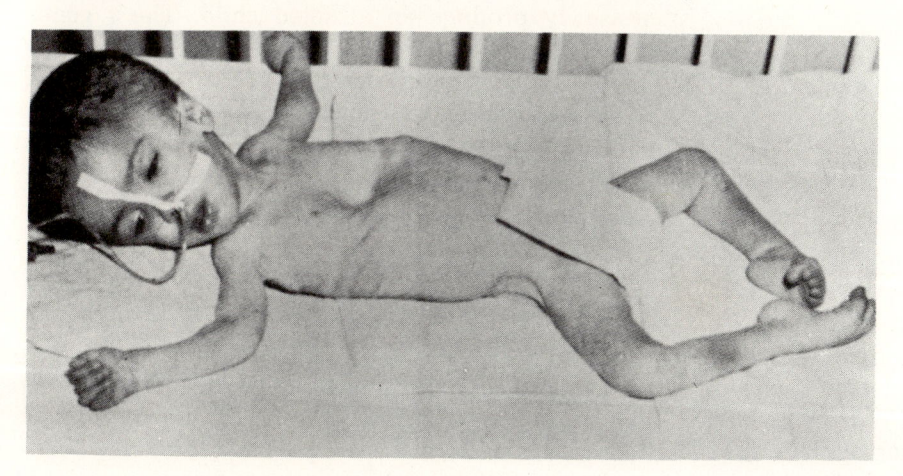

we can see (1) that p^2 represents the frequency of normal individuals in the population who do not carry the Tay-Sachs gene, (2) that $2pq$ is the frequency of those who are not afflicted with the disease but carry the defective gene, and (3) that q^2 represents the frequency of afflicted children (tt). Since we have estimated that afflicted children occur in the population with a frequency of 1/8100, and since this is represented by q^2, it follows that

$$q^2 = 1/8100$$
$$q = \sqrt{1/8100}$$
$$q = 1/90$$

Having thus found the value of q, and since q represents the gene t, we now have an estimate of the frequency with which the defective gene occurs in the population; this frequency is 1/90. Stating it differently, out of every 90 sperm or egg cells drawn randomly from the population, one sperm or egg cell carries the gene for Tay-Sachs disease.

Now, if we know the value of q, can we find the value of $2pq$? This is quite simple, since

$$p + q = 1$$
$$(T) \quad (t)$$

Therefore,

$$p = 1 - q$$
$$p = 1 - 1/90$$
$$p = 89/90$$

and finally,

$$2pq = 2 \times 89/90 \times 1/90$$
$$= 1/45$$

Since the value of $2pq$ represents the frequency with which carriers of the Tay-Sachs gene are found in the population, it is obvious that we now have some very useful information. We can now state that one out of every 45 randomly selected Jewish individuals will be a carrier and therefore a potential parent of a child afflicted with Tay-Sachs disease. This assumes, of course, that he or she marries another carrier. Putting it in another way, we can say that if an individual is randomly selected from the Jewish population of the United States, the probability that he or she is a carrier (Tt) of Tay-Sachs disease is 1/45.

Our next step is the application of this information to a specific example. Suppose that Henry and Ruth are of Ashkenazi Jewish ancestry but they have no known history of Tay-Sachs disease in their families. What is the probability that any given child that they may have will be afflicted?

Since we have estimated the frequency of carriers to be 1/45, and since Henry and Ruth have no known history of Tay-Sachs disease in their families, we have to assume that the probability that Henry is a carrier is 1/45 and that the probability that Ruth is a carrier is also 1/45. Since they will not produce an afflicted child unless they are *both* carriers, we need to know the probability that they both carry the Tay-Sachs gene. This is very much like the problem of two dimes coming up heads at the same time, and so we use the product rule. Thus,

$$\begin{array}{cc} 1/45 & \times & 1/45 \\ \text{Henry} & & \text{Ruth} \end{array}$$

which gives us a probability of 1/2025. On the other hand, even if they *are* both carriers, they will not necessarily produce an afflicted child. The t genes that they carry must come together; the probability that this will occur in any given fertilization is 1/4. Again, the product rule comes into play. Since we must compute the probability that two events will occur in a row, (1) both parents must be carriers, and (2) the defective genes that they each carry must come together in the formation of the zygote. Therefore,

$$1/2025 \times 1/4 = 1/8100$$

which not surprisingly turns out to be the estimated incidence of Tay-Sachs births in the population of which Henry and Ruth are a part.

This time let us change the problem rather drastically by assuming that Henry had a brother that died of Tay-Sachs disease. This new information tells us that Henry's parents must have been carriers, and that Henry was produced by the following cross:

Parents: $Tt \times Tt$
Offspring: TT, Tt, Tt, tt

Henry's afflicted brother is represented by tt, but since Henry is *not* afflicted, it follows that he must be one of the other three possibilities—TT, Tt, or Tt. Therefore, the probability that he is a carrier (Tt) is two out of three, or 2/3. Now, the probability that Henry and Ruth will produce an afflicted child in any given birth is

$$2/3 \times 1/45 \times 1/4 = 1/270$$

which is a significant increase over the previously computed probability that they will produce a Tay-Sachs child.

Now suppose that Henry and Ruth happen to be second cousins. Because of their close common ancestry, second cousins have 1/32 of their genes in common. In other words, if Henry and Ruth are second cousins, it follows that if Henry possesses a given gene, the probability that Ruth also carries that gene is 1/32. Therefore, the probability that Henry and Ruth will produce an afflicted child in any given birth now becomes

$$2/3 \times 1/32 \times 1/4 = 1/192$$

If they were first cousins, their relationship would be even closer and they would have 1/8 of their genes in common. The probability of having a Tay-Sachs child in any given birth would then be

$$2/3 \times 1/8 \times 1/4 = 1/48$$

It can be seen from this example that marriage between close relatives increases the probability that "bad" recessive genes will come together. How important this is depends on the closeness of the relationship. For example, first-cousin marriages are frowned upon by the Catholic church and are forbidden by law in over half the states in the United States. Objections to second-cousin marriages are less common.

Carrier Identification

Fortunately, modern knowledge of the biochemistry and hereditary mechanism of Tay-Sachs disease has made possible programs of carrier identification that can take much of the guesswork out of the foregoing situations. This carrier identification procedure is based on the fact that Tay-Sachs disease is associated with a lack of the enzyme called *hexosaminidase-A*. It has been discovered that *carriers* of the defective gene (Tt) do not produce the same levels of this enzyme as normal individuals (TT). Furthermore, the reduced level of the enzyme found in carriers can be detected by a simple blood test.

In the light of this new knowledge, we can now suppose that Henry and Ruth are tested for the presence of the Tay-Sachs gene. There would be three possible outcomes:

1. Neither is identified as a carrier. The probability that they will produce an afflicted child is therefore

$$0 \times 0 \times 1/4 = 0$$

2. One or the other, but not both, is identified as a carrier. This time, the probability would be

$$1 \times 0 \times 1/4 = 0$$

3. Both are identified as carriers. The probability that any given child would be afflicted is now

$$1 \times 1 \times 1/4 = 1/4$$

In the first situation there is no problem; Henry and Ruth can have children without any fear of Tay-

Sachs disease. This is also true of the second situation, but in that case the cross would be

$$\text{Parents: } TT \times Tt$$
$$\text{Offspring: } TT, TT, Tt, Tt$$
$$\qquad\qquad\qquad \uparrow \quad \uparrow$$

This indicates that with each birth, the probability is 1/2 that the child would be a *carrier*. This is useful information because Henry and Ruth would want to make their children aware of the problem and urge them to seek appropriate testing.

In the third situation, in which the tests show that both are carriers, the probability is 1/4 that any given child would inherit the dread legacy of Tay-Sachs disease. Considering the consequences, a probability this high would appear to make any plans to have children unwise, but it is still the right of the couple involved to decide this according to their own personal beliefs and philosophy. A possible solution to this dilemma will be discussed in the next chapter.

It is likely that the future will bring more and better programs for identification of carriers of various severe genetic defects. Tay-Sachs is a good illustration because it meets the basic criteria of such a program. These criteria are: (1) the population that contains the gene must be easily identifiable (Tay-Sachs disease is very largely confined to those people of Ashkenazi Jewish ancestry), and (2) there must be a simple test that will identify the carrier with a reasonably high degree of accuracy. In addition, there must be widespread information concerning any testing program throughout the group that is most affected by a specific genetic defect. This once again points up the practical importance of becoming familiar with modern biological principles.

Sickle-Cell Anemia

We said in Chapter 7 that sickle-cell anemia is caused by a "wrong" form of hemoglobin that is produced under the control of a mutant gene. Like Tay-Sachs disease, sickle-cell anemia is an obviously "ethnic" disease because it is found almost exclusively among those people of African ancestry. While it is not as quickly and inexorably fatal as Tay-Sachs disease, sickle-cell anemia is nevertheless a significant source of worry and heartbreak in the black population. Afflicted children look forward to shortened lives that are filled with frequent bouts of pain and physical disabilities which keep them from many normal activities (see Figure 8.8).

A variety of hemoglobin called *hemoglobin-S* is produced. This differs slightly in structure from normal *hemoglobin-A* and is due to a mutant gene that originated in Africa. We will see in Chapter 15 that when hemoglobin-S exists in combination with hemoglobin-A, it has a certain survival value in areas which are infested with malaria. As with Tay-Sachs disease, children with sickle-cell anemia (SS) are produced by two carriers of the mutant gene as follows:

$$\text{Parents: } AS \times AS$$
$$\text{Offspring: } AA, AS, AS, SS$$

This gives a probability of 1/4 that any given child will be afflicted with the condition. Also note that the probability is 1/2 that any given child will be a carrier of the sickle-cell gene and therefore produce both hemoglobin-A and hemoglobin-S.

Most estimates place the occurrence of sickle-cell anemia among American blacks at one afflicted child for approximately every 625 live births. This means that q^2 is 1/625. If you apply the principle used in the preceding section, you can calculate that about 8 percent of American blacks are carriers. This is a high birth rate and a high frequency of carriers, especially if one considers the tragic and difficult consequences of the disease.

Like Tay-Sachs disease, sickle-cell anemia is an excellent subject for genetic counseling programs which are based on carrier identification. A primary characteristic of sickle-cell anemia is the "sickling" of a large percentage of the patient's red blood cells (see Figure 8.9). These malformed cells tend to block cap-

Gene frequencies 147

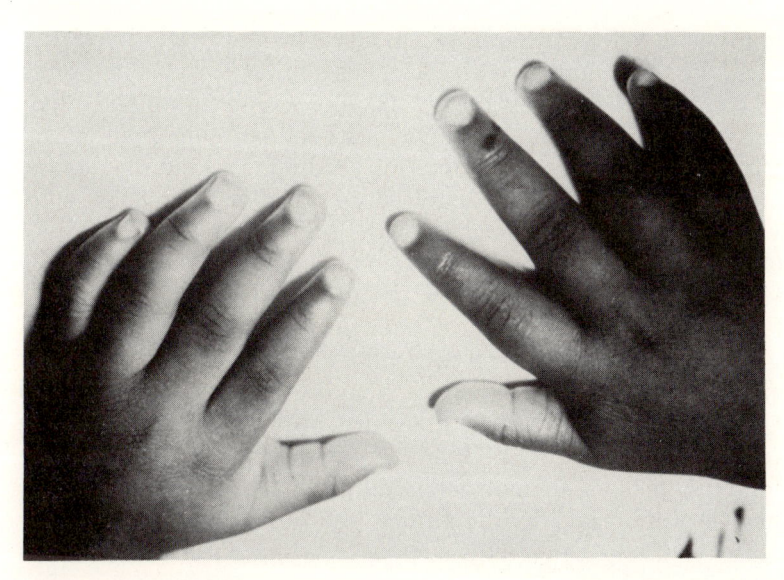

Figure 8.8
In this photograph, the swollen hands are those of a child afflicted with sickle-cell anemia. This hand-foot syndrome is a common complication of sickle-cell anemia during infancy and early childhood. (Courtesy of Roland B. Scott, M.D., Center for Sickle-Cell Disease, Howard University.)

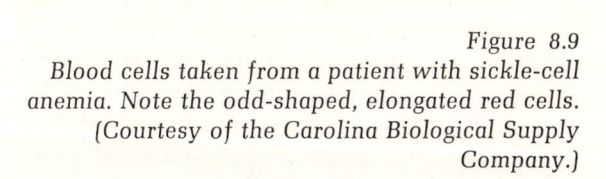

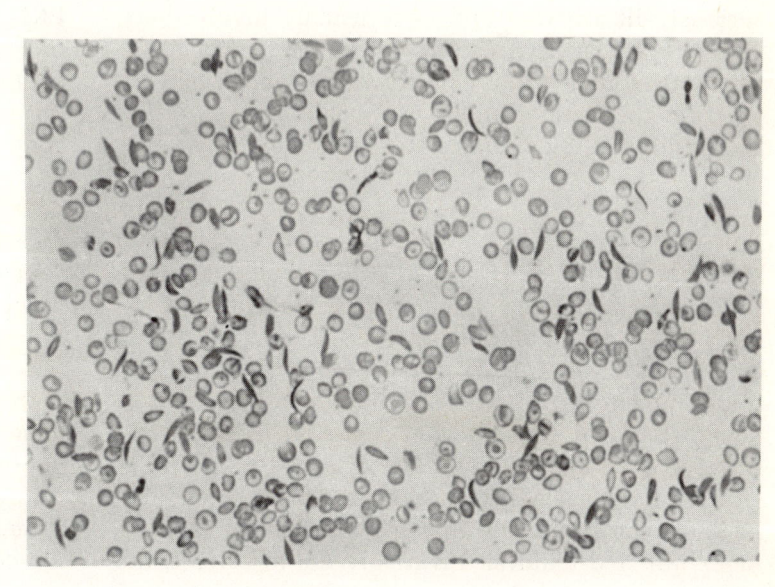

Figure 8.9
Blood cells taken from a patient with sickle-cell anemia. Note the odd-shaped, elongated red cells. (Courtesy of the Carolina Biological Supply Company.)

illary blood vessels, producing the distressing symptoms of the disease. The red cells of the carrier will also show some degree of sickling when subjected to decreased oxygen pressure, and simple tests have been devised that can identify carriers of the so-called sickle-cell "trait." Since the disease occurs mainly among blacks, the population of interest is easily defined, and increasing numbers of American blacks are taking advantage of the services offered by identification centers found in the larger cities of the United States. The significance of the test results are similar to those described in the preceding section for Tay-Sachs disease.

An unfortunate side effect of sickle-cell carrier identification programs involves a lack of understanding on the part of some prospective employers, insurance companies, and others that mistakenly regard the carrier as having some kind of mysterious illness. This is decidedly not true; the carrier of a single gene for sickle-cell anemia is not afflicted with the disease, any more than someone who carries a single gene for Tay-Sachs disease, PKU, or cystic fibrosis can be considered ill or a "poor risk" for employment or insurance. Cases where carriers of the trait have become seriously ill are rare and have usually involved exposure to unusually low oxygen pressure. This unfortunate misunderstanding demonstrates once again the need for the general public to become more aware of the basic principles related to genetic defects.

Phenylketonuria

You will recall from Chapter 7 that phenylketonuria (PKU) is a metabolic block disorder that usually results in severe mental retardation. Like albinism and Tay-Sachs disease, PKU is caused by the absence of a specific enzyme; in this case it is the enzyme necessary to convert the amino acid *phenylalanine* to another amino acid, *tyrosine*.

We have seen that children with Tay-Sachs disease can be produced only by two normal carriers, since afflicted children invariably die in early childhood and therefore cannot reproduce. Until recently the situation with PKU was similar, because although PKU children do not usually die in childhood, their condition has effectively prevented them from engaging in reproduction, thus producing a kind of "genetic death." Their "double dose" of defective genes were thereby removed from the population gene pool.

Now a new dimension has been added to the PKU problem. First, there are now widespread and apparently effective programs for the detection of PKU in newborn babies. Secondly, children so identified may be placed on the low-phenylalanine diet mentioned in Chapter 7. It is therefore probable that increasing numbers of PKU children will grow to maturity with reasonably normal intelligence, marry, and produce offspring. In the future, therefore, there could very likely be matings such as $PP \times pp$, $Pp \times pp$, and even $pp \times pp$.

Now, we must consider that in the case of a gene such as PKU, the gene P is mutating to gene p at some rate which is generally the same as the rate by which gene p is removed from the population by the "genetic death" of the afflicted children (pp). By decreasing this rate of removal through dietary treatment of PKU children, and assuming that the mutation rate of P to p remains the same or even increases because of greater radiation hazards, we are obviously creating a situation in which the frequency of a harmful gene will increase as time goes on.

Another case in point is *retinoblastoma,* a cancerous tumor of the eye that is caused by a dominant gene. The patient who would normally not survive to reproductive age may be saved by early surgical treatment which results in blindness in one or both eyes. Medical treatment of individuals with these and other severe genetic defects saves their "bad" genes from extinction and preserves them as part of the gene pool, something that would not be true in the "natural" situation. This raises some obvious ethical questions. Should this practice be continued even though it theoretically jeopardizes future generations? Or can we refuse to give a chance for a reasonably normal life to a child that would otherwise be severely retarded or die of retinoblastoma? Before you give a dispassionate, objective answer, think of what your answer would be if it were your own child!

Quite obviously, the problem of resolving the question of society's responsibility to the individual versus the question of society's responsibility toward future generations is one that requires a modern-day Solomon, but so far none has come forward. It has been suggested that society should pick up the medical bills for afflicted persons but at the same time insist that they be sterilized. Do you find this to be an acceptable solution? Meanwhile geneticists argue pro and con, some claiming that continued medical treatment of PKU and other severe genetic defects will increase the load of "bad" genes in the population to the point where future generations may consist largely of physical wrecks!

At present, our society is certainly not convinced that those living now should be sacrificed for the benefit of those who will live in the future. Will such decisions require an Orwellian society of the future, one that is willing to view its members in the same way that we now regard our breeding stocks of animals? None of the alternatives is appealing, but nevertheless it is a dilemma that must someday be resolved.

Inheritance of Blood Groups

Different blood types are due to a variety of specific proteins that are found on the outside surfaces of red-cell membranes. By far the most familiar are the A–B–O blood groups because of their long-recognized importance in blood transfusions. In addition, however, there are a number of other blood types that are much less well known. All appear to be related to specific proteins which in turn are determined by specific gene combinations.

A–B–O Blood Types

We have been dealing up to now with gene pairs that are carried at specific locations on a certain pair of chromosomes. We now know that if we let one member of a gene pair be represented by *A* and its partner by *B*, there will then exist in the population the following three combinations:

Suppose, however, that a third gene *O* exists in the population gene pool and is carried on the chromosome at the same locations as *A* and *B*. Since these three genes—*O*, *A*, and *B*—can combine *only two at a time,* the following six combinations are possible:

This type of inheritance is the mechanism by which the A–B–O blood groups are inherited in man. There is a further complication, however, because the *A* and *B* genes are *codominant;* that is, neither is dominant to the other. At the same time they are both dominant to gene *O*. Thus the actual blood types are determined by the various possible gene combinations as follows:

Type A	Type B	Type AB	Type O
AA	BB	AB	OO
AO	BO		

Genes *A* and *B* control the production of proteins A and B, respectively, and these proteins are carried on the outside surfaces of the red blood cells. Gene *O* evidently does not cause a protein to be synthesized, and therefore the red cells of individuals with type O blood do not carry either the A or B protein. This situation is summarized by the following diagrams (the circles represent red blood cells):

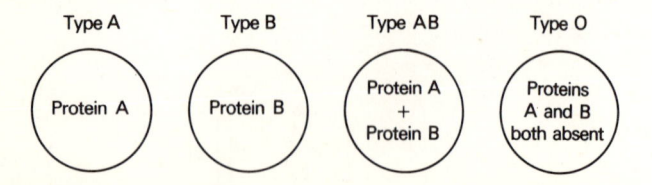

Type A	Type B	Type AB	Type O
Protein A	Protein B	Protein A + Protein B	Proteins A and B both absent

When the body is invaded by a protein that does not "belong," or in other words is not called for by the individual's genetic makeup, this "foreign" protein is called an *antigen*. Usually, the body provides specific *antibodies* which combine with the antigen in an attempt to destroy it. This is called an *antigen-antibody reaction*. Thus if a person with type A blood is given a transfusion with type B blood, protein B will be a foreign protein or *antigen* to the person with type A. The antibodies which are free in solution in Mr. A's blood plasma will therefore cause the red cells given to him by Mr. B to clump together. This clumping effect tends to block the small capillary vessels; this is especially dangerous in the kidney where it may cause kidney shutdown and the consequent death of Mr. A. You can see that a similar thing will happen if we give type A to Mr. B, or if we give type AB to Mr. A, Mr. B, or Mr. O. Since Mr. O's red cells do not carry either proteins A or B, his blood may theoretically be given to any of the other three. You can also see that types O, A, and B can theoretically be given to Mr. AB, since Mr. AB's plasma cannot contain either antibody A or antibody B. For these reasons, types O and AB are sometimes called the *universal donor* and the *universal recipient*, respectively. All of this is summarized in the table below:

Type (antigen)	Antibodies in plasma
O	Anti-A and anti-B
A	Anti-B
B	Anti-A
AB	None

In practice, however, transfusions require that the blood of the donor is carefully *cross-matched* with that of the recipient in order to avoid any possible disastrous complications that might arise from incompatibility of blood types.

The A–B–O blood groups are also useful in certain aspects of medical-legal practice. The identification of blood types is of obvious value in police laboratory work. In addition, they may be used as a part of a pattern of hereditary factors to help settle cases of disputed maternity or paternity. For example, suppose that a baby has type A blood and the mother has type O blood, and the latter accuses a man with type O blood of being the father. Since the O gene is recessive to both genes A and B, the accused cannot be the father because a mating between him and the mother could produce only type O children:

Parents: OO × OO
Offspring: All OO

Therefore, the no doubt vigorous denials of the accused would appear to be justified.

In another case, suppose the mother is type B, the reluctant father type A, and the child type O. The following is one of the *possible* crosses that would fit this situation:

Parents: AO × BO
Offspring: AB, AO, BO, OO

The results show that the accused *could* be the father. By itself, this evidence does not furnish positive proof, since any number of other candidates for the honor of fatherhood with either *AO* or *OO* genotypes could equally well fit the picture. We can see, therefore, that this approach can establish innocence but not necessarily prove guilt. It must be pointed out, however, that at least fifteen different kinds of blood groups have been discovered, and they can all contribute toward a pattern of heredity that could be used as evidence.

Recently, more attention has been given to the relationships between the A–B–O blood groups and possible complications during pregnancy. Suppose, for example, that a mother is type O and the fetus is type A. The mother's blood contains antibodies against the foreign protein on the fetal red blood cells. Furthermore, these antibodies pass through the placenta and enter the fetal bloodstream where they attack the red blood cells. This may result in a mild form of a condition called *erythroblastosis*. More importantly, however, there is evidence that this mother/fetus A–B–O incompatibility may be potentially more serious during the earlier stages of fetal development. In fact, statistical evidence suggests that spontaneous abortion occurs more often in early pregnancy when the mother is type O and the father is type A, B, or AB.

The Rh Factor

The discovery of the *Rhesus factor*, or *Rh factor*, has explained a previously puzzling tendency for some infants to be stillborn or die shortly after birth from *erythroblastosis*, which is a condition involving the destruction of a high proportion of red blood cells. Approximately 85 percent of the white population in the United States is *Rh positive*; that is, they carry the Rh protein factor on their red blood cells. The remaining 15 percent do not carry the Rh factor on their red cells and are therefore *Rh negative*.

If Rh+ red cells are introduced into the bloodstream of an Rh− person, the body will respond to this "foreign protein" by producing specific antibodies as part of the typical antigen-antibody reaction. Subsequent transfusions with Rh+ blood may therefore result in adverse reactions; for this reason the Rh factor is carefully typed along with the A–B–O blood groups before transfusions are given.

During pregnancy, the problem is likely to arise after the first or second child is produced by an Rh− mother and an Rh+ father. The gene for the Rh factor (*R*) is dominant to the gene for its absence (*r*). Therefore, the following crosses are possible when the father is positive and the mother is negative:

Parents: $RR \times rr$
Offspring: All Rr

or

Parents: $Rr \times rr$
Offspring: Rr, rr

In the first case, the father is RR for the Rh gene. Consequently, all fetuses will be Rh+ and during pregnancy problems could arise. In the second case, the father is Rr and the probability of a problem during any given later pregnancy is 1/2.

Figure 8.10 illustrates the mechanism of Rh incompatibility arising during pregnancy. During the later stages of pregnancy, red blood cells from an Rh+ fetus may escape into the maternal bloodstream through small tears that develop in the placenta. Then at birth, the placenta is torn from the uterine wall, resulting in a massive invasion of the maternal bloodstream by the fetal Rh+ cells. At that point, the Rh− mother begins to build antibodies against the foreign Rh factor, and the level of antibodies increases through subsequent pregnancies in which Rh+ babies are involved. After the first or second pregnancy, therefore, there is increasing danger that massive amounts of anti-Rh antibodies will move across the placenta and attack the fetal red blood cells, with severe erythroblastosis as the result. It is important to note, however, that this becomes a problem only when the mother is Rh-negative and the baby is Rh-positive. Whether or not a woman who is Rh− will actually have problems depends upon (1) the gene combinations carried by her Rh+ husband, and (2) the number of pregnancies that she goes through. Because the antibodies against the Rh factor need time to build up, problems are not likely to occur in the first or even in the second pregnancy, except in cases where the mother was exposed to the Rh antigen through a transfusion or a miscarriage.

*Figure 8.10
Diagram showing development of Rh antibodies in an Rh-negative mother as a result of pregnancy with an Rh-positive fetus. Note that in subsequent pregnancies, the mother's antibodies may destroy Rh+ fetal red blood cells.*

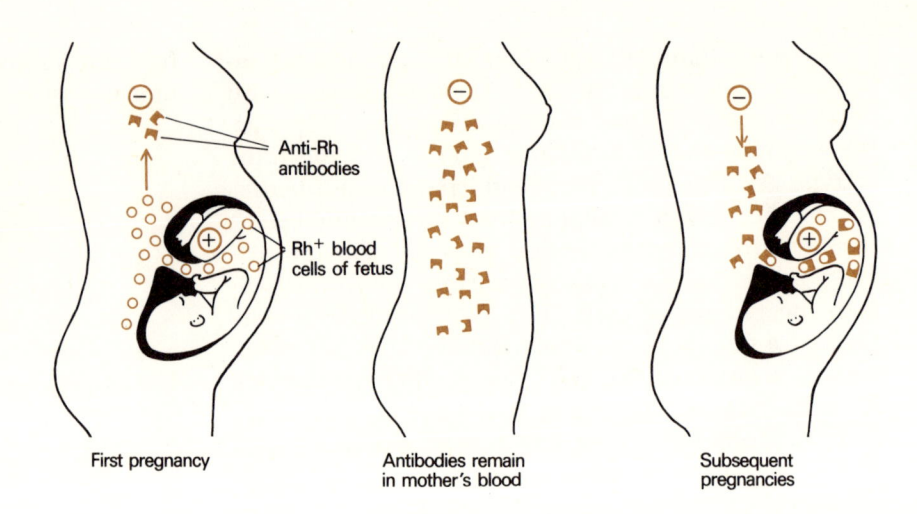

Medical management of erythroblastosis is based upon the principles we have discussed here. If it is determined by a simple blood test that a woman is Rh− and her husband is Rh+, then her antibody levels can be carefully monitored during each pregnancy. If the possibility of an erythroblastotic baby is suspected, a complete blood transfusion can be given immediately following birth. A more recently developed preventive approach consists of injecting the mother with massive doses of anti-Rh antibodies (*Rhogam*) following each childbirth. This temporary, artificially produced antibody level destroys fetal red blood cells that have entered her bloodstream when the placenta was torn from the uterine wall, and thereby removes the stimulus that causes antibody production.

Sex-Linked Genes

You will recall that the human male normally carries an XY chromosome pair, while the female normally has a pair of X chromosomes. The normal male and female karyotypes on page 54 show that the Y chromosome is much smaller than the X; we can therefore assume that the Y chromosome does not carry at least some of the genes that are carried by its X counterpart. It was once thought that the X and Y chromosomes carry gene pairs on part of their structures, but this is now regarded as questionable.

The female carries on her sex chromosomes genes that exist as usual in pairs, and a given gene at a given location on one chromosome can be expected to have a partner at the same location on the other chromosome. On the other hand, a gene found at a given locus on the X chromosome carried by a male may not have a partner on the Y.

Undoubtedly the best known condition associated with sex-linked or X-linked genes is *hemophilia*. Hemophilia is a serious "bleeder" disease caused by a lack of one of the factors required in the chain of events leading to normal blood-clot formation. The disease appears almost exclusively in males, although a very few female hemophiliacs have supposedly been identified.

The gene for hemophilia (*h*) is recessive to the gene (*H*) for normal clotting of the blood. We therefore have the following possible gene combinations in males and females:

Females		Males	
X	X	X	Y
H	H	H	none
H	h	h	none
h	h		

Note that when the female carries the recessive gene for hemophilia on one of her X chromosomes, there is, as it turns out, a high probability that she carries the dominant gene for normal blood clotting on her other X chromosome. This *masks* the effect of the hemophilia gene and she is therefore normal, but a *carrier*. On the other hand, if a male carries the hemophilia gene (h) on his single X chromosome, his affliction with the disease is assured because his Y chromosome will not carry the normal, dominant gene.

Since the male always receives his Y chromosome from his father and his X chromosome from his mother, it is apparent that hemophilia, like all X-linked genes, is passed from mothers to their sons. This was recognized by ancient Hebrew laws regarding circumcision. The Talmud stipulated that a male

Figure 8.11
History of hemophilia in the descendants of Queen Victoria.

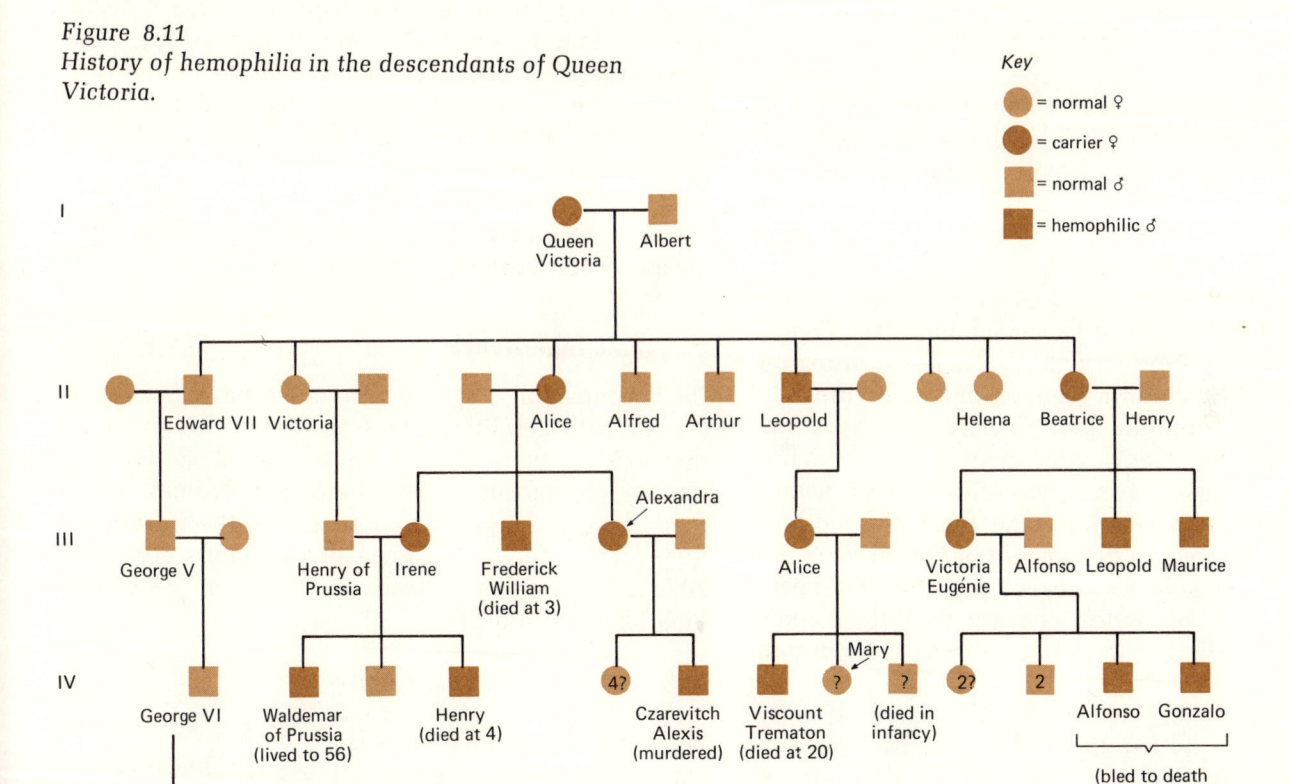

child was exempt from this ritual if two brothers had already died from bleeding connected with the operation. It is even more significant that the male children of the mother's sisters were also exempt, indicating an awareness that her sisters could also be carriers.

Males who are afflicted with hemophilia generally are invalids and are unlikely to marry and produce families. Therefore, the typical cross involves a normal male and a carrier female. We can represent this by the following, keeping in mind that the Y chromosome carries neither gene H nor its partner h:

$$\text{Parents: } YH \times Hh$$
$$\text{Offspring: } \underbrace{YH, Yh,}_{\text{Males}} \underbrace{HH, Hh}_{\text{Females}}$$

Our probability model shows that *if* a given child is a male, then the probability that he is a *hemophiliac* is 1/2. *If* a given child is a female, then the probability that she is a normal *carrier* is also 1/2. We can also say that the probability that any given *birth* will produce a hemophilic boy is 1/4.

The most famous family history of hemophilia involves the descendants of Queen Victoria (see Figure 8.11). From all appearances, the mutant gene probably originated with Victoria herself and according to some it was a mutation that may have changed the course of history. Alexis, who was the son of Nicholas II, the last Russian Czar, was afflicted with hemophilia, having received the gene from his mother, Alexandria, who was a descendant of Queen Victoria. According to some historical accounts, the Czar's preoccupation with his wife's concern over their son's condition contributed to the success of the Russian Revolution!

While hemophilia is the classic example, it is by no means the only condition that is caused by sex-linked genes. More than a hundred have been identified, many of them having tragic consequences for those afflicted. They include an X-linked form of muscular dystrophy, blindness due to degeneration of the optic nerve, a scaly skin condition called *ichthyosis*, and other, less serious afflictions such as color-blindness and night-blindness.

In discussing sex-linked genes, we have emphasized those carried on the X chromosome. These are usually referred to as *X-linked*. It follows, however, that genes carried on the Y chromosome are also sex-linked, and that they are passed exclusively from father to son. At this point, however, the significance of Y-linked genes has not been definitely established.

Sex-linked conditions are those that are controlled by genes carried on the sex chromosomes. They should not be confused with traits such as beard development, pattern baldness, breast development, and others which are controlled by autosomal genes but are expressed only in the appropriate hormonal environment. Women, for example, carry genes for a beard, but they are suppressed by the presence of female hormones. The determination of sex itself is not simply a matter of the presence of an XX or an XY pair, since it can also be greatly influenced by the hormonal environment.

Polygenic Inheritance

During most of our discussion, we have been concerned with conditions that are controlled by genes that exist in pairs and are found at a single location on the chromosome. The various combinations of these gene pairs can be compared to the results obtained from repeated tosses of two coins. The possible combinations of heads and tails, or "gene combinations," would be:

2 heads (HH)
1 head, 1 tail (HT)
2 tails, (TT)

The proportion of HH to HT to TT is therefore 1:2:1, which with the exception of X-linked conditions is the typical offspring ratio that we have observed throughout this chapter.

Many inherited traits, indeed some of the most common traits, follow a different pattern because they are controlled by *groups* of genes rather than by single pairs. Skin color, eye color, hair color, height, and intelligence are among those traits that appear to fit the pattern of *polygenic inheritance*. A characteristic common to such traits is a tendency to be expressed in gradations from one extreme to the other. Skin, hair, and eye colors, for example, run a gamut of shades from dark brown or black to very light. Scores obtained on intelligence tests also appear to reflect a range from low to high, with most individuals clustering somewhere around the midpoint.

We can illustrate the principle of polygenic inheritance by considering the results obtained from repeated tosses of four coins. This would result in five possible gene combinations:

4 heads (HHHH)
3 heads, 1 tail (HHHT)
2 heads, 2 tails (HHTT)
1 head, 3 tails (HTTT)
4 tails (TTTT)

These five gene combinations would occur in proportions given by the so-called *binomial distribution*. This yields:

$$(p + q)^4 = p^4 + 4p^3q + 6p^2q^2 + 4pq^3 + q^4$$

This shows that the "gene" combinations occur in a proportion of 1:4:6:4:1. This is illustrated in Figure 8.12 by a bar graph with a so-called *normal curve* superimposed upon it. This normal curve is typical of polygenic inheritance and illustrates how traits that are controlled by groups of genes tend to vary in their expression from one extreme to the other, with most individuals located somewhere near the midpoint or *mean* of the population curve.

We will return to polygenic inheritance in Chapter 16 when we discuss the relationship of this pattern of gene action to traits such as skin color and intelligence.

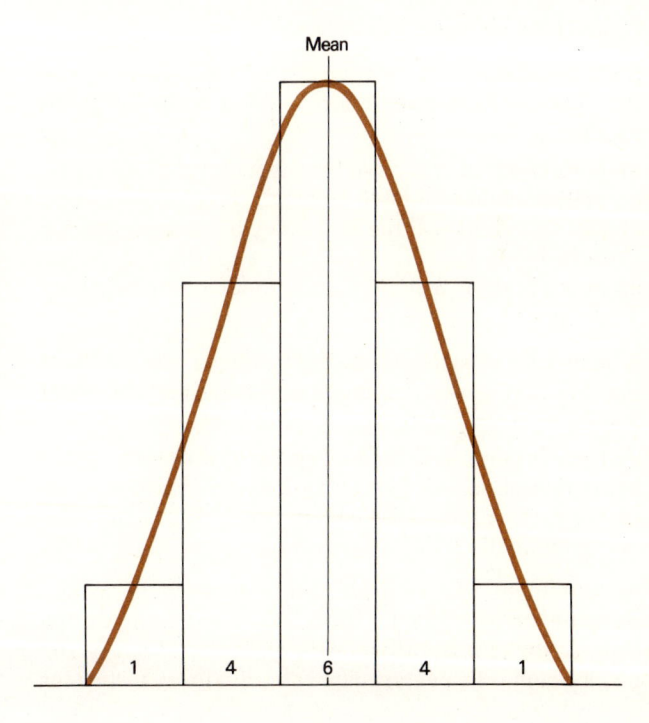

Figure 8.12
Diagram showing the relative frequencies of the five possible gene combinations found in a population when a given trait is controlled by two pairs of genes instead of one pair. These frequencies are similar to those of the head-tail combinations obtained from repeated tosses of four coins (see text).

Summary

In this chapter, our discussion has centered mainly around problems that occur when genes do not produce the kinds of results they are supposed to. Basically, in the language of Chapter 7, we have been concerned with the consequences of minute changes in DNA structure. Unfortunately, the examples that we have chosen are only a few in a long list of hereditary defects that plague mankind. In the next chapter, we will consider problems that result when whole chromosomes go astray.

Questions

1. A couple plan to have three children. Assuming that the probabilities of producing a boy or a girl are 1/2, respectively, what is the probability that
   a) the first child will be a boy?
   b) the third child will be a girl?
   c) all three children will be boys?

2. Both members of a couple have brachydactyly and they plan to have two children. What is the probability that
   a) both children will have brachydactyly?
   b) neither child will have brachydactyly?
   c) the first child will die in infancy of serious skeletal defects?
   d) both children will die in infancy of serious skeletal defects?

3. A man whose sister died of Tay-Sachs disease marries his first cousin. If they have two children, what is the probability that
   a) both children will die of Tay-Sachs disease?
   b) neither child will die of Tay-Sachs disease?
   c) the first child will die of Tay-Sachs disease *and* the second child will be normal?

4. According to one estimate, the frequency of sickle-cell anemia carriers in the black population of the United States is approximately eight out of every 100 black people. In one specific case, a wife is tested for the sickle-cell "trait" and is determined to be a carrier. Her husband has not been tested, but there is no known history of the disease in his family. If they have two children, what is the probability that
   a) both children will have sickle-cell anemia?
   b) the first child will be normal and the second child will be afflicted?
   c) *at least one* of the two children will be afflicted with sickle-cell anemia?

5. A couple have four children. The first child was normal, the second was normal, the third suffered severe erythroblastosis, and the fourth was normal. Thinking in terms of the Rh factor, what are the genotypes of the father and mother? What is the probability that a fifth child would be afflicted with erythroblastosis?

6. As a result of a baby mixup in the maternity ward, two mothers both claim the same baby. The blood groups of the mothers, their husbands, and the baby are as follows:

   > First mother: A
   > Second mother: A
   > First mother's husband: O
   > Second mother's husband: AB
   > Child: O

   On the basis of the above evidence, to which of the two mothers would you award the child? Why?

7. Discuss the ethical and practical pros and cons of each of the following statements:
   a) We should continue our present medical practice of keeping people alive long enough to pass on their defective genes and thus increase the frequency of birth defects.
   b) We should stop treating those with severe genetic defects in order to stop "polluting" the gene pool.

8. If you knew that you and your husband (or wife) were both carriers of PKU, how would you evaluate your chances of having a PKU child? Do you consider this to be a high probability? How would you handle the situation, and why?

9. Referring back to question (8) above, suppose you were assured that you would not have a child actually afflicted with PKU, but that the probability was 2/3 that it would be a carrier. Would this change your previous attitude? If so, how?

10. A normal woman whose brother is afflicted with hemophilia is married to a normal man. She is concerned about the possibility that her first child will be a hemophiliac. What advice would you give her? Why?

9

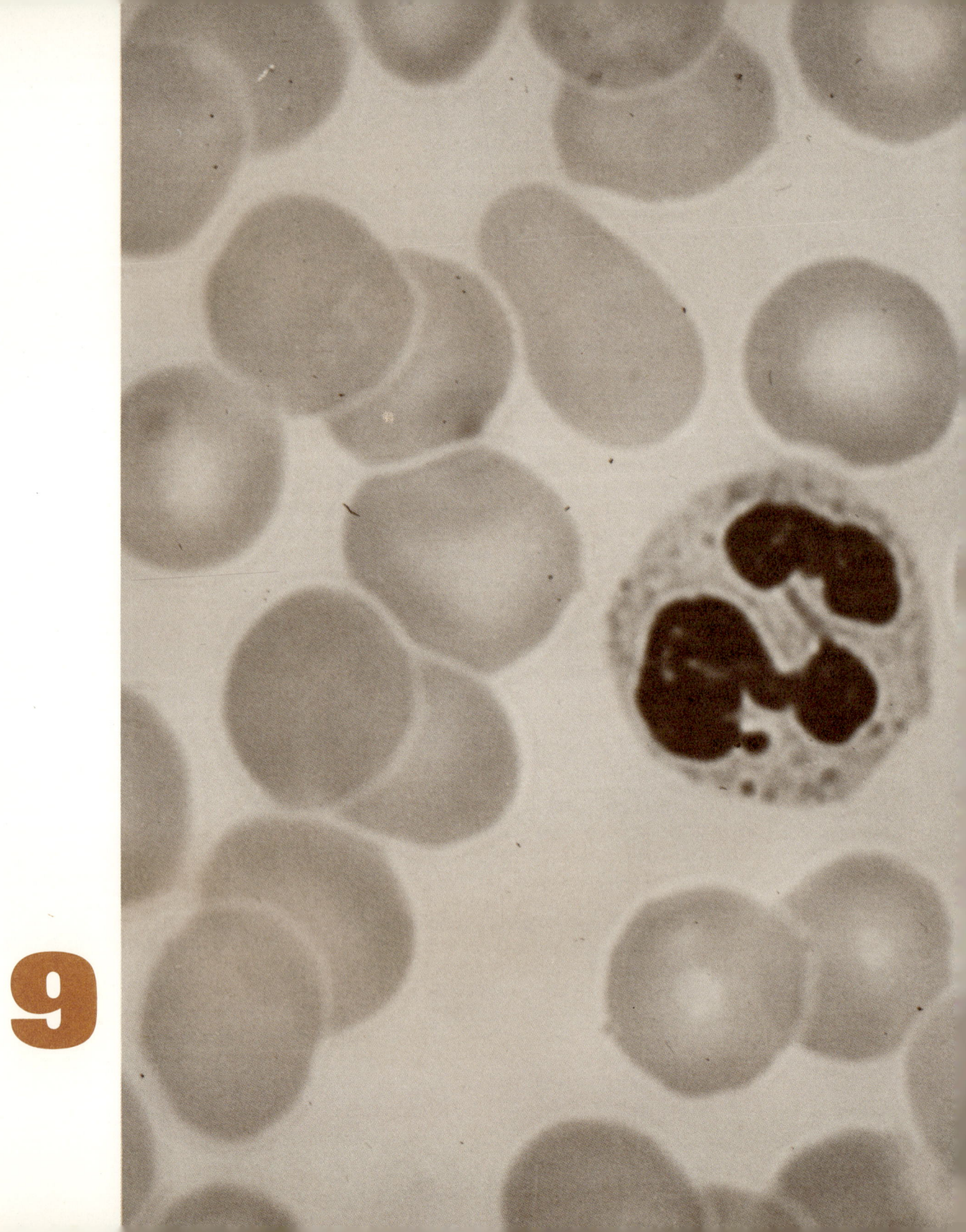

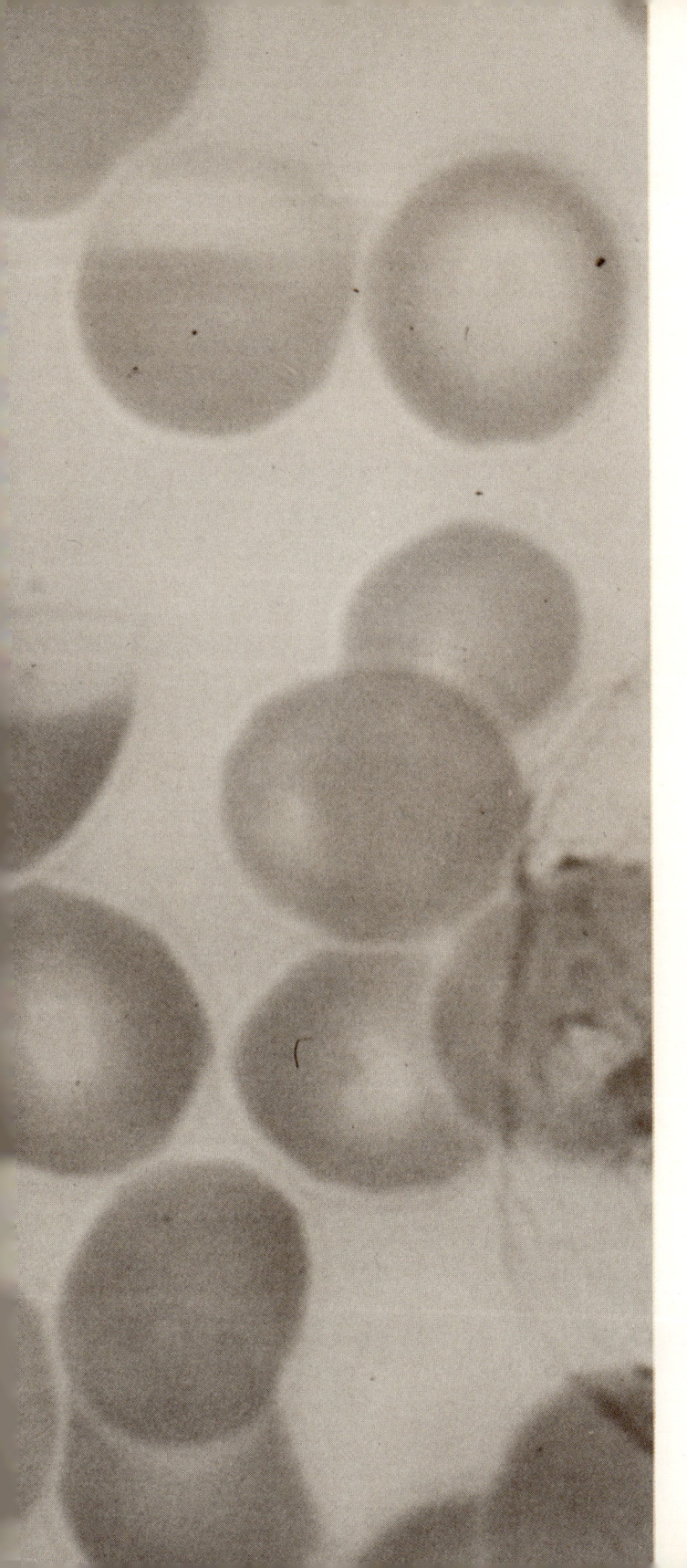

A Problem of Choice

Nondisjunction

As we begin this chapter, it might help to review the discussion of meiosis in Chapter 4. We must recall that the primary function of meiotic cell division is to produce sperm and egg cells that will carry exactly one-half the parental chromosome number. Thus in man, one member of each of the 23 pairs of parental chromosomes finds its way into the sex cell. Therefore it follows that when two gametes unite at fertilization, the zygote will have 46 chromosomes.

For example, when an egg cell is formed, a given chromosome pair normally separates in such a way that only one member of that pair ends up in the egg cell. This is shown by the following diagram, which summarizes the steps that actually occur with only one chromosome pair represented:

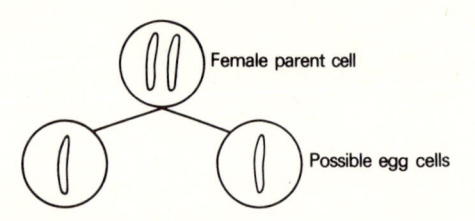

Unfortunately things do not always happen this way. In some cases a chromosome pair may fail to separate, and *both* members of that pair may find their way into the egg cell as follows:

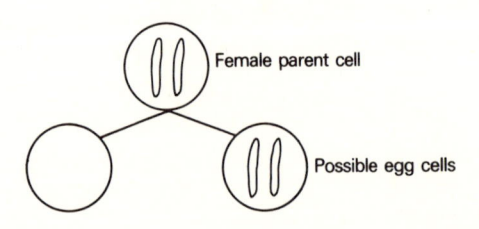

This failure of a chromosome pair to separate, resulting in a *doublet* in the egg cell, is called *nondisjunction* and is the primary cause of a number of very serious genetic defects. You can easily see that an egg cell containing a doublet actually carries 24 chromosomes instead of the normal 23. If it should now unite with a sperm cell containing 23 chromosomes, the resulting zygote will have 47 chromosomes, which is one more than the normal complement of 46. This abnormal variation in chromosome number almost invariably creates problems, the kind of problem depending on which chromosome pair is involved. One might think that having an extra chromosome would be a good thing, but for reasons not completely understood, having more than 46 chromosomes appears to be definitely not a case where more is better. Having *less* than 46 chromosomes also creates difficulties; this situation results from an egg or sperm cell that has only 22 chromosomes because of nondisjunction.

Down's Syndrome

Down's syndrome, also known as "mongolism," is one of the better-known of the genetic defects associated with nondisjunction. This condition was first described in 1866 by Dr. John Langdon Down, who called it "mongoloid idiocy" because a skin fold at the inner corners of the eyes gives the afflicted child a vaguely oriental look. Actually it is not the same fold that characterizes the true oriental eye. It should be stressed that Down's syndrome is not associated with oriental groups to any greater degree than with Caucasians.

A *syndrome* is a medical term for a collection of symptoms typically associated with a given disease. Not all the symptoms are necessarily always present, but the physician looks for a certain pattern when making a diagnosis. The child that is typical of Down's syndrome (see Figure 9.1) presents a number of characteristics including the skin fold at the inner corners of the eyes, a flat nose bridge, open mouth, protruding tongue, and a short neck. The hands tend to be short and stubby with a prominent crease across the palm. Some mongoloids have a ring of whitish spots on the iris of the eye that are called "Brushfield's spots."

A significant number of Down's syndrome children suffer from congenital heart defects, and they are usually highly susceptible to infection. In the past, therefore, many mongoloid children did not survive for long, but antibiotics and improved surgical procedures have increased the life expectancy of mongoloids to the point where the total number has quadrupled within a generation's time. Even now, however, more than half do not survive beyond age ten, and only about one-third live beyond age thirty. Dr. G. A. Jervis of the Institute for Research in Mental Retarda-

Figure 9.1
A child with Down's syndrome. Her karyotype is similar to the one shown in Figure 9.2. (Courtesy of the National Foundation–March of Dimes.)

tion has even suggested that patients with Down's syndrome may tend to age faster than normal. Normally they do not reproduce, but there have been occasional cases where females with Down's syndrome have given birth to children.

The physical development of mongoloids tends to be much slower than that of normal children. Their mental abilities vary, to some extent with the degree of attention which they receive and with patient efforts to train them to do simple tasks. In any event, they must be considered severely retarded. E. Peter Volpe notes that mongoloids account for fully 15 percent of all those in institutions for the mentally retarded. Given a birth rate of approximately 1:700 and considering the essentially hopeless nature of the condition, it may be seen that Down's syndrome exacts a terrible cost which must be paid with the vast sums of money required to institutionalize these unfortunate children and, more importantly, with the mental anguish of the families involved. All of this because of one extra chromosome!

The child with Down's syndrome has 47 chromosomes instead of the normal 46. The karyotype of a mongoloid (shown in Figure 9.2) reveals that the extra chromosome is due to the presence of three chromosomes 21 instead of the normal pair. Incidentally, in some sources this is described as chromosome 22, but this apparent confusion results from what has been a more or less arbitrary method of numbering the chromosome pairs and need not concern us here. At any rate, when *three* of any given chromosome are found in the karyotype instead of the normal pair, it is called a *trisomy*. Figure 9.2 presents the "trisomy-21" chromosome picture that is typical of approximately 95 percent of all mongoloids.

As stated previously, Down's syndrome is caused by nondisjunction. Furthermore, there is a striking correlation between maternal age and the occurrence of trisomy-21, strongly suggesting that the nondisjunction responsible for mongolism occurs only in the egg cell. For example, in women under age 30, the incidence of children with Down's syndrome is approximately one in 1500 live births. The birth rate for mongolism rises dramatically with maternal age, becoming as high as one in 30 for women in the 45-year-old age group. So far there appears to be no evidence that the age of the father bears any relationship to mongolism.

Exactly why there is a correlation between the incidence of mongolism and maternal age remains an open question. Logically it would appear to be related to the aging process in the ovary, possibly making the

Figure 9.2
Karyotype of a female child with trisomy-21, or Down's syndrome. Note the "triplet" at chromosome 21. (From "Birth Defects Article Series," National Foundation–March of Dimes, Vol. 4, No. 4.)

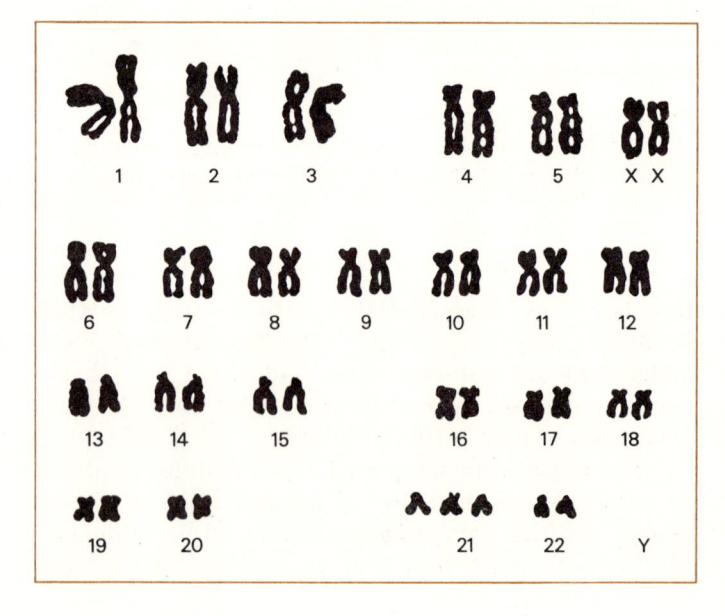

older egg cells more prone to "mistakes" in meiotic division than those produced by a younger woman.

The vast majority of Down's syndrome cases, therefore, result from fertilization of an egg cell containing *two* chromosomes 21 by a sperm carrying a single 21, producing a zygote containing a *triplet* at number 21, as shown below:

The relationship of maternal age to certain genetic defects has produced an interesting side effect in areas where population control is emphasized. In modern Japan, for example, women are terminating their childbearing periods at an earlier age as part of an overall effort to reduce population. As might be expected, this has led to a decrease in the incidence of Down's syndrome and other genetic defects associated with nondisjunction in the older mother.

Stray Sex Chromosomes

Nondisjunction can also give rise to problems which involve the sex chromosomes. For example, when an egg cell is produced, nondisjunction can affect the XX pair normally carried by the female, resulting in the following possibilities:

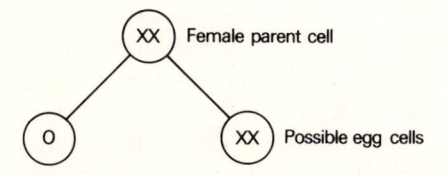

One possibility is an egg cell that carries two X chromosomes instead of a single X chromosome. The other possibility is an egg cell that does not carry even one X chromosome; this situation is usually symbolized by O.

What then are the possibilities that could result from fertilization? If we assume normal sperm cells, they will carry either a single X or a single Y chromosome. Therefore, the various possible combinations of sex chromosomes in the zygote are as follows:

Combination XXY is a triplet as in mongolism, except that this triplet involves different chromosomes. The individual has 47 instead of 46 chromosomes, and we may expect difficulties of some kind. In this case, the patient has *Klinefelter's syndrome,* a condition that occurs approximately once in every 500 live male births (Figure 9.3). Despite the two X chromosomes found in the XXY condition, the afflicted person is always a male; this suggests the effects of the "maleness" genes that are carried on the Y chromosome. Like Down's syndrome, the incidence of Klinefelter's syndrome appears to be associated with maternal age.

The Klinefelter's syndrome patient tends to be tall with poorly developed testes and breast development similar to that found in a normal female. For some unknown reason, it appears that brain development is especially sensitive to the presence of too many or too few chromosomes, and a significant number of Klinefelter's patients are retarded to some degree.

The second combination (XXX) produces a "superfemale." Such individuals tend to have fewer physical

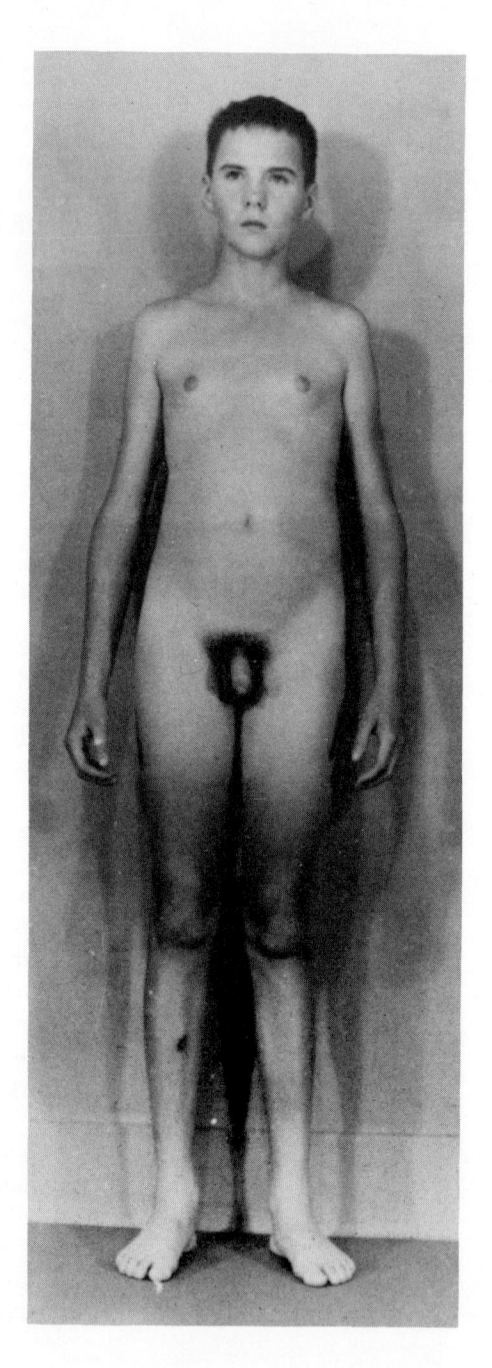

Figure 9.3
The boy in this photograph shows the breast development and long limbs typically associated with Klinefelter's (XXY) syndrome. (Courtesy of the National Foundation–March of Dimes.)

problems, and they may in some cases be fertile and produce normal offspring. Once again, however, the presence of the extra chromosome results in a tendency to be mentally retarded.

The third combination (XO) is found in those afflicted with *Turner's syndrome* (see Figure 9.4). This condition occurs once in about 2500 live births, and there is evidence that it may result from nondisjunction in the male parent as well as in the female parent. Actually, many more XO combinations are probably produced than the live-birth incidence would suggest, since it has been estimated that only about 3 percent of XO zygotes survive until birth. This tendency to abort is very possibly due to the fact that a part of the normal amount of chromosome material is missing. As might be expected, the Turner's syndrome patient is a female, but the ovaries do not develop, she fails to menstruate, and there is an absence of normal breast development. Typically, the Turner's syndrome patient is rarely over five feet in height, and a significant number are mentally retarded to varying degrees.

The fourth of the possible combinations (YO) can be disposed of quickly, because there is no evidence that a YO zygote ever develops. This failure to develop can probably be explained by the fact that a considerable amount of hereditary material normally provided by the X chromosome is missing.

Finally, an XYY combination has been noted as occurring in a number of males, apparently the result of nondisjunction during sperm formation. First described in 1962, the significance of the XYY deviation has been a subject of controversy. The XYY male has been described as tall with a tendency toward low

Figure 9.4
This girl is afflicted with Turner's syndrome (XO). Patients with this condition typically have short stature and do not attain full sexual maturity. (Courtesy of the National Foundation–March of Dimes.)

intelligence and violent behavior. Those observations, coupled with surveys that found a significant percentage of XYY males among prison inmates, led to speculation that the "XYY syndrome" is associated with criminal behavior. On the other hand, it has been pointed out that a prison is an excellent place to look for people who exhibit criminal behavior! Further doubt has been cast on the "criminal tendencies" theory by the fact that males with the XYY combination have been found who show no such tendencies. At this point the issue is still in question, and more evidence from careful experimentation is needed before definite conclusions can be drawn. On the other hand, even such experimentation may pose problems. For example, in a recent study, researchers screened newborn babies for chromosomal abnormalities including the XYY karyotype. Their objective was to follow the development of the XYY children in an effort to learn if there is a true association with behavioral problems. Interestingly, this study has been sharply criticized by other scientists who contend that when parents are informed that the child has the XYY karyotype, it could lead to a "self-fulfilling prophecy" situation. In other words, unusual parental anxieties concerning the child's behavior might possibly create behavioral problems which might not otherwise develop. On the other hand, the critics maintain that if parents are *not* told that their children are part of an experimental study, it constitutes fraud and deceit. This is an example of the ethical problems that always haunt experiments in which human subjects are involved.

Sex Chromatin

Normally, the female carries a pair of X chromosomes. According to the *Lyon Hypothesis,* proposed originally by Dr. Mary Lyon, only one member of the X pair is active in any given body cell of the female. Which member of the pair is active in any given cell is apparently determined randomly during early embryological development. Therefore, according to this hypothesis, if we let the X pair be represented by X_1 and X_2, X_1 will be active in part of the female's body cells and X_2 will be the active chromosome in the remainder of her cells.

The *inactive* X chromosome apparently coils into a deep-staining body that lies near the boundary of the cell nucleus and is visible during interphase. This structure, called a *Barr body,* was first described in 1949 by Dr. Murray L. Barr. It forms the basis for *sex chromatin* studies that have proved to be valuable tools for the diagnosis of conditions such as nondisjunction that involve the sex chromosomes. Figure 9.5 shows a diagram of a cell nucleus containing a Barr body.

The sex chromatin test is a simple one. A smear is obtained by lightly scraping the inside of the cheek to obtain a sample of the cells lining the mouth. The cells are then stained and examined under the microscope for the presence or absence of Barr bodies. If no Barr bodies are found, the patient is termed *chromatin negative.* This would be true in the case of the normal XY male. Since he carries only one X chromosome, it will be active in all cells; therefore no

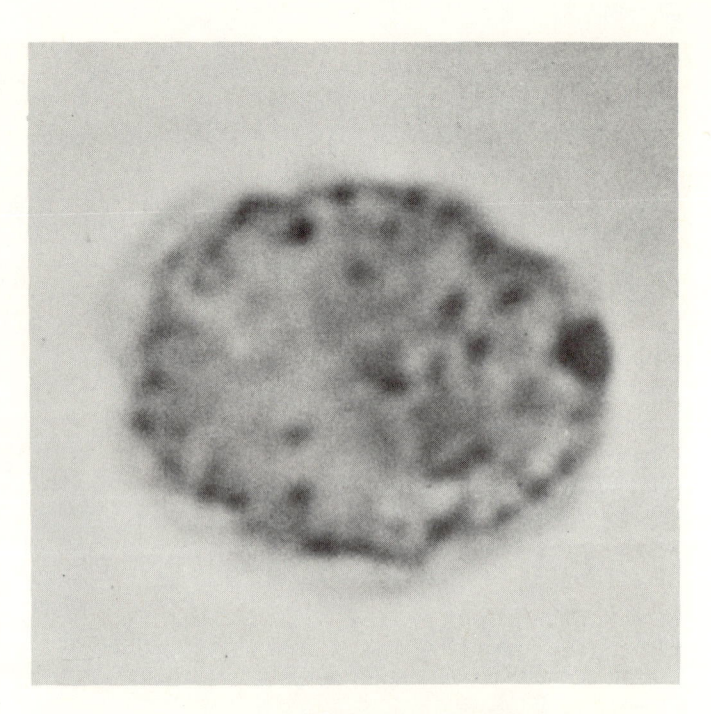

Figure 9.5
The small, dense area near the edge of the cell nucleus is called a Barr body. According to the Lyon hypothesis, the Barr body is associated with one of the two X-chromosomes found in the normal female.

inactive X chromosomes will be present. The same would be true, however, of a patient with Turner's syndrome, since you will recall that she has an XO karyotype and therefore, like a normal male, carries only one X chromosome. A chromatin-negative result can therefore help confirm a diagnosis of Turner's syndrome which has been tentatively made on the basis of clinical symptoms.

The "superfemale" carries three X chromosomes. Two of the three will be inactive; the superfemale will therefore be *chromatin positive* with *two* Barr bodies found in the cells. The normal female will also be chromatin positive, but her cells will contain only *one*

Barr body in their nuclei because she possesses only one inactive sex chromosome. The individual with Klinefelter's syndrome (XXY), although a male, will also be chromatin positive and like the normal XX female, his cell nuclei will also carry one Barr body. Quite obviously, however, a patient with Klinefelter's syndrome will be distinguished from the normal female by other characteristics. Generally the rule is this: The number of Barr bodies found in the nucleus of a cell will be one less than the number of X chromosomes found in that cell.

Another "sex symbol" is the *drumstick* that is found on the nuclei of a certain percentage of the white blood cells of a normal female but not found on those of the normal male (see Figure 9.6). Following the same pattern established with Barr bodies, the drumstick is present in the male with Klinefelter's syndrome but is absent in the female with Turner's syndrome.

Chromosome Translocations

Deviations from the normal chromosome picture may also occur as a result of *translocations*. A translocation involves a physical attachment of a chromosome belonging to one group to a chromosome belonging to another group. This produces an abnormal chromosome that may then be passed on from parent to offspring.

Previously we saw that the primary cause of Down's syndrome is an extra chromosome number 21, or a triplet at the 21 number. We also saw that this

Figure 9.6
A white blood cell from a human female. The small projection from the nucleus is called a drumstick. (Courtesy of the Carolina Biological Supply Company.)

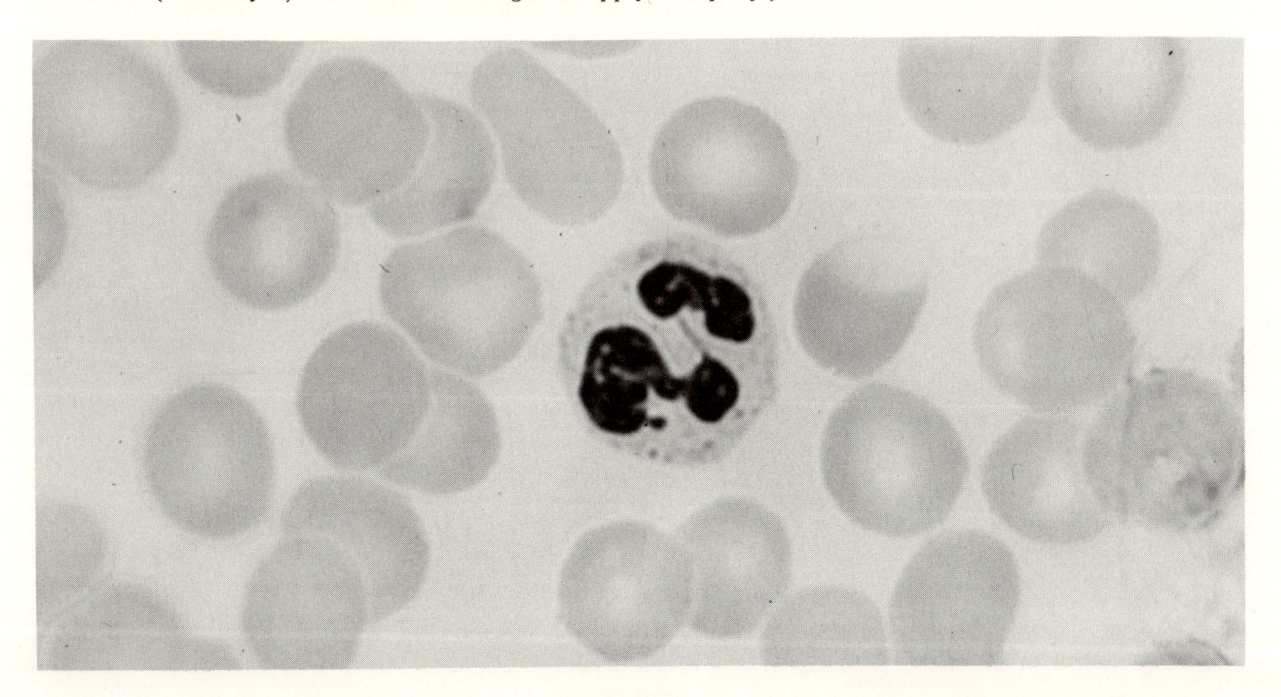

"trisomy-21" condition is produced by nondisjunction during egg-cell formation. There is, however, a small percentage of mongoloids in which the extra chromosome 21 results from the translocation mechanism.

In this case, the translocation occurs between number 15 and number 21. It involves the attachment of a large segment of chromosome 21 to a large segment of chromosome 15 to produce a translocation chromosome 15/21. Figure 9.7 shows the karyotype of a carrier of a 15/21 translocation. Note that the carrier's total chromosome content includes a 15/21 translocation, a normal 15, and a normal 21. The carrier has two 15's and two 21's and is therefore normal. Now, if we suppose that the carrier happens to be the female parent, she will produce four possible kinds of egg cells with respect to the 15 and 21 chromosome numbers:

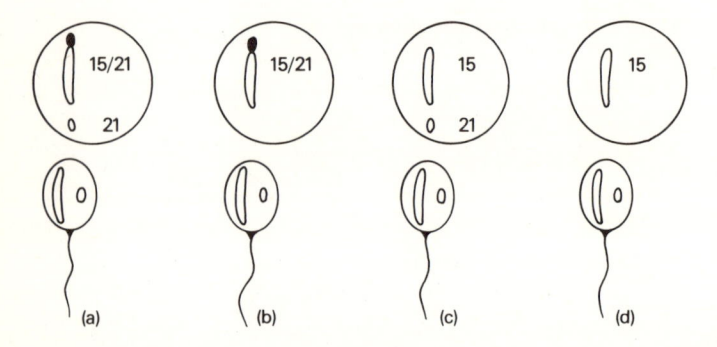

We can also see that a normal sperm cell carries a single 15 and a single 21 toward each of the four possible egg cells. Fertilization would therefore result in one of four possible zygotes, again in terms of their 15 and 21 chromosome contents:

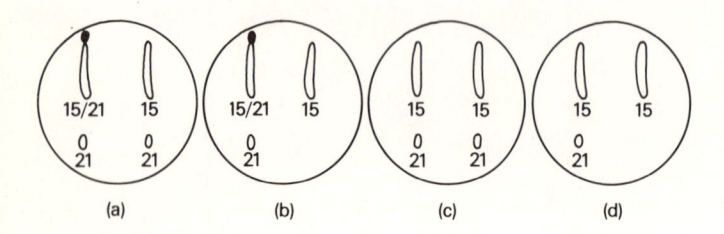

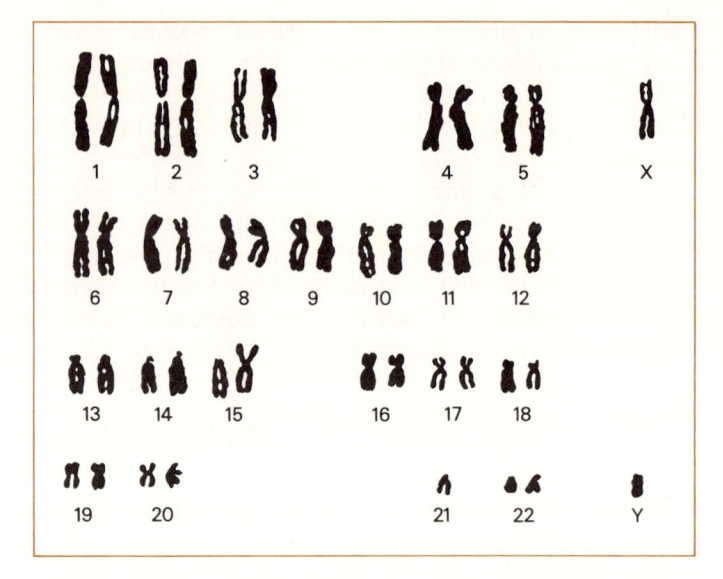

Figure 9.7
Karyotype of a 15/21 translocation carrier. Note the 15/21 combination and that only one "regular" 21 is present. The net result is two 15's and two 21's.

Note that the first possibility has two 15's and an actual total of *three* chromosome 21's. This is equivalent to the trisomy-21 that occurs because of nondisjunction; it therefore produces the usual symptoms and characteristics of mongolism.

The second possibility contains the 15/21 translocation, one 15, and one 21. This individual will carry the equivalent of two 15's and two 21's and will be normal, though a carrier. The third possibility is a normal noncarrier, and the fourth possibility is missing a chromosome 21. This fourth possibility does not develop due to the lack of genetic information carried on the missing chromosome 21.

Because the last possibility does not develop, we can eliminate case (d). We are left with an *a priori* probability model that tells us that the probability of a Down's syndrome child in any given birth is 1/3, the probability of a normal carrier is 1/3, and the probability of a normal noncarrier is also 1/3. The em-

pirical model, however, does not conform to the *a priori* model. Experience has shown that the probability that Down's syndrome will occur is actually more like 1/5 when the translocation is carried by the mother, and it may be as low as 1/20 when it is carried by the father. It should be emphasized at this point that translocation mongolism can be carried by either parent. One explanation for this difference between the *a priori* and empirical models suggests that the sperm cells that carry the translocation chromosome are less viable than those which carry the normal chromosomes. The translocation sperm cells are therefore less likely to fertilize the egg cell. To this should be added the fact that mongoloids, like many fetuses with serious genetic defects, tend to spontaneously abort during early development.

Translocation mongolism is called "familial" mongolism because it occurs in certain families with a definite hereditary pattern. The trisomy-21 condition that produces most cases of Down's syndrome is not, strictly speaking, inherited in the sense of being passed from one generation to the next, since it is essentially a result of nondisjunction in the ovary of the older mother. Translocation mongolism, on the other hand, is unrelated to the age of the mother, and as indicated before, the 15/21 chromosome translocation can be carried by either parent.

Chromosome Deletions

An abnormal chromosome picture is occasionally due to the presence of a *deletion,* a term that refers to a specific chromosome that has a portion missing.

One serious form of mental and physical retardation that is associated with a chromosome deletion is *cri du chat,* or "cry of the cat." It is so called because the cry of an infant afflicted with this condition sounds like the mewing of a cat. Figure 9.8 shows a child suffering from this syndrome, and Figure 9.9 shows the karyotype that is typically associated with the cri du chat condition. Note the deletion, or loss, of a portion of the short arm of chromosome 5.

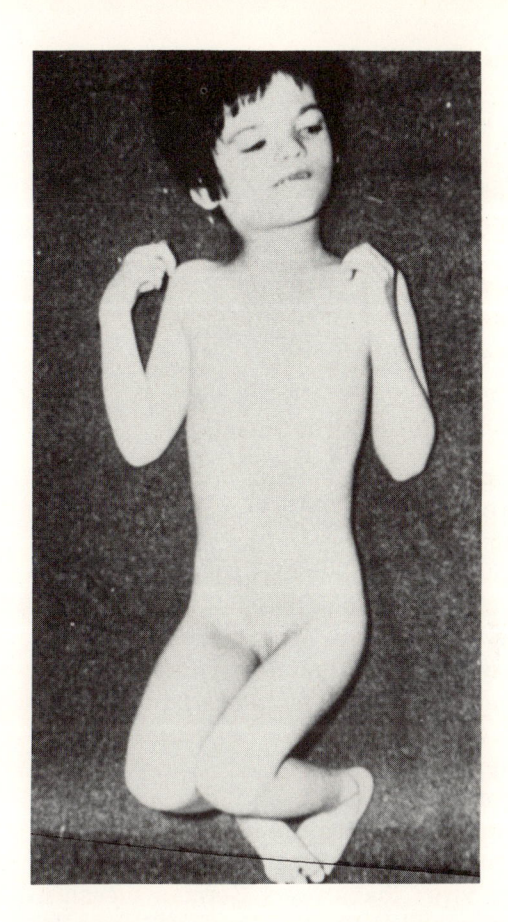

Figure 9.8
A patient afflicted with cri du chat syndrome. This condition is caused by a deletion of part of the short arm of chromosome number 5. The karyotype associated with cri du chat is shown in Figure 9.9. (From R. A. Boolootian, Human Genetics. New York: Wiley, 1971. By permission.)

Figure 9.9
Karyotype of a patient with cri du chat syndrome. Note that part of the short arm of one member of the number 5 chromosome pair is missing. This is called a deletion.

A form of leukemia, or cancer of the blood-producing cells, has also been linked to a chromosome deletion. Patients with this specific variety of leukemia have a karyotype that shows a deletion on chromosome 21. This abnormal 21 is called the "Philadelphia chromosome" because it was first described by investigators working in that city.

Environmentally Produced Birth Defects

Up to now we have been talking primarily about birth defects that are associated with specific errors in genetic makeup. We have seen that these genetic errors mainly involve either a mutant gene, as in Tay-Sachs disease, or a deviation from the normal chromosome picture, as in Down's syndrome.

There are a variety of birth defects, however, that are caused by environmental factors, or perhaps in some instances, by an interaction between environmental and genetic factors.

Teratogens are agents such as viruses or drugs that cross the placenta and produce defects in the developing embryo. Perhaps the most well-known teratogen is the virus that causes German measles, also known as *rubella*. Unlike "regular" measles (known as *rubeola*), German measles is a very mild disease in the adult, so mild that it often goes unnoticed or is mistaken for a slight cold. When rubella occurs in a pregnant woman, however, the virus crosses the placental barrier and may have devastating effects on the developing baby. The resulting birth defects can be extremely serious; they may include blindness due to cataracts, severe hearing loss, heart defects, and mental retardation. It should be stressed, however, that the effects of the rubella virus are most serious when exposure to the disease occurs in early pregnancy, generally within the first three months. After that, the probability of injury to the fetus drops off considerably, and exposure during late pregnancy apparently presents little or no problem.

The connection between rubella and birth defects was first noted by an Australian eye physician, Dr. N. McAllister Gregg, who noticed an unusually sharp rise in the incidence of eye defects among children born shortly after a widespread rubella epidemic in 1940. This observation was further confirmed by a rubella epidemic in the United States in 1964–1965 which resulted in an estimated 30,000 handicapped children. Fortunately, since then a preventive vaccine against rubella has been developed, and mass inoculations in the United States will undoubtedly bring about a drastic reduction in rubella-induced birth defects in the future. Quite obviously, it is especially important that every girl be immunized at some point before her potential childbearing period begins. Mass inoculations of both sexes will further help prevent future epidemics of this deceptively mild disease that can produce so much tragedy and sorrow.

Drugs also may have teratogenic effects. One of the best known is *thalidomide,* a drug that was taken as a sedative by pregnant women in Germany during the 1950's. This supposedly harmless drug affected limb development in the early embryo and resulted in thousands of tragic cases of children born with incomplete arms resembling seal flippers. Fortunately, in the United States a cautious attitude on the part of Dr. Frances O. Kelsey, a woman scientist with the Food and Drug Administration, held up the release of thalidomide for general use and thereby held the number of defective children to a minimum. As with rubella, the effects of thalidomide appeared to be most serious when it was taken during early pregnancy, and the exact nature of the birth defect seemed to depend upon the specific stage of development that was taking place.

Because of the tragic thalidomide episode, doctors should be extremely cautious when prescribing drugs for pregnant women. No one is really sure just what effects any drug might have on the developing embryo, and pregnant women are now encouraged to avoid any and all drugs unless they are absolutely necessary.

There is evidence that smoking during pregnancy may have harmful effects on the developing baby; it has also been suggested that alcohol may injure the baby's nervous system, especially during the later stages of pregnancy. Also, while the evidence is considered by some to be inconclusive, there are indications that the use of LSD may affect the chromosome patterns of its user, resulting in broken and bizarre chromosomes and thereby setting the stage for future birth defects.

A significant number of birth defects have been related to abnormally difficult births, especially those in which the baby was deprived of oxygen for a critical length of time. Such lack of oxygen may produce brain damage, ranging from so-called "minimal" brain damage to severe cases of cerebral palsy. Fortunately, the frequency of this type of birth defect is decreasing where better prenatal care and well-trained obstetricians are available. Unfortunately, as we look toward the end of the twentieth century from our new vantage point on the moon, we can see that adequate prenatal and obstetrical care is still not available to the vast majority of the nearly four billion people living on earth.

Amniocentesis—A Problem of Choice

It is clear from this and the two previous chapters that our main emphasis has been on birth defects and their hereditary and environmental causes. It is estimated that a child with a serious abnormality is born every two minutes, and that the total number of such births amounts to as many as 250,000 annually in the United States alone. Therefore, this is a significant problem which exacts an enormous toll in human suffering and which constitutes a drain on the resources of society.

At this point, we may begin to wonder if there is any possibility of having a normal baby! This is indeed a source of worry to many an expectant mother; the new mother's first concern is likely to be whether her baby is normal and healthy with all the necessary fingers and toes. Fortunately, the probability that any given couple will produce a defective child is actually very small, and most couples have little need for concern unless they happen to have knowledge of a special problem that presents an unusually high risk.

Still, while we should not emphasize the possibilities of birth defects to the point of hysteria, neither should the problem be ignored and simply left to chance or to the will of the Almighty. This is especially true if we consider that in many cases at least, there are now tools and procedures that enable the enlightened couple to do something about a problem that may arise. For example, in Chapter 8 we discussed carrier identification programs related to genetic disorders such as Tay-Sachs disease and sickle-cell anemia. It would certainly make sense for a Jewish couple or a black couple to take advantage of such programs, just as they would avail themselves of immunization against polio or tetanus.

A recently developed weapon in the war against birth defects is a procedure called *amniocentesis*. Amniocentesis provides a basis for making a *prenatal* diagnosis of a growing number of birth defects early in pregnancy. The procedure has now been performed many times with no serious problems for the mother or the fetus, and it has been shown to be a highly reliable method of prenatal diagnosis.

Amniocentesis involves the withdrawal of a small amount of amniotic fluid from the uterus; this is done somewhere around the 14th to the 16th week of pregnancy (Figure 9.10). It is accomplished by inserting a needle directly through the mother's abdominal wall and into the uterus, after which a small amount of the fluid that bathes the fetus is drawn into a syringe. Although it sounds painful, it actually presents no more discomfort to the mother than a blood test and no anesthetic is required. The procedure is done under sterile operating-room conditions, and it must be performed by a physician highly skilled in the technique.

The amniotic fluid thus collected contains cells that have been lost by the fetus and that therefore have the same genetic makeup as the fetus. These cells are grown in laboratory culture and analyzed

172 A problem of choice

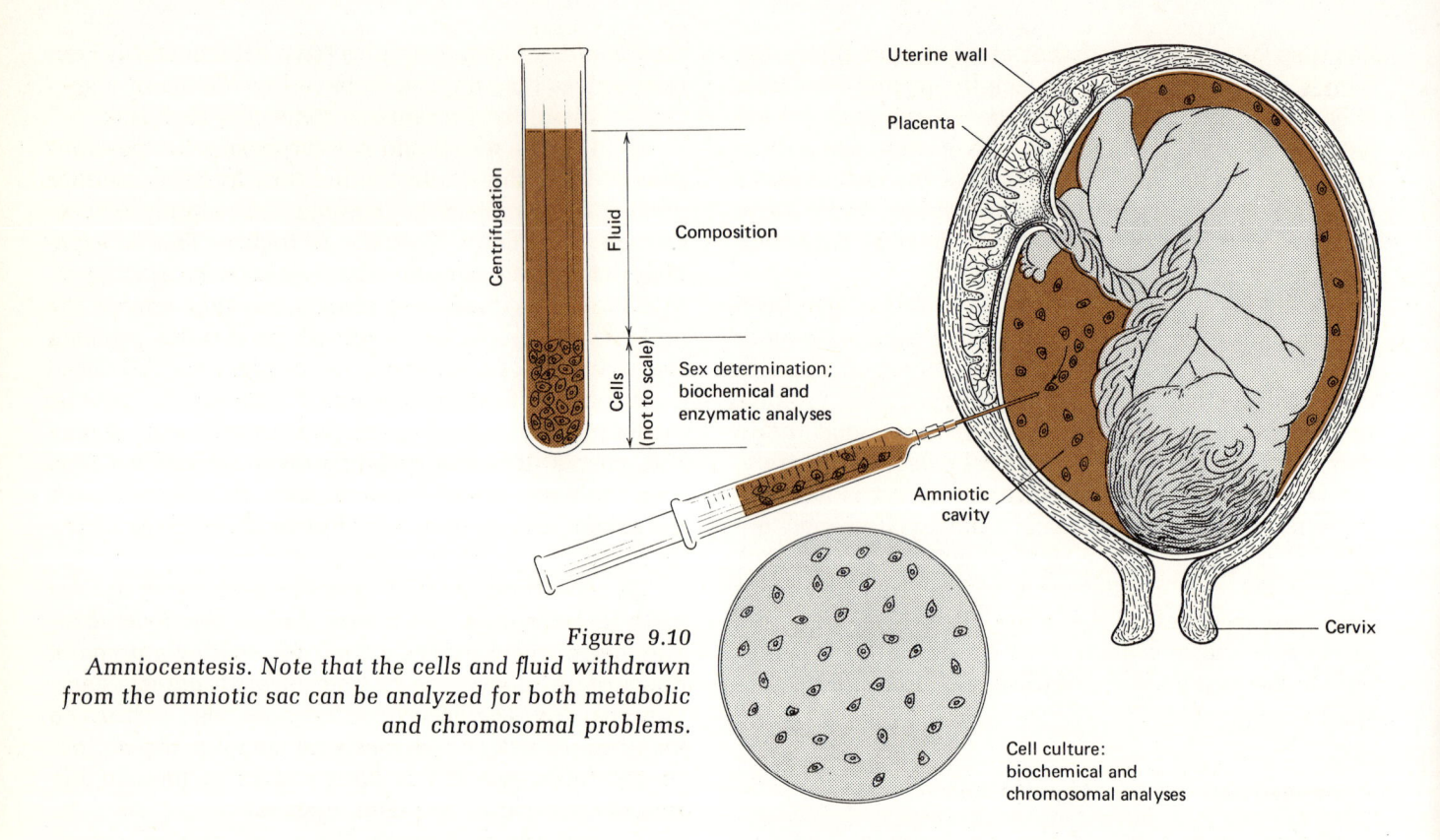

Figure 9.10
Amniocentesis. Note that the cells and fluid withdrawn from the amniotic sac can be analyzed for both metabolic and chromosomal problems.

for the fetal chromosome picture, or karyotype, and for the presence and level of certain enzymes, depending on the situation. Thus a chromosome-related disorder such as Down's syndrome may be detected, or enzyme levels may be evaluated to diagnose a metabolic block disorder such as Tay-Sachs disease. If a karyotype is developed from the fetal cells, it is also possible to accurately predict the sex of the unborn child. This latter prediction is more than merely a help to the parents in deciding between pink and blue receiving blankets, since the sex of the fetus is an important consideration when X-linked genes are involved.

It must be stressed that amniocentesis is definitely not a routine procedure to which every expectant mother can or should be subjected. While there is no evidence of significant immediate risk to mother or fetus, there is still much to be learned about long-term risks associated with removing amniotic fluid, although even this risk is generally discounted. At any rate, doctors are reluctant to recommend this procedure unless there is a specific and compelling reason why it is indicated. Some typical situations in which amniocentesis would be useful are described by the following cases.

Situation 1: A Jewish couple already has one Tay-Sachs child and the wife is in the early stages of a second pregnancy. The presence of the first afflicted child shows that both parents are carriers, and the risk that the second pregnancy will also terminate with a Tay-Sachs child is one in four. Amniocentesis is obviously indicated in this case.

Situation 2: A husband and wife have learned through a carrier identification program that they are both carriers of Tay-Sachs disease, even though they have no afflicted children as yet. The wife is in the early stages of pregnancy. Amniocentesis is also indicated in this situation.

Situation 3: A couple has a child with Down's syndrome, and an analysis of the child's chromosomes shows that it possesses the 15/21 translocation. Further chromosome studies of the parents show that the father is a carrier of the translocation. The wife is pregnant again; amniocentesis is again indicated.

Situation 4: A woman in the early stage of menopause becomes pregnant at age 44. Although her previous children were normal, amniocentesis is indicated because of the drastically increased probability of a nondisjunction trisomy-21 Down's syndrome child.

Situation 5: A woman whose husband is normal already has a child with hemophilia and she is pregnant again. In this case, amniocentesis could be done to detect the sex of the fetus. If it is a male, the probability that it will be a hemophiliac is 1/2; if the fetus is female, there is no possibility that the child will have hemophilia, although she could be a carrier like her mother.

In the foregoing sample situations, we again used Tay-Sachs disease and Down's syndrome as examples, but there is actually a growing list of genetic defects that are detectable by amniocentesis and which includes both metabolic block disorders and chromosomal abnormalities. Unfortunately, amniocentesis does not permit detection of abnormalities that are due to teratogens such as thalidomide or the rubella virus. Only certain *genetic* defects are detectable through chromosome or enzyme analysis.

It is important that we understand that a favorable report obtained from amniocentesis does not guarantee that the baby will necessarily be normal in all respects. Usually the procedure is done to evaluate the fetus for a *specific* abnormality for which a high risk value has been shown to exist in the family. It is true that the child could conceivably have some other unrelated defect, but the probability of this would be very small.

Suppose that a fetus is tested for a serious genetic defect and the report is *not* favorable, meaning that with almost absolute certainty, the child will be born with a severe, incapacitating abnormality. What now? At this point, one of two things must happen. Either the defective fetus is allowed to continue development or the pregnancy is terminated by abortion. Actually, couples that seek prenatal diagnosis by amniocentesis are generally committed beforehand to a termination of pregnancy if the fetus proves defective. Otherwise there would be no point in requesting the procedure, and the physician would probably not consent to expose the mother and fetus to even the minimal risk involved if there were not at least an intent to seek an abortion if the diagnosis proved unfavorable.

As we pointed out way back in the first chapter, a choice now exists in many cases of severe genetic defects where it did not exist before. The young parents who watched their first child deteriorate before their eyes as it suffered the lingering death of Tay-Sachs disease do not have to repeat that experience or permit another child to go through it, provided they are willing to accept abortion as a solution. A young mother who finds that she was exposed to rubella during early pregnancy may have that pregnancy terminated rather than take the chance of bringing a severely handicapped child into the world.

The U.S. Supreme Court has lowered the legal barriers to abortion. To have it done or not done is now largely the decision of the woman who must bear the child and the major responsibility for its care.

On the other hand, many people still refuse to accept abortion as an alternative even if it should mean that the most seriously handicapped child imaginable is knowingly brought into the world. There are those who believe that a mongoloid is a part of God's plan, and some parents of mongoloids have said that a mongoloid child has made their lives fuller and richer than would otherwise have been possible. Still others maintain that it is infinitely more acceptable, morally and otherwise, to terminate a pregnancy rather than to knowingly permit the development and birth of a child that will live out its years scarcely aware of its own humanity, a burden to itself and to others.

A choice is now possible, at least in certain situations. Who will make that choice? Shall it be made by society or by the individual? It has been suggested that society should make the choice in order to protect the interests of the majority, but most of us probably instinctively rebel at this concept and what it could lead to. At this point at least, we are still free to choose according to individual beliefs, backgrounds, and individual views of morality.

Questions

1. Distinguish between nondisjunction and the 15/21 translocation situation as they apply to Down's syndrome.

2. Explain how you would use the sex chromatin test to help in the diagnosis of (1) Klinefelter's syndrome, (2) Turner's syndrome, and (3) a "superfemale."

3. In recent years, a great deal of attention has been called to the "battered child" syndrome. If it can be shown that a certain couple repeatedly beats their child to the point of inflicting crippling injuries, would you advocate that society force the parents to give up that child? What role do the rights of the individual play in this case?

4. The results of an amniocentesis test show conclusively that a woman is pregnant with a child that will be severely defective, both mentally and physically. Should the authorities force the woman and her husband to consent to an abortion? Why or why not? How does this situation compare to that of question 3 above?

5. Chromosome studies done on a certain couple show that the wife is apparently normal but the husband is a carrier of the 15/21 translocation chromosome for Down's syndrome. Evaluate and discuss each of the following alternatives that might be considered open to this couple:

    a) The couple could have children without regard to the above findings on the basis that the chances of producing a mongoloid in this situation are empirically only about one in twenty.
    b) The couple could proceed to have children with the attitude that whatever happens must be accepted as the will of God, and if a child with Down's syndrome is born, it should be considered as part of God's plan.
    c) The husband could have a vasectomy and the couple could then adopt children.
    d) Since the wife is chromosomally normal, the couple could have children by artificial insemination, using sperm from anonymous donors.
    e) The couple could choose to have each pregnancy monitored by amniocentesis, terminating a given pregnancy by abortion if it is found that the fetus will be a mongoloid.

6. What differences, if any, would you say exist in the "right-to-life" antiabortion concept as applied to an apparently normal fetus and as applied to a fetus with Tay-Sachs syndrome?

7. According to Garrett Hardin, our concept of what is moral and what is immoral has changed throughout history as circumstances have changed. In what ways could increasing knowledge of genetics affect moral and ethical values now and in the future?

8. Suppose that you (or your wife) were exposed to rubella during early pregnancy. Would you feel that an abortion would be justified? Why or why not?

9. In the United States, the current trend is toward smaller families. How might this trend affect the number of births of children with chromosome abnormalities that result from nondisjunction in the egg cell?

10. One point of view suggests that the termination of a pregnancy by abortion should be regarded differently if the fetus has Down's syndrome than if it has Tay-Sachs disease. This point of view is based on the fact that with proper medical attention and patient training, a child with Down's syndrome may have a reasonable lifespan and may learn to do simple tasks. Do you agree with this point of view? Why or why not?

10

Human Nutrition

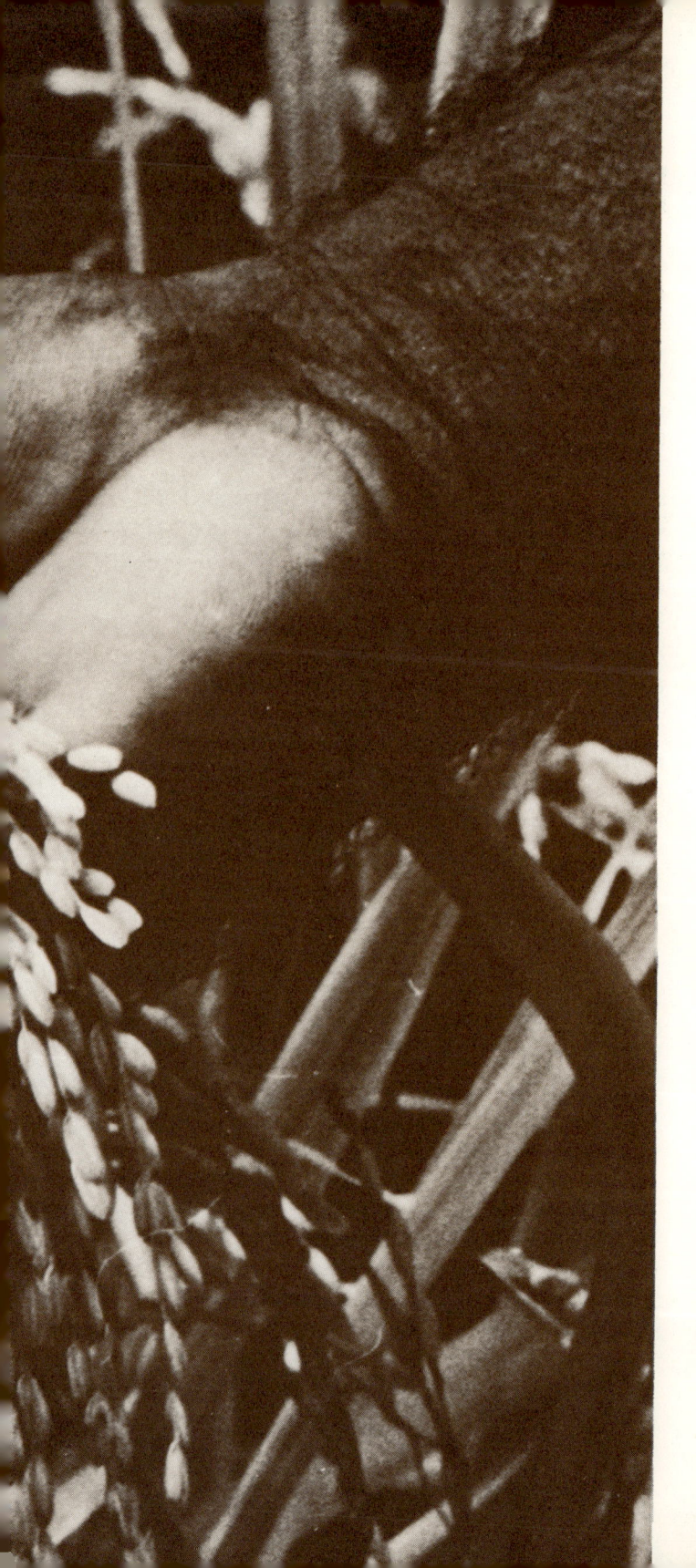

The Importance of Food

Whether mankind is more preoccupied with sex or with food probably depends upon the circumstances of the moment. Men and nations have fought over both at one time or another. The face of Helen of Troy may have launched a thousand ships, but Napoleon observed in his practical way that an army travels on its stomach. As with sex, we have made food and the eating of it the center of elaborate rituals which are part of our social customs. Elegant restaurants raise the preparation and serving of food to a ritualized art. Our most cherished traditions include picnics, banquets, the Thanksgiving turkey, the seder at Passover, the Easter ham, the wedding breakfast—and who likes to dine alone?

To the billions who face daily hunger, however, the question of food is even more fundamental. Food is always more fundamental when there is a lack of it or when constant malnutrition opens the door to disease, degrades human existence, and shortens life.

Like those ancient nucleic acid molecules in the primeval seas, we depend upon the environment for the materials necessary for our existence. Oxygen is one such material; you will recall its role in the energy-releasing process (Chapter 3). In addition to oxygen, the environment must provide other materials which we call *food*. Why do we need food? What major roles do foods play in maintaining life and health?

First, living organisms must have a source of *energy*. Energy is generally defined as the ability to do work, but in biology we think of it as the capacity to maintain life, to support growth and movement, and to drive the thousands of complex chemical reactions that take place within the cell. Recalling the "energy scheme" from Chapter 3, you will remember that green plants store the sun's energy in the bonds of food molecules. All animal life is therefore indirectly dependent on the radiant energy from the sun.

In addition to energy, the body requires a constant supply of *building materials*. During the period of rapid growth from conception to about 18 years of age, there is a significant increase in body size, and part of what is taken into the body is retained as part of its structure. The body, however, is never a totally "finished product." It is constantly renewing itself, so that even after the growth period is over there is a lifetime of tearing-down and rebuilding processes which require a supply of structural materials. The growth of hair and fingernails are more obvious examples of this renewal process, but it also takes place in other parts of the body such as muscles, blood, and the intestinal lining. The brain is the exception, however. Apparently, upon completion of growth there is no renewal of brain cells. In fact, as the body grows older there tends to be a progressive loss of brain cells.

Finally, the growth of new tissue and the release of energy for body activities are regulated in part by food substances called *vitamins*. Small amounts of these vitamins are required to keep the "machinery" of the body operating smoothly. We can very roughly compare their role to that which lubricating oil plays in the operation of a machine. The oil is neither a structural part of the machine nor is it fuel for the machine. Also like vitamins, it is required in very small quantity. Yet without the lubricating oil, the machine will not operate properly and may soon grind to a halt.

To summarize, the food materials that we must continually take in from our environment have three essential functions. First, they provide energy for body activities. Second, they furnish structural materials for growth and repair of tissues. And third, they play a regulatory role in cell activities.

Classes of Nutrients

There are five broad categories of nutrients that are necessary to an active, healthy life. These five categories consist of (1) carbohydrates, (2) lipids, (3) proteins, (4) minerals, and (5) vitamins. In Chapter 2 we discussed the chemical structures of carbohydrates, lipids, and proteins, so you may wish to quickly review that material at this time.

In general, when we speak of a "balanced" diet, we mean a diet which contains at least some of each of the above five categories of nutrients. Table 10.1 shows the primary role or roles that each class plays in general nutrition.

Table 10.1
Roles played by the five classes of nutrients.

	Energy	Construction	Regulation
Fats	+++	+	0
Carbohydrates	++	+	0
Proteins	+	+++	0
Minerals	0	++	+
Vitamins	0	0	+++

Carbohydrates

Carbohydrates, which include starches and sugars, are excellent sources of energy. Although their energy content is not as high as that contained in fats, they are the "first choice" of the cell as material for the respiration process. As long as sufficient carbohydrates are present in the diet, fats will not be used to any appreciable extent, which is one reason why it is much easier to gain weight than to lose it. In the United States, carbohydrates account for about 50 percent of the average diet, but in poorer countries, starches and sugars may amount to 80 percent of the total intake.

In addition to providing energy, carbohydrates, especially *cellulose,* supply bulk to the diet, thus helping to prevent constipation. You will recall that cellulose is the material of the plant cell wall. Although cellulose is a carbohydrate, the human organism cannot derive energy from cellulose because we lack the necessary chemical machinery (enzymes) to break it down. The presence of carbohydrates also promotes the development of intestinal bacteria. These bacteria are normally always present in our intestines, and they have the capacity to manufacture certain materials needed by the human organism. Vitamin K, which plays an important role in blood clotting, is one such factor. Carbohydrates are therefore an important part of a balanced diet.

Lipids

There are three kinds of lipids which are important in nutrition. These are the *triglycerides, the phospholipids,* and *cholesterol* (Figure 10.1). Triglycerides are the so-called neutral fats, and they consist of a glycerol molecule to which three fatty-acid molecules are attached. Fats supply the most concentrated source of energy, and they also represent a stored form of energy which is drawn upon when the carbohydrate supply is too small to maintain energy requirements.

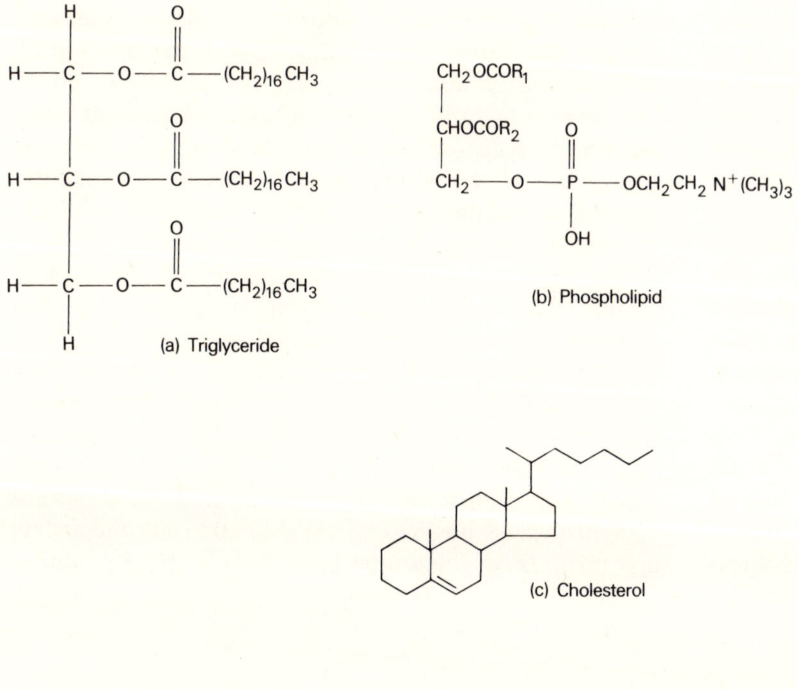

Figure 10.1
Structures of the three classes of lipids important in nutrition. In the phospholipid, R_1 and R_2 are fatty-acid chains.

Phospholipids consist of a glycerol molecule to which two fatty acids and a phosphate group are attached, the latter being substituted for the third fatty acid that is found in triglycerides. Phospholipids are important structural components in cell membranes, in the nervous system, and in a substance called *thromboplastin*, which functions in the blood clotting process.

Cholesterol has a structure which is quite different from the other two forms of lipids. In fact, its basic molecular structure closely resembles that of the sex hormones and certain hormones secreted by the adrenal glands. Cholesterol enters the body through the diet, especially one that is rich in eggs, butter, and animal fat. In addition, however, cholesterol is manufactured in body cells, especially in the liver. Along with phospholipids, cholesterol is an important structural component of cell membranes. It is also deposited in the skin, making the skin resistant to the action of various chemical agents and helping to prevent excessive water loss by evaporation. In recent years, a great deal of publicity has been given to the relationship between this important nutrient and heart disease. We will discuss this problem in a later section.

From 40 to 60 percent of the total energy intake of the average American consists of lipids; it has been found that lipids tend to make the diet more palatable and to help satisfy the appetite. In addition, certain vitamins are fat-soluble; that is, they will dissolve in fats but not in water. Fats must therefore be present if such vitamins are to be used effectively. At one time it was assumed that lipids were not an essential part of a balanced diet, but the varied and important roles which they play in body nutrition establish beyond question that they are indeed necessary.

Proteins

Since a balanced diet is one which contains appropriate quantities of all five classes of nutrients, it may be incorrect to say that any single group is more important than the rest, just as it is illogical to say that any one of the three legs of a tripod is more important than the other two. If, on the other hand, any one group were to be given that honor, it would be the proteins. The protein molecule has been called the "noblest piece of architecture produced by nature," which is consistent with the varied and critical roles that it plays in living systems. Table 10.1 shows that proteins are important structural components of the body. They are used in the growth and repair of tissues; they form the structure of many hormones; they form hemoglobin; they play an indispensable role as enzymes (Chapter 7). They can also be broken down to release energy if necessary, but since they are a poor source of energy, we cannot depend solely upon protein for this important aspect of nutrition.

The Food and Nutrition Board of The National Research Council recommends a daily intake for adults of 0.9 gram of protein per kilogram of body weight. This amounts to about 60 grams per day for a 150-pound man. An additional 10 grams per day is recommended during pregnancy, and an additional 20 grams from nursing mothers. It is also recommended that the daily intake for children under one year be 2.2 grams per kilogram of body weight, which is significantly above the daily intake of the adult. The average daily protein intake for the average middle-class American is estimated to be about 200 grams per day, which is well above the recommended minimum.

Unfortunately, a large percentage of the peoples of the world, including some Americans, take in less protein than is needed to maintain body health. Not only is the amount of protein in the diet important, but the kind or quality of protein is also critical. Protein molecules are made up of various amino acids. There are twenty such amino acids that are important to living systems (see Box 7.1). Eight of the twenty are called *essential* amino acids because they cannot be synthesized by the cells (Table 10.2). In one sense, the term is unfortunate because *all twenty* amino

Table 10.2
List of essential amino acids.

Isoleucine
Leucine
Lysine
Methionine
Phenylalanine
Threonine
Tryptophan
Valine
Histidine (essential for infants)

acids are "essential" (required) for good health. In this case, however, essential refers to the fact that certain amino acids *must* be taken in with the diet. For children there are nine, possibly ten, essential amino acids, apparently because children have not yet developed the same capacity as adults for synthesizing certain amino acids. This is significant because people in many parts of the world who depend upon plants as the major source of protein are lacking in certain essential amino acids, especially *lysine*. Protein deficiency is therefore an important aspect of malnutrition and will be discussed further in a later section.

Minerals

Iron is perhaps the most familiar of the minerals that are important to human nutrition, since we are bombarded almost daily by advertisements for products that promise to cure "iron-deficiency" anemia. Since iron is an important part of the hemoglobin molecule, its deficiency will result in a form of *anemia*, which is a general term for failure of the red cells to carry sufficient oxygen to the cells. Iron is also an important factor in the respiration process in the mitochondria. Foods high in iron include egg yolk, liver, meat, and legumes such as peas and beans.

Calcium and *phosphorus* are important constituents of bones and teeth. High intake of both calcium and phosphorus are recommended for children during the years of skeletal growth, and increased intake of these minerals is also important during pregnancy and for the nursing mother. Calcium also aids in blood clotting, in the transmission of nerve impulses, and in muscle contraction. You will also recall from Chapter 7 that phosphorus is an important building block of both DNA and RNA. Foods containing calcium and phosphorus include milk and milk products, meat, fish, legumes, and nuts.

Sodium and *potassium* help to maintain a balance in body fluids, and they both have an important function in the transmission of nerve impulses. Most of the sodium in our diet comes from sodium chloride (table salt). Potassium is found in grain cereals, bananas, meat, poultry, fish, and nuts.

Iodine is a necessary constituent of *thyroxin*, which is the hormone secreted by the thyroid gland that helps regulate energy metabolism. A deficiency of iodine may therefore cause a deficiency of thyroxin. The negative feedback system which we considered in Chapter 5 brings about increased activity by the pituitary gland, which stimulates growth of the thyroid in an attempt to produce more thyroxin. This enlargement of the thyroid gland is called *simple goiter*. The full, well-rounded neck produced by simple goiter was considered a sign of beauty during Renaissance times, and it is said that Napoleon was upset because many recruits from the Alpine region of Europe had to be rejected because their necks were too large to fit into the uniform. Iodine is scarce in the soil of several regions of the world, including the Great Lakes of North America. Today, simple goiter is largely unknown in areas where iodized table salt is used.

Other minerals important to nutrition include *magnesium, fluorine,* and *sulfur*. Magnesium functions as a catalyst in energy metabolism; fluorine is found in the teeth and skeletal system; sulfur is a constituent of a number of amino acids. Fluoridation of drinking water supplies has been shown to be valuable in reducing the prevalence of tooth decay in children, since fluorides tend to harden the enamel or outer coating of the teeth. Contrary to the fears expressed when the fluoridation of drinking water was first proposed, the addition of fluorides in small amounts is not harmful.

Vitamins

Table 10.3 provides a summary of the vitamins that are important to human nutrition. Vitamin K, which is necessary for proper blood clotting, is synthesized by bacteria in the large intestine, but most vitamins must be taken in with the diet. Furthermore, except for vitamins A and D, vitamins are not stored in the body to an appreciable extent; they must therefore be taken in regularly.

In recent years there has been an interest in using certain vitamins as *drugs*, that is, as therapeutic agents in dosages far in excess of recommended nutritional requirements. For example, a large daily intake of *ascorbic acid* (vitamin C) has been recommended as a prevention of the common cold. Although there is some evidence that this may be true in some cases, many physicians remain skeptical. Another series of studies suggests that excess vitamin C may play a role in preventing the accumulation of cholesterol in the arteries, a condition commonly related to heart and blood-vessel disease. Various claims have been made for extra dosages of vitamin E. For example, it has been suggested that vitamin E promotes healing, and claims are even made that it slows down the aging process. Unfortunately, there is no substantial evidence that these claims are generally valid for the human organism.

Folic acid, one of the B vitamins, is necessary for the production of red blood cells. Extra amounts of this vitamin are recommended during pregnancy because of the extreme demands placed upon the mother's circulatory system.

Vitamins are essential to the normal operation of body processes and are therefore an important part of the nutritional picture. Many vitamins, for example, act as *coenzymes*. That is, they form a part of the structures of certain enzymes, without which important cell reactions will not take place (Figure 10.2). We will have more to say about the effects of vitamin deficiencies in a later section of this chapter.

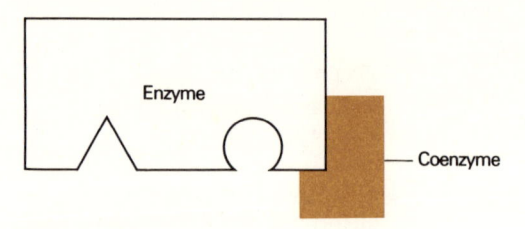

Figure 10.2
Combination of a coenzyme with an enzyme. The coenzyme is in this case a vitamin.

Water

Water is absolutely essential to nutrition. While it is possible to exist for a considerable time without food, the length of time that one can go without water is limited. You will recall that in Chapter 2 we pointed out that water is a necessary medium for the thousands of reactions that take place in the cell. Most reactions take place in solution, and water is the principal solvent of the body.

Digestion and Absorption

Suppose that you have a meal consisting of a steak, a baked potato with butter, a slice of bread with butter, and a cup of coffee with sugar and milk. From what we have discussed so far in this chapter and in others, it is obvious that in order for these foods to carry out their energy, structural, and regulatory functions, they must somehow get into the cells.

When you swallow foods, they are in a sense still "outside" the body because they are isolated in the tubelike *alimentary canal*, or *digestive tract* (Figure

Table 10.3
Summary of the principal vitamins. (From J. W. Kimball, Biology, 3rd ed. Reading, Mass.: Addison-Wesley, 1974, p. 63.)

Vitamin	Deficiency disease	Sources	Other information
A	Night blindness	Milk, butter, fish liver oils, carrots, other vegetables	Precursor in the synthesis of the light-absorbing pigments of the eye Stored in the liver Toxic in large doses
Thiamine (B$_1$)	Beriberi Damage to nerves and heart	Yeast, meat, unpolished cereal grains	Coenzyme in cellular respiration
Riboflavin (B$_2$)	Inflammation of the tongue Damage to the eyes General weakness	Liver, eggs, cheese, milk	Prosthetic group of flavo-protein enzymes used in cellular respiration
Nicotinic acid (niacin)	Pellagra (damage to skin, lining of intestine, and perhaps nerves)	Meat, yeast, milk	Converted into nicotinamide, a precursor of NAD and NADP—two important coenzymes for REDOX reactions in the cell
Folic acid	Anemia	Green leafy vegetables Synthesized by intestinal bacteria	Used in synthesis of coenzymes of nucleic acid metabolism
B$_{12}$	Pernicious anemia	Liver	Each molecule contains one atom of cobalt
Ascorbic acid (C)	Scurvy	Citrus fruits, tomatoes, green peppers	Coenzyme in synthesis of collagen
D	Rickets (abnormal Ca^{2+} and PO_4^{3-} metabolism resulting in abnormal bone and tooth development)	Fish liver oils, butter, steroid-containing foods irradiated with ultraviolet light	Synthesized in the human skin upon exposure to ultraviolet light Toxic in large doses
E	No deficiency disease known in humans	Egg yolk, salad greens, vegetable oils	—
K	Slow clotting of the blood	Spinach and other green leafy vegetables Synthesized by intestinal bacteria	Necessary for the synthesis of prothrombin, an essential agent in the clotting of blood

Figure 10.3
The human digestive system. (See text for details.)

10.3). It is only after they enter the cells in a form suitable for carrying on cellular functions that foods can be considered to be "inside" the body. Therefore, in order to get "inside" the body, the food must enter the "metabolic mill," the first phase of which is the digestive process.

Digestion is the process by which the large molecules of the foods that we eat are broken down by chemical and mechanical means to smaller molecules. *Absorption* is the process by which the products of digestion pass through the lining of the digestive tract and enter the circulatory system. Finally, the molecules that result from digestion are transported by the blood to all cells in the body.

Figure 10.3 illustrates the general plan of the digestive system. Beginning with the mouth, food materials are progressively broken down by enzymes as they go through the stomach and small intestine. This chemical digestion is aided by a mechanical breakdown of food materials. This includes chewing and the churning motions of the stomach and intestines. This mechanical breakdown increases the surface area that is exposed to digestive enzymes, thus making the chemical action more rapid and more complete.

Digestive enzymes are secreted by the salivary glands in the mouth, by the stomach lining, and by many glands lining the small intestine. Most of these glands are found in the first ten inches of the small intestine, which is called the *duodenum*. In addition, there are two important accessory organs that secrete substances which are carried by ducts into the duodenum. One of these, the *liver*, secretes *bile*, which aids in the digestion of fats by breaking up larger masses of fat into smaller droplets so that they can be more effectively broken down by enzymes. The other organ, called the *pancreas*, secretes enzymes which aid in the chemical breakdown of carbohydrates, proteins, and fats.

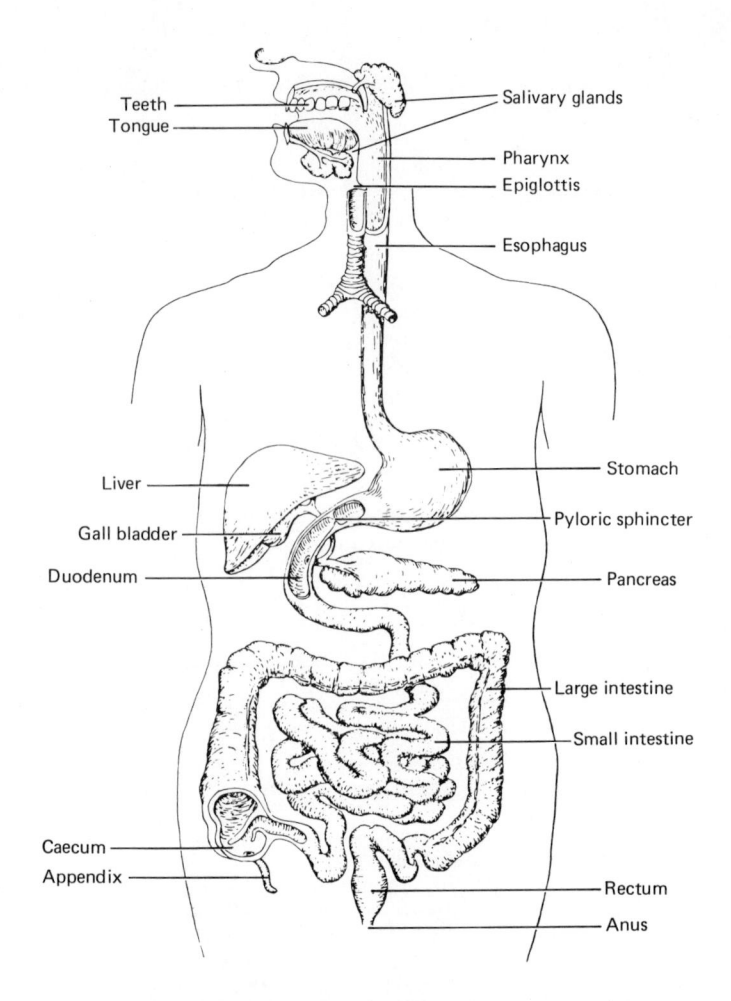

Going back to our sample meal, what are some of the major constituents of these foods and what happens to them in the digestive process? The steak is high in proteins and fats; the butter contains fats and cholesterol; the bread and potato are high in starches, and the milk in the coffee contains milk sugar, or lactose, in addition to a number of other nutrients. Finally, the sugar that was added to the coffee is table sugar, or sucrose. The following is a summary of what happens to these representative foods as large molecules are broken down by the action of enzymes into smaller molecules.

1. Each molecule of sucrose is broken down into simple sugars consisting of one molecule of *glucose* and one molecule of *fructose.*
2. Each molecule of lactose is broken down into simple sugars consisting of one molecule of *glucose* and one molecule of *galactose.*
3. Starches are first digested into *maltose,* a double sugar. This is then broken down into two molecules of *glucose.*
4. Fats (triglycerides) are broken down into *glycerol* and *fatty acids.*
5. Proteins are progressively broken down into *amino acids.*

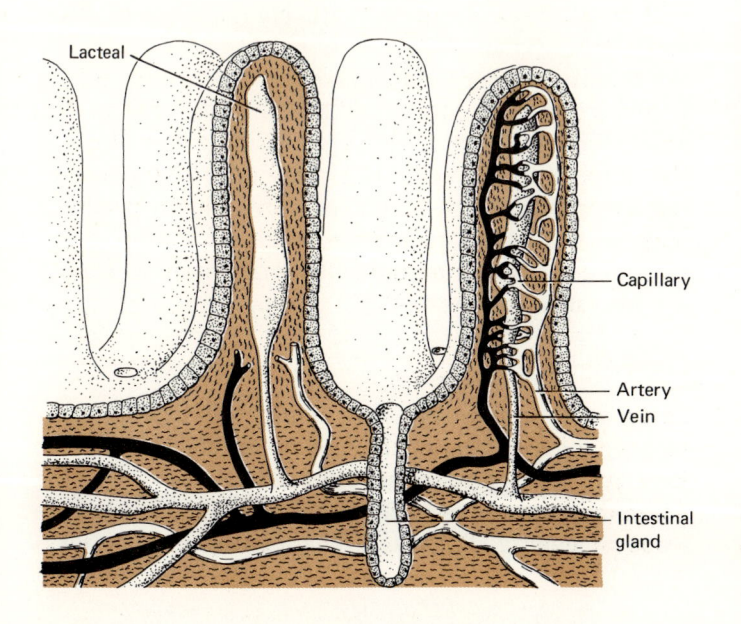

We are assuming, of course, that vitamins and minerals of various kinds would also be among the products of digestion.

In the next step, the usable products of digestion must be absorbed through the lining of the digestive tract into the body's transport system. Most absorption takes place in the small intestine. One of the few substances that is readily absorbed through the stomach lining is alcohol, which accounts for the rapid effects of a before-dinner cocktail taken on an empty stomach. Absorption through the small intestine is aided by tiny, fingerlike projections called *villi* (singular: *villus*) which line the intestinal wall (Figure 10.4). Both the shape and the large number of these villi significantly add to the total absorptive area. Usable nutrients such as simple sugars and amino acids pass into the blood capillaries of the villus, from which they are carried by the bloodstream to the cells of the body. Fatty acids and glycerol apparently recombine to form triglycerides, which take the form of small droplets that also contain phospholipids and cholesterol. They then enter the central *lacteal* of the villus and are transported by the lymphatic system to the bloodstream.

The mass of material which is left after the usable nutrients are absorbed continues into the large intestine, or *colon.* In the first part of the colon, most of the water content is absorbed into the blood, leaving the residue to pass into the rectum where it is eliminated in the form of the *feces.*

Absorption is an obviously important part of the whole nutritional picture. Sometimes nutrients are

Figure 10.4
Diagram of an intestinal villus. Note the capillaries and the lacteal.

not adequately absorbed from the small intestine even though the food is successfully digested. Conditions that involve decreased absorption by the intestinal wall are generally classified as *sprue*. In general, the causes of sprue are not well known, though in some cases a cure will be effected by removing wheat and rye flour from the food. It is known, however, that vitamin D is required for the absorption of calcium and that fats are necessary to the absorption of fat-soluble vitamins.

More on the Metabolic Mill

After the usable products of digestion are absorbed and transported to the various parts of the body, they cross the cell membranes and enter the cell's metabolic machinery. It is this machinery which uses the products of digestion for energy production and for the manufacture of structural materials. In this section, we will see how cells deal with the various classes of molecules.

Carbohydrate Metabolism

Carbohydrates supply the cells with their main source of energy. The breakdown of glucose and the capture of energy in the form of ATP was described in Chapter 3. Simple sugars including glucose, fructose, and galactose are the major products of carbohydrate digestion; the most important of these is glucose. Both fructose and galactose can be converted by the cell's machinery to glucose. In fact, you will recall from Chapter 7 that failure of the cell to convert galactose to glucose is part of an inherited condition called galactosemia.

After a meal that contains carbohydrates, any glucose that does not enter the energy pathway is converted to *glycogen*, which is stored in the liver and muscle cells. Glycogen, also called animal starch, consists of long chains of glucose molecules which are built with the expenditure of energy. As glucose in the cells is depleted and more is needed for energy production, glycogen is broken down again to glucose, which reenters the bloodstream and is carried to the cells (Figure 10.5).

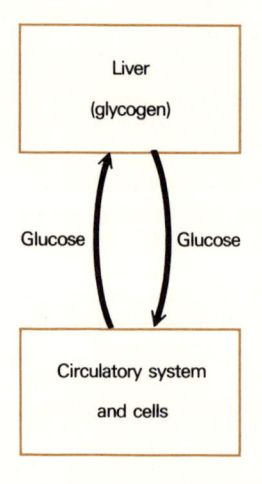

*Figure 10.5
Relationship between glucose and glycogen in the body.*

Glucose may also be converted to five-carbon sugars such as ribose and deoxyribose, which you will recall from Chapter 7 are components of RNA and DNA.

Finally, glucose may be converted by the cell to triglycerides and stored in the form of fat in the *adipose* cells. Most of this conversion takes place in the adipose or fat cells, but some may also occur in the liver. Alternatively, glucose breakdown products may be converted to certain kinds of amino acids, provided that other amino acids are present to supply the necessary amine (—NH$_2$) group.

Lipid Metabolism

Fatty acids are broken down into two-carbon fragments which can enter the energy-producing cycle or take part in synthesis reactions. Glycerol may also enter the respiratory pathway and be used for the release of energy. Fatty-acid chains may also be used as a basis for the synthesis of cholesterol, or they may be used in the synthesis of certain amino acids. Fats cannot be converted to carbohydrates, but both glycerol and fatty acids can be used as sources of energy.

Protein Metabolism

In Chapter 7 we described the process of protein synthesis. You will recall that DNA acts as the ultimate pattern whereby messenger RNA (mRNA) orders amino acids into the special sequences called for by the genetic material of the individual. One of the principal aspects of protein metabolism, therefore, is the breakdown of animal or plant protein into amino acids, which are then rearranged into specific *human* structural proteins and enzymes.

Amino acids may also undergo a *deamination* process in which the amine group is removed as follows.

$$\boxed{NH_2} - \underset{\underset{R}{|}}{\overset{\overset{H}{|}}{C}} - COOH \rightleftharpoons \underset{\underset{R}{|}}{\overset{\overset{O}{\|}}{C}} - COOH + \boxed{NH_3}$$

Amino acid   Keto acid   Ammonia

This process produces a *keto acid* and a molecule of *ammonia*. The keto acid can then enter the energy pathway; in this way, proteins can be used for energy production, although the energy available is comparatively small. Generally, we would see a high utilization of protein for energy during periods of starvation or during high protein diets when little or no carbohydrates are taken in. In fact, during periods of extreme starvation, the body will break down its own muscle protein to use as a source of energy. The ammonia that results from deamination of amino acids is highly toxic and must be converted in the liver to *urea,* which is eventually excreted through the kidneys. For this reason, it has been suggested that prolonged use of a weight-reduction diet consisting exclusively of protein may place an undue burden on the liver and kidneys.

In the process of *transamination,* an amino acid may be synthesized by using a breakdown product of glucose and the amine (—NH_2) group from another amino acid. For example, *alpha-ketoglutaric acid* is a breakdown product of glucose, produced as it goes through the energy pathway. By reacting alpha-ketoglutaric acid with the amino acid alanine, we may synthesize another amino acid called *glutamic acid* as follows.

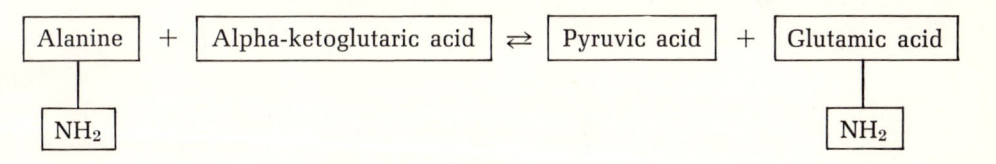

$$\boxed{\underset{\boxed{NH_2}}{Alanine}} + \boxed{Alpha\text{-}ketoglutaric\ acid} \rightleftharpoons \boxed{Pyruvic\ acid} + \boxed{\underset{\boxed{NH_2}}{Glutamic\ acid}}$$

Thus, glutamic acid can be synthesized by combining a product of glucose metabolism with an amine group from another kind of amino acid. It is important to note that not all twenty amino acids can be synthesized in this way. The necessary molecules for such synthesis are not present in every case. Those that cannot be synthesized by the cell are the *essential* amino acids that we mentioned earlier.

188 Human nutrition

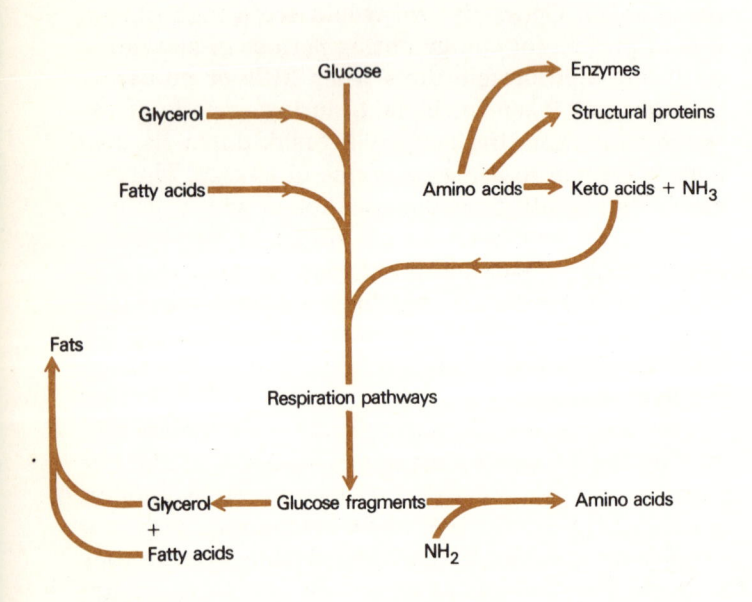

Figure 10.6
Diagram summarizing some of the important relationships among nutrients in the metabolic mill.

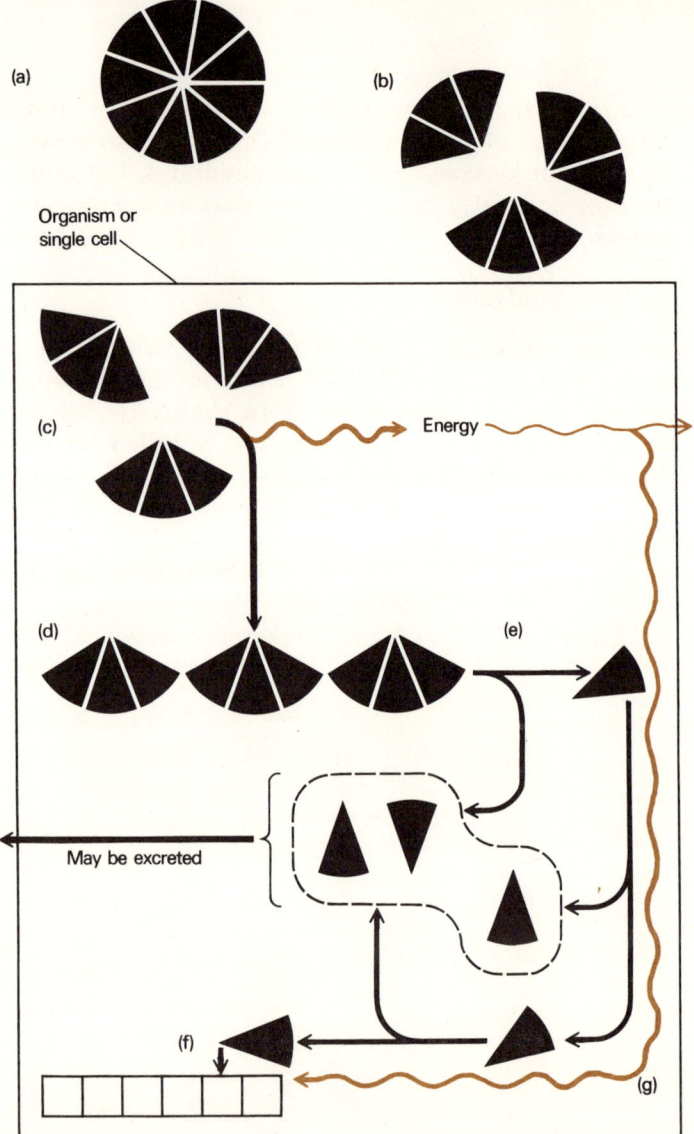

Figure 10.7
The metabolic mill. A large food molecule (a) such as a fat or protein is partly broken down (b). Some fragments enter cell (c), where they are broken down with release of energy. Other fragments are changed into other units (e), and portions may be excreted. Then large units are assembled for construction and storage (f) using energy (g).

The foregoing description of the metabolism of the major nutrients is only a small part of the complex reactions that take place in the metabolic mill. They are presented here mainly to show the versatility of the cell in the way it handles fundamental food molecules. Figure 10.6 attempts to summarize the various changes that food materials undergo in the process of metabolism. Figure 10.7 summarizes the "metabolic mill."

Earth—The Hungry Planet

As the world enters the final quarter of the twentieth century, the century of progress, at least one billion people are underfed. At least 500 million suffer from chronic hunger, and hundreds of thousands more face actual starvation. Persistent undernourishment affects 400 million others, and more than half of these are children.

In Bangladesh during 1974, 400,000 people fled to the capitol city of Decca because of the famine in the countryside. But food was no more plentiful in the city, and each morning the street sweepers cleaned the gutters of scores of bodies. The Director-General of the United Nations Food and Agricultural Organization went to Bangladesh to see the effects of starvation firsthand. He reported,

> The sight of small children pitifully clinging to life, surrounded by dead bodies, gives one an angry sense that we are still too far away from the frightening reality of hunger and malnutrition which millions of persons suffer day after day while diplomats . . . talk far into the night.

A 130-nation World Food Conference was held in Rome in late 1974. The American Secretary of State told the conference, "It is clear that population cannot continue indefinitely to double every generation. At some point we will inevitably exceed the earth's capacity to sustain human life." On the other hand, in his message to the same conference the Pope said,

"It is inadmissible that those who have control of the wealth and resources of mankind should try to resolve the problems of hunger by forbidding the poor to be born." In the end, the World Food Conference advocated a "balance between population and food supply," but refused to actually come out in favor of birth control.

Malnutrition constitutes the single greatest health problem on our planet today. Even the most conservative estimates place annual deaths from starvation in the hundreds of thousands. In most cases, these deaths are officially blamed on infections or parasites, but it is poor nutrition that has destroyed the individual's resistance to the final invasion by disease organisms. Paul Ehrlich defines death by starvation as "any death that would not have occurred if the individual had been properly nourished, regardless of the ultimate agent."[*]

As might be expected, nutrient-deficiency diseases are rampant in the underdeveloped countries of the world, where large numbers of people are forced to exist on substandard diets. As might also be expected, the major portion of the victims are children, who are especially susceptible during the years of rapid growth.

Kwashiorkor is a west-African word that refers to "the sickness that develops when another baby is born." It is a protein deficiency disease that develops even if a diet that is adequate in carbohydrates and lipids is provided (Figure 10.8). In children with kwashiorkor, there is retardation of physical growth and the child develops a characteristic potbelly. The hair pulls out in patches, fluids collect in the tissues, and if the child is not treated, death will ultimately follow. It is estimated that kwashiorkor affects as many as 50 percent of the children in some of the world's poorer areas where the main diet staples are starches and sugars.

---

[*] Paul R. and Anne H. Ehrlich, *Population, Resources, and Environment*. Philadelphia: W. H. Freeman, 1972, p. 87.

Marasmus is another common deficiency disease that appears to result from general undernutrition, including a deficiency of proteins and other sources of calories. The child with marasmus becomes thin and wasted, with wrinkled skin and enormous eyes that are reminiscent of the children in a Moppets painting. Marasmus is on the increase largely because of a decrease in the practice of long-term breast feeding among women in the underdeveloped countries. Mother's milk appears to contain the necessary dietary ingredients, but early weaning places the child at the mercy of the deficient diets of the region.

It should be pointed out, however, that hunger and its consequences are not confined solely to the poorer areas of the world. Public Health Service studies have found that even in affluent America, there are between *ten and fifteen million chronically hungry people,* with children making up a large proportion of those with deficient diets. It is of particular interest that in recent years, a number of extremely severe cases of marasmus and kwashiorkor have been identified in low-income areas of the United States! This indicates that there could be many more cases, not so severe, that have not been discovered.

These cases of severe protein malnutrition complicate childhood diseases such as measles. The body loses its capacity to fight infections and to recover rapidly when it is malnourished. There is therefore increased susceptibility to all infectious or parasitic agents—bacteria, viruses, fungi, and protozoa.

Vitamin A deficiency is also widespread in the underdeveloped countries. This vitamin is found mainly in yellow vegetables such as carrots, and its deficiency causes changes in the eye that eventually lead to blindness.

Beriberi, another disease caused by a vitamin deficiency (this time *thiamine*), is common in Southeast Asia and in the Philippines. In those areas, many people live on a high-carbohydrate diet and prefer polished rice, or rice with the husks removed. The husks, which are rich in thiamine, are fed to stock animals.

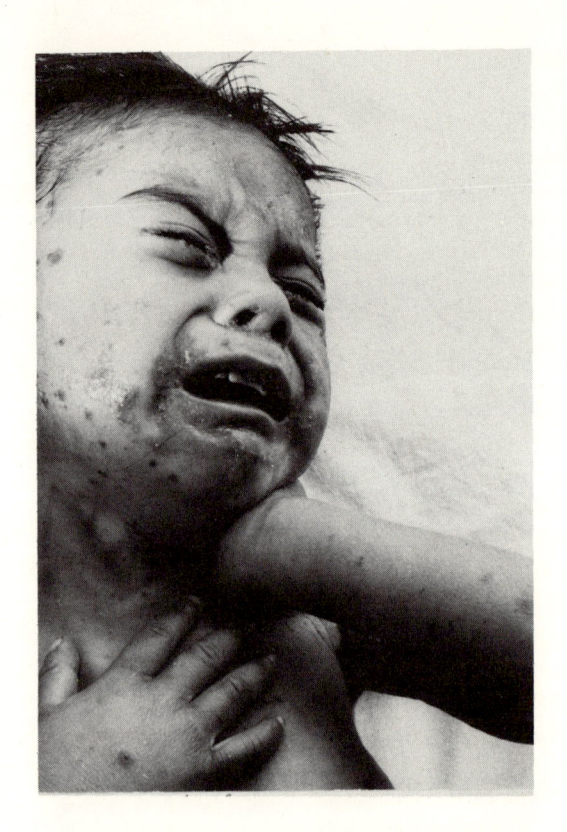

Figure 10.8
This child from Guatemala City is a victim of kwashiorkor. This is a disease of malnutrition that results from a protein deficiency. (Courtesy of the World Health Organization.)

Anemia is common among people in the poor areas of the world. Women are especially affected by the lack of protein, vitamin B_{12}, and folic acid. During pregnancy, the fetus absorbs large supplies of iron from the mother's stores. It is little wonder that women who are subjected to one pregnancy after another often develop severe anemia along with the chronic weakness, apathy, and loss of appetite that are typical symptoms of this condition.

Nutrition and the Brain

Hidden behind the thin armor of the skull lies the delicate and incredibly complex organ that is the very essence of humanity. The human brain is unique; it alone accounts for the wide gulf that separates the human animal from his nearest relatives on the evolutionary family tree. We have already seen how easily the complex patterns of brain development can be disturbed by metabolic block disorders and chromosome abnormalities or by environmental agents such as the rubella virus. We have also seen how the victims of such disorders are usually deprived of ever reaching their full potential as human beings.

It is therefore understandable that one of the most disturbing aspects of widespread malnutrition is the possibility that severe and chronic malnutrition may interfere with the development of a properly functioning brain. The impairment of brain development by any means in even an occasional child is certainly a cause for concern. Malnutrition, on the other hand, is like some insidious epidemic that affects whole populations and thus has the potential for creating problems in thousands of cases. It is therefore little wonder that the biological problem of malnutrition and brain development takes on significant social and even political dimensions.

Research with experimental animals such as rats has clearly established that protein deficiency during brain development has long-lasting effects on the animal's ability to learn, and that such effects appear to be irreversible even when the deficient diet is replaced by one that is nutritionally adequate. Unfortunately, research on the effects of malnutrition on human brain development and learning ability presents a number of difficulties. First, there are both practical and ethical considerations which make it difficult, if not impossible, to carry out planned, controlled experimentation with human subjects. Second, environmental factors other than nutrition tend to play a significant role in the development of human intellectual capacity, and it is difficult to isolate the effects of nutrition alone. And third, there are many kinds and degrees of malnutrition, which contributes to the difficulty of isolating specific deficiencies which might impair brain development. Even considering these difficulties, however, there is growing evidence that the quality of the human brain and its all-important intellectual performance may be irreversibly affected by malnutrition during fetal development or during the critical first years following birth.

The nervous system begins to form very early in human fetal development. There is evidence that most of the earlier development consists of cell divisions that produce the *neurons,* or nerve cells. These are the basic functioning units of the brain, and all *ten billion* neurons which are normally found in the adult brain are apparently formed by the time the child is born. This means that during fetal development, an average of over 300 neurons must be produced each second! There is evidence that two spurts of brain growth occur during early development. The first takes place from the 10th to the 18th week of fetal development, and the second from the 20th week until four months following birth. It has been suggested that prenatal brain development is most vulnerable to the effects of malnutrition during these periods of rapid growth, when each of the parts of the total structure must be completed in a given time and in a given sequence.

Consider the following analogy. If you are building a house, specific building materials must be available when you need them. If they are not, then important structures will be missing from the finished house. Also, certain items must be installed at definite times. For example, the electrical wiring must be finished *before* the walls are closed in. If we assume that you cannot add such features *after* the house is completed, the deficiencies in the structure would be irreversible and the house would never "function" in the way that it should. Now, the course of fetal development of the brain quite obviously cannot be held up (as might the construction of a house) until the proper building materials (nutrients) are brought in. If they are not there when needed, the finished brain will likely be deficient, and it is also likely that the deficiency will be irreversible. Prenatal nutrition, namely, the nutrition of the mother during pregnancy, is therefore of utmost importance.

At birth, the human brain weighs about three-quarters of a pound in comparison to the adult weight of over three pounds. Following birth the brain grows rapidly, doubling its weight during the first two years of life and then continuing relatively rapid growth for approximately the next ten years.

As we mentioned previously, the approximately ten billion neurons that make up the brain's basic functioning units are apparently present at birth. Most of the brain growth following birth apparently consists of the addition of proteins, cholesterol, and *glial cells*. Physiologists are not certain of the exact functions of the glial tissue, but it is assumed to be necessary for the proper functioning of the neurons. In addition, postnatal brain growth involves a significant increase in the number of connections among the neurons, thus completing the complex circuitry that is so essential to the brain's function as a super computer. We can now go back to our house-building analogy to point out that proper nutrition *following* birth is as important as fetal nutrition, since if the building materials are not there at the appropriate time, deficiencies in brain structure are likely to result. Authorities tend to disagree on the extent to which the effects of malnutrition on the postnatal phase of brain growth are irreversible. Some maintain, however, that permanent damage can result from malnutrition at any stage of brain growth, but is most likely to occur at times when brain growth is most rapid.

An adequate supply of protein seems to be especially critical to proper brain development, which is not surprising in view of the general importance of protein as a structural component of tissue. This critical need for protein is especially significant if one considers that the poverty areas of the world show varying degrees of protein deficiency.

It has been recognized for some time that protein-calorie malnourished infants are extremely apathetic and listless. It goes without saying that the performance of such children in school will be substandard. Although in some cases a return to adequate nutrition may bring about improvement, research studies strongly suggest that the learning ability of children who have recovered from malnutrition is generally lower than the level of that of children who have had continuously adequate nutrition. At any rate, in the poverty-ridden areas of the world where the spectre of starvation is a constant dinner guest, there is little chance of recovery because there is small likelihood that the children will suddenly be given an adequate diet.

While it is true that the intellectual performance of a given individual is related to both known and unknown social and environmental factors other than nutrition, the basic biological relationship between brain development and the availability of adequate building materials (nutrients) cannot be ignored. As a practical matter, it must be assumed that chronic malnutrition in pregnant women and in young children will very likely affect the quality of the next generation of humanity, forming a vicious cycle of malnutrition, apathy, and continued poverty from which it will be difficult for millions of earth's inhabitants to escape.

Nutrition and Aging

As people grow older, they tend to develop special nutritional problems. Due to a variety of causes, the aged often show decreased food intake. In some cases this may be due to poverty; instances of older people subsisting on social security checks and dog food are far from unknown. In other cases, loss of teeth and ill-fitting dentures make eating difficult. Some aged persons fail to eat regularly because food is too much trouble to prepare, or they may simply forget to eat. In addition, it is thought that the need for lysine and thiamine increases with age. Low-protein diets cause a decreased intake of thiamine, which tends to further depress the appetite.

In general, older people tend to suffer from either lack of information or misinformation concerning nutrition. They may think, for example, that milk is a food needed by children, not realizing that the skeleton is constantly renewing itself and must be maintained by sufficient calcium intake. It is estimated that the loss of 100 milligrams of calcium per day could reduce the skeletal structure by 30 percent over a ten-year period. This calcium deficiency may lead to a decrease in bone mass, resulting in a condition called *osteoporosis*.

It is also advisable that older people restrict their food intake, since there is a possible link between fats and blood-vessel disease. In spite of this, there is a tendency for the aged to favor high-fat foods such as eggs and bacon.

It is considered likely that poor diets tend to increase the health problems of the aged. This is a growing societal problem which has been and continues to be largely ignored except by local and sporadic programs. Better ways must be found to bring adequate nutrition to the ever-increasing numbers of "golden-agers"(!) who are living out their lives in an isolation enforced by poverty and infirmity.

Nutrition and the Pill

It is estimated that well over ten million women in the United States alone regularly use oral contraceptives. This number constitutes about 20 percent of all American women who are of childbearing age.

Recent studies have shown that more attention should be given to the nutritional requirements of women who are "on the Pill." There is evidence, for example, that oral contraceptives interfere with the absorption of folic acid, a vitamin which is necessary for red blood-cell formation. A deficiency of folic acid may lead to the clinical symptoms of anemia, the "tired, rundown" feeling that results when inadequate amounts of oxygen are transported to the cells. It is estimated that as much as 50 percent of the dietary folic acid may fail to be absorbed from the intestine when oral contraceptives are present in the system, and since American women tend to take barely adequate amounts of folic acid to begin with, it is important that the diets of Pill users be supplemented with this B vitamin.

A possible link between oral contraceptives and a deficiency of vitamin B_6 (pyridoxine) is also currently under study. Pyridoxine is involved in the metabolism of amino acids; this might create special problems for women who are taking oral contraceptives along with low-protein diets.

There is evidence that oral contraceptives actually help with the metabolism of iron. This is probably due to an increased absorption of iron from the intestinal tract. It may also be a factor that women who take the Pill have a tendency to lose smaller amounts of blood during the menstrual period.

The foregoing discussion points up the need for further research into the effects of oral contraceptives on nutrition. More information is essential so that appropriate measures can be taken to supplement the diet with folic acid, B_6, and other nutrients that may prove to be deficient in women who use oral contraceptives.

Eating for Two

During the nine-month period from conception to birth, the fetus develops from a microscopic cell to a full-term baby weighing several pounds. The pregnant woman therefore harbors within her own body a parasite that makes enormous demands for the nutrients it needs for energy and to sustain its fantastic rate of growth. It is little wonder, then, that pregnancy presents special nutritional needs that must be met to ensure the health of both mother and child. Pregnant teenage girls present even more problems, since their nutrition must meet both the additional needs of pregnancy and their own growth requirements.

During pregnancy, the fetus absorbs large quantities of iron, which it stores against the period after birth when it will subsist on mother's milk, which is naturally low in iron. The developing fetal skeleton requires extra amounts of calcium and phosphorus, and if the mother's diet is deficient in these minerals, the fetus will rob her of these materials from her own teeth and bones. The fetus also demands extra supplies of amino acids which it incorporates into the structural proteins of its tissues.

All of this requires that the mother increase her intake of nutrients, with special attention given to protein, iodine, folic acid, iron, and vitamin D. Increased vitamin D will tend to ensure that maximum amounts of calcium will be absorbed from the intestine. In addition, it is necessary that she increase calcium and phosphorus intake to protect the integrity of her own teeth and bones. The old saying "a tooth for every child" has, like many old sayings, a considerable basis in truth, especially when a pregnant woman's diet is not adequate.

Malnutrition during pregnancy has been linked to spontaneous abortion and appears to be a factor in premature births. Studies related to famine situations in both Holland and Russia during World War II suggest, however, that good nutrition *prior* to pregnancy increases the chances of a normal pregnancy even under poor nutritional conditions. In Holland, for example, there were only half as many stillbirths as there were in certain areas of Russia, where the women were poorly nourished *before* becoming pregnant.

Once the baby is born and has left its parasitic existence, the next problem is to feed it. In more affluent societies, there is a choice between breast-feeding and bottle-feeding with supplements of cow's milk or commercial formulas. There is general agreement that mother's milk is more likely to contain all the necessary nutrients which the baby needs in order to double its weight in the first five months after birth. Mother's milk is also likely to be free from substances that might produce allergic reactions, and it also contains antibodies. Cow's milk, on the other hand, may be lacking in thiamine and ascorbic acid because these are usually destroyed by pasteurizing. Various commercial formulas are available, however, which provide the essential nutrients and which are allergy-free.

Breast-feeding has been highly praised as a "wonderful experience" for the mother, but whether it is or not depends upon her attitudes and whether it is convenient. Mothers who bottle-feed, either because of choice or insufficient milk secretion, should not be made to feel guilty, since there is no evidence whatsoever that bottle-fed babies do not grow up just as healthy and happy as those who are breast-fed. This assumes, of course, that the bottle-fed baby is given all the necessary nutrients.

If the mother chooses to nurse her baby, then her own special nutritional needs continue to be as important as they were during pregnancy. She still requires extra protein, for example, and it is recommended by the Food and Nutrition Board of the National Research Council that she increase her calcium intake to at least 50 percent above normal to avoid losing calcium from her own stores.

In those areas of the world where constant malnutrition is the rule, the mother does not have the luxury of choosing between bottle-feeding and nursing her baby. In this situation, prolonged breast-feeding may be the only hope for the child, since in most cases it is the only source of adequate nutrition during the critical six months or so after birth. Early weaning places the child on a sugar-starch diet which is deficient in protein, and as we have seen, this leads to nutritional problems that adversely affect both physical and mental development.

Too Much Food

Even while the spectre of malnutrition hovers over the greater part of the world, those who live in the affluent Western countries are spending much of their time fighting the "battle of the bulge." The "Fat American" is an outstanding example of what happens when we have too much of a good thing. It is estimated that there are at least 25 million overweight people in the United States alone.

It is estimated that the mortality rate for overweight people is at least 50 percent higher than for those of normal weight. A number of specific diseases are associated with obesity (a polite word for "fat"). Notorious among these are heart and circulatory disease, kidney disease, diabetes, and cirrhosis of the liver.

What makes people gain weight to the point of obesity? Perhaps the simplest answer is that people *like* to eat, and when food is abundant they will eat it! Our supermarkets overflow; candy stores and soda fountains flourish; pizzerias have sprung up on every corner. The food is there, so we eat it. Then we get into our automobiles to visit a friend who lives two blocks away!

Fundamentally, whether one gains weight, loses weight, or remains the same depends on (1) how much food is taken in, and (2) how much food is used for energy. The energy available from food is measured in Calories, as all weight-conscious Americans know. A Calorie is defined as the amount of heat energy required to raise the temperature of a little more than a quart of water one degree on the Celsius scale. To use a simple analogy, when we eat, we are putting money (Calories) in the bank. When we exercise, we draw out money (Calories) from the bank. Just staying alive also requires energy, so this adds to the withdrawal made by exercising. If our daily deposits exceed our daily withdrawals, quite obviously our bank account (waistline) will grow. Therefore, if our energy requirements are less, our food intake must be less. A lumberjack can afford to eat more—indeed he should eat more than someone who sits behind a desk all week and then rushes off to the local health spa to frantically exercise on Saturday. Also, a child or teenager who is growing needs more food than an adult who no longer needs extra nutrition for growth. Unfortunately, we tend to carry over our teenage eating habits into adulthood, often with disastrous results for our waistlines. It should be stressed, however, that reduced food intake should not involve an inadequate supply of the necessary nutrients. A diet that is lower in Calories must still be a balanced diet. It is partly for this reason that many physicians and nutritionists violently object to the "fad" diets that have been advocated in recent years. We hear talk of high-fat diets, high-protein diets, yoghurt diets, "drinking-man" diets, and many more, some of which violate the basic principles of nutrition with which you are hopefully now familiar. Such diets have great appeal because they promise us that we can eat like the proverbial pig and still have the figure of a model. They often provide a quick loss of weight, which we just as quickly gain back when we return to the eating habits that caused us to gain weight in the first place. We simply cannot escape the fact that money put into the bank must not exceed the amount withdrawn.

It would be wrong, however, to assume that all cases of obesity are caused by overexercise with a knife and fork. It has been demonstrated, for example, that malfunction of the appetite control center in the hypothalamus (Chapter 12) may be at fault in some instances, although such cases are probably rare. It has also been found that some people have more adipose tissue (fat cells) than others. One theory proposes that an overabundance of fat cells may be produced by overfeeding during infancy. Also, although there is no clear-cut evidence, some studies and even casual observation suggest a genetic tendency toward obesity in certain families. A definite genetic factor has been demonstrated in mice, where there is an inherited tendency to produce more fat. Even when these mice are put on a restricted diet, they tend to lose protein rather than their fat stores. Finally, in recent years there has been a tendency to blame some cases of overeating on psychological factors. Whether a person is hungry or not, the act of eating may be a source of comfort for the person who is worried, sad, lonely, frustrated, or even distressed by the fact that he or she is overweight!

While eating too much may cause health problems, eating too much of the wrong thing may also lead to problems. Currently there is a great deal of concern and publicity related to the high consumption of cholesterol in the United States and in other countries where food is relatively abundant. Cholesterol has a tendency to form deposits called *plaques* on the inner surfaces of arteries. Because of this, it has been linked to heart and blood-vessel disease, which is one of the leading causes of death in the United States. Cholesterol deposits in the *coronary arteries* which feed the heart muscle can lead to a blocking of one of those arteries, resulting in *coronary thrombosis,* or the classic heart attack. Excess triglycerides have also been indicated as a contributing factor. In addition, cholesterol plaques which are deposited on the inner walls of larger arteries may cause the artery to weaken (Figure 10.9). The pressure of the blood causes the weakened blood vessel wall to balloon outward, a condition called an *aneurysm.* When the aneurysm breaks, death from massive hemorrhage quickly follows.

The formation of cholesterol and triglyceride de-

*Figure 10.9
Photograph of an artery showing cholesterol plaques on inner wall. (Courtesy of Coral Reef Photographs, Inc., Hollywood, Florida/Dalton, Georgia.)*

posits on arterial walls is a widespread condition called *atherosclerosis*. It definitely affects more men than women, and men are much more prone to heart attacks before the age of 50. After menopause, women begin to catch up; therefore it is assumed that the high levels of female hormones in premenopausal women protect them from atherosclerosis. Generally it is recommended that both men and women reduce their intake of eggs, milk, butter, and other foods that contain high levels of cholesterol. This reduction is especially important for people who obtain little or no exercise. Exercise apparently helps prevent high blood-cholesterol levels. It is also quite possible, though not proven, that heredity may play a role in how well cholesterol and triglycerides are metabolized.

Summary

"Man is what he eats" is an old German proverb that sums up in a few words what we have tried to relate about the importance of nutrition. Nutrition is not only a personal problem that affects each of us individually, but it is rapidly taking shape as one of the greatest peacetime problems to confront the world in modern times. The number of mouths to feed and stomachs to fill has *doubled* in the world since the end of World War II. Those in high positions of influence continue to stand silent on the issue of birth control while world population continues to expand, especially in the poor countries where famine has already begun.

In the fourteenth century, the Great Plague spread across Europe, killing one-third of the world's population. Unless the impossible is done, unless the richer nations are willing to make sacrifices, and unless there are drastic *worldwide* changes in attitudes toward birth control, the great famines that will result may rival the Great Plague in their destruction of human lives—while the rest of the world watches on color television.

Questions

1. Discuss the three principal uses of food in the body.
2. Discuss the interrelationships among carbohydrates, proteins, and lipids in the "metabolic mill."
3. How are digestion and absorption related to the total picture of nutrition?
4. What is meant by "essential" amino acids? Why are they essential?
5. What is the relationship between growth and the need for protein?
6. What does breast-feeding in poor areas have to do with the first six months of the child's life?
7. What are some possible causes of obesity? What is probably the most common cause?
8. In what ways can you relate the material from Chapter 7 to the material of this chapter?
9. What far-reaching social and political consequences may result from the effects of widespread malnutrition on brain development?

1

"...And Replenish the Earth"

The Problem

Of all the Biblical injunctions, none has been obeyed with more enthusiasm than the command to "be fruitful, and multiply, and replenish the Earth...." According to the Population Reference Bureau in Washington, D.C., the earth's total population at mid-1975 was close to 3.97 billion and increasing at a rate of over one million every five days. At that rate, an additional 73 million passengers crowd aboard Spaceship Earth with each passing year, and each three-year period sees an increase in world population that is equivalent to the addition of a new United States! If the present rate of increase continues, the number of people on earth will double every 36 years or so; it will reach over 7 billion by the time today's college student is in his mid-50's. By the time *his* children reach age 65, the world will be inhabited by 15 billion

"... And replenish the earth"

Figure 11.1
Graph showing the increase of human population through history. Note the extreme increase since about 1850. (Courtesy of the Population Reference Bureau, Inc., Washington, D.C.)

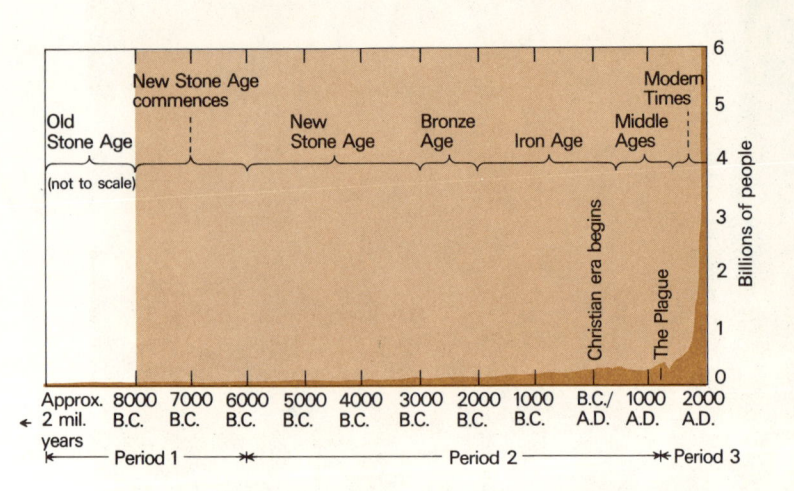

souls—assuming that the present rate of increase does not change appreciably in the meantime.

Looking at Figure 11.1, we can see that the "population explosion" is a relatively recent phenomenon when viewed as part of man's total history. You will note that the human population changed very little during the Old Stone Age, a period when man still depended on hunting and food gathering for subsistence. With the development of primitive agriculture and animal husbandry, man, though still dependent on the natural environment, exerted more control over it. Beginning around 8000 B.C., the population began a slow rise through the beginning of the Christian era and up to modern times. At that point, the graph shows a sudden, explosive increase, marking the birth of the population problem.

The population explosion is fundamentally a biological problem, since it involves the consequences of human fertility coupled with lowered death rates brought about by biomedical advances. Of added interest to the biologist is the relationship that cannot help but exist between the increasing mass of humans and the deteriorating natural environment. Of course, the population problem is enormously complicated by a tangled confusion of other factors, including the social aspects of human sexuality and the human family, as well as the traditions and morals imposed by the varied cultures of mankind. It also has definite political overtones, but vote-conscious politicians generally shrink from association with population-control measures and may possibly continue to do so well beyond the point of no return.

With increasingly few exceptions, it is generally agreed that the problem of exploding population must be solved while there is still food and space and before the quality of life on this planet is irreversibly degraded. In announcing plans for 1974 World Population Year, United Nations Secretary-General Kurt Waldheim said,

> It is impossible to think of solutions to the major problems confronting the world economic development, environmental pollution, improvement in the quality of life, even disarmament—without some reference to population trends. The evidence is all around us.*

---

* 1972 Annual Report, Planned Parenthood Federation of America.

The Arithmetic of Population Growth

At any given moment, new babies are coming into the world while some of the babies of yesteryear are leaving it. Life is a one-way road, its beginning marked by birth and its end by death. This simple and obvious fact that people are born and die is the fundamental basis of the arithmetic of population growth.

At a party, we have a room containing 100 people. If two people come in through the front door while two other people leave through the back door, quite obviously we still have 100 party-goers in the room. On the other hand, if *five* new people enter the room while only two leave through the back exit, we now have a net gain of three people, and the population number has become 103. Equal numbers of births and deaths therefore result in no growth and, of course, in no decline. An *excess* of births over deaths, however, will cause an increase in population size.

Let us begin by defining the so-called "crude" birth and death rates. The *crude birth rate* is the number of births per year per each group of 1000 individuals making up the total population. The *crude death rate* is measured by the number of deaths per 1000 individuals found in the population. Now if we let b and d represent the crude birth and death rates, respectively, it follows that

$$r = b - d$$

where r represents an annual rate of increase. In the unlikely event that d is greater than b, r would then be a rate of *decrease,* but this is not the problem that haunts us at the moment. It may also be seen that if b is *equal* to d, then the population number will remain constant. The problem at hand stems from the persistent *positive* value of r that results from the fact that b consistently exceeds d!

As an example, in mid-1975 it was estimated that the crude birth rate was 31.5 per each 1000 people making up the earth's population. The corresponding death rate was estimated at 12.8 deaths per 1000. Therefore, for every 1000 people living on earth in mid-1975, 31.5 new people were coming aboard annually, while only 12.8 people were leaving. This resulted in a *net gain* of 18.7 people per 1000. Therefore,

$$r = b - d$$
$$= \frac{31.5}{1000} - \frac{12.8}{1000}$$
$$= \frac{18.7}{1000}$$
$$= \frac{1.87}{100}$$
$$\approx 1.9 \text{ percent}$$

Note that r is expressed as a percent; in other words, a gain of 18.7 per thousand is equivalent to 1.87 per hundred, which is rounded to 1.9 percent.

This rate of growth (r) is comparable to the interest rate granted by a savings bank to its depositors. Suppose, for example, that you deposit $100.00 in the bank at 6 percent *simple* interest per year. In that case, your money will earn $6.00 per year, and the principal will therefore increase at a rate of $6.00 per year. At the end of five years, your original $100.00 would grow to $130.00.

More than likely, however, the bank will offer *compound* interest. This means that as the interest is credited to your account at the end of the first year, it will join with the original principal in earning interest during the second year, and so on. Interest, in other words, earns more interest. Under this plan, at the end of the first year you would have $106.00, *all* of which would then earn 6 percent interest during the second year, yielding a new principal of $112.36 at the close of the second year. By the end of five years, your original $100.00 will have grown to

$141.85, or considerably more than you would have received from a simple interest plan. To demonstrate the power of compound interest, suppose that you forgot that you had deposited the $100.00 and it remained in the bank for 200 years earning 6 percent interest compounded annually. At that time, some lucky descendant could collect over 12 million dollars! If only life were not so short we could all eventually become millionaires.

Leaving our dreams of riches and getting back to population problems, it may be seen that if births exceed deaths, the resulting annual rate of increase is comparable to an interest rate. Furthermore, since those people who are added to the population number will in turn produce more people, this rate of increase is equivalent to the concept of "interest earning interest," and is therefore *compounded*. Also, since reproduction goes on continuously in a human population, the rate of increase is compounded continuously. If we now apply a little algebra and calculus to the basic formula for compound interest, we can derive a formula for projecting future population numbers that is based on continuous compounding of a certain annual rate of increase. This projection formula turns out to be as follows.

$$N_t = N_0 e^{rt}$$

where N_t is the projected number in a given population t years from now; N_0 is the size of that population as of now; e is 2.718, a constant; r is the annual rate of increase $(b - d)$, and t is the elapsed time in years.

We can now illustrate the use of the above formula using data from the 1975 "data sheet" published by the Population Reference Bureau. The total population on earth as of 1975 was estimated at 3.967 billion, and the rate of increase as discussed earlier in this section was 1.9 per 100 individuals, or 1.9 percent. Now, if we wish to project the earth's population size in the year 2000 with 1975 as our starting date (a 25-year time span), we can apply our formula as follows:

$$\begin{aligned} N_t &= N_0 e^{rt} \\ N_t &= 3.967 \times 2.718^{(0.019)(25)} \\ &= 3.967 \times 2.718^{0.475} \\ &= 3.967 \times 1.61 \\ &= 6.387 \text{ billion} \end{aligned}$$

where N_0 is 3.967 billion, r is 1.9 percent, or 0.019, t is 25 years, and e is 2.718.

At this point, we should note that we have been using the word *project*, and that we have been avoiding the use of the word *predict*. There is an important difference between these terms. For example, given an honest coin we can *predict* that if we toss that coin 100 times, heads will come up approximately half the time and tails the other half. We can make such a prediction because the probability model based on an honest coin does not change, but remains forever the same. We can *not*, however, predict that a population will be at some definite size at a given future point in time, because r will not necessarily remain the same. The annual rate of increase, which is an important part of our formula, is dependent on a number of variables that may change in unpredictable ways during the next 10, 20, or 100 years. Our figure of 6.387 billion people by the year 2000 is therefore a *projection* based on an assumption that birth and death rates will not appreciably change during the period between 1975 and 2000. We are therefore obliged to admit that our projection could be drastically affected by factors such as shifting attitudes toward birth control or major medical breakthroughs that could significantly lower death rates.

By further manipulating the projection formula, we can determine the time that it would take for a population to double at a given rate of increase. If you happen to like mathematics, you might like to follow the steps below. If not, don't panic because the important thing is the application of the end result.

Our basic formula is

$$N_t = N_0 e^{rt}$$

and the idea of doubling can be represented by

$$\frac{N_t}{N_0} = 2$$

Now, since $N_t/N_0 = 2$ and $N_t/N_0 = e^{rt}$, it follows that

$$e^{rt} = 2$$

Also, since e, or 2.718, is the base of the natural logarithm, then

$$\ln 2 = rt$$

which means that

$$t = \frac{0.70}{r}$$

where t is the elapsed time in years, 0.70 is the approximate natural log of 2, and r is the annual rate of increase of the population in question.

The doubling rate of a population provides the most dramatic illustration of unlimited population growth because of the almost unbelievable effects of repeated doubling. As an example, suppose that you put a dollar in the bank on September 1. Then on September 2 you deposit two dollars, on September 3 you add four dollars, then eight, then sixteen, and so on each day for the month of September. On September 15 you would deposit $16,384. On September 25, you would add $16,777,216, and on September 30 your deposit for the day would be $536,870,912. Adding all of your daily deposits, the total amount that you would have in the bank on September 30 would be over one *billion* dollars!

If we now apply this principle of exponential doubling to the earth's human population, we obtain equally astonishing results. For example, using the current estimate of a 1.9 percent annual rate of increase, the present doubling time is:

$$t = \frac{0.70}{0.019}$$

$$= 36 \text{ years}$$

Therefore, assuming a 1975 world population figure of approximately 4 billion, 36 years later we will have 8 billion people on earth; 72 years from now we will have 16 billion. Projecting ahead (note that we are *projecting* because we are assuming that the present r will remain at 1.9 percent), the grandchildren of today's college student will be living on a planet crowded with 32 billion people, or 593 people for every square mile of the earth's *entire* land surface!

In case anyone remains unconvinced that Spaceship Earth does not have unlimited room, let us carry our projections further. If we project ahead for the same time that has elapsed since Columbus voyaged to America, we arrive at approximately the year 2450. At the current rate of doubling, we would have 65,536 billion people on earth. This breaks down to over one million people for every square mile of the entire land surface of the earth, and almost four people per square yard. Naturally we are including the Sahara Desert, the top of Mount Everest, the South Pole, and other equally unlikely places.

If that seems a trifle crowded, consider what we would have as far ahead in time as the Battle of Hastings (1066) is behind us. Just before the end of the twenty-seventh century, there would be over 15,000 people on every square *yard* of the earth's total land surface.

At least we won't be lonely!

Why Explosive Growth?

One clue to the origins of the population explosion can be found among the gravestones of an old cemetery. Stone after stone tell of those who died "ere half their years," and the graves of a husband and wife may be flanked by small stones that mark those of children who never had a chance to grow up (Figure 11.2). To the observant, the gravestones are weatherworn reminders of long-forgotten scenes—parents grieving for a lost child or a young husband mourning his bride of a year, dead in childbirth before she was twenty. They remind us that a horseman called Pestilence once rode among us unopposed, sweeping away many of those whose lives had hardly begun and before they could add their own children to the population. Death was never far away; it could come with the next case of diphtheria or pneumonia, or in the form of a typhoid organism that might lurk in a drink of sparkling, unchlorinated water. And across the world at large, malaria, the greatest killer of all, took a staggering annual toll.

Today, those of us in the developed industrial countries of the Western world take for granted the vaccines for diphtheria, whooping cough, and typhoid. We take for granted the wonder drugs which, among other things, have all but eliminated the necessity for the tuberculosis sanitarium, and the techniques of modern surgery that make light of appendicitis. DDT, cursed and maligned by the environmentalist, has significantly lowered death rates due to malaria by destroying the carrier mosquito. Because of the "medical revolution," many young physicians today have never seen a case of diphtheria. Armed with penicillin, they no longer sit helplessly at the bedside of a little girl while pneumonia takes the life-giving breath from her lungs. None of this is to say that Death has been conquered; it still waits patiently at the end of the road. For large numbers of people, however, modern medical science has put off the inevitable and drastically increased the chances that a given child will grow to maturity, reproduce, and thereby add to the population.

By lowering death rates, we have decreased the number leaving through the back door each year. Meanwhile, new party-goers crowd into the room as

Figure 11.2
Gravestones of a Revolutionary War veteran and his wife. Note the smaller monuments which mark the graves of their children who died before age 18.

birth rates in many parts of the world remain high. The value of r therefore remains positive, since $r = b - d$, and the fantastic power of exponential doubling takes the world closer to the brink of disaster.

Looking at the above formula ($r = b - d$), we can see two possible solutions to the population problem. First, the birth rate (b) could be lowered, thus reducing the annual rate of increase to zero. Or second, the death rate (d) could be increased; this would also decrease the value of r. Now, it is unthinkable that even the most ardent advocate of zero population growth would want to throw away the vaccines and miracle drugs and watch the cemeteries again fill with small gravestones. The "death rate" solution is simply not acceptable, and we must therefore turn to the "birth rate" solution as our hope for the future.

It has proven to be extremely difficult to bring birth rates down to a level compatible with zero population growth. It is generally true, however, that birth rates tend to go down in populations that acquire economic and educational advantages. In the United States, for example, there appears to have been a significant reduction in fertility in recent years; this has occurred despite opposition by the Church and with little or no help from the government. On the contrary, in 1972 the Presidential Commission on Population made an urgent appeal for action that was based on two years work at a cost of two million dollars. The report was largely ignored—a political potato obviously too hot to handle. It can probably be assumed that politicians up to and including the President of the United States will avoid significant involvement unless the public demands action.

Even in sophisticated and economically advanced societies, subtle pressures may operate to keep birth rates above replacement level. For example, a couple's desire for a boy (or girl) may lead to the birth of several children before the child of the desired sex is born. Or in the case of the unhappy marriage, instead of producing fewer children as one might logically expect, sometimes the wife has more children because she seeks solace in the nursery, knowing that she will always be needed and loved by her baby. In such situations, it is not unusual to find a new baby on the way by the time the youngest starts school.

The attitude still persists in our society that there is something "wrong" with a couple that has no children. Needless to say, this attitude is notorious among would-be grandparents, but it extends further. If the master of ceremonies of a television show should pass among the audience and interview John Doe, there is an embarrassed silence if Mr. Doe admits to no children after 15 years of marriage. On the other hand, Mr. Smith's proud declaration of 10 children is greeted by loud applause, in which the government graciously joins by granting the fertile Mr. Smith a healthy reduction in income tax! Now, there is no doubt that applause is well deserved by any parent who successfully performs the difficult task of raising healthy and happy children, but if the mere production of large numbers of offspring is all that counts, then the lowly fruit fly should bring down the house.

In some areas such as Latin America, the father of many children is still looked upon by some as having *machismo*, or manhood, with masculinity measured by the number of children produced. The traditional male attitude toward pregnancy and the problems of child care, usually expressed in more subtle terms than machismo, has no doubt contributed significantly to the present population problem. Statutory laws related to birth control issues have been made in the past by male-dominated legislatures, and strict Church laws that forbid more effective forms of birth control continue to flow from an all-male Church hierarchy.

To some people, children represent a form of self-fulfillment as well as a kind of immortality, a chance to see their own lost dreams come true through the accomplishments of their offspring. To others, a large family may be a psychological defense against a life of dull employment and social contacts. Thus as economic advantages increase and the "good life" becomes available through better education, there is a tendency for family size to diminish.

The present situation in the United States is an illustration of what can happen if higher economic and educational levels are combined with opportunities for family reduction in the forms of better contraceptive measures and relaxed abortion laws. In 1967, birth expectation data showed that wives in the 18-to-24 age bracket expected an average of 2.9 births before completing their childbearing years. Allowing for the factor of infertility, this would suggest that all women in the 18-to-24 age category would have completed their childbearing years with an average of about 2.6 births per woman. Referring to Figure 11.3, it may be seen that a projection based upon a fertility rate of about 2.5 births per woman (series D) would result in a total U.S. population of almost 350 million by the year 2020, when the average college student of today will be about 70 years old. This represents an increase of approximately 140 million people over the present number.

Similar data collected in 1971 and 1972, however, show a drop in birth expectations by the 18-to-24 age group to 2.1 births per woman. This fertility rate, considered the "replacement level," shows the effects of changing attitudes on the part of young couples toward family size. At this point, we need to inject some words of caution lest we celebrate too quickly. First of all, the birth expectations of young women may not match their actual fertility rate during their childbearing years, since good intentions are not always matched by results. Secondly, a "replacement level" fertility rate does not mean automatic zero population growth (ZPG), as some have rushed to conclude. The effects of replacement-level fertility depend on the *number* of fertile women producing offspring; Figure 11.3 (series E) shows that even this fertility rate will result in a total of about 275 million people by the year 2020, although this is admittedly an improvement over the series D projections. Still, it is important to emphasize that even if the fertility rate should remain at "replacement level," the brakes cannot be effectively applied to U.S. population growth for at least 65 to 70 years.

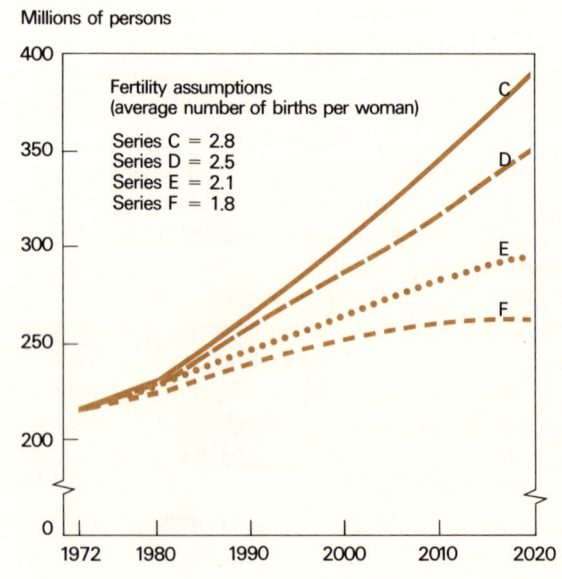

Figure 11.3
Total population projection for the United States, 1972 to 2020.

In addition, the natural rate of increase that results from an excess of births over deaths is not the only source of population growth in the United States. In 1886, the Statue of Liberty became a symbol of freedom and economic opportunity in the New World, and Ellis Island became the first stop for the "homeless and tempest-tossed" who came streaming in from Europe's "teeming shores." At that time, the resources of the United States appeared to be limitless, and there was plenty of room for all as the immigrants came to help fashion the superpower of today. As of now, 400,000 new immigrants still legally enter the United States each year, and some estimates place the total of legal *and* illegal entries closer to one million per year. The seemingly vast area of the country is deceptive, because most of the new immigrants travel to the cities, thus adding to the already insurmountable urban problems and in some cases swelling the welfare rolls. Immigration evidently accounts for about 20 percent of the normal annual growth of the U.S. population, and some Americans are wondering why they should be asked to limit their own families when four million or more new immigrants will enter over the next ten years. The President's Commission on Population did recommend more stringent quotas, but pressures from a variety of sources continue to be applied in the direction of maintaining or even increasing current quotas.

While there is hope of progress toward lowered birth rates in the United States and other countries where the economic and education levels are relatively high, there is little or no progress in large areas of the world where poverty and illiteracy contribute to continued high birth rates.

The situation in India is a classic case in point. As of now, India's population has surpassed 600 million people, all of them confined within an area that is one-third that of the United States. Hunger and malnutrition are commonplace, and 600,000 people live, sleep, and die in the streets of Calcutta. Deaths from malaria and cholera, once the leading killers of the people of India, have been significantly reduced, but the birth rate remains high. In 1973, the birth rate was 42 per thousand and the death rate was 17 per thousand, yielding an annual rate of increase of 2.5 percent applied to an already staggering total. The population increases by 14 million people every year, and at the present rate it will double every 28 years. At the present rate of increase, India's population will reach the impossible figure of one *billion* by the end of the present century. Meanwhile, attempts to bring significant birth control programs to the great masses of India's people have thus far met with little success, since they founder in a morass of poverty, illiteracy, apathy, and long-established traditions that tend to oppose voluntary limits on family size.

Latin America provides another example of uncontrolled population growth taking place in a setting of poverty and illiteracy. In 1973, the rate of increase in Latin America as a whole was 2.8 percent per year, while the rates of several individual countries were estimated at 3.4 percent per year. In Colombia poverty is a way of life, and desperate women often try anything from self-induced abortion to infanticide. Meanwhile, contraceptives are generally opposed on grounds of religious beliefs.

There have been several attempts to tie U.S. foreign aid to population control. This approach is politically unpopular due to pressures brought to bear by groups that are opposed to birth control. This approach would probably be ineffective anyway, since the success of an "exported" birth control program depends in large measure on the motivations, education level, and traditions of the target country. Even in the United States, for example, there are members of the black population who look with suspicion upon birth control programs, regarding them as a conspiracy to keep the black population from increasing. This attitude is understandable in the light of history, but it is also extremely unfortunate, since the large numbers of blacks who live at poverty levels in the large-city ghettos have the most to lose by uncontrolled population growth.

Consequences of Population Growth

Let us begin with a useful though admittedly oversimplified analogy. Suppose we have a pasture that can provide adequate food for as many as ten cows. We therefore say that ten cows represent the *carrying capacity* of the pasture; in other words, the pasture can provide sufficient nutrition for ten cows and no more. Now, it follows that if we put *five* cows in the pasture, the "quality of life" of those five cows will be high, and they will be well fed and contented. At this point, we might decide to increase our herd to *ten* cows, but since the carrying capacity of the pasture has still not been exceeded, the food supply should still be adequate.

If, on the other hand, we now increase the herd to *fifteen* cows, we will exceed the carrying capacity of the pasture and our cows will no longer be well fed and contented. At this point, we can rush in with fertilizers and extra irrigation, thereby increasing the grass yield and raising the carrying capacity. Thus encouraged, we continue to increase our herd, but soon find to our dismay that there is a limit beyond which we cannot increase the carrying capacity of the pasture. As the herd continues to increase, the cows compete for available food; the weaker animals die of malnutrition and its associated illnesses. Death has therefore intervened, bringing the numbers down to the carrying capacity of the pasture, which we are now likely to find will support even fewer than the original ten cows due to the damage done by overgrazing.

In one sense, the earth is a pasture that will support a "herd" of just so many people. It is difficult to say exactly what the maximum carrying capacity of the earth might be, since it depends on so many factors including man's ability to "rush in with fertilizers and irrigation." At any rate, we can see that the inexorable arithmetic of population growth must eventually take human populations well beyond the earth's carrying capacity despite efforts to bolster it through technology.

Even now, like too many cows in a pasture, mankind is by no means well fed and contented. Generally, those living in the industrialized Western nations enjoy a more-than-adequate diet, but even in the affluent United States, there are those who border on malnutrition. If we consider the world at large, however, *probably no more than a third of the four billion people on earth receive sufficient nutrition.* For the majority, therefore, malnutrition and associated protein and vitamin deficiency diseases are part of a way of life.

Despite the evidence that population growth is out of control, there are still individuals and groups, some with considerable political influence, that oppose broad applications of birth control measures. This opposition is usually rationalized by claims that (1) the population crisis is exaggerated by "doomsayers"; (2) there is a huge area of potentially arable land that can be brought under cultivation; (3) food production can be enhanced by the use of fertilizers, irrigation, and new high-yield varieties of grains, and (4) there are practically limitless resources from the sea.

All of the above alternatives to population control have been carefully examined by scientists and found wanting. For example, Paul Ehrlich has pointed out that most of the land that can be cultivated under present economic conditions is already under cultivation. He estimates that it would cost 28 billion dollars per year to open new land to feed just those people that are annually added to the population. Much of the now uncultivated land is in the tropics, but these apparently lush areas prove to be useless for agriculture because the soil becomes hard and bricklike when exposed to the sun.

It is generally agreed that it is cheaper and easier to increase production on land which is already under cultivation. The "green revolution," which is a term applied to the techniques that increase food supply by using fertilizers, irrigation, and new genetic varieties of cereals, can be helpful to a degree, but scientists are careful to point out that the green revolution is *not* a substitute for population control. Instead it is regarded only as a way to "buy time" while the brakes are being applied to population growth.

The food that is available from the sea is by no means limitless. Additional fish protein would also be a useful way of buying time, but the fishing fleets of the various nations of the world have already made significant inroads into some of the more desirable species, and the future will undoubtedly see more international squabbling over fishing rights such as occurred between Britain and Iceland in 1973. To those who would have the world subsist on algae from the sea, it should be pointed out that one cubic *meter* of sea water will yield on the average only one cubic *centimeter* of algae. Add to this the fact that most of the open sea is a "biological desert" and that only a relatively small portion of it can be counted on as a source of even such unappetizing foods as algae.

Malnutrition and starvation are by no means the only consequences of uncontrolled population growth. People are the most active and prolific polluters of the environment, and it therefore follows that more people means more pollution. This is especially true in areas where affluence brings more automobiles, more automatic washers, more factories, and a greater need for sewage treatment facilities that somehow never seem to keep pace with population increases. Even the benefits derived from the green revolution are mixed with the polluting effects of pesticides and fertilizer runoff. In Chapter 15, we will explore the problems of environmental deterioration more fully, and we will see that the "energy crisis" is also in part a result of overpopulation.

Those who oppose measures of population control often point out, and correctly, that plenty of wide-open space is still left on planet Earth. But people do not live in Antarctica, or in the Sahara Desert, or on the slopes of Mount Everest. One can travel through the Great Plains of the midwestern United States and see very few people living on that vast acreage. Instead, people tend to congregate in the large cities and suburbs that have combined to form the *megalopolis*. Megalopolitan living can be found along the eastern seaboard and in southern California, where one can travel for many hours through areas of extremely high population density. Figure 11.4 shows how the distribution of people in the United States shifted from 1940 to 1970. This illustrates the tendency to move to the cities and suburbs. We can probably assume that the vast majority of the 50 million or so *additional* Americans that will exist by the year 2000 will add to the already insurmountable problems faced by the large megalopolitan areas. As more and more people are added to already overcrowded areas, more city blight, more pollution, greater tension, and even higher crime rates appear inevitable.

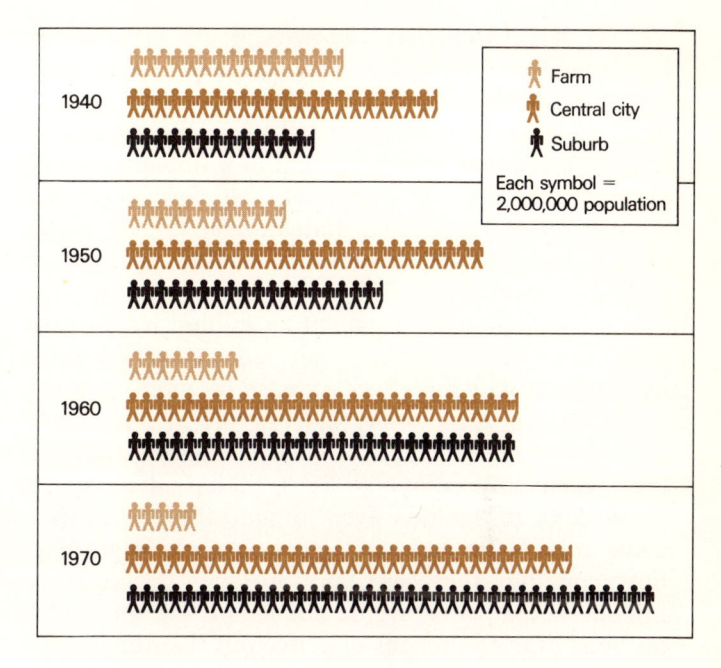

Figure 11.4
Shifts in U.S. population from 1940 to 1970. Note the tendency of people to concentrate in cities and suburbs. (Courtesy of the Population Reference Bureau, Inc., Washington, D.C.)

We have seen that decreased death rates resulting from the "medical revolution" have contributed to rapid population growth. In addition, advances in medical science have produced a side effect that is too often ignored when population problems are discussed. In 1974, approximately 10 percent of the population of the United States, or about 21 million people, were age 65 or older. Assuming *replacement* fertility rates, the number in the 65-or-older age bracket will increase to about 29 million by the year 2000. The "dependency load" consisting of those over age 65 will therefore increase by approximately 9 million over a relatively short period of time.

Today, many of our so-called "senior citizens" lead reasonably enjoyable lives, taking advantage of the variety of "golden ager" organizations that have come into being in recent years. For the most part, however, they represent the tip of the iceberg sticking out of the water; we do not see (or we ignore) the millions of aged persons who live out their final years in lonely rooms or substandard nursing homes, forgotten by society and too often ignored by the medical science that helped them attain the dubious joys of old age. To these people, the term "golden ager" is a cruel joke.

If we go through the doors of the best of nursing homes, we enter another world. It is a world of wheelchairs, of loneliness, of the forgotten. We see a former corporation executive shuffle down the hall, satisfying a need to be useful by delivering mail to his fellow patients. We see an old man tearfully complain that his children never visit him because his aged brain denies him the pleasure of knowing that they left only five minutes before. We see a woman painfully move around her room in a wheelchair, packing her belongings as part of a daily ritual in which she prepares to go back to a home that no longer exists. We see the boredom, the loneliness, and the isolation that hasten deterioration of the mind. What may be worst of all, we see the loss of dignity that is perhaps the greatest curse of the aged. This is the other side of the coin of the miracle drugs and it constitutes one of the major challenges of modern society.

When we take our leave of the home, it is with a sigh of relief because these people from another time make us strangely uncomfortable. Perhaps it is because we are sorry that we cannot really help, or perhaps we see in them what we may someday become. As we go through the door, we are disquieted by the feeling that we may someday return.

What About the Future?

To most biologists, it is strange that any controversy over the critical need to stop population growth should even exist. The arithmetic of population growth says it all. For example, can we possibly arrive at a situation 500 years from now when four people live on every square yard of the entire land surface of the earth? This is obviously absurd, but just as obviously, if we do not drastically reduce birth rates, then death rates will rise to solve the problem long before 500 years have elapsed.

The present situation in the United States is encouraging. A Gallup poll in 1968 revealed that 41 percent of Americans regarded the "ideal" family as consisting of four or more children. This figure has since dramatically decreased, probably as a result of a combination of factors including economic conditions, better forms of contraception, changing attitudes of women toward childbearing and childrearing responsibilities, and an increasing public awareness of the problems related to the population explosion. As of this writing, the fertility rate in the United States appears to be down to replacement level, and if this holds, it should at least make a small dent in the problem.

Even so, the United States is not an island. We are all passengers on Spaceship Earth and we must ultimately share in the total problem that is sure to result from continued worldwide population growth. The annual rate of increase must be reduced without delay, hopefully by bringing about a reduction in birth rates which can only be achieved by worldwide application of birth-control measures. Otherwise the "death-rate solution" will be the only alternative, and those dread horsemen of the Apocalypse—Pestilence, War, and Famine—will once again ride roughshod over the earth.

Man must come to realize that even a minimum quality of life on this planet is in peril. If the human population is limited, we will at least have a chance to live with ourselves and with the other forms of life that evolved with us from those primeval seas of long ago. On the other hand, if we continue to allow the population to double and redouble, we will destroy the environment. We will deplete our sources of food and energy, and it may not be an exaggeration to say that we will set the stage for our own extinction.

Questions

1. Identify and discuss some of the social pressures and traditions in American society that influence our personal population-related decisions. What are some of these decisions that affect population?

2. Discuss the good and the bad effects of population control on minority groups such as Puerto Ricans, Mexican Americans, and black Americans.

3. Show how births, deaths, and immigration combine to produce a population growth condition.

4. Define and carefully differentiate between the terms *projection* and *prediction*. What are some of the factors that prevent us from predicting future population levels?

5. Some Americans resent their being asked to limit their own families while nearly half a million immigrants are permitted to enter the country each year. Do you advocate cutting off immigration, or do you think the present policy should be continued? Defend your position.

6. Why are the problems of population distribution almost as important as the problems of population growth itself? In what ways are the two related?

12

Psychobiology

Introduction

Imagine for a moment that medical science has perfected the art of transplanting the brain from one person to another. Now let your imagination go further and suppose that you have been in a serious accident that resulted in the loss of both legs and irreparable internal injuries. Your brain is intact and healthy but the doctors agree that survival is impossible.

Fortunately for you, in the same hospital there is another person of the same sex who has suffered severe head injuries. His (or her) brain function is destroyed and he is considered clinically dead even though heart and lung functions are being temporarily maintained by mechanical life supports. You are both rushed to the operating room where a medical team is preparing to transplant your healthy brain to the body of the donor. The operation is successful, and when the mists clear you are back in your hospital bed. Or is it the other person who has come back from the operating room? Who are you?

When you are well enough to move around, you stand before a mirror and look at an apparition that stares back with different eyes from a face that is not yours. You touch the strange face with wondering fingers that are not your own and you look down at a body that you do not recognize. Is the person in the mirror *you*, or is it the other person? Or are you now some kind of mixed-up Frankenstein creation made from two people whose lives briefly crossed under the operating room lights? Who died, you or the donor of the body in the mirror? The body that was *yours* has already been buried, but were *you* buried with it? Were the usual prayers and eulogies of a funeral service necessary or even appropriate in this case?

You close your eyes as you ask again, "Who am I?" Then you realize that in spite of the fact that you would be unrecognizable by family and friends, it is indeed *your* personality that survives. Your thoughts, memories, dreams, plans, desires, loves, and hates are there as before. It was the other personality who ceased to be, dying with the 3.5-pound, pinkish-gray mass of protein, fat, and water that had lain hidden within the bony crypt of his or her skull.

At this point we must regard the foregoing as just a bit of science fiction; but then it is difficult to know what the future might bring. After all, not so long ago a similar account of a heart transplant would have been considered just as fanciful. At any rate, the point of the story is clear. The brain is the seat of personality and behavior, and it is the single greatest contribution to the very concept of being human. More than any other organ, it is responsible for the great gulf that separates man from his closest relative in the animal world, and like the gift of fire, it can be servant or tyrannical master, a mixture of blessing and curse.

Considering the intimate relationship of the brain to the human condition, it might seem incredible that we know less about its functioning than we know about the depths of the sea or a star light-years away in space. Actually it is only recently that man has turned loose his brain's great powers on the problem of understanding itself. One of the great problems has been and continues to be the unbelievable complexity of the "supercomputer" that we carry around in our heads. Sir John Eccles, Nobel Laureate and one of the foremost of the brain researchers, feels that the human brain represents the most highly and most completely organized matter in the entire universe.

It is little wonder, therefore, that the brain has become one of the most exciting and challenging of the new frontiers of scientific research. Vast amounts of literature are produced annually as scientists from a variety of disciplines step up the attack, but even so, our knowledge is far from complete. It is already apparent, however, that by increasing our understanding of the structure and function of this uniquely human organ, present and future brain research will have profound significance for our problems and behavior.

The Neuron—The Basic Unit of Behavior

The fundamental unit of the nervous system is a highly specialized cell called a *neuron*. Neurons, or nerve cells, exist in a variety of shapes and sizes (Figure 12.1), but the general plan of the typical neuron is illustrated in Figure 12.2.

The cell body contains the usual cell organelles that were described in Chapter 3, including mitochondria, Golgi bodies, lysosomes, an endoplasmic reticulum, and a nucleus. A centriole is rarely found in mature neurons. This is significant because once the nervous system is fully developed, nerve cells do not reproduce by mitotic division. Neurons are therefore not replaced as they are lost through injury, the

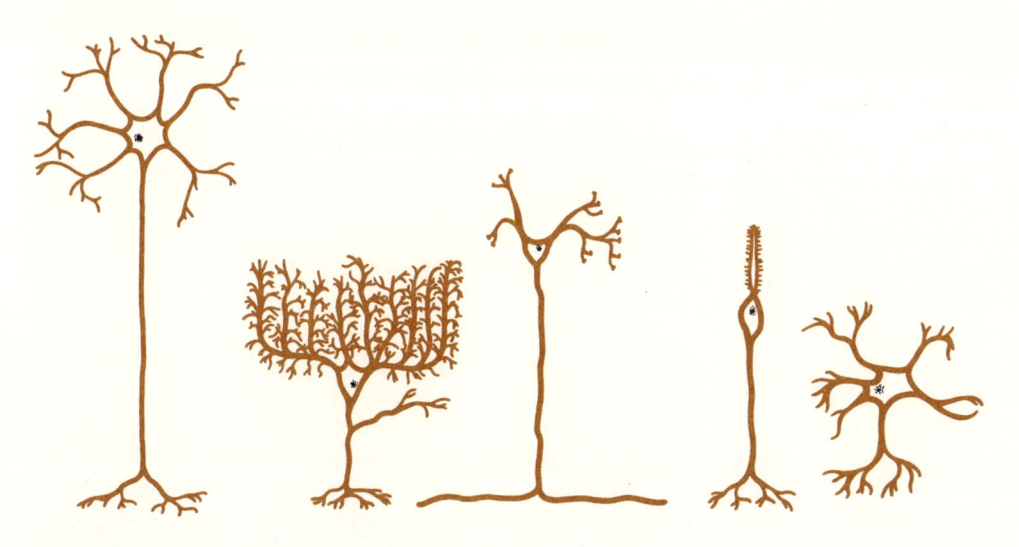

Figure 12.1
Diagram showing various shapes and sizes of neurons.

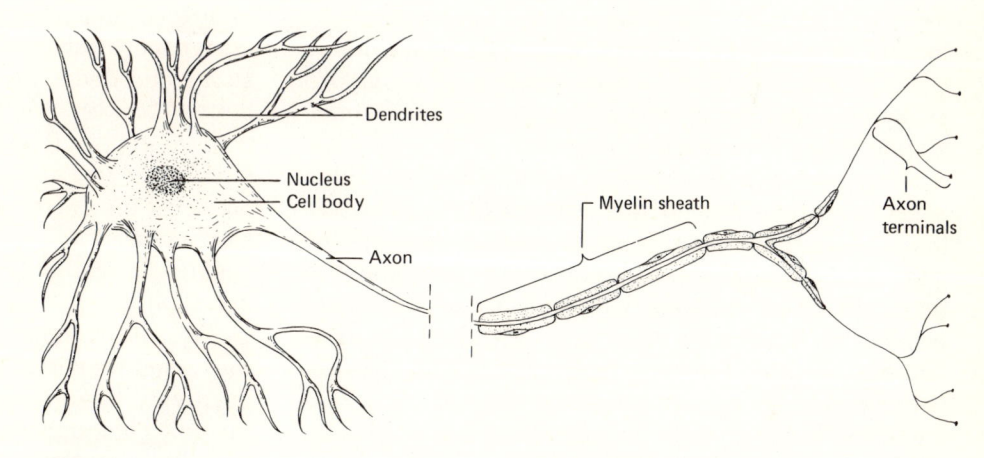

Figure 12.2
Diagram of a typical neuron, showing the cell body with nucleus, dendrites, and axon.

effects of disease or drugs, or through the aging process.

The cell body membrane continues outward into branches called *dendrites* that extend a short distance from the cell body. Dendrites carry nerve impulses *toward* the cell body. The long, sheath-covered, more sparsely branched process that extends from the cell body is the *axon*; it carries impulses *away* from the cell body to muscles, glands, or to other neurons. The axon of a given neuron may be very short or it may extend up to several feet, as in the case of those which carry impulses to and from the lower extremities.

A *nerve* is a bundle of axons surrounded by connective tissue, much like a coaxial cable consisting of separate wires bound together and wrapped with insulation. Just as a telephone cable contains wires that carry messages in both directions, a nerve consists of both sensory and motor fibers. *Sensory fibers* carry impulses from *receptors* to the spinal cord and brain. Receptors include those cells for pain, heat, cold, and pressure which are located under the skin, as well as such major receptor organs as the eye, the ear, and odor detectors in the nose. Meanwhile, motor fibers carry impulses from the brain and spinal cord to *effectors* such as muscles and glands. The typical nerve is a white, stringy structure, an example of which can easily be seen running along the muscles of a Thanksgiving turkey drumstick.

The human brain contains an estimated 10 billion neurons. In addition to the neurons themselves, the brain tissue is composed of another 100 billion *glial* cells. The glia appears to be the structural "glue" of the brain. Glial cells are wrapped around the axons of neurons, especially those that extend up the arms and legs and into the body trunk. These glial cells appear to play an important role in the conduction of impulses, because their loss may be associated with severe disorders of the nervous system such as multiple sclerosis. Multiple sclerosis (MS) is a very serious disease that occurs when the axon sheath (*myelin*) degenerates in patches throughout the brain or spinal cord, or both. This interferes with the axon's ability to conduct nerve impulses, resulting in a number of symptoms including severe weakness and jerking movements of the legs and arms. Multiple sclerosis generally attacks young adults, and unfortunately there is no cure. The MS patient usually lives a considerable number of years, confined much of the time to a wheelchair.

Other glial cells are wrapped around blood vessels in the brain, forming a barrier to the diffusion of materials from the blood into the brain tissue. This "blood brain barrier" quite possibly affords the brain an extra degree of protection against invasion by microorganisms. Unfortunately, it also makes it more difficult to treat brain infections with antibiotic drugs. In addition, the white blood cells that are normally a natural line of defense against disease organisms have difficulty crossing the blood brain barrier to attack the infection at its site.

The Nerve Impulse

The nerve impulse is both electrical and chemical in nature. It depends primarily on the existence of a difference in charge potential that extends along the axons and dendrites as they branch out from the cell body. This difference in potential is due to a concentration of positively charged particles outside the membrane and a concentration of negatively charged particles inside the membrane. The neuron can thus be compared to a small battery in which the inside is negatively charged with respect to the outside (Figure 12.3).

When a neuron is stimulated, a tiny segment of the membrane suddenly becomes permeable to the positively charged particles, which then rush to the inside. This appears to affect the permeability of the adjacent segment, and so on. Thus the impulse flashes along the nerve fiber like a flame traveling along the fuse on a stick of dynamite (Figure 12.4).

Figure 12.3
An axon at resting potential. There is a greater concentration of positively charged particles outside the membrane than inside. Large negatively charged particles provide the negative charge inside the membrane.

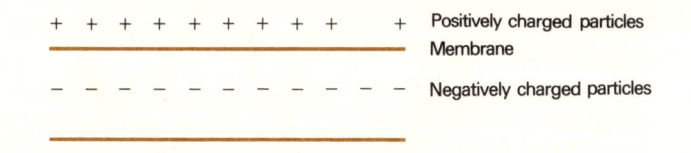

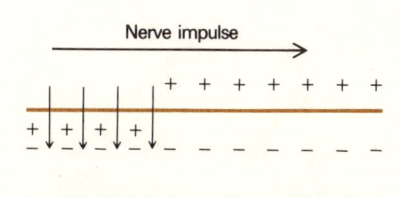

Figure 12.4
Transmission of an impulse along an axon. The positively charged particles enter, depolarizing the membrane. This depolarization effect moves swiftly along the axon.

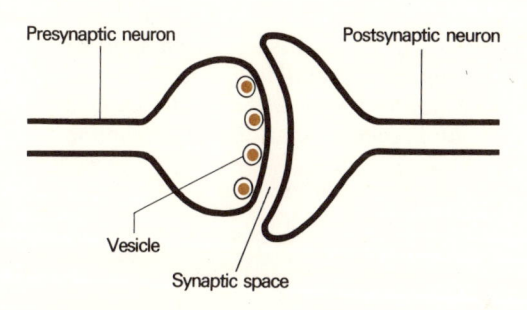

Figure 12.5
General structure of a synapse. The vesicles contain a neurotransmitter.

Such impulses may travel along axons at speeds great enough to cover the length of a football field in one second!

The Neuron Synapse—The Switch

A neuron does not exist in isolation. Along with other neurons, it is part of a very complex "circuitry" system through which impulses travel in an infinite variety of patterns that produce thoughts, memory, and the various aspects of behavior. A single neuron, for example, may be acted upon by hundreds of others.

Nerve impulses travel from one neuron to the next through a system called a *synapse*. Synapses may be roughly compared to the switches in a complex system of electrical circuits that permits the operator to make the circuits perform a variety of functions. Figure 12.5 illustrates the more general structures of a synapse. Note that the axon of the first neuron does not actually touch the dendrite of the second neuron. For many years, the transmission of a nerve impulse across this synaptic space was a matter of great mystery. At one time, for example, it was thought that the impulse "sparked" across the space much as a high-voltage electrical current leaps from one electrode to another like a bolt of lightning. Later research, however, revealed that the impulse travels across the synapse by chemical action (Figure 12.6). The end of the axon secretes a chemical called a *neurotransmitter* which moves across the space and stimulates the membrane of the dendrite or cell body of the adjacent neuron. The most common and best known neurotransmitter is *acetylcholine*, but others found especially in the brain include *epinephrine, norepinephrine, serotonin,* and *dopamine*.

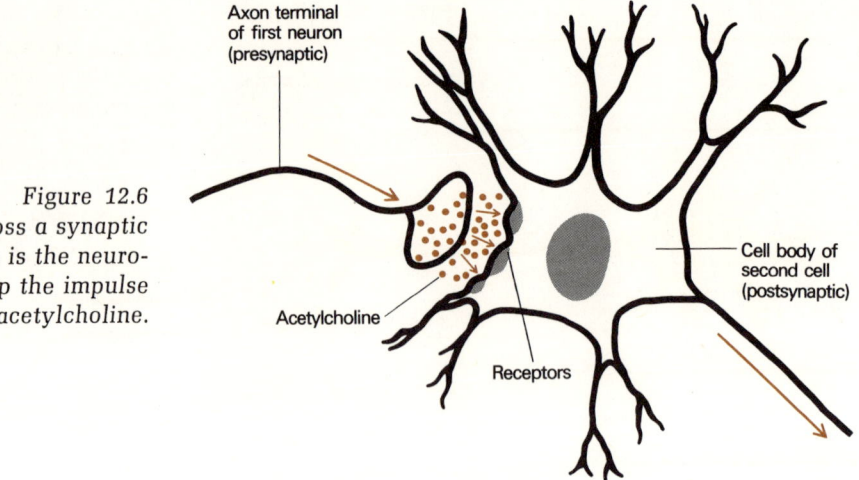

Figure 12.6
Transmission of an impulse across a synaptic space. In this case, acetylcholine is the neurotransmitter. Cholinesterase will stop the impulse by breaking down the acetylcholine.

A few years ago, scientists discovered that an inadequate supply of dopamine in the brain is associated with *Parkinson's disease,* which is a severe degenerative condition that usually begins at middle age, becoming progressively more serious as the patient grows older. The symptoms of Parkinson's disease include severe tremor of the hands and progressive stiffening of the limbs and neck. Speech and swallowing may be affected, and the patient develops a characteristic blank facial expression. Although the patient's mind usually remains clear, he becomes progressively more helpless and may spend years confined to a wheelchair, unable to attend to his simplest needs without help.

It was found that the brain of the Parkinson patient is unable to synthesize a chemical called L-Dopa, which is one of the stages in the production of dopamine. It was then discovered that when patients were given L-Dopa, there was in some cases dramatic improvement. A few even improved to the point where they were able to resume reasonably normal activities. Unfortunately, further experience has shown that L-Dopa is not effective in all cases of Parkinson's and that in many cases the improvement seems to be only temporary. In addition, a persistent problem has been the difficulty in getting L-Dopa to cross the blood brain barrier in sufficient quantities to do any good and still not produce side effects such as nausea.

Once the neurotransmitter has initiated an impulse in an adjacent neuron, the chemical must be quickly destroyed. Otherwise, quite obviously, the "switch" would be left on, causing unwanted impulses to continue to travel along the neuron. In the case of *acetylcholine,* this problem is solved by an enzyme called *cholinesterase* which breaks down acetylcholine into acetic acid and choline in a small fraction of a second, thus preventing further impulses from crossing the synaptic space.

A structure similar to the synapse between neurons is found at the *neuromuscular junction* (Figure 12.7). This is the point at which a nerve impulse is transmitted to a muscle fiber, causing the fiber to contract. As before, acetylcholine is released at the axon terminal, but in the neuromuscular junction it initiates an

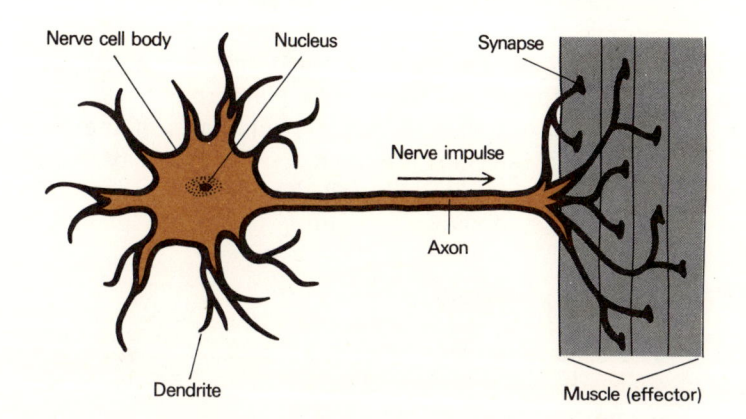

Figure 12.7
A neuromuscular junction. Nerve impulses cause acetylcholine to be secreted at each of the synapses in the muscle fiber, resulting in contraction of the fiber.

"action potential" over the surface of a muscle fiber. As in the interneuron synapse, the acetylcholine is rapidly broken down by cholinesterase.

Myasthenia gravis is a severely weakening disease that is associated with the neuromuscular synapse. In this condition, the effect of acetylcholine on the muscle fiber is significantly reduced. As a result, the patient suffers from extreme weakness and even paralysis. Treatment with *neostigmine,* a drug that inhibits the breakdown of acetylcholine by cholinesterase, greatly increases the concentration of acetylcholine at the muscle fiber and often brings a dramatic improvement within minutes after the drug is injected.

The Command Post—The Brain

The brain lies at the upper end of the spinal cord, cushioned by membranes and fluid within the protective vault of the skull. No other organ, not even the heart, is so jealously guarded against injury, which is fortunate since the brain's fragility matches its importance.

The living brain is a pinkish-gray mass of jellylike material consisting largely of water, fats, and proteins. It weighs about 3.5 pounds in the adult human male and somewhat less in the female. Lest male chauvinism should receive a boost from this fact, we should hasten to add that the female brain tends to be smaller because her overall stature is less than that of the male. Her brain weight relative to her body weight matches that of the male, and her intelligence most assuredly does. The subordinate role that women have traditionally played in the business, political, and professional worlds has never reflected a lesser intelligence, nor even a different kind of intelligence than that of males. Her role was determined long ago in the caves of primal man, when her inescapable function of childbearing held her by the cookfire and later on in the kitchen. In fact, up to relatively recent times, marriage automatically meant years of childbearing with no choice but to bear the attendant responsibilities. Now, with effective methods of contraception, changing social attitudes, and labor-saving household devices, the modern woman has a choice. She is rapidly coming into her own in competition with what was in the past assumed to be the "superior" male intellect.

The brain (Figure 12.8) is separated by a deep fissure into left and right halves which are connected by the nerve fibers making up the *corpus callosum.* The *brain stem* is continuous with the spinal cord and consists of the *medulla oblongata* and the *pons.* Directly behind the pons is the *cerebellum.* Above and in front of the brain stem are situated the *thalamus* and *hypothalamus* areas, which are essentially masses of gray matter or neuron cell bodies.

In the human being the greater part of the brain mass consists of the *cerebrum,* which is most highly developed in man and widely separates us from other animals in terms of intellect and behavior. Figure 12.9 shows the differences in cerebral development that exist among various groups. In general, the cerebrum follows the general contour of the thalamus and appears to have developed from it and around it like a giant puffball (see Figure 12.8). Numerous nerve fibers connect the thalamus with various parts of the cerebrum, and the thalamus appears to be a relay station between the brain stem and the cerebrum, especially with regard to sensory stimuli.

Cerebral Functions

The brain has often been compared by frustrated researchers to a "black box" that mysteriously performs functions so complex as to defy imagination. It has been said that man finds his own computers

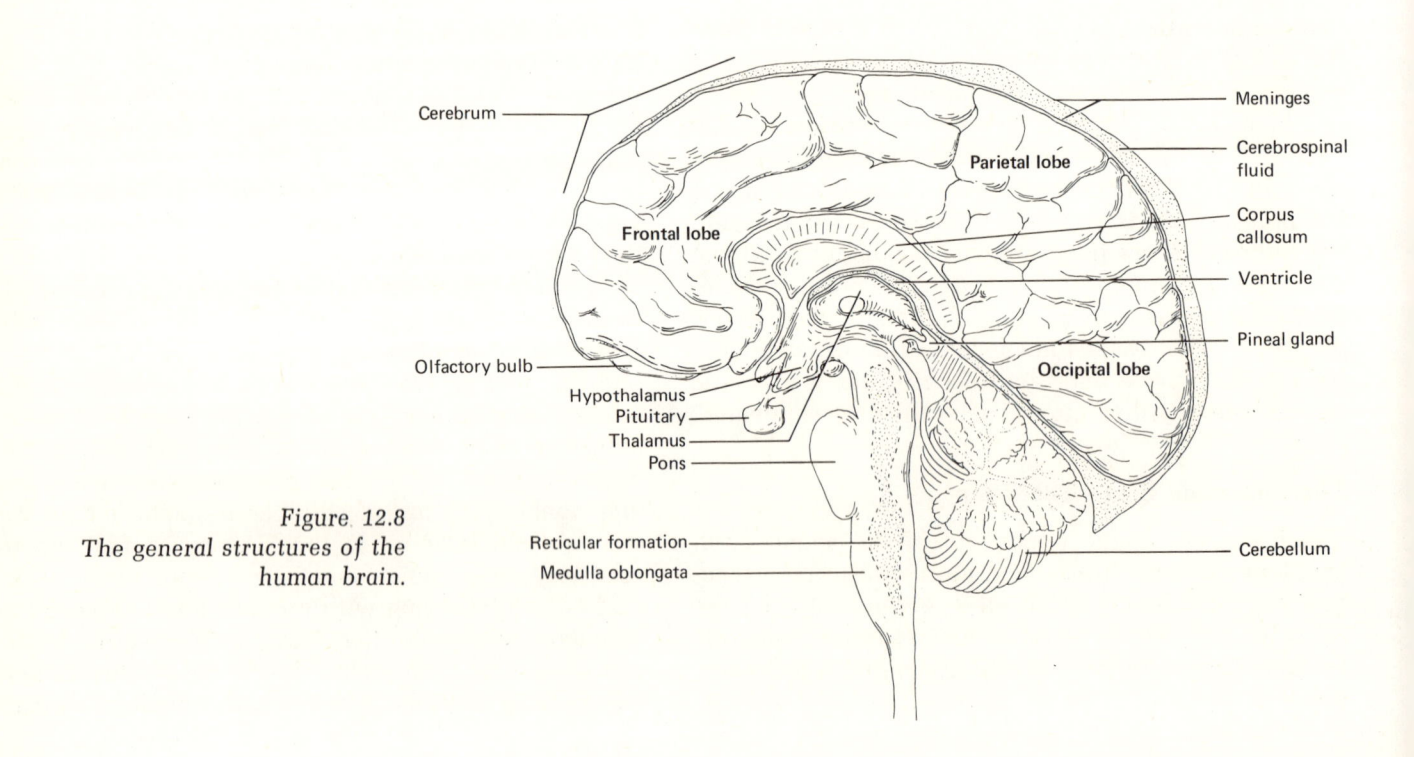

*Figure 12.8
The general structures of the human brain.*

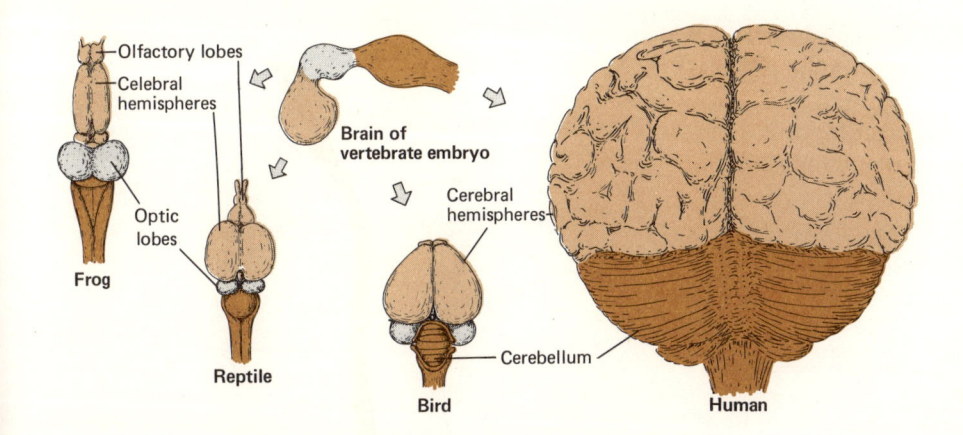

Figure 12.9
Diagrams of various animal brains. Note especially the differences in cerebrum development.

easy to understand because he *made* them, but he has only scratched the surface of the workings of a brain that allows him to build giant, sophisticated computers that are like primitive toys when compared to that "supercomputer" produced by the random and aimless processes of evolution.

The special and distinctly human functions of the cerebrum are certainly among the least well understood, and the importance of gaining further understanding of its functions is obvious when we consider that such knowledge could give us a tremendous amount of insight into the problems of the human personality. This is reason enough why brain research must be given a priority comparable to those assigned to cancer and heart research.

The split-brain studies done by Robert Sperry and his associates have shown, for example, that the cerebrum is actually two brains connected by the bundle of 200 million or so fibers that make up the corpus callosum. When these fibers were cut in the brains of cats and monkeys, it was discovered that any tasks learned while the right eye was covered had to be completely relearned when the left eye was covered instead. In fact, it was even possible to "teach" each of the two separated hemispheres a different solution to the same problem! The Sperry group therefore concluded that the two halves of the cerebrum are capable of independent learning. Later, while working with human patients whose corpus callosums had been cut in an effort to treat severe epileptic seizures, it was found that these patients were able to orally describe objects felt with the right hand but were unable to describe those felt with the left hand. Since the right side of the body is connected to the left brain and the left side to the right hemisphere, Sperry and his group cited this as evidence that the speech centers are located in the left cerebral hemisphere.

Psychologist Robert Ornstein found that the left brain appears to be associated with speech and analytical tasks, while the right brain appears to rule over intuitive behavior and spatial relationships. Normally, of course, the two halves work together as information is fed back and forth through the corpus callosum. In keeping with the supposed different functions of the two hemispheres, Ornstein found that lawyers have more active left brains, while in artists the right brain seems to play the greater role. Ornstein suggests that Western educational systems tend to emphasize the left hemisphere and too often ignore the right half with its flashes of inspiration which are perhaps the true basis of genius.

222 Psychobiology

One of the better known aspects of the cerebral cortex is its role in motor activities. These include activities which result from muscle contractions and they include walking, general body movements, and speech. Figure 12.10 shows the motor area located immediately in front of a deep groove or fissure called the *fissure of Rolando*. This area has been extensively mapped by the procedure of stimulating various

(a)

(b)

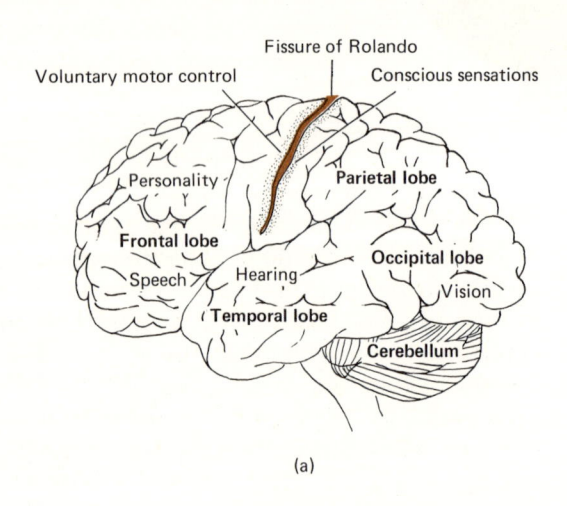

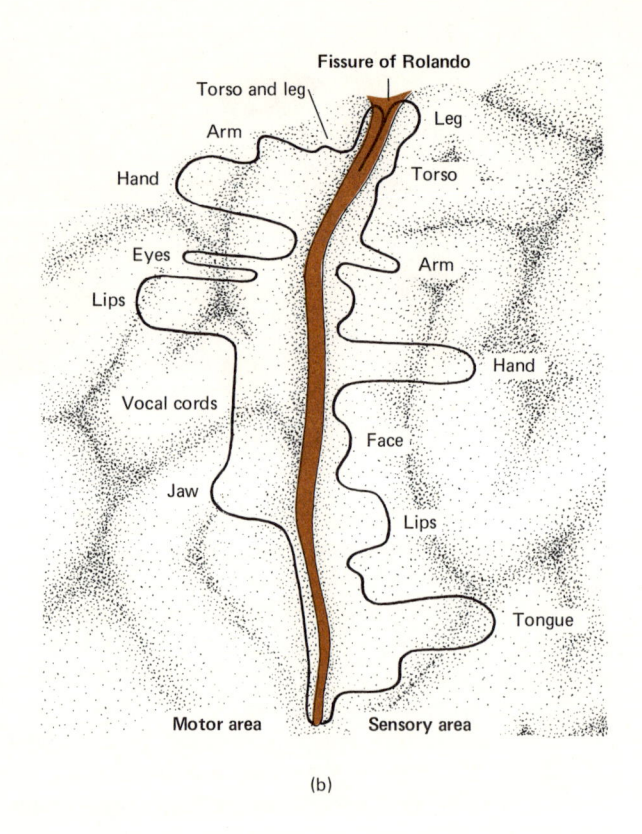

Figure 12.10
Motor areas of the cerebrum are located in front of the fissure of Rolando; sensory areas are found behind the fissure of Rolando.

points of the human brain with electrodes and noting the part of the body that responds. The diagram of the distorted little man (Figure 12.11) shows the relationships between divisions of the motor area and the associated body localities. Note especially how the mouth, tongue, and other speech parts are assigned an unusually large area of the motor cortex. It is believed that this ability to control the necessary motor functions of speech is unique to the human brain, since even if lower animals can formulate an idea, they still are not able to perform the motor activities necessary to put the idea into words.

The cerebral cortex also contains sensory areas that interpret impulses carried up the spinal cord by way of sensory nerve pathways. Sensory areas are therefore those in which sensations such as light, heat, cold, and pain are interpreted. These sensory impulses reach the thalamus first, and it is the thalamus that initiates the interpretation of specific sensations. Other pathways, however, lead to that portion of the

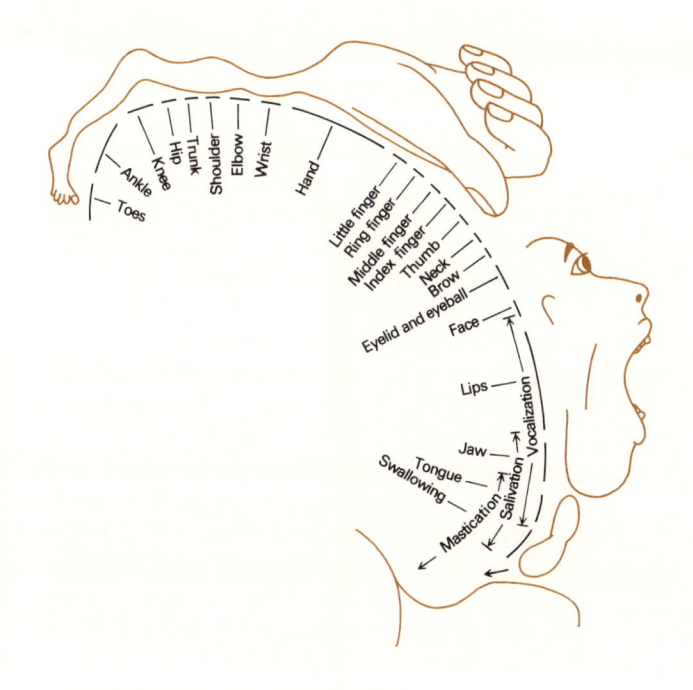

Figure 12.11
"Distorted man" diagram showing areas of the body that are associated with areas of motor control in the cerebrum. Note the large area of the human brain that is involved with hand and mouth movements. (Adapted from J. E. Crouch and J. R. McClintock, Human Anatomy and Physiology, *New York: Wiley, 1971, p. 547.)*

cerebral cortex just behind the fissure of Rolando, and it is in the cortex that finer interpretations of the sensory signals are made.

What about the rest of the cerebral cortex? What functions are performed by, say, the frontal lobes? These are called the "silent" areas and their functions are still for the most part unknown. Some authors associate the frontal lobes with "moral values" or "moral decisions." Unfortunately, the term "moral" is rather difficult to handle and it might be better to speculate that the frontal lobes store concepts of behavior that are generally required of us by that great mass of other people we call society. We learn early in life the do's and don'ts of our particular culture and we develop inhibitions against behaving in certain ways. In recent years, "inhibition" has become a bad word in some circles, but while it is true that too many or the wrong kind can be harmful to a person's mental health or personality, it is probably well to remember that a healthy set of behavioral inhibitions is the lubricant of the machinery of human relationships.

Memory. A middle-aged man who has spent his adult life in the city is driving down a country road when the scent of new-mown hay enters the open windows of the car. Suddenly the years fall away and a long-forgotten scene from his farm boyhood flashes through his mind. He is once again in the hayfield watching his father pitch the sweet alfalfa onto the wagon, while the ancient team patiently waits for the command to move on to the next mound of the hay that is to be their winter feed. One scene gives way to another with the vivid detail of a slide show, and he is perched atop the swaying wagonload as the horses plod through the apple orchard, its trees loaded with their promise of the fall harvest to come.

How can a seemingly insignificant clue such as the scent of new-mown hay, the distant voice of a crow, the sound of a Christmas carol, or the taste of a certain food open the floodgates of memory and bring to life scenes of forty years ago? Why is it that you cannot recall what you had for lunch two days ago, yet you can remember the details of your first day at school, your first date, or the death of a friend or member of your family. What happens to the sensory experiences that are carried to the cerebral cortex? Are most of them fleeting impressions to be lost within minutes, or are they stored there like old movies ready to be rerun when the right clue brings them to mind? Experiments performed during brain operations suggest that the latter is true, at least in part. When certain areas of the cortex were stimulated, patients (fully awake because no pain results from cutting into the brain itself) have reported voices and scenes from the past that were supposedly long forgotten.

In the very elderly, where there is often a problem of failing memory, there is sometimes an inability to recall one's own name or the name of a son or daughter. This loss of memory in the aged is often a result of changes in the blood vessels that supply the brain, changes that cause neurons to fail due to a lack of oxygen and nutrients. In addition, a substance called *lipofuchsin* has been noted to collect in neuron cell bodies of the elderly; whether this signals degenerative changes due to the aging process is still unknown. An especially puzzling phenomenon of the aging brain is a tendency for the very elderly to recall in vivid and minute detail events that occurred seventy years ago but be unable to remember even important events that happened only minutes before. One elderly nursing home resident, a former musician, can regale his listener with detailed and fascinating stories of his experiences on the show circuit that go back to the turn of the century. He cannot, however, play the piano in the nursing home because he cannot remember where it is! Evidently the aging brain tends to retain old and well-established long-term memories but is unable to form new memories due to death of neurons or changes in their ability to function.

Memory appears to be a function of the cerebral cortex. Unfortunately, our knowledge of how the cortex processes and stores information is relatively limited, although intensive research in this area has produced some interesting if often conflicting results.

Current theory favors the separation of memory into short-term and long-term varieties. For example, it is suggested that *short-term memory* involves the routing of impulses around and around a circuit of neurons—a so-called reverberating circuit—lasting from a few seconds up to an hour or so. For example, if you look up the telephone number of a local store, you will probably remember it long enough to put the call through, but afterward you will not be able to recall it. Some experts think that short-term memory is a kind of protective mechanism that keeps the brain from becoming cluttered with trivia.

How does short-term memory move to the level of long-term memory, and how is the information stored in the brain's ten billion neurons? And by what mechanism is this information called out by an appropriate stimulus like a computer program at the touch of a button?

One school of thought holds that *long-term memory* somehow involves permanent changes in the neuron circuits, probably as a result of strengthening certain synaptic connections and thus forming specific pathways which represent memories. Some scientists feel that this strengthening of synaptic connections is accomplished by the release of chemicals; the results of a number of well-publicized research studies appear to support such a chemical basis for memory. One study done in the early 1960's by the Swedish scientist Halger Hyden showed that the concentration of RNA in brain cells changes significantly during learning activities. Recalling that RNA acts as a pattern for protein synthesis, Hyden's work suggests the possibility that "memory proteins" are synthesized during learning. In 1970, Dr. Georges Ungar reported that he had isolated a specific substance from the brains of rats that he had trained to be afraid of the dark. Ungar identified the substance as a polypeptide consisting of fifteen amino acids, which he named "scotophobin," meaning "fear of the dark." The most interesting part of Ungar's work, however, is his claim that when "scotophobin" was injected into mice, they almost immediately developed a similar fear of the dark, supposedly indicating a direct transfer of memory in the form of "memory molecules." Ungar's work has come under strong attack by other scientists who have criticized both his methods and his findings. His results, though interesting, are therefore not considered conclusive.

The chemical theory of memory sounds logical but it runs into difficulty when we consider the findings of Dr. Karl Lashley. Lashley trained rats to run mazes, and then he removed various parts of the animals' cortexes in an attempt to isolate the part of the cortex that contained the memory. To his surprise he found that no matter what part of the cortex was removed, the animal retained at least part of the memory. He went on to find that the degree of memory loss depended on the total amount of cortex removed, not on the location. Since then, Dr. E. Roy John has criticized the idea that learning establishes new pathways among the neurons. Dr. John favors a "statistical" theory of memory which holds that the learning process involves the *average* firing pattern of a group of neurons rather than what happens to individual neurons. Therefore, according to Dr. John, the recall of a specific memory involves a unique wave pattern throughout various regions of the cerebral cortex. Unfortunately, such electrophysiological theories still do not account for the way by which long-term memory is established.

Scientists therefore do not agree as to whether memory is chemical, electrical, or both. At this point, the final answer to this fascinating mystery still lies hidden in the crumpled gray matter of the cerebral cortex.

The Cerebellum

The *cerebellum* (Figure 12.12) has long been known to play an important role in balance and coordination. Essentially, we are able to stand upright and maintain that position without "thinking about it" because the cerebellum constantly monitors the stretch of muscles and the position of the body through receptors located in the muscles and in the inner ear. Thus the cerebellum constantly sends impulses to appropriate muscles in order to maintain balance. In a similar

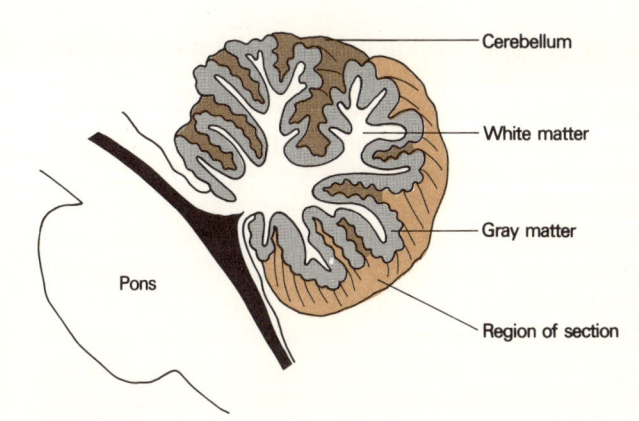

Figure 12.12
The cerebellum. (See text for details.)

fashion, the cerebellum controls coordination by exerting a "damping" effect on muscle action. As an example, when you reach for a cup of coffee, it is essential that your "reach does not exceed your grasp," causing your hand to overshoot its mark. As you reach for the cup, which itself represents a decision originating in the cerebral cortex, the cerebellum monitors impulses from the muscle receptors and sends impulses through the motor area to make certain that your hand stops exactly at the cup handle. Tumors and other problems with the cerebellum are relatively easy to diagnose because a primary symptom is often a loss of coordination and balance.

The Brain Stem

The brain stem extends from the spinal cord into the brain, ending at the *thalamus,* which is a rounded area at the top of the stem. In the brain stem we find the *medulla oblongata* as well as the bulging portion above the medulla called the *pons* (see Figure 12.8).

The medulla contains critical vital centers that participate in the control of heart rate, blood vessel size, and respiration. For example, if there should be a sudden drop in blood pressure, receptors located in the large artery leaving the heart will send impulses to the appropriate center in the medulla. The medulla in turn sends impulses to the heart and blood vessels, causing the heart rate to increase and the arteries to constrict, thus raising the blood pressure. The rate of respiration is also controlled by the medulla through sensitivity to the level of carbon dioxide in the blood. Thus we can see that the medulla controls extremely vital body functions and why injury to the medulla can quickly cause death.

The *reticular activating system (RAS)* is another structure that is found in the brain stem (Figure 12.13). The RAS has come under intense investigation in recent years, and it appears to have at least two important functions. First, it filters out sensory impulses that come into the central nervous system from the various receptors. It has been suggested that by acting as a selective filter, the RAS helps us to ignore much of the irrelevant stimuli that bombard us most of the time. Otherwise we would find it difficult to concentrate on any one task if our attention could be diverted every time a clock strikes or a car goes by on the street.

The RAS also has a function related to the sleep and waking processes. Evidently during sleep, the RAS refuses to pass stimuli to the cortex, thus permitting the cortex to "sleep." On the other hand, if a strong stimulus should occur during sleep, such as the unwelcome ringing of an alarm or the smell of smoke, impulses are allowed to go through the RAS and the cortex is activated. Even here it should be noted that the RAS is selective in terms of what stimuli are permitted to awaken the cortex. The city dweller, for example, may sleep through the sounds of heavy traffic and fire sirens going by on the street, and yet a mother in that same situation will awaken instantly at the first sound of her baby crying in the night. Her husband, on the other hand, will probably sleep blissfully on, unless they share responsibility for the two o'clock feeding.

It has also been discovered that the RAS must be bombarded by a minimum of stimuli if wakefulness of the cortex is to be maintained. Dullness of one's

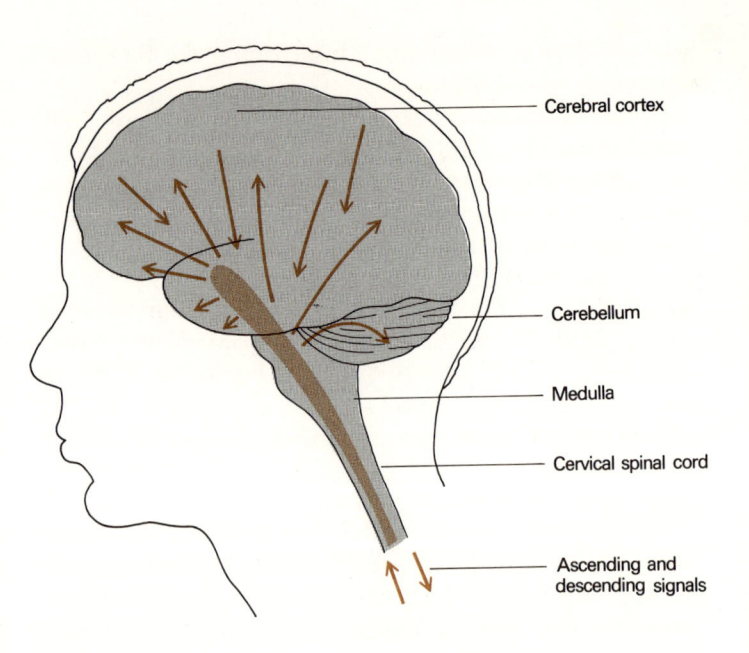

Figure 12.13
The reticular activating system (RAS). (See text for details.)

surroundings (or company) may send one to sleep. The professor whose students are dozing before his eyes might well take the hint that he is not bombarding their brains with enough interesting stimuli! At the opposite extreme, if the cerebral cortex is unusually stimulated by events of the day or by worries, then sleep may come with great difficulty even under the right kinds of conditions.

Some interesting studies with student volunteers who were subjected to long periods of isolation and boredom suggest that if the brain is to function properly, the RAS must be more or less continuously bombarded with sensory stimuli. Subjects who were deprived of stimuli for long periods suffered intense boredom that led to confusion, hallucinations, and other abnormal forms of behavior. It can therefore be seen how the effects of isolation on the human mind have been used in the past to prepare prisoners of war for brainwashing sessions. Boredom has also been suggested as one of the hazards of retirement, especially for a person who has developed few or no interests outside his or her work. It has in fact been long recognized that people whose jobs permit them to work well beyond the usual retirement age may tend to postpone the loss of memory and confusion that is typical of the aged.

Sleep. If you are now approaching the age of 21, this means that you have slept about 7 years. By the time you reach age 60, you will have slept a total of 20 years, or about one-third of your life. Considering the amount of time we spend sleeping, it is understandable that scientists have been giving increasing attention to this pleasurable and important activity.

Sleep has always been a source of wonder. It has been called the "little death," and Shakespeare referred to it as "death's counterfeit." Cervantes extolled the virtues of sleep saying,

> Now blessings light on him that first invented this same sleep! It covers a man all over, thoughts and all, like a cloak; 'tis meat for the hungry, heat for the cold, and cold for the hot. 'Tis the coin that purchases all the pleasures of the world cheap.

Today the television commercials extol its virtues in a different way by promising all variety of potions guaranteed to bring sweet oblivion to the insomniac.

Scientists have discovered that sleep occurs in several phases, but in general the average sleeper passes through phases of dreamless sleep and so-called REM sleep. REM stands for "rapid eye movement," and it was first discovered when researchers noted that changes in sleeping subjects' brainwave patterns from the "sleep" pattern to an "active" pattern were accompanied by rapid movements of the eyes under their lids. Further investigation showed that in most cases the subjects were dreaming during the REM period. It soon became apparent that normal sleep is marked by alternate periods of dreaming (REM sleep) and deeper sleep without dreams. Furthermore, investigations have shown that everyone probably experiences these dream periods even though dreams are usually forgotten upon awakening, leading some people to claim that they "never dream."

An interesting sidelight of sleep research is the fact that when subjects were awakened at the start of each REM period, or in other words, when their dream periods were interrupted, they became irritable and nervous and some even developed temporary disorientation. When allowed to resume normal sleep patterns, they then spent an unusual amount of time dreaming and some reported having nightmares. It appears, therefore, that we need to dream in order for the brain to function normally, although no one has yet offered a satisfactory explanation for this need to dream.

The Limbic System

The *limbic system* of the brain has been intensely studied by researchers in recent years. Although textbooks disagree on what structures make up the limbic system, it is generally considered to include the *thalamus,* the *hypothalamus,* the *amygdala,* and the *hippocampus* (Figure 12.14). The amygdala, hippocampus, and thalamus are paired structures, one member of each pair located on either side of the brain.

The limbic system is part of the "old brain"; it appears to be associated with more primitive forms of behavior such as eating, sexual activities, and basic emotions such as fear and rage. In addition, some authorities suspect that the hippocampus may be involved in the formation of memory traces, especially long-term memory that results from stimuli accompanied by strong emotional reactions such as fear or deep sorrow.

The thalamus, you will recall, is located at the upper end of the brain stem and acts as a relay station for impulses to other parts of the brain such as the cerebral cortex. In addition, the thalamus appears to be associated with pain sensations.

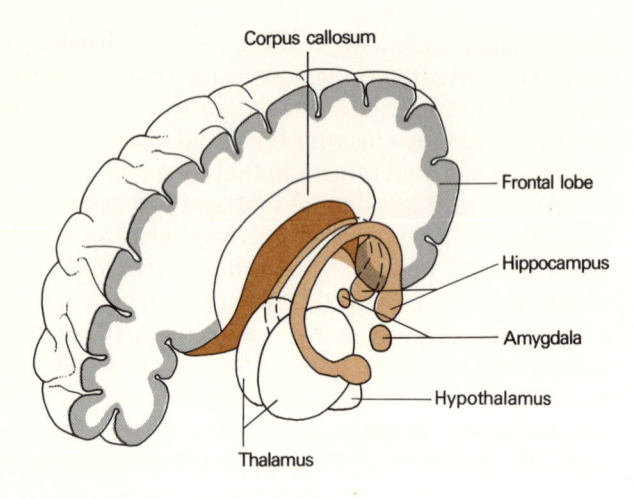

*Figure 12.14
Diagram showing portions of the limbic system.*

The hypothalamus, as the prefix "hypo" implies, lies below the thalamus and appears to have several interesting and basic functions. First, you may recall from Chapter 5 that the hypothalamus is involved in sexual function through its control over the anterior lobe of the pituitary gland. Of special interest is its role as the "biological clock" which turns on periodically to control the menstrual cycle. In addition, the hypothalamus contains nuclei (groups of cell bodies) that have been called the *feeding* and *satiety* centers. These nuclei evidently act as appetite control centers, probably responding to the level of glucose in the blood. Destruction of the satiety center in rats, for example, resulted in the rats eating themselves to the point of obesity. On the other hand, when the feeding center is destroyed, a rat will starve to death unless force-fed. All of this may shed some new light on why many people have great difficulty controlling their appetites and weights. In some situations at least, a voracious appetite might be due to malfunction of the satiety center rather than to "psychological" factors as so often assumed.

Other nuclei in the hypothalamus control at least in part the maintenance of body temperature at a constant level. These "thermostat" nuclei apparently respond to blood temperature. As the body temperature drops, signals from the hypothalamus cause an increase in muscle vibration, reaching the ultimate in the shivering reaction so common in swimmers coming from warm water into colder air. This muscular vibration produces heat, thereby raising the body temperature. On the other hand, if the body temperature increases, the hypothalamus signals evidently produce sweating and dilation of blood vessels near the skin surface, thus helping to lower the body temperature. There is also speculation that the hypothalamic thermostat responds to infection by raising the body temperature, the resulting fever thus providing a more hostile environment for the invading bacteria. An interesting sidelight is the recent claim by researchers that if the hypothalamic thermostat could be permanently lowered somewhat through the continued use of an appropriate drug, man's lifespan would be increased to as much as 150 years! This, of course, is definitely only in the realm of speculation.

An interesting discovery was made quite by accident by researchers who found a so-called *pleasure center* located near the hypothalamus. Rats with electrodes implanted in their pleasure centers repeatedly stimulated this area by pressing a lever; in fact, some rats stimulated the pleasure center more than 2000 times per hour for 24 consecutive hours! So far, the pleasure-center situation has not been satisfactorily explained. For one thing, we are not certain exactly why rats repeatedly self-stimulate this area, since up to now no rat has come forward with testimony describing his exact feelings when the pleasure center is stimulated. It is a *human* assumption that the rat repeatedly self-stimulates the center because it produces "pleasure," but it could possibly be no more than a compulsive behavior pattern that is somehow associated with the stimulus. Another possibility is that stimulation of this particular brain area stimulates neurons that would otherwise be stimulated by satisfaction of such basic drives as sex and hunger.

Reports on the functions of the amygdala have proven to be most confusing, but research has so far suggested that this portion of the limbic system is indeed important. The amygdala appears to be associated with a variety of emotional responses, the particular response depending on which part of the structure is stimulated. An ordinarily tranquil and friendly cat, for example, may fly into a rage and attack the experimenter when the amygdala is stimulated. In other cases, the same animal may exhibit fear. In addition, stimulation of various parts of the amygdala may produce activities associated with eating or evidences of sexual excitement. Having linked this structure with basic emotional responses, researchers in the brain field are now looking for ways to use their knowledge concerning amygdala functions in the treatment of severe emotional disorders.

The Control of Behavior

It is clear that a human society must impose certain standards of behavior on its individual members. It is essential that we do not steal from one another, kill one another, inflict bodily harm on one another, or endanger others by driving down Main Street at 80 miles per hour. Societies have developed a variety of behavioral controls which include religious and cultural traditions, peer pressures, and statutory laws enforced by police. Generally we tend to accept a degree of difference in the behavior of others as long as it is within tolerable limits. But when an individual's behavior patterns exceed certain limits, society may step in and in extreme cases remove the offending person by imprisonment or by confinement to a mental institution.

Since the brain is essentially the seat of behavior, it was inevitable that as man learned more about its structure and functions, he would attempt to control behavior through the brain itself. Consequently, the more time-honored methods of behavior control by torture rack and dungeon have been supplemented by drugs which affect brain function and most recently by direct surgical intervention. In many instances, these new drugs and techniques are undeniably beneficial to both the individual and society in a way that is no different from, say, the development of the Salk polio vaccine. There is, however, one aspect of direct behavioral control that many people find worrisome. The brain is, after all, the seat of the personality, of the *person,* and we tend to instinctively feel that manipulating the brain is somehow very different from removing an appendix or even a leg. We accept the idea of a surgeon entering our skulls if his purpose is to remove a brain tumor or repair a broken blood vessel, but we feel uneasy if he does so to alter our behavior, possibly to make it conform to what someone else's idea of what "acceptable" behavior should be.

The practice of *psychosurgery* involves surgical destruction of certain portions of the brain in an attempt to treat psychiatric conditions, that is, behavior that deviates too far from what we consider as "normal." Psychosurgery was first introduced in the 1930's when surgeons began performing an operation known as a *prefrontal lobotomy* on mental patients with schizophrenia. Schizophrenia is a severe and all too common mental illness that occurs in several forms and involves extreme, bizarre behavior patterns. The schizophrenic is typically disoriented; he may be hostile, belligerent, and severely agitated, or he may sink into the depths of depression and guilt. He withdraws from reality and enters a world filled with visions and voices that only he can see and hear. With his mind totally crippled and unable to function in society, the schizophrenic in the past was generally doomed to live out his life confined in a mental hospital.

The prefrontal lobotomy is a rather crude operation which breaks connections leading to the frontal lobes of the brain. In many cases, lobotomy makes patients more docile and easy to manage. Unfortunately, the operation also tends to blunt the patient's normal emotional responsiveness to the point where he becomes something of a zombie. It is estimated that during the 1930's and 1940's, as many as 50,000 lobotomies were performed in the United States alone. Some lobotomies are still done today, but they have been largely replaced by drug therapy as a treatment for schizophrenics.

In the mid-1940's, a French surgeon named Henri Laborit began giving a drug called *promethazine* to preoperative patients, hoping it would reduce the chance of surgical shock. Surgical shock involves a sudden and often dangerous fall in blood pressure following an operation and is apparently due to the secretion of a chemical called *histamine*. Since promethazine was known to be an *antihistamine,* Laborit reasoned that it might help to keep blood pressure at normal levels.

He then noticed an unexpected side effect. It is said that the anticipation of an operation is often worse than the surgery itself, and Laborit noted that patients given promethazine were unusually calm and much less apprehensive than was usual when facing the unknown terrors of surgery. When it was thus recognized that promethazine, which is a derivative of *phenothiazine,* had possibilities for psychiatric use, a search began for other derivatives which led to the discovery of *chlorpromazine*.

It was found that chlorpromazine appeared to be specific for schizophrenia, since it tended to both calm the agitated patient and bring the withdrawn, depressed patient back to reality. Thus beginning in the mid-1950's, chlorpromazine and similar drugs have revolutionized the treatment of schizophrenia and radically changed the climate of mental hospitals. Thousands of schizophrenic patients have been able to return home and function reasonably well, and the rate of admission of such patients has dramatically decreased.

On the other hand, the issues that have been raised concerning the treatment of hyperkinetic children with amphetamine drugs are not as clear. The hyperkinetic child is excessively active and may show other symptoms such as lack of coordination, confusion between left and right, and difficulties with short-term memory. These children are often diagnosed as having *minimal brain damage (MBD)*.* Such damage is thought to be due to problems in fetal development, difficult births, or brain infections. More recent evidence suggests, however, that in some cases genetic factors could be involved. It has been estimated that as much as five to ten percent of all school children may be suffering from MBD to some degree. It is important to note that minimally brain-damaged children are not necessarily retarded; in fact, their I.Q.'s tend to follow the same distribution as those of the general population.

The MBD child is likely to be a discipline problem at home and especially at school. He is fidgety, impulsive, and has an extremely short attention span. In spite of normal or even above-normal intelligence, he may have difficulty learning to read and write. In general, he is difficult to manage and is a disruptive influence in the classroom. Also, due to a tendency to dominate and bully other children, he is usually a social failure and has few if any friends.

As long ago as 1937, it was discovered that hyperkinetic children significantly improved when given amphetamine drugs. They were less fidgety, their attention spans increased, and they required less discipline both at home and in school. Their school work tended to improve, perhaps partially as a result of an improved attention span and increased ability to concentrate on one thing at a time. Amphetamines are stimulant drugs, and why a *stimulant* should have a *calming* effect on a hyperkinetic child is still not known for certain. One possibility is that the drug stimulates a part of the brain which in turn inhibits the reticular activating system (RAS), thus preventing excess stimuli from entering the cortex. Another pos-

---

* Today, MBD is becoming more commonly known as *minimal brain dysfunction.*

sibility is that amphetamines act directly on the cortex in some way, perhaps to increase attention span.

On the other hand, the treatment of hyperkinetic children with amphetamines should be approached with caution. First of all, the diagnosis of minimal brain damage in a child is often extremely difficult. Symptoms typical of MBD may also indicate a different type of psychiatric disorder. Furthermore, not all behavioral problems indicate that something is organically wrong; it has been suggested that misbehavior in school may in some cases simply reflect a deterioration of the school's general disciplinary climate. We are therefore faced with the possibility that amphetamines might be given to children whose physical conditions do not warrant it.

Psychosurgery as a method of behavior control has made considerable progress since the days of the crude prefrontal lobotomy. Increased knowledge of brain structure and surgical techniques have provided a potential for control by either directly stimulating or destroying certain parts of the brain. For example, it has been found that removal of the amygdala may greatly reduce overly aggressive behavior in children. By implanting electrodes and stimulating certain areas, the surgeon can alternately create pleasure, fear, or depression. In some instances he can stimulate various portions of the brain until he finds the area of "unwanted" behavior, which he can then destroy with electricity. Or, patients may have electrodes implanted in their pleasure centers and wear self-stimulation units on their belts. Thus equipped, they may stimulate their own pleasure centers as often as they please!

Dr. José M. R. Delgado even speculates on the possibility of computer control when he says,

> A two-way radio communication system could be established between the brain of the subject and a computer . . . anxiety, depression, or rage could be recognized [by the computer] in order to trigger stimulation of specific inhibitory structures.

Such speculations quickly raise an image of the ultimate dictatorship—a population of brain-mutilated zombies controlled through their electrodes by a favored few operating from giant computer centers. This frightening picture probably exaggerates the potential of psychosurgery for evil, since it has been demonstrated more than once in this century alone that the control of whole peoples by a few can be accomplished much more easily by the more traditional methods of propaganda, with imprisonment or execution for those who refuse to be brainwashed!

Still, there are practical questions that must be asked. What kinds of behavior should be changed by surgical intervention? What *is* "unacceptable" behavior anyway? To whom is it unacceptable? To the patient? To his family? To the neurosurgeon? To whatever governmental authorities happen to be in power?

Psychiatrist Willard M. Gaylin raises the practical question of "informed consent." By law as well as by the rules of medical ethics, the patient must be informed as to the need for an operation as well as its possible consequences and must consent to such an operation. If the surgery happens to be a gall bladder operation, it is assumed that the patient can make a rational decision, but as Dr. Gaylin points out, in the case of psychosurgery "the diseased organ is the consenting organ." Therefore, the question is whether such a patient could ever give rational consent to an operation that might change his behavior and thereby alter his personality.

On the other hand, just because some aspects of psychosurgery may make us uneasy, its potential for good should not be overlooked. The most traditional forms of surgery, even tonsillectomies, carry a potential for unforeseen complications, and we do not see this as a reason for "throwing out the baby with the bath water." How would we feel, for example, about a psychosurgical operation that would free a patient from agonizing and uncontrollable pain? Or suppose an operation could change a patient's behavior in such a way that he would be lifted from the depths of a depression that keeps him from functioning? Or still further, suppose that psychosurgery would enable a man to avoid spending his adult life in a prison or in a mental institution? If we consider cases such as these, the possibilities of psychosurgery for good may well be enormous, unless you wish to take the point of view that it is a person's right to spend his life in prison if he wishes rather than have his behavior altered by a surgeon. And so it goes. One thing is clear, however. As man's knowledge of the brain and his ability to control its functions continue to expand, we must keep pace with an awareness of the important ethical issues that are involved.

Drugs and the Brain

To the average middle-class American, the term "drug user" means hippie pot parties, a heroin addict searching desperately for his next fix, a speed freak mugging an old lady for money to support his habit, or someone on a bad LSD trip suffering the tortures of the damned. In fact, the thought of such unlovely scenes may so upset our middle-class pillar of the community that he gulps down his cocktail, puts out his cigarette, and goes to his well-stocked medicine cabinet for an aspirin, an antacid, or a tranquilizer!

Drug abuse is not found only on ghetto streets or in back alleys. It is a national blight, a cancer that spreads throughout the organism of society, invading every segment, every level. We are a nation of users, misusers, and abusers ranging from those who depend on their stocks of nonprescription painkillers, tranquilizers, sleeping pills, and stimulants to those whose very humanity has been sacrificed to drug addiction and alcoholism.

In this section, we will consider some of the major problem drugs and, where possible, relate their effects to the material presented earlier in this chapter. It is obvious that a full treatment of drugs and the drug problem is not within the scope of this book, but considering its importance, the reader is urged to examine more extensive treatments of this subject. One particularly excellent source is a book by Dorothy and Daniel Girdano entitled *Drugs, A Factual Account* (refer to the Annotated Reading List beginning on page 361).

The drugs of major abuse can be divided in a general way into the *depressants* (alcohol, barbiturates, tranquilizers), the *stimulants* (amphetamines, cocaine), and the *hallucinogens* (LSD, mescaline). There are also the true *narcotics* which include opium and its derivatives (morphine, heroin, codeine).

The physiological effects of drugs are, of course, on the nervous system, and the specific nature of those effects are related to the purpose for which a particular drug is taken. Depressants, for example, appear to depress the RAS, thus promoting drowsiness or sleep by decreasing the number of impulses sent to the cortex. Depressants may also inhibit nerve impulses at the synapse either by preventing the secretion of the neurotransmitter or by causing it to break down faster than normal. Stimulants, on the other hand, have the opposite effect because they increase nerve impulses across the synapse by mimicking the action of neurotransmitters and by stimulating the RAS to bombard the cerebral cortex with greater frequency.

Still other effects that drugs may have on the nervous system include direct stimulation of the appetite control and pleasure centers of the hypothalamus. The methamphetamine (speed) user, for example, repeatedly injects the drug for a period of 72 hours or more until he suffers complete exhaustion. This behavior is similar to the rat that repeatedly self-stimulated its so-called pleasure center over a 24-hour period.

Some drugs such as alcohol and barbiturates cause the user to build a *tolerance* so that more and more of the drug is needed to produce the desired effect. The physiological basis of tolerance formation is not well understood, but it has great significance for the user. The heroin addict, for example, may start out with one injection per day, but tolerance to the drug quickly develops and he will find himself needing five or six injections per day in order to ward off the unbearable symptoms of withdrawal. The heroin addict thereby builds a habit that costs up to $100 per day to support. Since most addicts are not wealthy nor even gainfully employed, it is little wonder that the addict turns to crime in order to get the drug that his body so desperately needs.

The foregoing is a case of *physical* dependency on a drug. Physical dependency involves withdrawal symptoms that are so painful and distressing that the user will do anything to acquire more of the drug in order to avoid withdrawal. Contrary to popular belief, however, not all addiction involves such physical dependency. Drugs such as marijuana, for example, and even some of the so-called "hard" drugs like cocaine and the amphetamines do not produce physical dependency. In these cases the dependency is *psychological*. Actually, however, even with drugs that create physical dependency, psychological dependency is also likely to be present.

Depressants

Ethyl alcohol, or *ethanol,* is certainly one of the oldest and most widely used depressants. Most people find it difficult to believe that alcohol is a depressant, since they tend to associate it with feelings of exhilaration. But the sight of the overenthusiastic "life of the party" passed out on the livingroom floor should be enough to convince anyone that alcohol is indeed a depressant. Alcohol is absorbed into the bloodstream primarily through the first part of the small intestine; it is absorbed more rapidly if the drinker has not had a heavy meal shortly before. When it reaches the brain, its first action is to depress neuron activity in the frontal lobes. Since the frontal lobes evidently control inhibitions, it is apparently this frontal-lobe depression that influences behavioral changes. Generally these changes depend on the individual's basic personality and on the mood he is in when he starts drinking. He may become loud and aggressive, silly, overamorous with the wrong lady such as the boss's wife, or perhaps just more quiet than usual. As the intake of alcohol continues and the blood-alcohol content rises, other areas of the cerebral cortex such as the motor and visual areas are involved. By this time the functions of the cerebellum are depressed, and a loss of balance and coordination results. Finally, when large amounts of alcohol are ingested, the brain stem will be affected and death can result from depression of the respiratory center in the medulla. Fortunately for our "life of the party," he will usually pass out before this occurs, thus preventing the ingestion of more alcohol.

In our society, there is a curious tendency to treat excessive drinking as a joke. Even during the Prohibition Era of the 1920's and early 1930's, the law which forbade the sale of liquor was laughed at and held in contempt by vast numbers of otherwise law-abiding citizens. Illegal bars called "speakeasys" flourished, serving people from all strata of society. Meanwhile the bootlegger, the entrepreneur of Prohibition, made huge profits from illegal sales of liquor while enjoying the respect and admiration of many of his fellow citizens. The occasional fine that he had to pay was counted as a business expense. The law was finally repealed in 1933 mainly because it had never worked in the first place and relatively few people really wanted it.

Alcohol is definitely not a joke. Certainly the abuse of it is not. To the unnecessary highway tragedies that occur due to drunken drivers we must add at least five to six million Americans who are physically and mentally sick because of a total dependence on alcohol. They are the alcoholics. They come from every walk of life and every level of society. Alcoholism has no respect for race, color, creed, sex, age, or financial condition. Its victims include the poor, the rich, teachers, physicians, lawyers, farmers, clergymen, and politicians. They are not to be compared with the tired businessman and his occasional before-dinner martini or even with the "life of the party" who invariably fails to hear the chimes usher in the New Year.

The typical alcoholic follows a long and tortuous road which is downhill all the way. Generally, he begins as a casual social drinker but changes over a period of time to a solitary drinker, often using alcohol to help him through trying periods. It may take a number of years, but ultimately his whole life revolves around alcohol; he lives only for the next drink while he deteriorates both physically and mentally. Without doubt, alcoholism is one of the leading causes of misery in our society.

The family of depressant drugs known as *barbiturates* is also subject to high abuse. The barbiturates, often in the form of sleeping pills, are prescription drugs which include *Nembutal, Seconal, Amytal,* and *phenobarbital*. The effects of the first three are generally of short duration, but those of phenobarbital are longer lasting. This fact combined with phenobarbital's antispasmodic action makes it a useful treatment for epilepsy when it is taken with another drug called *Dilantin*. In addition, phenobarbital may be prescribed in small, regular doses for patients with high blood pressure.

The barbiturates depress the function of the reticular activating system so that fewer impulses are sent to the cerebral cortex, thus promoting drowsiness when taken in sedative doses as sleeping pills. Unfortunately the body develops a tolerance to barbiturates, and an individual who comes to depend on barbiturate sleeping pills may find himself requiring larger and larger doses in order to achieve the desired effect. In addition, the sedative effect evidently interferes with REM sleep, and when the drug is discontinued the patient may have to pay the "REM sleep debt" with excessive dreams and restless sleep. This may induce him to return to the drug, thus sowing the seeds of a vicious cycle that becomes more and more difficult to break. Still another danger from long-term use of barbiturate sleeping pills is the fact that it causes a physical dependence that creates severe withdrawal symptoms when the drug is stopped. This withdrawal process may be severe, dangerous, and even fatal if not carried out under close medical supervision.

Stimulants

The *amphetamines, cocaine, nicotine,* and *caffeine* are some stimulant drugs that are most subject to widespread use, misuse, and abuse.

The amphetamines include *Dexedrine, Dexamyl, Benzydrine,* and *methamphetamine,* also known as "meth" or "speed." The amphetamines appear to mimic the action of a neurotransmitter that affects the rate of impulses in the area of the hypothalamus. In addition, amphetamines bring about a release of adrenalin from the adrenal glands. Amphetamines act as appetite depressants and "pep pills," and they have been widely prescribed for patients who need to reduce their weight by a large amount. Dexedrine has been extensively used for weight reduction, since it not only depresses the appetite but also gives the patient a sense of well-being in spite of the reduction in food intake. Unfortunately the patient often develops a psychological dependency on the pill as much for the peppy feeling that it produces as for its role in appetite control. Also, unfortunately for those of us who like to eat and still stay slim, there is no magic drug that will permit us to do so. In the vast majority of cases, as soon as an appetite depressant is discontinued the pounds return as rapidly as the dieter goes back to his old eating habits.

Methamphetamine, or *speed,* may be "mainlined," that is, injected directly into a vein. Following injection of the drug, the user experiences a variety of physical sensations that soon culminate in a "flash" or "hit." This consists of an extreme sense of euphoria, or pleasure, and has been described as a "whole-body orgasm." As soon as the euphoria wears off, the user injects himself again, repeating the process for 72 hours or longer without sleep or food.

It is easy to see why the compulsively repeated self-stimulation of the hypothalamus with speed has been compared to the repeated self-stimulation of the pleasure center carried on by the rat. Evidently speed affects a pleasure center or centers in the limbic system in much the same way. Unfortunately, all good things must end, it seems, and so it is with the "speed freak." Toward the end of the injection cycle, called a "run," the user tends to become disoriented and irritable with paranoid feelings of suspicion. This is why the speed user may often be dangerous to himself and to others as well.

Much has been written in recent years about cigarettes (nicotine) and their association with fatal illnesses such as lung cancer, emphysema, and heart disease. Extensive campaigns against smoking urge older smokers to quit and younger people to make certain they never start. Without going into detail, it is enough to say here that the evidence against the cigarette is overwhelming. Considering this, it is puzzling why so many young people start smoking and why cigarette consumption in the United States remains so high.

Caffeine, usually taken in coffee and sometimes in nonprescription pep pills, is a mild stimulant, but excessive use can lead to problems, especially sleeplessness or diarrhea due to overstimulation of the intestinal tract. For example, one physician commented that his business showed a significant increase when coffee machines were installed in a local industrial plant.

Hallucinogenic Drugs

Hallucinogenic drugs are those that distort perception and produce hallucinations that are sometimes pleasing to the user but at times may bring an opposite reaction such as a "bad trip" on LSD. The hallucinogens include *LSD, mescaline,* and possibly *marijuana,* although marijuana is more difficult to classify.

LSD (*lysergic acid diethylamide*) is one of the strongest and most unpredictable of the hallucinogenic drugs. Its ingestion produces time distortion, visual distortion, and hallucinations involving psychedelic colors that only an LSD user can describe. LSD users like to think that the drug expands the mind, increases perception, enhances creativity, and helps to solve the secrets of the universe. There is no evidence that any of this is true; moreover, experiments have shown that the LSD user *thinks* he is more creative as a result of the drug, but in reality he is not.

On the other hand, there is considerable evidence that LSD is potentially a very dangerous drug that in some cases can lead its user into a permanent psychosis. Its action on the nervous system is not well understood, but it is suspected that all hallucinogens interfere with the function of serotonin, which is one of the neurotransmitters important to healthy brain function. LSD has been found to concentrate in the limbic system as well as in the auditory and visual areas of the cortex.

LSD users are subject to panic reactions known as "bad trips." Evidently the panic reaction occurs when the user finds the extreme visual and auditory distortions of reality too difficult to take, or possibly the common physiological reactions such as nausea and increased heart rate may contribute to the feeling of panic. Another serious side effect of LSD is the "flashback." This seems to involve a sudden recurrence of LSD symptoms long after the drug has been excreted from the body. Flashbacks can in fact occur well over a year following the last LSD experience. Flashbacks are said to occur most often in individuals who have experienced a panic reaction with LSD, and use of other drugs such as marijuana have been known to set off an LSD flashback. Flashbacks are usually if not always unpleasant experiences, leading to reactions ranging from remorse to hysteria. Suicide is a possible response to the emotional lows and fears common to the flashback phenomenon.

Mescaline, a drug obtained from the peyote cactus, has been used at various times by American Indians as an important part of religious ceremonies. It is the only hallucinogenic drug to have any degree of legal sanction in the United States, and this is limited to the ritual ceremonies of the Native American Church of North America. Its members believe that the drug permits them to communicate more meaningfully with God. As one member put it, "The white man goes to his church and talks *about* Jesus, but the Indian goes into his tipi and talks *to* Jesus."

Marijuana

Marijuana, also called "pot" or "grass," is regarded by some as a hallucinogen, although others think of it as a depressant similar to alcohol. It is obtained from the hemp plant *Cannabis sativa,* which grows wild and can also be cultivated in most parts of the world.

The leaves and flowering tops of the plant are dried and used as marijuana, which is rolled in the form of a cigarette called a "joint" or "reefer." In very hot climates the flowers become coated with resin, which may be collected in brown cakes known as *hashish.* The active ingredient of both marijuana and hashish is *tetrahydrocannibinol (THC),* which is several times more concentrated in hashish than in marijuana. Unfortunately, the mechanism by which THC affects the nervous system is not well understood at this point.

The marijuana user inhales the smoke deeply into the lungs and holds it there so the drug may be absorbed through the lung capillaries into the bloodstream. This produces what has been described as a mild euphoria, or "high," with an easing of anxieties and tensions. Some users also report a sharpening of the senses and "mind-expanding" sensations similar to that ascribed to LSD.

Marijuana is not addicting in the sense that it produces a physical dependency complete with withdrawal symptoms. On the other hand, neither are cocaine and speed physically addicting. Indeed it must be emphasized that *psychological* dependence can be just as addicting as physical dependence. Ask the veteran cigarette smoker who will tell you that quitting is easy; he has done it thousands of times!

Is marijuana a harmful drug or is it merely a pleasant relaxant that should be legalized? Opinions on this question are conflicting and are too often based on emotion instead of hard scientific evidence. There are research studies which suggest that *excessive* use of marijuana may well be harmful. There is evidence, for example, that its use may interfere with cellular production of DNA. Still other research results suggest the possibilities of lung damage, brain damage, and chromosome damage. It has also been shown that like alcohol, marijuana may seriously impair driving skill and judgment.

On the other hand, there is no overwhelming body of hard evidence that the occasional use of marijuana produces long-term *harmful* effects, although it must be said that neither is there conclusive evidence that it is *not* harmful. There is definitely a need for extensive further research to find out once and for all if the moderate use of marijuana is harmful or not.

The Opiates

The opiates include *opium* and its derivatives: *morphine, heroin,* and *codeine.* These are the true narcotics, producing severe withdrawal symptoms as the effects of the drug wear off.

Obtained from poppy seeds, opium is one of the world's oldest narcotics, but its use in the United States today is minimal. In 1805, morphine was introduced as a pain reliever and is still widely used for this purpose, especially in terminally ill cancer patients in whom addiction is secondary to the relief of pain. During the Civil War when morphine was used liberally on the battlefields, it was noted that soldiers who were given repeated doses of morphine became physically dependent upon it and suffered withdrawal symptoms when the drug was stopped.

In 1898, heroin was introduced and was hailed as a cure for morphine addiction. Today heroin has no medical use but is the narcotic of choice on the streets, where addicts build habits costing up to $100 per day. Heroin, also called "horse" or "smack," is administered directly into a vein. It apparently affects the pleasure center of the hypothalamus, producing intense euphoria and driving away the user's cares and problems—temporarily, at least. Heroin rapidly produces physical addiction; the victim is "hooked" before he realizes that he cannot use just a little bit of heroin and then stop. The body also forms a tolerance to the drug, and ever-larger doses are required to attain the euphoric state that the user seeks.

Treatment of heroin addicts continues to be a huge problem. Generally, those who "kick the habit" by going through withdrawal under medical supervision are free from the drug for only a short time until the problems that originally led them to drug addiction bring them back to it again. *Methadone,* a synthetic narcotic, has shown some promise as a form of treatment. Methadone maintenance does not require the addict to go through withdrawal. Instead it appears to block the need for heroin. While methadone is itself a narcotic, patients under methadone treatment are often able to lead useful lives.

Why Drug Use?

The rapidly spreading drug culture in this country has been explained in a variety of ways. Experts speak of "alienation" as a root cause. The generation gap is often blamed, which is curious since a generation gap has existed for thousands of years! Poverty has received its share of the blame, but poverty is not new nor does it explain the wide use of drugs by children of the affluent. Drugs are said to offer a way out of one's troubles, a way to overcome lack of success and a lack of self-esteem. Whatever the cause in the individual case, drug addiction is an intolerable, life-wasting situation for the addict and a growing problem for society. Certainly it has become one of the most important and difficult issues of our time.

Questions

1. Describe the functions of (1) the cerebral cortex, (2) the medulla, (3) the cerebellum, and (4) the RAS.

2. Explain how the synapse functions, and show how drugs might alter that function.

3. Describe the functions of the limbic system. Suggest ways in which future research on this system could be of practical importance.

4. Explain why alcohol initially produces a feeling of exhilaration even though it is classified as a depressant.

5. What similarities do you see between the problems of law enforcement during Prohibition and the drug abuse problems of today?

6. If you had complete power to make laws concerning drug abuse, what kinds of penalties would you set for users? for pushers? for dealers? Would you make the penalties less severe in the case of one drug versus another? Explain.

7. In your opinion, how large a role does *curiosity* play in the drug problem? Explain.

8. What are some of the ethical problems associated with the treatment of abnormal behavior patterns with drugs or psychosurgery?

13

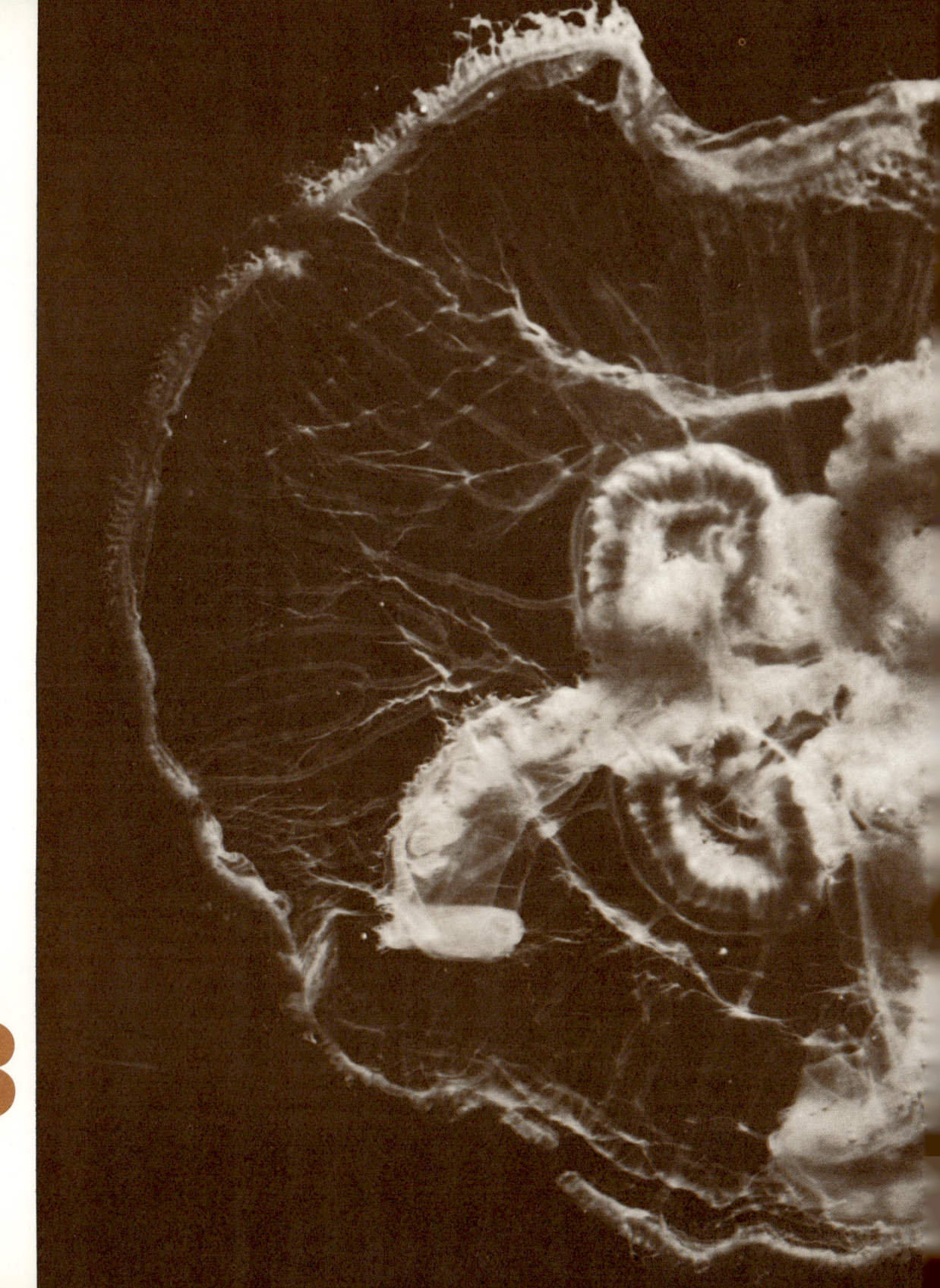

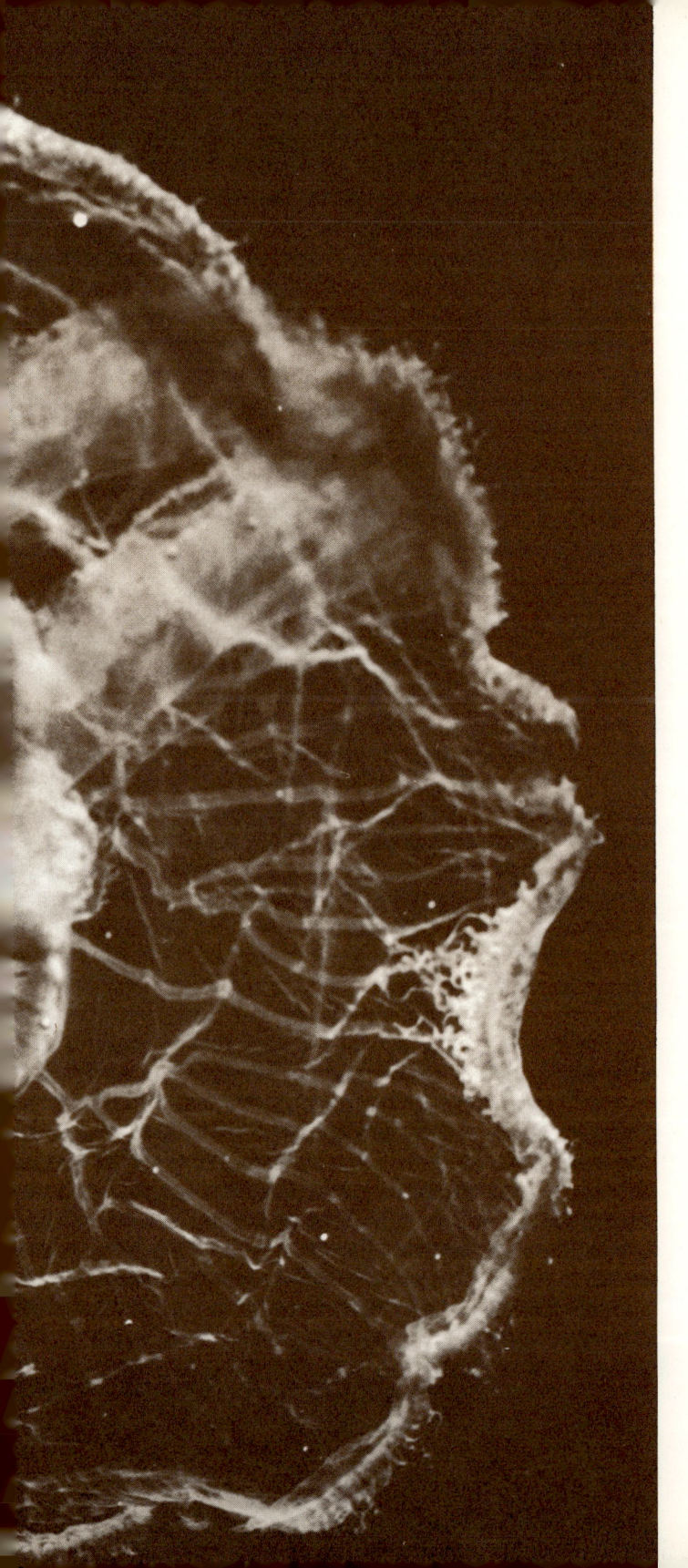

Noah's Ark

Order Out of Chaos

From its simple and tenuous beginnings in the primeval seas of long ago, life evolved through eons of time, filling the land, air, and waters of the earth with millions upon millions of organisms. Biologists estimate that today there are over two million species, ranging from the tiniest bacteria to the great whales that roam the seas unchallenged by all except man.

Biologists have continually struggled to bring order out of this chaotic diversity of life by devising various classification schemes. Such systems of classification go back as far as Aristotle, the Greek philosopher who lived about 300 B.C. Since Aristotle's time, many systems have been proposed, but even today there is no one scheme that is universally accepted by biologists. Biologists who specialize in classification are called *taxonomists,* and those who attempt to use evolutionary relationships as a basis for classification are called *systematists*. Taxonomists like to spend their spare time bickering with other taxonomists over the details of classification. It is therefore well to admit before we begin that no matter what system we choose to discuss here, there will be those who will disagree. In fairness to the taxonomists, we should point out that the reason for the bickering is the fact that it is probably impossible to devise a completely logical classification scheme into which every single species will neatly fit.

All present-day classification schemes are based in part on the work of the Swedish naturalist Carl von Linné, more commonly known as Linnaeus. In 1758, Linnaeus described his basic scheme of classification in a classic work of biology called *Systema Naturae*. This system is based on the idea that a *species* is a group of organisms which look alike or very similar, and which can interbreed and produce *fertile* offspring. Thus despite racial and ethnic differences, all human beings belong to the same species because interbreeding among the various groups produces fertile offspring. Similar species are in turn grouped into a large category called a *genus*; from this genus-species classification, the "scientific names" of organisms are derived. The wolf, for example, is *Canis lupus*. *Canis* is the name of the genus to which the wolf belongs and *lupus* refers to the species. The dog, on the other hand, is *Canis familiaris*. As you can see, the dog and the wolf belong to the same genus but to different species. All dogs, whether they be terriers or St. Bernards, belong to the same species, *Canis familiaris*. To use further examples, *Felis leo* is the scientific name for the lion, and *Felis catus* (or *domesticus*) is the name given to the domestic cat. (The genus *Felis*, by the way, includes cats that purr.) Man is *Homo sapiens* (*sapiens*: "wise"). All human beings presently on earth belong to the same species and therefore also to the same genus.

Different species ordinarily do not interbreed in their natural state but sometimes do in captivity. The most notable example of interbreeding between two species involves the horse (*Equus caballus*) and the donkey (*Equus asinus*). The offspring is a mule, and mules are sterile. Mules therefore do not produce more mules; they must be obtained by breeding horses with donkeys.

Continuing with Linnaeus' system, genera (plural for genus) are combined into *families,* which in turn make up *orders*. Orders are further combined into *classes*. Classes are combined into *phyla* (singular: *phylum*), which together make up a *kingdom*. Table 13.1 shows how a domestic cat and a man (or woman) would be classified according to this system.

Table 13.1

	Man	Domestic cat
Kingdom	*Animalia*	*Animalia*
Phylum	*Chordata*	*Chordata*
Class	*Mammalia*	*Mammalia*
Order	*Primates*	*Carnivora*
Family	*Hominidae*	*Felidae*
Genus	*Homo*	*Felis*
Species	*Sapiens*	*Catus*

Looking at Table 13.1, you can see that both the cat and the man belong to the kingdom *Animalia,* the Animal Kingdom. They also both belong in the phylum *Chordata,* which includes all animals with backbones as well as those with primitive beginnings of backbones. In addition, both the cat and the man are placed in the class *Mammalia,* the mammals. In other words, they are both animals that have backbones, but in addition, they both have hair and feed their young with milk secreted by the *mammary glands* of the female. Up to this point, therefore, the cat and the man are closely related, but when we come to the order, their relationship becomes more distant. The cat belongs to the order *Carnivora,* which includes predatory animals that exist primarily on meat. Man, on the other hand, is placed in the order called the *Primates*. This order includes mammals that have features such as stereoscopic vision, opposable thumb and fingers, and four limbs with five digits bearing nails. In this order along with man are the great apes, the New World monkeys, the lemurs, and the tarsiers. The cat is placed in the family *Felidae,* which includes all cats such as lions, tigers, leopards, and cougars. Man belongs to the family *Hominidae,* which includes only ourselves, since all other groups of manlike creatures have long been extinct (see Chapter 16). The genus *Felis* includes cats that purr, and *catus* refers to the domestic cat. As we mentioned before, modern man is the only example of the species *Homo sapiens.*

In 1859, Charles Darwin proposed that present-day species were *not* present at the "dawn of Creation," but are descended from ancient ancestors through a process of slow change which produced the diversity of life on earth today. These ancient ancestors were similar to, yet different from, their own modern descendants; today's species show evidence of relationships which imply a common ancestry. This concept of *organic evolution* created a sensation in Darwin's time, as man was forced to change his thinking as to his own origins as well as the origins of all other forms of life. Evolution is one of the most fundamental principles of modern biology, and we will take a closer look at this important concept in Chapter 16.

Linnaeus lived about 100 years before Darwin introduced his Theory of Evolution. It is doubtful, therefore, that Linnaeus thought in terms of classifying organisms according to common ancestry, since he probably shared the common belief of his time that all species on earth were the original and unchanging works of the Biblical special Creation. Linnaeus' classification scheme was a "natural" system. In other words, he and his followers simply grouped living organisms according to their most obvious similarities of body structure, or what biologists call *morphological* characteristics. Linnaeus, therefore, was probably unaware that he was actually classifying plants and animals into groups according to their places on the family tree of evolution.

After Darwin, however, biologists began to realize that the Linnaean system worked suprisingly well as a device to show similarities in evolutionary backgrounds. In other words, it became apparent that various species show close structural relationships to one another because of their descent from a common ancestry. Those species that show the most obvious structural similarities share a *less remote* common ancestor. Classification then became a study of *phylogenetic* relationships, meaning the *evolutionary* relationships between groups of organisms.

Like most systems, phylogenetic classification is not without its pitfalls and problems. Some species, for example, may show extremely close structural similarities simply because they evolved in similar habitats and not because they are closely related in an evolutionary sense. Also, not all groups of organisms fit neatly into appropriate pigeonholes. This is especially true as we go up the classification ladder to the larger groupings such as classes and phyla where common ancestry may not be as obvious. In general, however, the evolutionary family tree of life on earth is quite apparent to the practiced eye of the systematist as he scans the classification scheme. Furthermore, one does not have to be a systematist to see that patterns of relationships tend to emerge when plants or animals are grouped according to specific characteristics.

Try thinking of it this way. With which of the following do you have the closest relationship: your brother or sister, your first cousin, or your second cousin? With which of these do you share the *least* remote common ancestor? the *most* remote? Now apply a similar line of reasoning to your relationship with a chimpanzee, a dog, and a fish.

Modern Classification

If you were to casually walk through a woods or field noting the various forms of life around you, you would probably see trees, bushes, flowering plants, birds, insects, and possibly a few chipmunks or squirrels. If you are lucky, a deer might cross your path. If you were then asked to broadly classify the living things that you saw, you would probably divide them into two groups—plants and animals. Historically, this is what happened when biologists first began to classify living organisms. They established two great kingdoms; one was the Animal Kingdom and the other was the Plant Kingdom. These two kingdoms were then each subdivided into phyla, classes, orders, and so on.

When the microscope was invented, it revealed a whole new world of millions of organisms, most of which did not seem to fit the old scheme. After numerous attempts to place them in one category or the other, it became obvious that the old scheme by which all life was divided into plants and animals would no longer work. A new scheme had to be developed.

Modern classification is therefore based upon the division of all organisms according to two fundamental cell types—those having *procaryotic* cells and those having *eucaryotic* cells. Procaryotic cells lack a cell nucleus bounded by a membrane, although there is a distinct region in the cytoplasm where genetic material is found. The genetic material (DNA) is not organized into complex chromosomes as it is in eucaryotic cells. Procaryotic cells also lack mitochondria, chloroplasts, an endoplasmic reticulum, Golgi bodies, and lysosomes. Procaryotes therefore have a very simple structure, which leads biologists to believe that a division of living organisms into these two basic cellular types occurred long before the evolutionary development of plants and animals as distinct groups.

In 1969, R. H. Whittaker proposed a five-kingdom scheme of classification. He placed the procaryotes in the kingdom *Monera* and divided the eucaryotes into four kingdoms: protists, plants, fungi, and animals. Whittaker's rationale for his scheme is far too complicated to go into here, but he based it on a number of factors including the different ways that organisms obtain nutrients. As you might suspect, not all biologists agree with the details of Whittaker's five-kingdom scheme, but we will use it here as a basis for a quick survey of the multitude of living organisms currently inhabiting the earth.

Procaryotes: The Monera

The Monera include all organisms having procaryotic cells. Several thousand species of procaryotes, or Monera, have been described and they are divided into two major groups, the *blue-green algae* and the *bacteria.*

The blue-green algae are widely distributed throughout the world, being found in both fresh and salt water as well as on rocks and in damp soil. The blue-greens are especially tolerant of pollution; they are likely to be found in abundance in polluted waters. There are at least 2000 species, and together they account for a high percentage of the total oxygen that is produced by photosynthesis. In addition, the blue-greens are important to aquatic life because of their ability to "fix" atmospheric nitrogen, that is, convert nitrogen into nitrates for use as nutrients. Like all procaryotes, the blue-green algae lack a nuclear membrane and cell organelles, such as chloroplasts, that are found in eucaryotes. They exist as single cells or as colonies. Examples of blue-green algae include *Nostoc* and *Oscillatoria* (Figure 13.1).

The bacteria are extremely small organisms without a distinct nucleus or organelles. There are somewhere around 1500 species which can be classified in part according to shape: *coccus* (spherical), *bacillus* (rodlike), and *spirillum* (spiral). This division of the Monera is much more familiar to the nonbiologist, since we tend to associate bacteria with diseases such as tuberculosis, pneumonia, diphtheria, and so on. Actually, disease-causing bacteria, called *pathogens,* are in the minority. Many species do not affect us one way or the other, but certain others are essential components of the entire scheme of life. Some kinds of bacteria, for example, play critically important roles in the nitrogen cycle, both by returning nitrates to

Procaryotes: the Monera   243

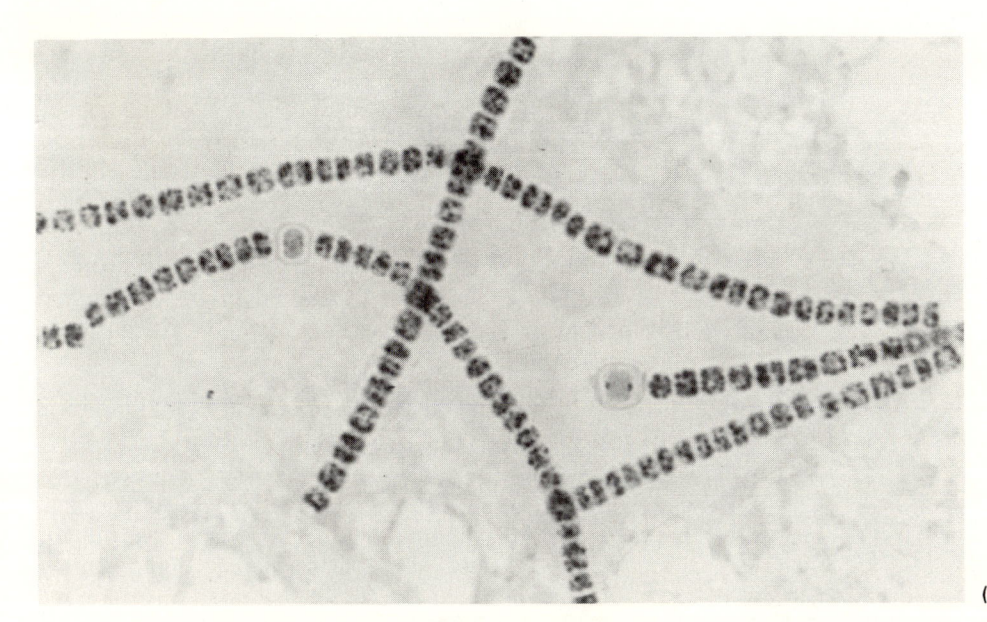

(a)

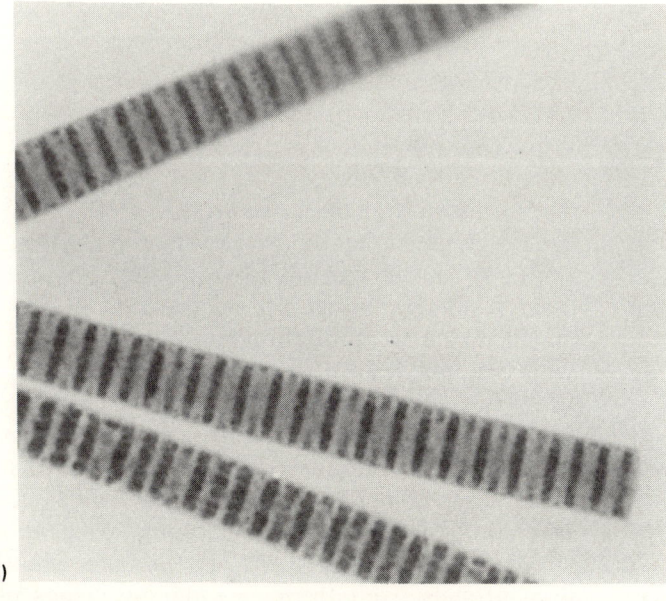

Figure 13.1
Nostoc (a) and Oscillatoria (b) are examples of blue-green algae. They do not have highly organized cell organelles and belong to the kingdom Monera. (Nostoc courtesy of the Carolina Biological Supply Company.)

(b)

(a)

(b)

(c)

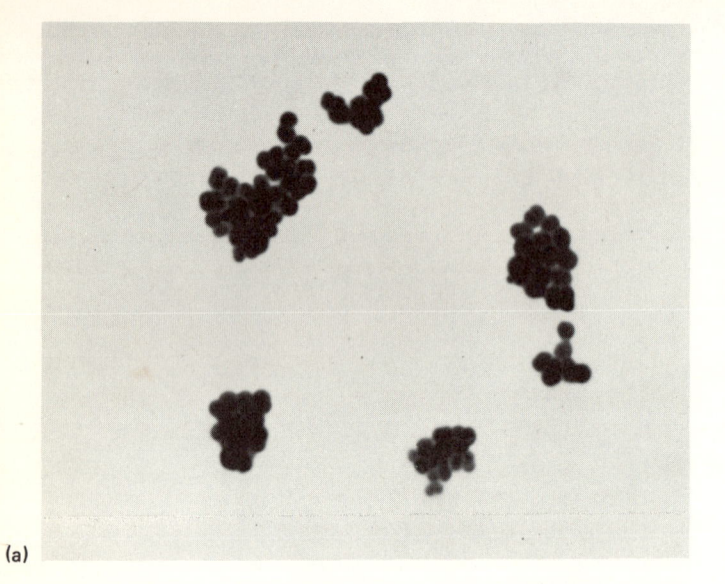

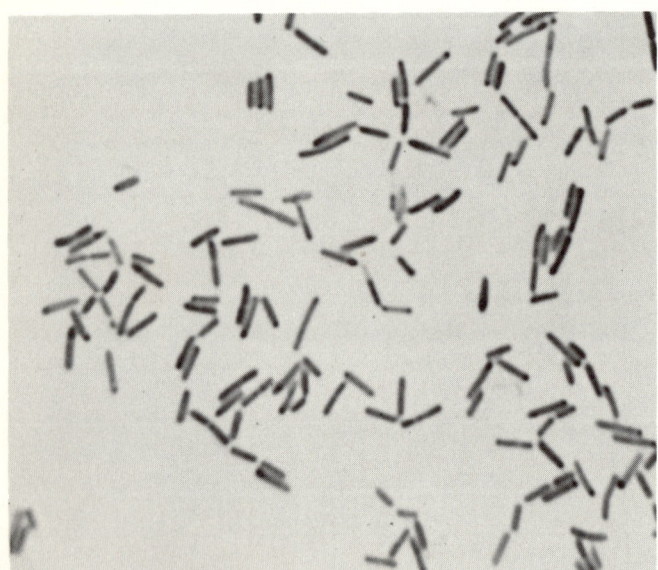

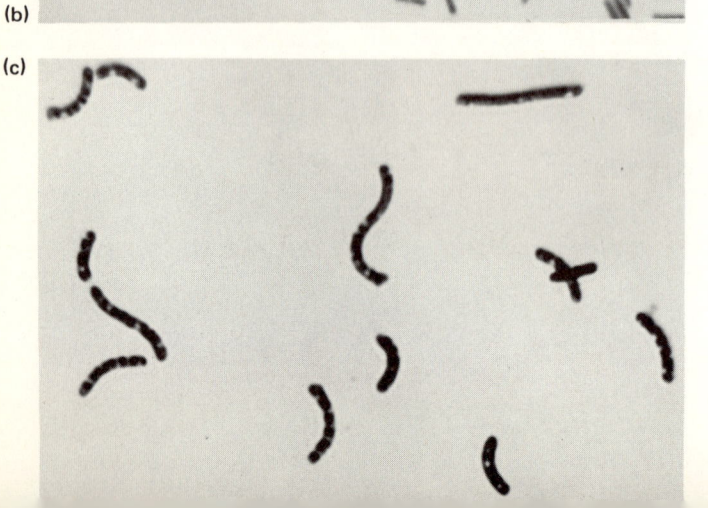

the soil from dead plant and animal material and by directly synthesizing nitrates from the nitrogen in the air. This latter group lives on the roots of certain plants such as clover and alfalfa. Other bacteria live in the intestinal tract where they aid in the production of vitamin K, which is needed in the bloodclotting process. Figure 13.2 shows some representative types of bacteria.

Figure 13.2
Bacteria exist in three major forms: (a) spherical (coccus), (b) rod-shaped (bacillus), and (c) spiral-shaped (spirillum). (Courtesy of the Carolina Biological Supply Company.)

Eucaryotes

The Protists

The protists are eucaryotes, meaning that they have cells with well-organized organelles such as mitochondria. The cell that we described in Chapter 3 is a eucaryotic cell. Protists are primarily single-celled organisms, although in some cases they appear to have aggregated together to form colonial types. Representative groups include the *protozoa*, *diatoms* (*golden algae*), and *euglenoid* organisms.

Protozoa are very abundant in both fresh and salt water. Some, like *Trypanosoma gambiense* (Figure 13.3), are parasitic. They produce diseases such as amoebic dysentery, sleeping sickness, and malaria. Most are free-living and have various means of locomotion. The *amoeba*, for example, moves about by extending its cytoplasm, forming *pseudopods* or "false feet" (Figure 13.4). The *paramecium* (Figure 13.5) propels itself through water by means of *cilia*, which are tiny extensions of the cell membrane that beat in unison, driving the organism in various directions. Protozoa are regarded as being closely related to animals, and it is possible that present-day multicellular animals evolved from protozoalike ancestors.

Eucaryotes 245

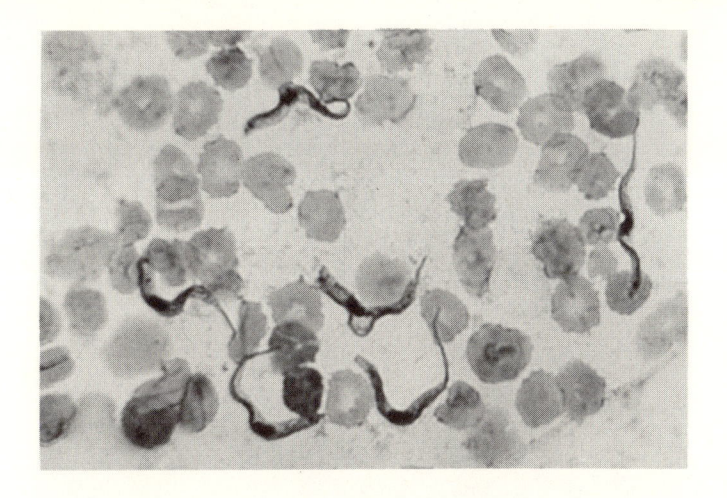

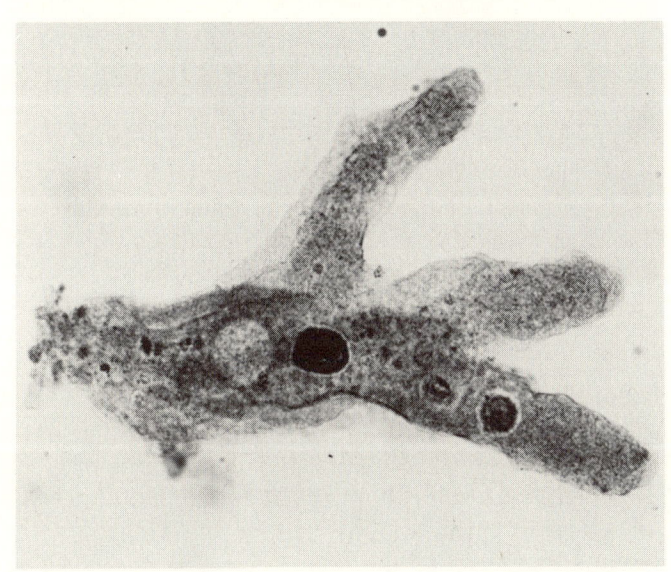

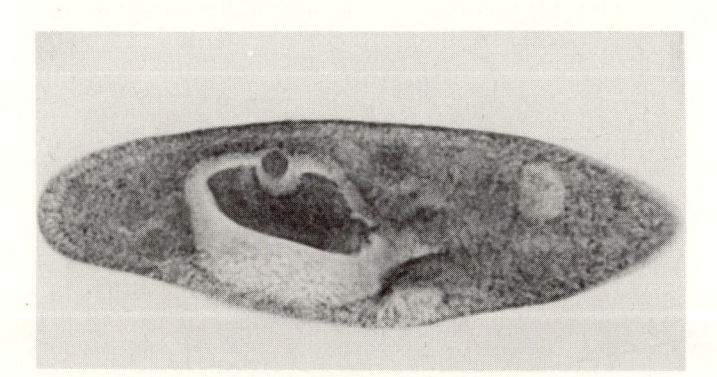

The *euglena* and its relatives comprise a group that was the subject of lively argument among taxonomists during the time when all organisms were divided into plants and animals. The euglena (Figure 13.6) is a protozoanlike organism that part of the time ingests its food like any animal, but it also contains chlorophyll, making it capable of carrying on photosynthesis. Is it plant or animal? Essentially, the problem was solved by placing it among the protists.

Figure 13.3 (top)
Trypanosoma gambiense *is a parasitic protozoan that causes a form of "sleeping sickness." (Courtesy of the Carolina Biological Supply Company.)*

Figure 13.4 (center)
The amoeba is a protozoan that moves about by extending portions of its cytoplasm to form pseudopods, or "false feet." (Courtesy of the Carolina Biological Supply Company.)

Figure 13.5 (bottom left)
The paramecium is a complex protozoan that moves through the water by beating tiny hairlike structures called cilia *which are located on its surface. Like the amoeba, the paramecium is a protist. (Courtesy of the Carolina Biological Supply Company.)*

Figure 13.6 (bottom right)
The euglena. Note the taillike flagellum that propels this organism through the water. The euglena is especially interesting because it contains chloroplasts and carries on photosynthesis. (Courtesy of the Carolina Biological Supply Company.)

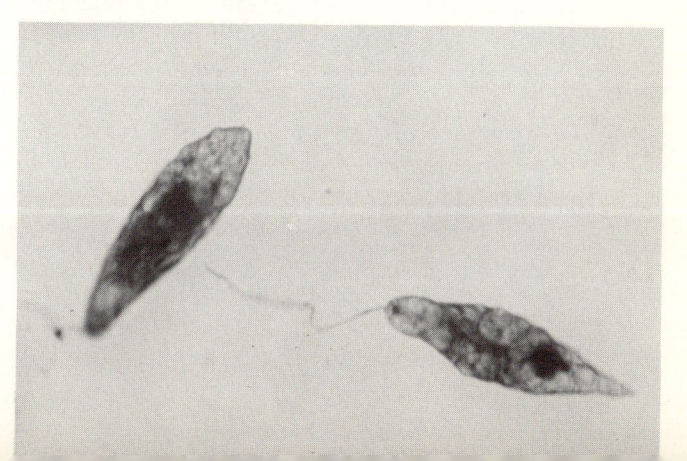

The *diatoms* (Figure 13.7) are among the most abundant organisms in the sea and comprise a large percentage of the *phytoplankton*. Phytoplankton consists of small plant or plantlike organisms which form the first link in many aquatic food chains (Chapter 14), and provide up to 70 percent of all the oxygen that is produced by photosynthesis. Diatoms are golden algae which deposit around themselves hard coats of silica, which is a sandlike substance. "Diatomaceous earth" consists of diatom shells and has a variety of uses, especially in filters and abrasives. In fact, next time you brush your teeth, you may or may not want to recall that your toothpaste consists largely of diatomaceous earth.

The Plants

The Plant Kingdom consists of multicellular organisms which have eucaryotic cells, each of which is surrounded by a rigid wall made largely of cellulose. The cells of plants also contain chlorophyll that is organized into chloroplasts (Chapter 3). When we discussed the "energy scheme" in Chapter 3, we emphasized the importance of the green plant to all life. We will return to this theme again in Chapter 14 where we consider the plant's major ecological role as the *producer*.

The Plant Kingdom includes the red, brown, and green *algae*, together with two groups of more complex plants known as the *bryophytes* and the *tracheophytes*.

The red and brown algae are seaweeds. One form of brown algae known as *kelp* may grow to lengths as great as 200 feet. The green algae are perhaps more familiar, especially to those who do not live near an ocean, since they are commonly seen floating on the surfaces of ponds or washing up on lake shores. Two common examples of green algae are *Spirogyra* and *Cladophora* (Figure 13.8).

Algae are simple forms of plants and are generally confined to water habitats. Forms of algae, probably similar to modern types, must have been abundant in the ancient seas. Biologists believe that land plants must have evolved from ancient forms of green algae which developed an ability to withstand periods of relative dryness, thus setting the stage for the escape of plants to the land.

The *bryophytes* are small plants without true roots, leaves, or stems. There are perhaps 20,000 or more species of bryophytes, including the *liverworts* (Figure 13.9) and the *mosses* (Figure 13.10). Bryophytes are *nonvascular* plants; that is, they do not have conducting systems to carry water and food throughout the plant structure. Our own arteries and veins, for example, make up our *vascular* system. In addition, sexual reproduction in the bryophytes requires the presence of a water medium through which sperm cells can swim to reach and fertilize the egg

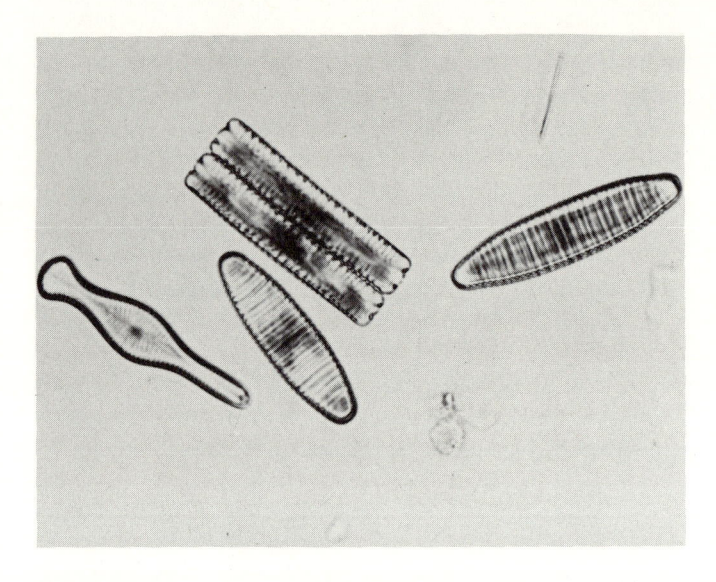

Figure 13.7
Mixed diatoms. Diatoms belong to a group of protists known as the golden algae. (Courtesy of the Carolina Biological Supply Company.)

Eucaryotes 247

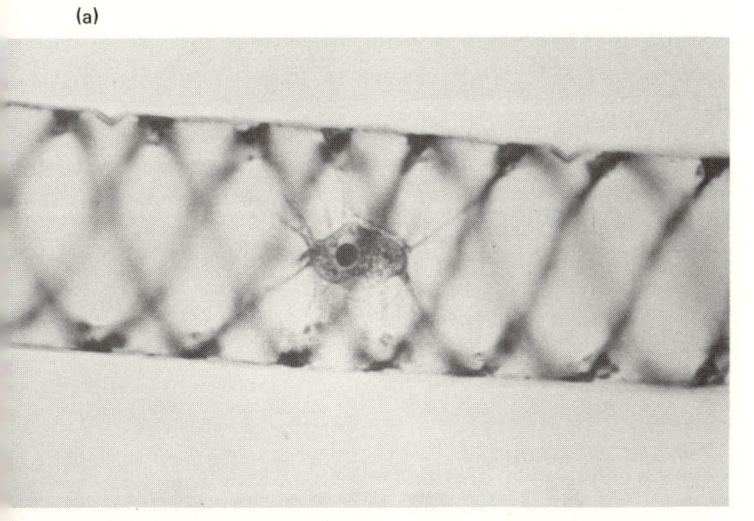

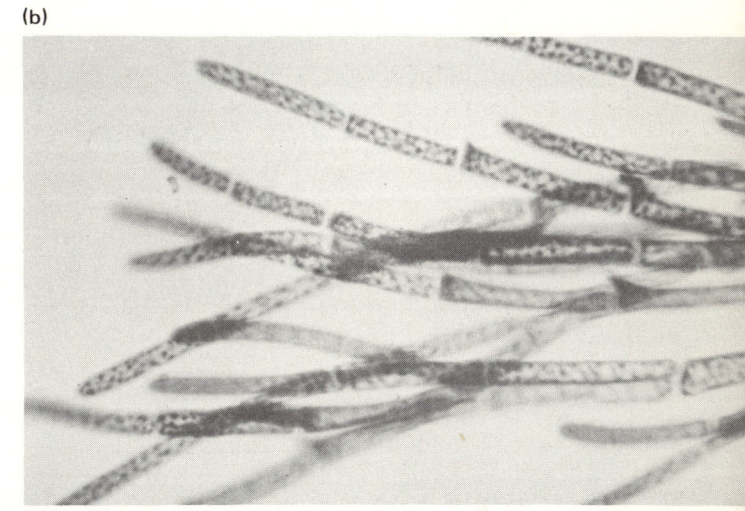

(a) (b)

Figure 13.8
Spirogyra (a) and Cladophora (b) are examples of green algae. Both types are abundant in ponds and lakes.

Figure 13.9
Liverworts belong to a group of plants called the bryophytes. They are generally found in moist wooded areas.

Figure 13.10
Mosses are also examples of bryophytes. They are especially abundant in the woods, where they grow on rocks and on fallen trees. (Courtesy of Dr. James D. Haynes, College at Buffalo, State University of New York.)

cell. This dependence on a more-or-less constant supply of moisture confines the bryophytes to wet areas, such as woodlands, and has limited their potential for spreading throughout the general land areas.

The *tracheophytes* dominate the land areas of the earth; indeed, the term "plant" usually brings to mind tulips, roses, strawberries, orchids, maple trees, pine trees, and other common plants which are members of this group. Both bryophytes and tracheophytes probably evolved from the green algae, but unlike the bryophytes, the tracheophytes are *vascular* plants, possessing conducting systems in their roots, stems, and leaves.

Members of one group of tracheophytes, which includes ferns and horsetails among others (Figure 13.11), are still restricted to moist areas, in part because of their mode of reproduction. Ferns, for example, have no seeds. Instead small brown areas, each of which consists of spore cases, develop on the underside of fern leaflets (Figure 13.12). When the

Figure 13.11
Ferns (a) and horsetails (b) belong to a group of plants called the tracheophytes.

(a) (b)

Eucaryotes 249

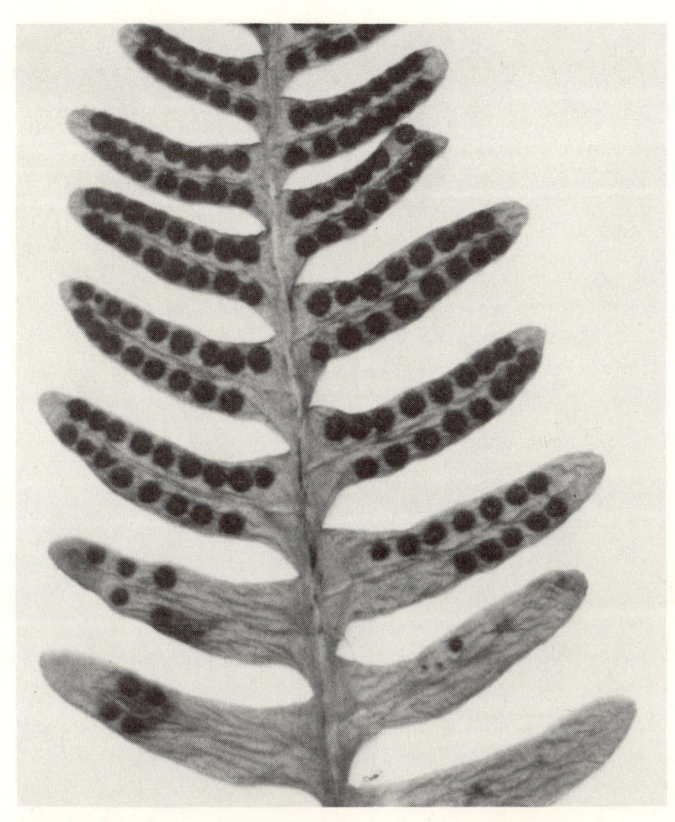

Figure 13.12
Spore cases, called sori, *which are found on the underside of fern leaflets.*

Figure 13.13
A group of evergreens which includes spruce *and* pine *trees.*

tiny spores are released, those that fall onto a moist area develop into small, heart-shaped plants so inconspicuous on the forest floor that they are rarely noticed except by botanists. Each small plant, called a *thallus*, contains two kinds of reproductive structures, one of which produces an egg cell and the other of which produces sperm cells. Sexual reproduction takes place as the sperm cells swim through water to the egg cell. The resulting fertilization of the egg cell produces a zygote which then develops into the more conspicuous fern.

The most familiar plants are found in a tracheophyte group which includes the various kinds of evergreens (Figure 13.13) and the flowering plants (Figure 13.14). These are the *seed plants* and are the most successful group on the evolutionary family tree of the Plant Kingdom. Their success is due in part to the efficient conducting systems in their roots, stems, and leaves, and in part to a system of sexual reproduction that does not require external moisture. Pollen grains containing the sperm nucleus are spread by insects or wind to the female portion of the flower

*Figure 13.14
Flowering plants are the most familiar of
the major plant groups. The flower itself
is the reproductive structure.*

called the *pistil*. A tube then grows from the pollen grain into the *ovary* of the pistil. The sperm nucleus enters the ovary through the *pollen tube* and fertilizes the egg cell. The resulting embryo then becomes part of a seed, which contains food material and has a tough coat that enables the embryo to survive for long periods of extreme cold or dryness until conditions are right for germination.

In addition, flowering plants have evolved various efficient methods of seed dispersal which ensure the spread of at least some of their offspring to fertile areas. The dandelion and maple tree furnish good examples of seed-dispersal adaptations (Figure 13.15). Other adaptations include the production of fruits and nuts which are carried by animals to areas away from the parent plant. Because of their adaptive features, seed plants have colonized most of the earth's land areas, occupying habitats that range from aquatic environments to the most arid desert areas.

The Fungi

In some systems of classification, the fungi are included with the plants, but in the Whittaker system,

Eucaryotes 251

the fungi are classified as a separate kingdom. The fungi include many thousands of different species, and as we will see in Chapter 14, they play an important ecological role as *decomposers*. They are plant-like organisms and have no true roots, stems, or leaves. Fungi are mainly *saprophytic*; that is, they gain nutrients from the environment by absorption from dead and decaying organic materials. Since they contain no chlorophyll, they are not capable of making their own food by photosynthesis.

The *club fungi* include mushrooms, bracket fungi, and puffballs (Figure 13.16). Some of the club fungi, especially certain mushrooms, are edible. Others, however, are extremely poisonous, and it is for this reason that amateurs should *never* gather mushrooms for a meal! It is important to note here that the so-called "certain" signs of edibility are by no means always certain.

Figure 13.15 (left)
Dandelion (a) and maple (b) seeds are carried by the wind away from the parent plant; this helps spread the species over a wide area.

Figure 13.16 (below)
A mushroom (a), a bracket fungus (b), and a puffball (c) are all examples of club fungi. They are saprophytes; that is, they absorb their food from dead and decaying organic materials. (Photo of puffball courtesy of Dr. James D. Haynes, College at Buffalo, State University of New York.)

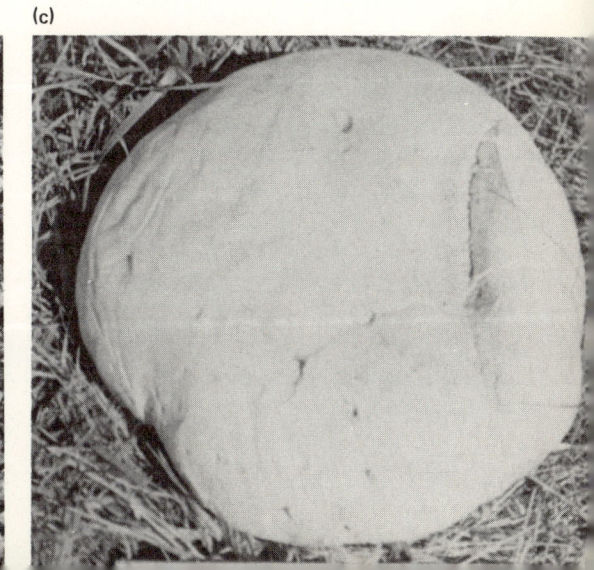

The *sac fungi* include the mold from which penicillin is derived, as well as several other molds that are used in the manufacture of cheese. Roquefort cheese, for example, owes its characteristic flavor and appearance to the mold *Penicillium roqueforti*.

Bread molds belong to a class of fungi that are algalike in appearance. This same group also includes the mold which caused the famous potato famine in Ireland during the mid-1840's that was responsible for the emigration of large numbers of Irish people to America. Still another group of fungi include those responsible for athlete's foot and "ringworm," which is not really a worm at all, but a fungus.

Slime molds are peculiar and fascinating organisms that grow among wet, decaying leaves on the forest floor. They are glistening sheets of orange, yellow, or red color that may grow up to two or three feet in diameter. At this stage of its life cycle, the slime mold looks like a colony of thousands of amoebas as it crawls about taking in bacteria and other bits of organic matter. After a period of feeding and growth, the slime mold crawls to a dry spot where it develops spore cases which eventually break, releasing the spores. Those spores which land in a wet area then develop into a new sheet of living material.

The Animals

Animals are multicellular organisms with eucaryotic cells that, unlike plant cells, do not have cell walls. Animal cells also lack chlorophyll. Because animals therefore cannot manufacture their own food, the major mode of nutrition is ingestive; that is, animals take in food materials from the environment and put it through a digestive process (see Chapter 10). Like the tapeworm, some absorb food nutrients directly from their environment; therefore they have no need for a digestive system.

It would take a very large zoology textbook to adequately describe even a major portion of the multitude of organisms that make up the Animal Kingdom. There are, for example, around 600,000 species of insects alone! We will therefore limit ourselves to a brief discussion of the more major phyla and some of their representative examples.

There are a number of invertebrate phyla, which are groups of animals without backbones. The simplest of these are the *sponges* (Figure 13.17), which are simple aquatic animals that usually consist of colonies of single cells. At one time, the skeletons of sponges were used as commercial sponges, but today

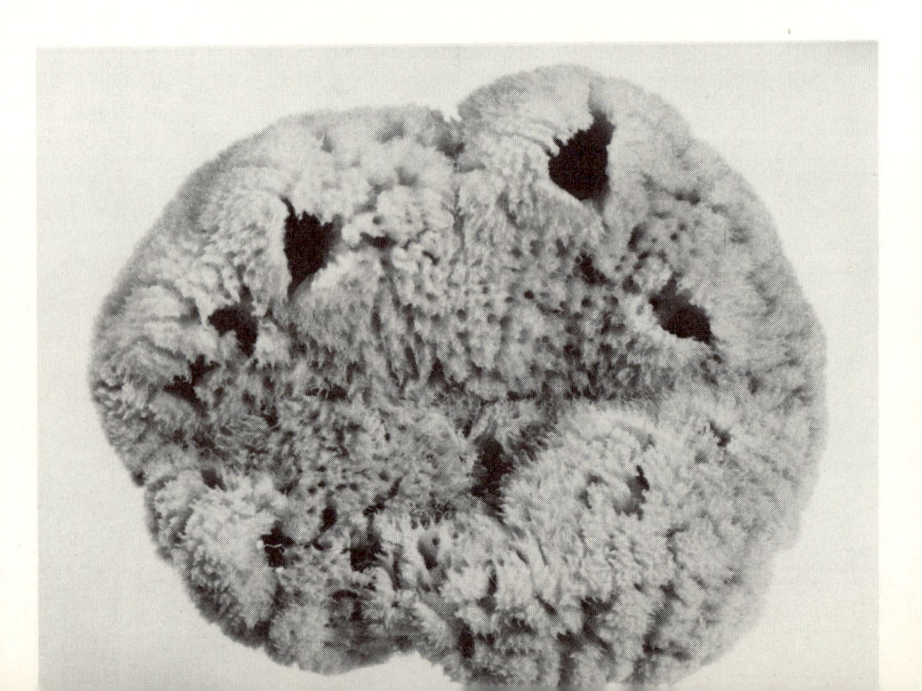

Figure 13.17
A commercial sponge. (Courtesy of the Carolina Biological Supply Company.)

Eucaryotes 253

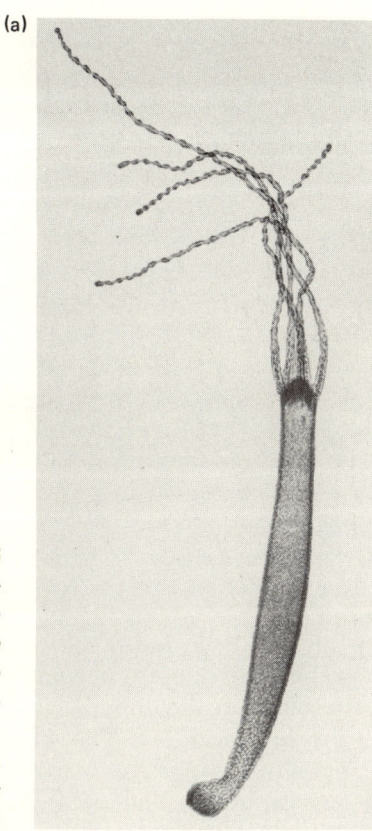

most of the latter are synthetic. Biologists generally regard the sponges as an evolutionary dead end from which no other group evolved.

The *coelenterates* are more complex than the sponges. They have a saclike body which usually has tentacles containing stinging cells (Figure 13.18). These stinging cells are used to paralyze any prey that is unlucky enough to come within the grasp of the tentacles. The Portuguese man of war (Figure 13.19) is a good example of the stinging power of some coelenterates. Its sting has been known to para-

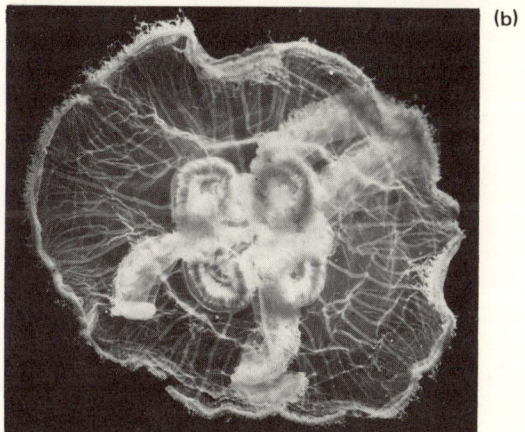

Figure 13.18
Hydra (a) and the jellyfish (b) are saclike animals belonging to the coelenterates. Note the tentacles which help the animals obtain food. (Courtesy of the Carolina Biological Supply Company.)

Figure 13.19
Portuguese man of war. (Courtesy of the Fisher Scientific Company, Educational Materials Division, Chicago, Illinois.)

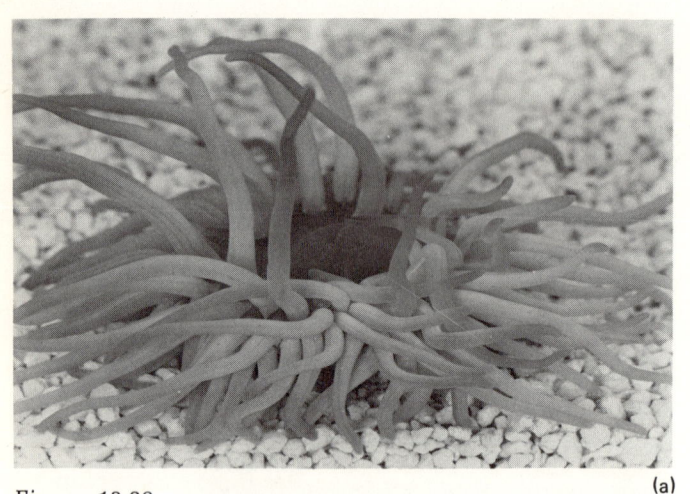

Figure 13.20
The sea anemone (a) looks like a flower but is actually an animal that belongs to the coelenterates. Corals (b) are also coelenterates. This particular variety is called a brain coral. (Courtesy of the Carolina Biological Supply Company.)

lyze swimmers in the ocean waters off the coast of southern United States. Sea anemones and corals (Figure 13.20) are other examples of coelenterates. Corals flourish in tropical seas. Their skeletons, which pile on top of each other over the years, build giant reefs off the beaches of tropical islands. One theory holds that the coelenterates may have given rise to the flatworms, but others believe that like the sponges, the coelenterates represent an evolutionary dead end.

Two invertebrate phyla, the *flatworms* and the *roundworms,* are significant because they contain a number of examples that are parasitic to man and other animals. Flatworms include the flukes (Figure 13.21) and the tapeworms (Figure 13.22), and the roundworms include trichina (Figure 13.23), hookworms, and the pinworms that often infest the intestinal tracts of young children.

The *molluscs* form another important invertebrate phylum. There are about 80,000 species of molluscs including clams, mussels, oysters, squid, and the octopus.

The *annelids* are segmented worms. Each external groove on the annelid body marks the location of an internal partition. The worm therefore consists of a series of compartments that are more or less similar to one another. The organization of the body into segments was apparently a significant evolutionary milestone. For one thing, the presence of segments permits an increase in the size of the organism by duplicating the internal organs in each segment, thus maintaining efficient organ function. The annelids also have a well-developed *closed* circulatory system, in which blood is pumped through the tissues, carrying oxygen and nutrients to the individual cells and taking waste materials away. A closed circulatory system was a necessary evolutionary development as animals left the watery medium of the sea and ventured onto land. The best known of the annelids is the earthworm, which is well adapted to living in moist, humus-filled soil. Actually, most annelids are aquatic. The better known of the aquatic annelids are the leeches, which are bloodsuckers that cling to their hosts with suction discs (Figure 13.24).

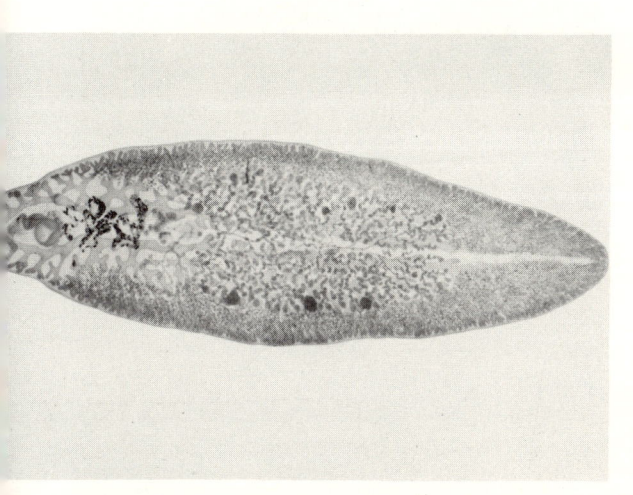

Figure 13.21
The fluke is a variety of flatworm. The type shown here is a parasite which invades the liver and bile duct. (Courtesy of the Carolina Biological Supply Company.)

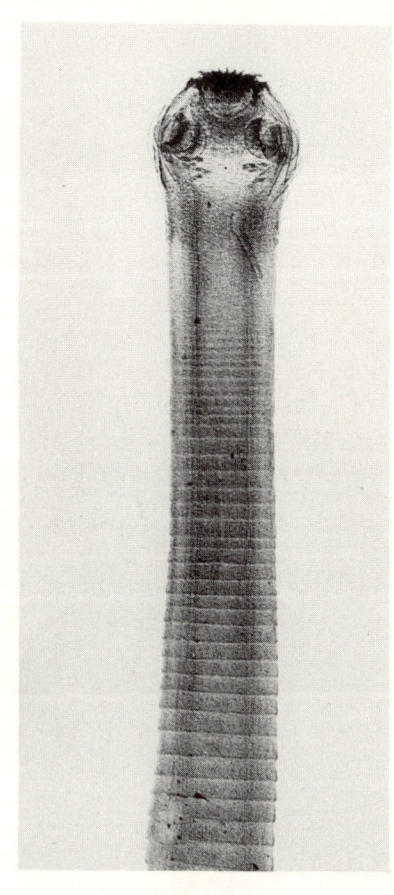

Figure 13.22
The tapeworm is a parasitic flatworm that lives in its host's digestive system, where it absorbs nutrients through the body wall. Note the suckers and hooks which help the tapeworm cling to the intestinal wall. (Courtesy of the Carolina Biological Supply Company.)

Figure 13.23
Trichina is a parasitic roundworm that may be transmitted to man from infested pork. This photograph shows the young stage of the worm encysted in muscle tissue. (Courtesy of the Carolina Biological Supply Company.)

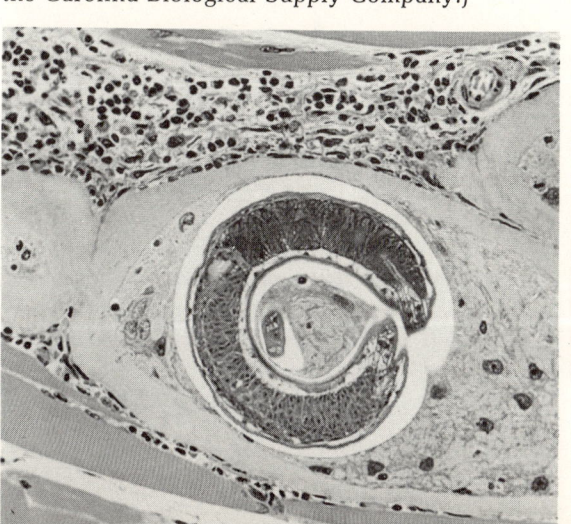

Figure 13.24
The leech is an example of a group of segmented worms called annelids. The leech is an aquatic animal that feeds on the blood of other animals. The common earthworm is another example of an annelid. (Courtesy of the Carolina Biological Supply Company.)

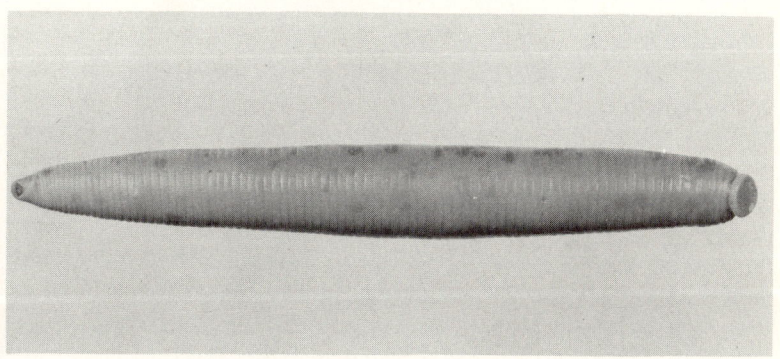

256 Noah's ark

The *arthropods* form one of the largest animal phyla. The group contains at least 650,000 species in all, most of which are insects. Arthropods are animals with outer skeletons, jointed legs, and definite body regions. The arthropods include insects, lobsters, spiders, crabs, crayfish, shrimp, centipedes, and millipedes. See Figure 13.25 for more examples of arthropods.

Figure 13.25
Some examples of arthropods are (a) the tarantula, (b) the grasshopper, (c) the horseshoe "crab," (d) the lobster, and (e) the cabbage butterfly. (Courtesy of the Carolina Biological Supply Company.)

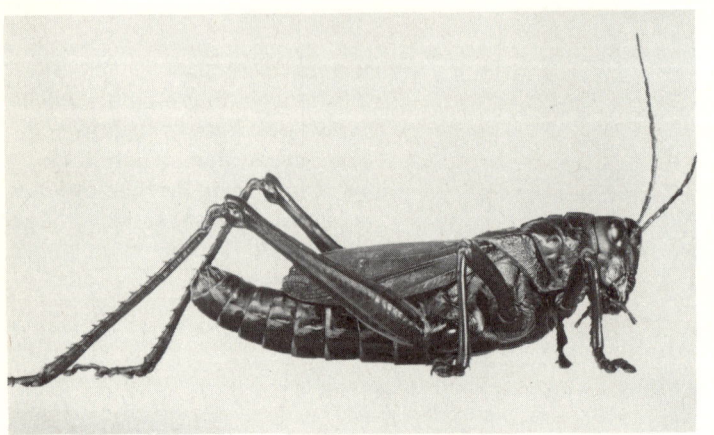

The *echinoderms* include animals such as the starfish and the sea urchin (Figure 13.26). The echinoderms are characterized by what biologists call *radial symmetry;* that is, the body plan is similar to that of a wheel in which the spokes radiate outward from the center in all directions. Evolutionists sometimes speculate that the echinoderm group may have given rise to the chordates because the larval stage of some echinoderms resembles some of the more simple chordates.

The chordates themselves are characterized by the presence of a backbone or a simple "prebackbone" structure called a *notochord* (Figure 13.27). Those in the latter group include acorn worms and tunicates, which may represent "links" in the evolution of the vertebrates.

The five classes of vertebrates are (1) the fishes, (2) the amphibians, (3) the reptiles, (4) the birds, and (5) the mammals.

Figure 13.26
Two examples of echinoderms: (a) the sea urchin, and (b) the starfish. (Courtesy of the Carolina Biological Supply Company.)

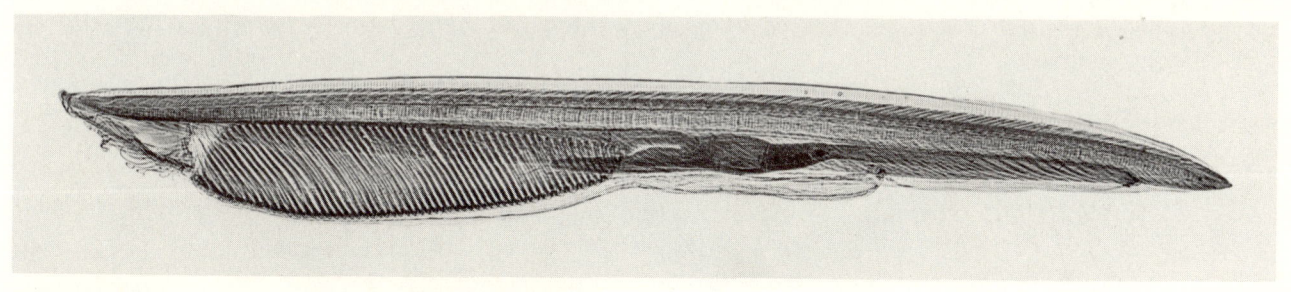

Figure 13.27
Amphioxus may represent an evolutionary link between the invertebrates and the vertebrates. It does not have a true backbone, but does have a simpler structure called a notochord. In true vertebrates, a notochord forms early in development, but is later replaced by a backbone. (Courtesy of the Carolina Biological Supply Company.)

The fish include the *jawless fishes* such as lampreys and other eel-like fishes which have suction disclike mouths and lack fins. The *cartilaginous fishes* are marine fish with jaws and fins and with skeletons that are formed of cartilage instead of bone. The sharks are well-known examples of this group (Figure 13.28). However, the vast majority of modern fishes such as trout and salmon belong to the group known as the *bony fishes* (Figure 13.29).

The *amphibians* are animals with moist skins which lack scales (Figure 13.30). Typically, their life cycle is spent partially in the water, where the larvae breathe by means of gills, and partially on land as adults which have lungs and are air-breathers. The frog, for example, lays its eggs in the water where they develop into tadpoles. The tadpoles maintain a fishlike existence until they change into the adult frog. In one sense, therefore, the frog appears to rep-

Figure 13.28
The shark belongs to a group of fishes which has a skeleton made of cartilage instead of bone. It is generally considered to be more "primitive" than the bony fishes. (Courtesy of the Carolina Biological Supply Company.)

Figure 13.29
The bony fishes include the more familiar game fish such as the perch shown here. (Courtesy of the Fisher Scientific Company, Educational Materials Division, Chicago, Illinois.)

Figure 13.30 (left)
The bullfrog is a common example of an amphibian. (Courtesy of the Carolina Biological Supply Company.)

resent a form of life that has only partially "escaped" the aquatic environment in which life itself was born. Toads and salamanders are other examples of amphibians.

The *reptiles* are lung-breathing animals with scaly, dry skins. They are believed to have evolved from amphibian ancestors, but unlike the amphibians, the reptiles have made good their escape to the land. The frog, for example, must maintain a moist skin because there is an exchange of gases through the skin to supplement the action of the frog's lungs. The reptile, on the other hand, has an improved lung function which removes this necessity to remain near the water. In addition, the reptiles "solved" the problems posed by reproduction without a watery medium through which the sperm can swim to the egg cell. The "solution" included the evolution of internal fertilization and a waterproof egg in which the embryo can develop on land. The reptiles include snakes, chameleons, lizards, turtles, and alligators (see Figure 13.31).

Figure 13.31
The cottonmouth snake (a) and the chameleon (b) belong to the reptiles. (Courtesy of the Carolina Biological Supply Company.)

Figure 13.32
Birds are believed to be descended from the reptile group. Shown here are (a) a common songbird, (b) a bird of prey, and (c) a waterfowl. (Courtesy of Dr. Olin S. Pettingill, Jr.)

The *birds* have feathers and light, hollow-boned skeletons which help them fly. Their scaly legs and beaks are marks of their reptilian ancestry (Figure 13.32). Birds are "warm-blooded"; that is, they maintain a constant body temperature regardless of the temperature of their environment. Fish, amphibians, and reptiles, on the other hand, are "cold-blooded" organisms; their body temperature tends to fluctuate with the environmental temperature.

Mammals, the class to which man belongs, have body hair and feed their young milk secreted by the mammary glands of the female. The young are born alive, with the exception of those of the egg-laying duckbilled platypus. In one group called the *marsupials,* the young are born at a very early stage of development and complete their development in a pouch located on the mother's body. Marsupials include the kangaroo and the opossum. Like the birds, mammals are "warm-blooded." They are divided into approximately fourteen orders, one of which (the primates) includes man and his closest relatives, the great apes. See Figure 13.33 for several examples of mammals.

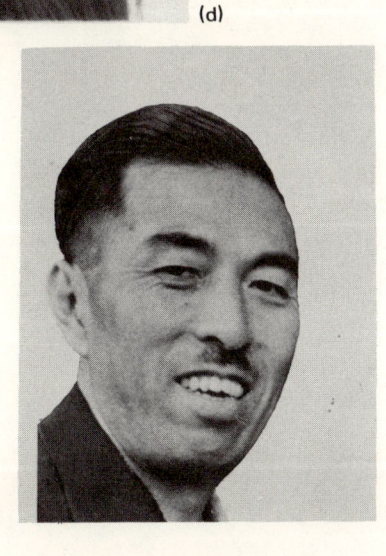

Figure 13.33
Examples of mammals are (a) the tiger, (b) the lion, (c) the dolphin, (d) the orang-utan, and (e) Homo sapiens. Which have the least remote common ancestor—the tiger and the lion, or the tiger and the dolphin? Homo sapiens and the orang-utan, or Homo sapiens and the lion? (Part (e) courtesy of the American Museum of Natural History.)

Summary

Despite the technical problems involved with classifying the multitude of earth's organisms, we can see that there is an element of unity and order in the wide diversity of life. On the cellular level, for example, the dog, the cat, the rat, the amoeba, and man all have much in common. They are all eucaryotes with similar patterns of cell organization. DNA, the basic genetic material, is universal throughout life even though the details of DNA structure may differ from species to species or from kingdom to kingdom. No matter what system of classification is used, one is struck by the obvious relationships which suggest the common origins of species.

How do these myriad organisms and species fit into the vast scheme of life? Do they exist separately, in the little pigeonholes of the taxonomist, or are they each only small parts of a vast and intricate pattern? In the next chapter, we will examine this pattern and see how all organisms live in a constant state of interdependence.

Questions

1. For purposes of classification, why is it better to divide living organisms into procaryotes and eucaryotes rather than into plants and animals?
2. Some classification systems group plants and fungi together. What justifications can be offered for dividing them into separate kingdoms?
3. Life shows both diversity and unity. Explain this statement.
4. What patterns of relationships emerge from classification systems? What is the significance of such relationships?

14

The Ecosystem

The Sun—The Ultimate Source

The earth spins on its axis like a gigantic roast on a spit, ever turning its seas and continents toward the solar radiation that streams through space from 93 million miles away. Without the sun, life on earth would have had no beginning; without it all life would cease to be. If some cosmic giant were to cast his shadow over the earth, he would watch it become a cold, dead planet, its oceans turn to ice, and its atmosphere solidify on a lifeless surface.

The sun is the ultimate source of earth's energy, including that which is needed to sustain the processes of life and that which is used to power the energy-consuming trappings of man's civilization. Energy is an all-important commodity, for when it is in

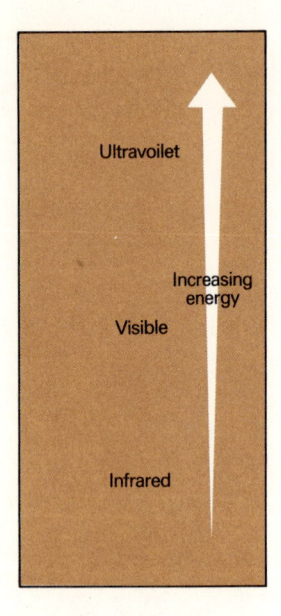

Figure 14.1
Spectrum of solar radiation. Note how the energy of radiation increases when going from infrared to ultraviolet.

short supply people starve, the automobile stays in the driveway, the house grows cold, industries collapse, the stockmarket tumbles, and nations quarrel.

The sun is 864,000 miles in diameter, over 100 times that of the earth. It is a luminous giant that dominates the solar system and dwarfs the nine planets that make up its family. Like a hydrogen bomb it combines hydrogen into helium, releasing vast amounts of energy that travels through space in all directions. Only the tiniest fraction of this solar energy is intercepted by the earth, the third planet from the sun and 93 million miles away.

The sun's energy comes to us as *radiant* energy, which travels through space at a speed of 186,000 miles per second. This energy consists of a spectrum of radiation that ranges from high-energy cosmic rays, x-rays, and ultraviolet through visible light to the lower-energy infrared and radio waves (Figure 14.1). The highest-energy portion of solar radiation is lethal to living organisms, but fortunately most of it is screened out by the atmosphere. In general, the atmosphere and cloud cover reduces the impact of solar radiation so that not more than 50 percent or so actually reaches the earth's surface.

The rotation of the earth on its axis results in daily variation of solar intensity at any given location as day alternates with night. Furthermore, the earth's annual revolution around the sun as well as the tilt of the planet's axis provide an additional source of variation in solar impact (Figure 14.2). The degree of solar radiation, therefore, varies with such factors as time of day, season of the year, and latitude. The effects of

Figure 14.2
Relationship of the earth to the sun. Due to the tilt of the earth's axis, the amount of radiant energy on different parts of the earth changes with the seasons.

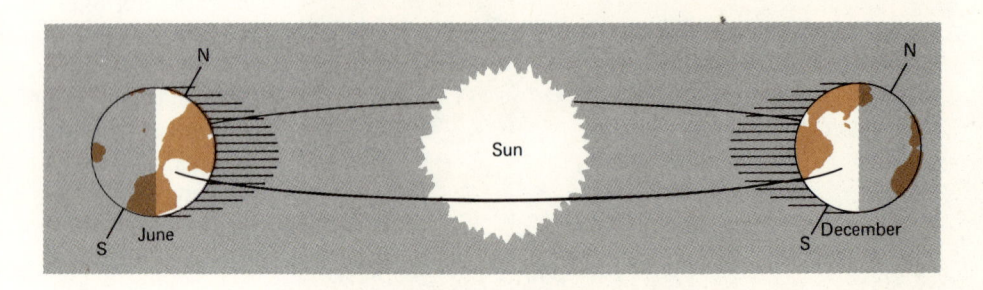

latitude on the living world are quite apparent if we think of the variation from the lush rain forests of the tropics to the sparse vegetation found on the frozen tundras of the north.

The ultraviolet portion of solar radiation is not visible to the eye, but it is well known to the swimmer or skier because of its tanning and burning effects on the skin. Ultraviolet radiation also plays a role in the synthesis of vitamin D, and in a later chapter we will see how this process may have been a factor in the development of variation in human skin color.

At the other end of the solar spectrum, infrared radiation is converted to heat as it strikes the surfaces of the earth. Some of this heat is lost back into space, but a great deal more would be lost if it were not for the insulating blanket provided by the atmosphere. Infrared radiation penetrates the atmosphere as it does the glass roof of a greenhouse, and like the greenhouse roof, atmospheric carbon dioxide and water vapor prevent an excessive loss of heat back into space. Otherwise, like the airless moon, the earth's surface at night would become unbearably cold. This "greenhouse effect" becomes more pronounced with higher concentrations of carbon dioxide and water vapor. You may already be aware that the first frosts of autumn are more likely to occur when nights are clear than when heat loss is reduced by a blanket of clouds.

That portion of solar radiation consisting of visible light may be dispersed in a rainbow of colors that includes (in order of increasing energy) red, orange, yellow, green, blue, and violet (Figure 14.3). These components of visible light are selectively reflected to the eye according to the nature of the pigments that give an object its color. For example, a red object *reflects* only the red portion of the spectrum and *absorbs* the remaining energies. A green plant, on the other hand, reflects the green portions of visible solar radiation and absorbs the red and the blue-violet portions. These red and blue-violet radiant energies are therefore absorbed and used by the plant as it converts light energy to the chemical energy that

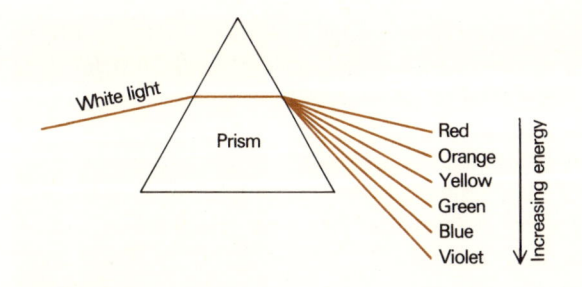

Figure 14.3
Diagram showing white light dispersed by a prism into the various colors (energies) of the spectrum.

turns the wheels of life. We might roughly compare the green plant to a television set that tunes in only the desired signal while rejecting those from other stations in the area.

The importance of the green plant to the grand scheme of life cannot be overemphasized. Looking back to Chapter 3, you will recall how the process of photosynthesis "traps" solar energy and stores it as chemical energy in the molecular bonds of foods such as carbohydrates. You will also recall that the green plant produces oxygen as a by-product of photosynthesis. You have also seen how cellular respiration releases the chemical energy that was stored in the food molecules, making it available to run the machinery of the living world. The energy relationships among living organisms are very important and also very complex, and the study of the energy flow from the sun throughout the world of life is a significant part of that branch of biology known as *ecology*.

Ecology

Our new-found concern with the natural environment has made *ecology* practically a household word. Like many so-called "scientific" terms that come into common use, it is often misunderstood and misused. Those who point out, for example, that polluting a lake by dumping into it municipal and industrial wastes "destroys the ecology" betray a lack of understanding of the true meaning of the word. In brief, the ecology of a lake or river is not "destroyed" by pollution, but *changed* in a way that may not conform to our values, both aesthetic and economic. For example, Lake Erie (Figure 14.4) is widely used as a classic illustration of a lake that has been "killed" by pollution. On the other hand, it is described by Dr. Robert Sweeney of the Great Lakes Laboratory at Buffalo, New York as a "very lively corpse." The lake still teems with life; in fact, it is rich in nutrients, but the polluted environment has drastically decreased or eliminated the game fish. Left in their place are greater numbers of "trash" fish such as carp and sheepshead, which are better adapted to survive under conditions of pollution.

The ecologist, meanwhile, is often equated with the environmentalist and is pictured as going about seeking injunctions against those who would pollute the environment. We should hasten to point out, of course, that most ecologists *are* environmentalists because of the nature of their training and interests. The reverse is not necessarily true, however; not all environmentalists are ecologists.

Modern ecology is a broad and complex science. It encompasses most of the other areas of biology as well as mathematics and the physical sciences. As such, it has gone far beyond its earlier beginnings which were largely associated with nature study and conservation. Since the word *ecology* is derived from the Greek word *oikos,* meaning *home,* it can be simply defined as the study of living organisms in their

Figure 14.4
Lake Erie (one of the Great Lakes) is widely used as a "model" of the effects of pollution on a body of water.

"homes," or in their natural environments. More specifically, however, modern ecology is concerned with the complex *energy relationships* that exist among various segments of the living, or *biotic,* world. This involves the interrelationships of plants and animals as well as the physical environments in which they live.

The ecologist is concerned with the *biosphere,* which may be defined as the total of all living organisms that exist on earth. It may also be said that the ecologist is concerned with the *ecosphere,* described by Dr. Lamont C. Cole of Cornell University as the sum total of all living things on earth *together* with the nonliving environment of which they are a part. As a practical matter, however, the ecologist usually confines his study to the *ecosystem,* which is the most fundamental unit of ecology.

The Ecosystem

An *ecosystem* is usually defined in terms of a relatively limited area. It may be a forest, or a pond, or even a fallen tree rotting on the forest floor (Figure 14.5). Given one of the foregoing situations, an ecosystem represents the living (biotic) portion together with the nonliving (abiotic) portion as they exist and interact with each other. Therefore, an ecosystem typically consists of (1) several populations of organisms making up a community, and (2) environmental factors such as temperature, light, and water. A study of such an ecosystem will reveal complex interaction and interdependence between the various populations as well as between the organisms and the physical environment.

A *population* is a geographically localized group of organisms belonging to the same species. A stand of oak trees, dandelions splashing a lawn with yellow in late spring, and red-winged blackbirds flying among the cattails at the edge of a creek all represent populations. For that matter, so does a field of wheat; a wheat field may also constitute an ecosystem if we include the physical environmental factors and various living organisms present in the field. A popula-

Figure 14.5
The forest and the pond (a) are different kinds of ecosystems. The log rotting on the forest floor (b) is still another type of ecosystem.

tion, therefore, consists of individuals with basically the same characteristics, although they vary among themselves in ways that are not always obvious. The members of a flock of sparrows may all look alike to us, but then we probably all look alike to the sparrows.

A typical ecosystem contains several different kinds of species-populations which together make up a *community*. Most natural communities contain representatives of the bacteria, fungi, algae, and higher forms of animals and plants. In addition, parasites of some kind almost inevitably live in association with other members of the community.

In a typical human community such as a small city, everyone has a place in the scheme of things, a job to do, a function to perform. The physician treats the sick; the police officer sees that law and order prevail. We could say that each individual or group of individuals has a place or *niche* in the community, contributing his or her efforts to the total operation and continued existence of that community. This situation might be roughly compared to that of the natural community, where the various species occupy special niches.

For example, the *producers* supply carbohydrates, proteins, fats, some vitamins, and oxygen. Therefore, we correctly conclude that the producers are the green plants which convert solar energy into the chemical energy that drives the living portion of the ecosystem. These indispensable members of the community range from the giant trees of a forest community, to the grasses of a meadow, to the tiny aquatic

Figure 14.6
Phytoplankton consists of varieties of microscopic plant life which act as major producers in many aquatic ecosystems. (Courtesy of Dr. James D. Haynes, College at Buffalo, State University of New York.)

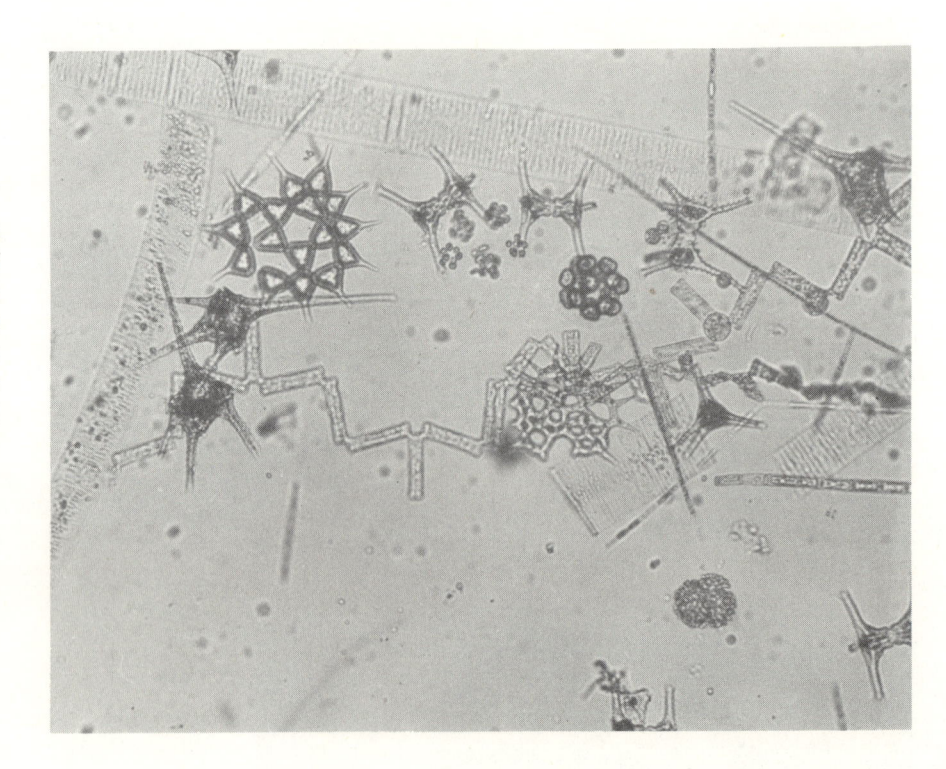

plants called *phytoplankton* (Figure 14.6) that live in a pond or an ocean. Generally, the name given to a community is taken from the *dominant* producer, which is the plant species that produces most of the energy for the community. We may therefore speak of a beech forest community or of a grassland community.

The typical community also contains *consumers*, which as the term implies consume or eat the producers and thus gain part of the energy that the producers captured from the sun. The first-level consumers are *herbivores*, or planteaters. The second-level consumers are *carnivores*, or meateaters, who prey upon the herbivores, thus deriving from the herbivores a portion of the energy which they in turn had gained from the producers. To add to this, in some communities there may be first- and second-level carnivores, where the second-level carnivore preys upon the first-level carnivore. Some animals, such as man, are *omnivores,* meaning that they eat both plants and other animals.

Meanwhile the *decomposers* break down the waste materials of the ecosystem, converting the dead bodies of other organisms to nitrogen compounds and other raw materials for recycling back to the producers. This is the "disassembly crew" of the community; they consist of bacteria, yeasts, molds, and other fungus plants (Figure 14.7). In one sense, they may be thought of as third-order consumers.

The *scavengers* are the "garbage men" of the community. They clean up debris consisting of dead animals and plants. A crow perched on the carcass of an animal killed on the highway is a familiar example of a scavenger. So is the vulture wheeling in the sky, patiently waiting for the western movie hero to run out of water as he painfully crosses the desert.

Most organisms in a community are hosts to several kinds of *parasites*. Parasites are organisms that live at the expense of other organisms; they include the tapeworm, the hookworm, the trichina worm that infests pork, as well as the bacteria that normally inhabit the large intestine. The best-adapted parasites are obviously those that live on or in their host without killing the host and thereby removing their source of food.

The foregoing represents the major niches that exist in a typical community. An important principle of ecology states that where two or more closely related species occupy the same niche, those species will compete with each other to the point where one or the other will either be driven to extinction or forced to another geographic location. Another possibility is that one or the other of the two species may be forced to occupy a different niche. To return to our human community analogy, it is somewhat like two supermarkets existing in a small town that can adequately support only one supermarket. The competition is se-

Figure 14.7
The fungus growing on this dead stump will help break it down to the point where it will become part of the soil.

vere, and only the "best-adapted" supermarket will survive. The other will either go out of business (become extinct) or move out of town to another location. Or one of the supermarkets might possibly convert to a department store and thus occupy a different niche.

On the other hand, if a community has no supermarket at all, there is obviously an excellent opportunity for someone to move in and take advantage of the fact that a portion of the community's wealth is available to buy groceries. In a similar fashion, all possible niches in a typical natural community tend to be occupied, because an empty niche represents an available portion of the ecosystem's energy. The pressures toward evolutionary change may therefore favor the movement of a species toward occupying an empty niche.

Energy Flow in an Ecosystem

The availability and use of energy constitute one of the most critical problems in the living world. A major role of the ecologist, therefore, involves the systematic study of the flow of energy through the ecosystem. This energy flow is illustrated most simply by the *food chain,* which in general terms might be described by the following.

Producers → Herbivore → First-level carnivore → Second-level carnivore → Decomposer

In more specific terms, the following is a possibility.

Green plants → Mouse → Snake → Hawk → Bacteria

A simple and familiar example of a shorter food chain is

Grass → Cattle → Man → Bacteria

in which man's niche is that of a first-level carnivore that "preys upon" the herbivore.

The food chain is an oversimplified version of what happens in the natural situation, since it implies a rigid, straight-line relationship in which a given animal will exploit one and only one source of food. Actually the appetites of most organisms (including man) are not that specialized, and they tend to vary their diets according to what is available. A fox, for example, may prefer mice, but it happily dines on a rabbit if one is available. In fact, when game is scarce a fox may supplement its diet with berries, apples, and insects. Wolves prefer small game that is easy to catch, but in times of famine they will band together in order to overpower a large animal such as an elk or moose. The mountain lion's taste may be for deer, but it will settle for an unwary cow or rabbit if deer are in short supply. A "man-eating" tiger is almost always old or lame and has turned to human prey because he is no longer able to kill the more fleet-footed game that makes up his normal diet. In the natural situation, therefore, animals normally eat several types of foods. Conversely, a specific food is eaten by several different animals. These predator-prey relationships are described by complex *food webs,* a simplified example of which is illustrated in Figure 14.8.

Perhaps the best illustration of the energy flow in an ecological community is the *food pyramid,* or *trophic pyramid.* Trophic means "to eat." The green plant, for example, is said to carry on *autotrophic* nutrition (self-feeder), while the nongreen plants and animals of an ecosystem are *heterotrophs* (other-feeders). You will recall that a conversion or transfer of energy is inevitably accompanied by a loss of part of that energy in the form of heat. For example, we pointed out in Chapter 3 that only a portion of the potential energy in a gallon of gasoline is available to drive the wheels of a car. In a similar fashion, only a portion of the potential energy of food molecules is converted to ATP during respiration. It therefore follows that not all of the potential energy contained in a given quantity of producer material is available to

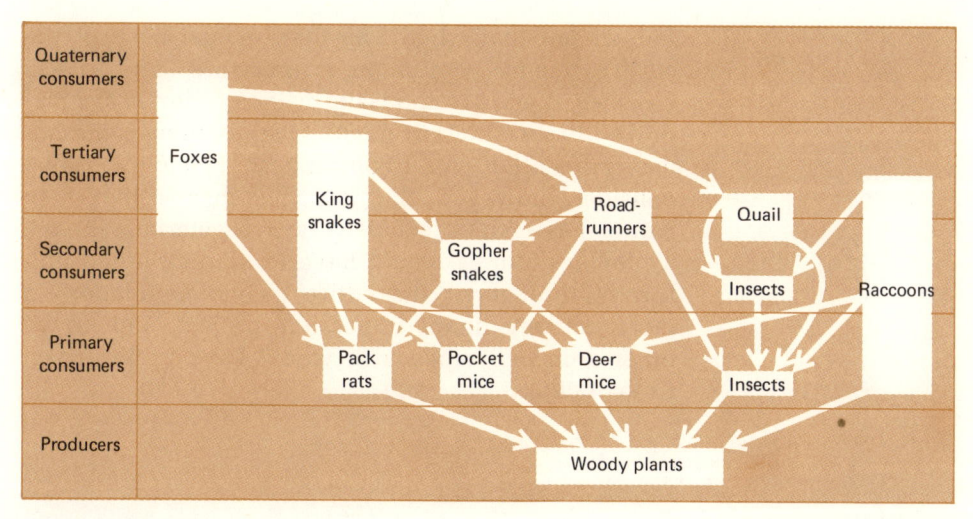

Figure 14.8
Diagram of a food web. The food sources of each population are indicated by arrows. For example, raccoons feed upon insects, deer mice, and woody plants.

the herbivores that feed upon it. Furthermore, not all the energy of the herbivores is available to the first-level carnivores, and so on up the food chain.

This important concept of decreasing available energy is illustrated by the food pyramid in Figure 14.9. It may be seen that as we go up from the base of the pyramid, each trophic level supports fewer individuals, because only a fraction of the energy contained in each trophic level can be transferred to the next.

There are some important and practical lessons to be learned from the realities of energy flow. Dr. Lamont C. Cole points out that each 1000 Calories of energy that are stored by the algae in Cayuga Lake are converted by small aquatic animals to protoplasm amounting to 150 Calories. In turn, a type of fish called smelt produces about 30 Calories of protoplasm from the 150. If man eats the smelt, he can convert each 30 Calories worth of smelt to 6 Calories of human fat and muscle. On the other hand, if he waits for trout to eat the smelt and then eats the trout, he can obtain only 1.2 Calories from the trout (see Figure 14.10). Dr. Cole goes on to point out that if man were really dependent on the lake for food, "he would do well to exterminate the trout and eat the smelt, or better yet, to exterminate the smelt and live on planktonburgers!"

At the present time, it is estimated that about one percent of the energy from *all* producers on earth is utilized for human nutrition. This is actually a tremendous amount, and yet fully two-thirds of the world population goes to bed hungry. Dr. Cole estimates that *all of the plant life on earth together* produces an annual yield of approximately 500 thousand trillion Calories of energy, of which about 50 percent is available to first-level consumers. Thus there are about 250 thousand trillion Calories available to first-level consumers per year. As of now there are about four billion people on earth, each of whom requires an *average* of 2200 Calories per day in order to main-

272   The ecosystem

tain an adequate diet. The total annual need of the earth's human population is therefore 3200 trillion Calories. Since 250 thousand trillion Calories are theoretically available, it might seem as though the human population could continue to expand. We must consider, however, that (1) this means that man would have to be exclusively a vegetarian, (2) the available Calories would have to be distributed uniformly throughout the entire human population—a most unlikely if not impossible project, and (3) all other animal life on earth would have to be driven to extinction!

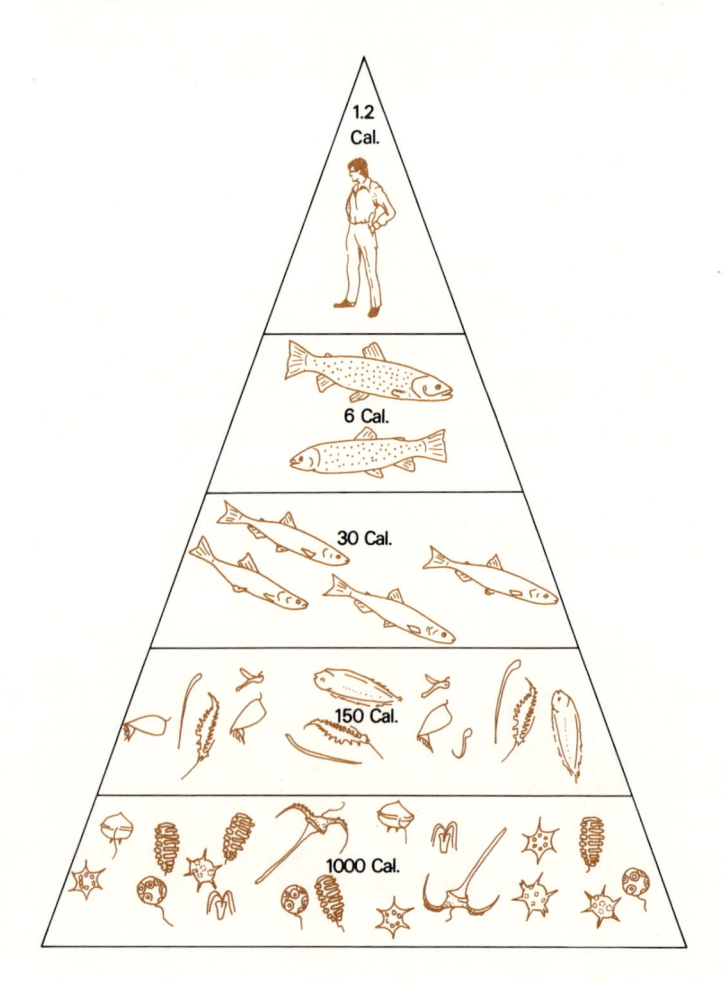

Figure 14.9
A food pyramid. Note that each higher trophic level supports fewer individuals due to the loss of energy from one level to the next.

Figure 14.10
A food pyramid illustrating the situation in Cayuga Lake. (See text for details.)

Environmental Factors in an Ecosystem

Every plant and animal that exists in a specific ecosystem must constantly interact with an environment that consists of other living organisms together with such physical factors as temperature, light, oxygen, and water. Thus each species lives in an environmental framework that consists of a variety of factors to which it has become *adapted* through the process of evolution. Also, when a species has become adapted to a set of environmental factors, these may become *limiting factors* in determining the distribution of the species.

The polar bear, for example, is well adapted by its fur and layers of fat to living in a very cold climate. At the same time, the polar bear will not voluntarily venture below the line that marks the southern boundary of the geographical area where the average annual temperature is no higher than 32°F. Man in his natural state would be limited to tropical climates, but man's brain has enabled him to use technology to live in areas where the temperature drops far below that to which he is naturally adapted. A fish cannot long survive on land because its gills are adapted to absorb dissolved oxygen from water. When the fish is removed from the water, the gill membranes dry out and become impervious to oxygen. Land animals, on the other hand, cannot survive in the water because their lungs are designed to absorb oxygen from air. All organisms, in fact, are adapted and limited to a specific set of environmental factors. Ecologists refer to this concept as the *principle of minimum conditions* under which the organism can exist.

The *substratum* is the medium or surface in or on which an organism lives. The substratum of the tapeworm, for example, is the intestinal tract of its host. The tapeworm is remarkably well adapted to its substratum by the suckers and hooks that enable it to cling tenaciously to the intestinal wall (Figure 14.11). A land plant, on the other hand, is adapted to a soil substratum in which it is firmly anchored by roots that also function to absorb water and nutrients.

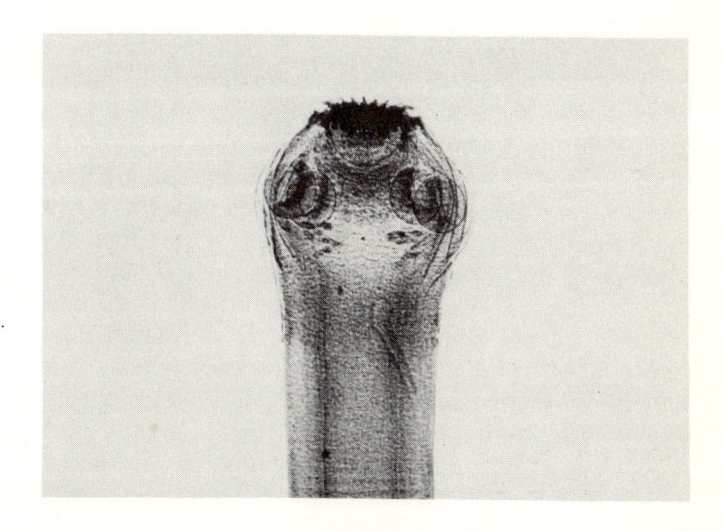

Figure 14.11
The suckers and hooks on the head of the tapeworm help it cling to the intestinal wall. (Courtesy of the Carolina Biological Supply Company.)

Other examples of substratum adaptation include the streamlined form of a fish which enables it to move easily through the water and the extra-large paws of the Canada lynx that provide natural "snowshoes" to keep it from sinking into the deep snow.

Living organisms vary quite considerably in their tolerance of *temperature* ranges, but for all species there are temperature extremes above and below which they cannot survive. Mammals and birds have temperature-regulating mechanisms that maintain body temperature within a narrow range regardless of the environmental temperature. On the other hand, groups such as the reptiles, amphibians, fish, and insects have body temperatures that tend to be the same as the temperature of their surroundings. This has a profound effect on their activity, for they tend to be quite active at higher temperatures, while activity is drastically reduced as the temperature drops. Temperature therefore becomes an important limiting factor for these so-called "coldblooded" animals.

If we consider that *light* is an essential factor in photosynthesis, it is not surprising that light, along with water, carbon dioxide, and soil nutrients, is a critical limiting factor in the geographic distribution of green plants. The depth to which sufficient light penetrates water, for example, determines the maximum depth at which plants will grow. Less than 0.1 percent of available light penetrates below 600 feet. For this reason, most of the photosynthesis that takes place in the sea occurs on the continental shelf relatively close to the shoreline. Contrary to popular belief, the vast areas of open sea do *not* contain an inexhaustible food supply; instead, for the greater part of each year the open sea is a "biological desert." Those animals that do live at great depths are largely scavengers, living on organic materials that settle down from above.

An adequate *water supply* is an obvious limiting factor, since no living organism can long survive without water. Some animals, however, are especially adapted to live in desert areas. The kangaroo rat and the camel are able to subsist on metabolic water, which is the water produced by the respiration process as shown below by the "overall" equation for respiration.

$$C_6H_{12}O_6 + 6O_2 \rightarrow 6CO_2 + 6H_2O$$

There are endless other examples of the ways organisms are adapted to the physical factors of the environment. This vital necessity to adapt to or "fit into" the environmental scheme is an important force behind the process of evolution. We will consider this further in a later chapter.

The Delicate Web

In addition to the effects of physical factors such as the substratum, temperature, light, and water, individual organisms and species within an ecosystem must constantly deal with the living or *biotic* environment. The biotic environment consists of all living organisms in the ecosystem, and the interdependence that exists among those that occupy the various niches forms a complex structure of delicately balanced food webs and energy relationships. The predator, for example, is obviously dependent upon its prey for its food supply, but in other not quite so obvious ways, the prey is also dependent upon the predator. Meanwhile, the destinies of both predator and prey are closely tied to those of the producers, decomposers, and other members of the community.

So much has been said and written about the "balance of nature" that the term has lost its meaning and has become a cliché. Sometimes it is interpreted to mean that an ecosystem is always orderly and unchanging. Actually, an ecosystem continually changes from time to time as populations fluctuate. This may result from shifting climatic conditions and other factors that affect food supplies, or from the extent to which the young survive to maturity. These fluctuations occur even without the intervention of man.

Of course, in a given community such as a beech-maple forest, we generally notice little or no overall change from one year to the next or even over a relatively long period of time. Yet there are parts of the community that are constantly changing. Animals are born, they live, and they die. Giant trees succumb to old age and windstorms, crashing to the ground where they wait for the decomposers to return them to the soil. Saplings spring up in their place, competing in a silent struggle for the light that filters through the forest canopy. One season gives way to another. There are wet summers and dry summers. There are mild winters and there are killing winters, when the woods are silent except for the bonelike rattling of bare branches and the creaking protest of the trees swaying in the chill wind.

In spite of this, however, populations are usually maintained in a relatively steady state by a negative feedback system that tends to operate within an ecological community. You may recall the term *negative feedback* from Chapter 5 when we discussed it in regard to hormone levels. At that time, we compared it to a furnace-thermostat system.

To illustrate how the negative feedback concept might operate in an ecosystem, let us consider a population of mice which has increased in size due to an unusually favorable summer or series of summers. First of all, as the mice population expands, it will tend to exceed the carrying capacity of the environment (Chapter 11) and food shortages will exert pressure against further expansion. Second, there will probably be a drop in fertility among the overcrowded mice. (Although it is not known why, animals such as mice and rats tend to show a significant decrease in fertility under conditions of overcrowding.) Females may produce smaller litters or no litters at all in some cases. Finally, predators such as owls and hawks will swoop down on the mice, illustrating the principle that predators tend to concentrate on prey that is most plentiful and easiest to obtain.

With the extra supply of energy available from the overabundant mice, there will be more room at the top of the trophic pyramid. Specifically, the numbers of hawks and owls will increase, thus bringing even more predation pressure on the mice. In time the combination of predation, lowered fertility, and food shortage will drive the mice population back down to the carrying capacity or even lower. As the number of mice decreases, the bonanza of energy will no longer be available to the hawks and owls and their numbers will also decrease (Figure 14.12).

The foregoing example is really an oversimplified version of a highly complex situation. However, it shows that the "balance of nature" is dynamic and constantly shifting, while usually maintaining the community at a steady state.

The various actors and their interrelationships in the ecological drama are part of a system created by a long, slow process of evolution. Furthermore, these relationships are so intricately interwoven that it is usually impossible to predict the long-term effects of tampering with any one part of the ecological machinery. In our example, what would happen if man should step in and kill the owls and hawks, as he is so fond of doing? What happens when we spray with insecticides that kill the predator insects along with the insects that we wish to eliminate?

It is important, therefore, that we understand the complex checks and balances that are part of the ecosystem structure. Removal of predators and competition, for example, may only permit a favored species to outgrow the all-important carrying capacity of its environment. Well-meaning persons who object to hunting deer and other animals should realize that controlled killing by hunters is often a contrived negative feedback operation that keeps the population down, thereby preventing mass deaths from winter famine or infectious disease.

Figure 14.12
A simplified version of the negative feedback system which helps maintain the "balance of nature." (See text for details.)

Symbiosis

The most obvious and dramatic relationship that exists between animals is that of predator-prey. The "law of fang and claw" brings to mind the sight of a lioness streaking across the African veldt, scattering a herd of zebras as she picks her victim and brings it to the ground. Or we may picture a graceful cougar crouched on a rock above an unsuspecting sheep, or a hawk swooping down on a hapless rabbit.

But there are other types of relationships between organisms in an ecosystem that are much less dramatic and, more often than not, go unnoticed by us. Of particular interest is a relationship known generally as *symbiosis,* which means "living together." Such symbiotic relationships may take one of several forms depending upon the benefits or harm that is brought to one or the other of the participants.

One type of symbiosis, called *commensalism,* is a kind of one-sided association in which one participant benefits from the relationship and the other participant is evidently neither benefited nor harmed. For example, the remora is a small fish that has a suction disc on the top of its head. It uses this disc to attach to the underside of a shark, thereby riding along with the shark and feeding on scraps from the shark's meal as they float by. Apparently the shark is not bothered by the remora, but the remora gets a free ride, a free meal, and who would not like to have a shark as a protector?

Another form of symbiosis, called *mutualism,* is a relationship in which both parties benefit. For example, the termite eats wood, as everyone knows. However, the termite cannot digest wood without a tiny protozoan that lives in the termite's gut. In the process, the protozoa receive their share of nutrients while the termite does all the work. If, on the other hand, the protozoa are killed by experimentally raising the temperature, the termites will starve even if they have a plentiful supply of wood.

The *lichen* (Figure 14.13) consists of two plant varieties that live together in a mutually beneficial

Figure 14.13
The lichen is a plant that consists of an alga and a fungus. (See text for details.)

manner. One participant, an alga, is a green plant and makes food by photosynthesis. The other, a fungus, does not contain chlorophyll and therefore benefits from the alga's foodmaking ability. In return, however, it helps to retain water, thereby keeping the alga from drying out.

The most widely known form of symbiosis is *parasitism.* In this relationship, one participant (the parasite) benefits but in the process brings harm to the other participant (the host). The amount of harm done may be relatively minor or, in some cases, it may end in the death of the host. There are innumerable examples of host-parasite relationships, and biologists regard parasites as one of the most successful groups of organisms.

Ectoparasites are those that live on the surface of the host. A leech attached to a turtle and slowly sucking the blood of the turtle is an example of an ectoparasite. So are lice, ticks, and the well-known fleas on a dog. *Endoparasites* live inside the host in places such as the intestinal tract, the lung, the bile duct, or the blood, to name a few. Some of the more commonly known endoparasites include tapeworms, hookworms, trichina worms, and malaria protozoa. There are many, many more.

A truly successful parasite does not kill the host, since if the host dies so does the parasite. In one sense, such a parasite burns down its own house. Some, such as the so-called tertiary malaria organism, do kill the host in significant numbers but make up for it by the efficient method by which their offspring are transmitted to other hosts.

The hookworm may be present in the human intestinal tract in significant numbers without killing the host, although symptoms of weakness and anemia are usually present. Trichina, a parasitic worm contracted from infested pork, will kill the host only if the infestation is especially heavy.

Circulation of Materials

An automobile will continue to run only if we keep replacing the gasoline which is lost while providing the necessary energy. When we are "out of gas" we are out of energy, and unfortunately to be sure, there is no possibility of recycling the gasoline from the engine back to the tank. In a somewhat comparable fashion, energy flows from the producers upward through the food pyramids of the ecosystem and is eventually lost. Consequently, the maintenance of life is always dependent upon the process of photosynthesis for the input of "new" energy from the sun.

Unlike energy, materials such as carbon, nitrogen, and water tend to circulate from the physical environment to living organisms and back to the environment again in a kind of never-ending "ashes-to-ashes, dust-to-dust" cycle. This movement of materials through the ecosystem, or *ecosphere* if one is referring to the total picture, is called a *biogeochemical cycle*. It must be stressed that the cycles shown in this section represent only a small part of the total picture. If we consider the large variety of materials that form the living organism, it is obvious that the total picture must be incredibly complex, so much so that there are many aspects which are not completely understood.

In the *carbon cycle* (Figure 14.14), the carbon contained in carbon dioxide is "fixed" in the form of carbohydrates by the process of photosynthesis. Decomposers release carbon from dead organisms in the form of carbon dioxide, and more carbon dioxide is produced by the respiration process and the combus-

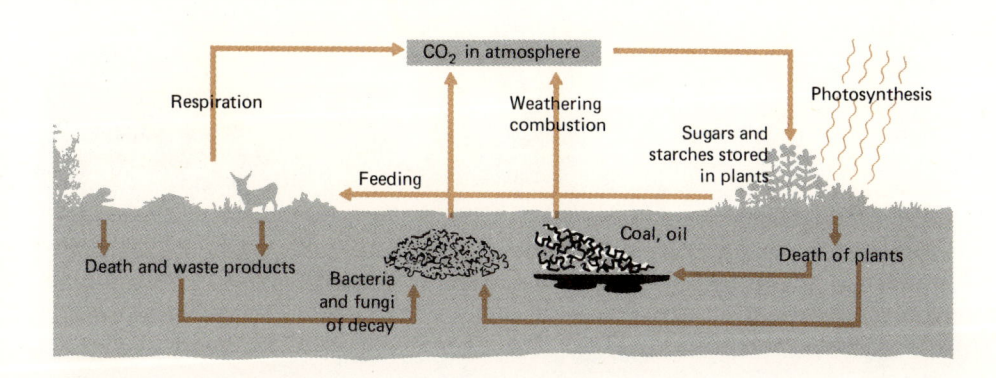

Figure 14.14
The carbon cycle.

tion of vast amounts of fossil fuels such as coal, oil, and natural gas. Under special conditions, large masses of carbon compounds may accumulate over millions of years as coal or oil, but the major pathway is from the atmosphere or from carbon dioxide dissolved in water to the living organism via photosynthesis, and then back to the atmosphere or water via respiration.

In the *water cycle* (Figure 14.15), water leaves the atmosphere in the forms of rain, snow, dew, and frost. It falls as rain or snow directly on the land and oceans. Part of it evaporates back into the atmosphere and part of it flows through streams and rivers into the oceans. Animals drink available water and plants obtain it through their root systems. Meanwhile, both plants and animals lose the excess water

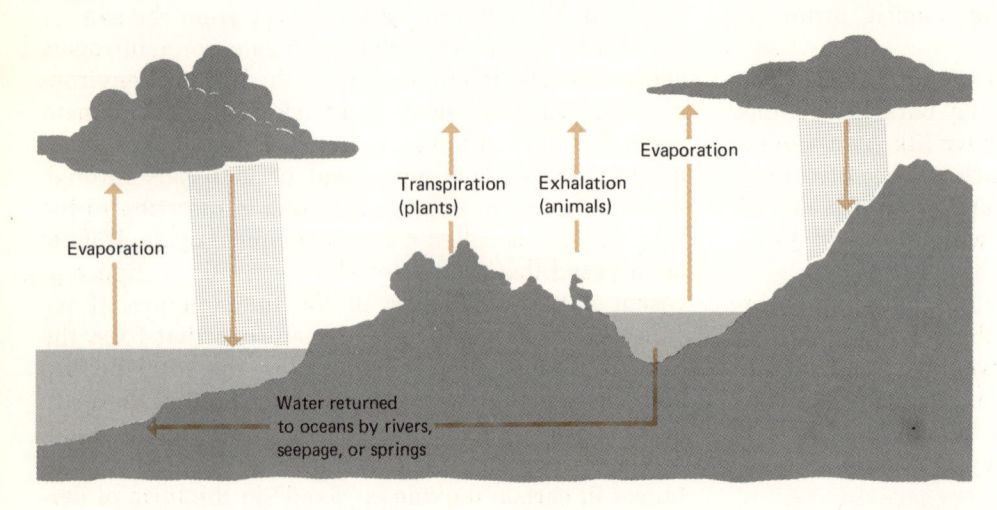

Figure 14.15
The water cycle.

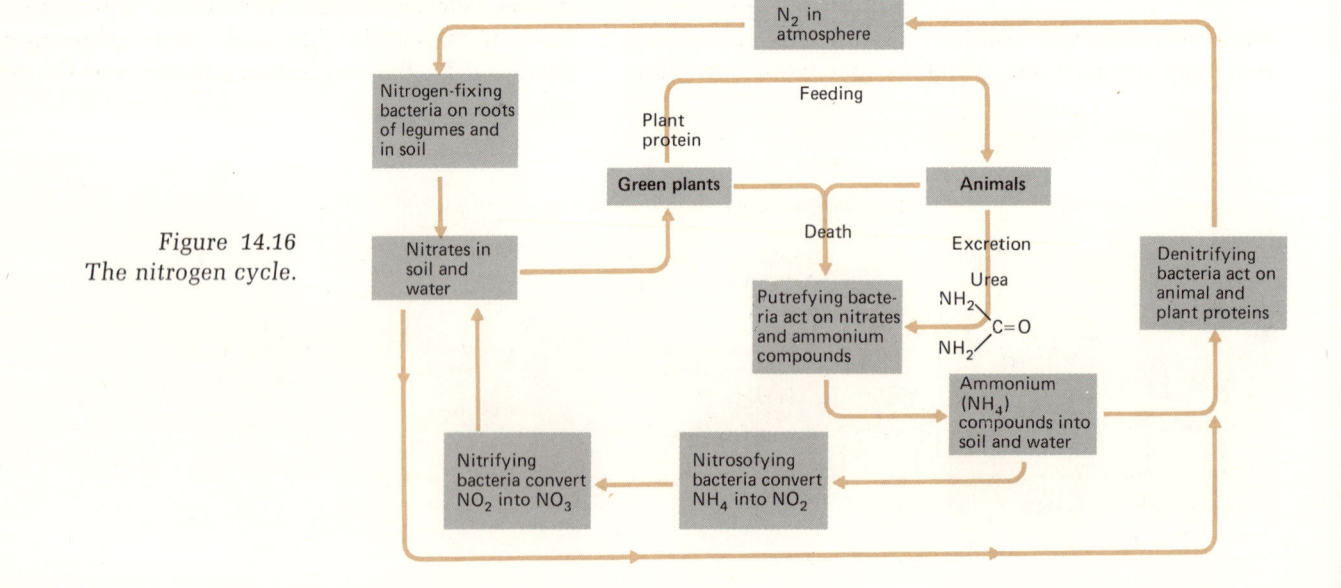

Figure 14.16
The nitrogen cycle.

that results from respiration. Water is a basic material of photosynthesis and it also serves as the universal medium in which the reactions of life take place.

Nitrogen is the basic element of protein, which you will recall is of "first importance" to the living system. In the *nitrogen cycle* (Figure 14.16), dead plants, animals, and their waste materials are decomposed, and nitrogen in the form of ammonia (NH_3) is released by the action of soil bacteria. Another group of bacteria changes the ammonia to nitrites (NO_2^-), and still another group changes the nitrites to nitrates (NO_3^-). The nitrates are then used by the green plant to synthesize protein. Still another group of bacteria, called nitrogen-fixing bacteria, are found in small nodules on the roots of legume plants such as clover and alfalfa. These nitrogen-fixing bacteria take free nitrogen from the atmosphere and convert it to a nitrate form which is directly usable by the plant in the manufacture of amino acids. Nitrogen-fixing bacteria are capable of synthesizing as much as 200 pounds of nitrates per acre of clover each year. In addition, some nitrogen compounds are synthesized by lightning and are washed into the soil by rainfall, but this contribution to the cycle is thought to be relatively insignificant.

Phosphorus, an important plant nutrient, presents a special problem which deserves mention here. Unlike nitrogen or carbon, phosphorus does not circulate from the atmosphere, but from sedimentary rocks. Part of the phosphorus that is eroded from bedrock is washed out to sea. Some is returned to the land in the form of wastes excreted by sea birds. Apparently, the amount of phosphorus returned by sea birds is not as significant as it once was, and ecologists feel that the stores of phosphorus in the land portions of the ecosphere may be diminishing at a faster rate than they are entering the system by rock erosion.

In theory, these natural cycles should be sufficient to maintain an adequate supply of soil nutrients, but the increasing demands made on agriculture by an ever-expanding human population have made it necessary to add artificial fertilizers containing both nitrates and phosphorus. This is itself an energy-consuming process, since energy is required for the manufacture of artificial fertilizers.

Ecosystem Development

In an earlier section, we said that the ecosystem is in a state of dynamic balance, a condition that is normally maintained by feedback systems. This description is typical of mature and *stable* ecosystems that show few overall changes despite constant fluctuations in the populations making up the community.

How do such mature ecosystems develop? How, for example, does an empty field become a forest if left alone long enough? What processes operate to convert a pond to a marsh and eventually to a woodland? Depending on the type of ecosystem we are dealing with, a series of changes occurs in a more-or-less predictable pattern in which one stage follows or *succeeds* another until a stable or *climax stage* is reached. This pattern is called *ecological succession.* In a "typical" succession, the first stage of development is the *pioneer stage,* which is succeeded by other stages until the climax community is established. During each stage, the living organisms present tend to modify the physical environment in a way that makes it possible for the organisms of the next stage to move in.

For example, a succession may take place on an expanse of bare rock. In this case, the likely pioneer organism is a crustose lichen (Figure 14.17) which gives way to the "leafy" form of lichen (Figure 14.18). Lichens are in turn succeeded by mosses (Figure 14.19). Ultimately, the *herbaceous stage* follows with plants such as wild strawberries, broomsedge grass, and goldenrod, as each stage of plant growth prepares the surface for invasion by the organisms of the next stage. If left undisturbed, this area may eventually develop into a climax stage that would likely be a beech-maple or oak-hickory forest.

280   The ecosystem

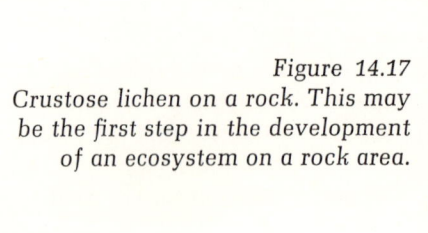

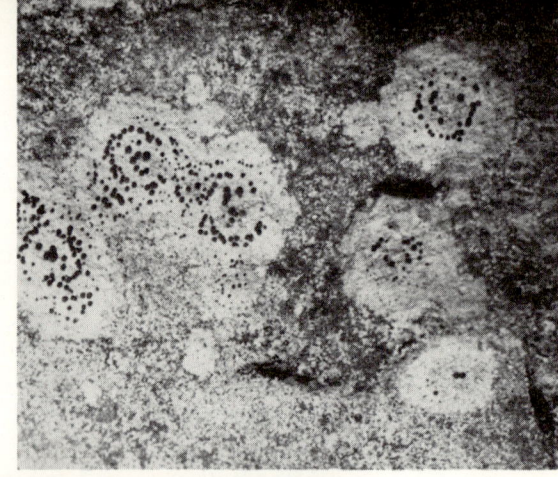

Figure 14.17
Crustose lichen on a rock. This may be the first step in the development of an ecosystem on a rock area.

Figure 14.18
The foliose lichen stage of a rock succession.

Figure 14.19
The next stage of a rock succession showing that lichens have been replaced by a moss. (Courtesy of Dr. Alden E. Smith, College at Buffalo, State University of New York.)

Ecosystem development 281

Newly formed pond
Planktonic algae (microscopic)
(a)

Submerged rooted plants
Chara   Pondweed
(b)

Emerging plants
Arrowhead Water lily Cattail Bulrush
(c)

Maple-elm-pine forest
White pine   Maple
(d)

(e)

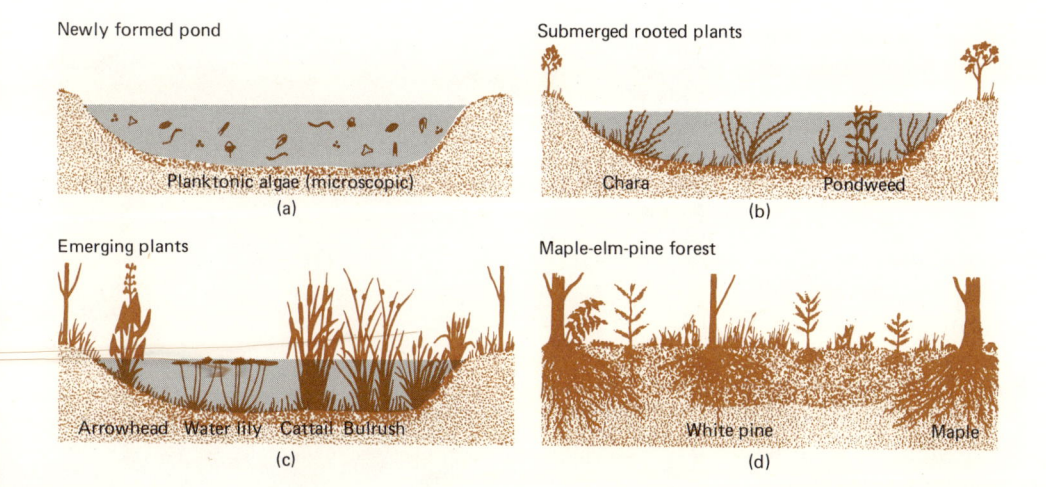

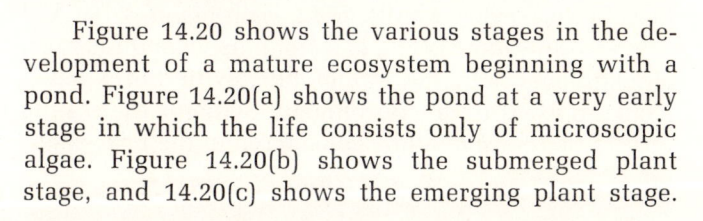

(f)

Figure 14.20 shows the various stages in the development of a mature ecosystem beginning with a pond. Figure 14.20(a) shows the pond at a very early stage in which the life consists only of microscopic algae. Figure 14.20(b) shows the submerged plant stage, and 14.20(c) shows the emerging plant stage.

Figure 14.20
Diagrams (a), (b), (c), and (d) show the stages in an ecological succession from a pond to a climax forest. Part (e) is a photograph of a pond that is rapidly filling in with cattails. Part (f) shows a pond that is completely filled in. Note that shrubs and trees are beginning to move in.

Finally in Figure 14.20(d), we see the establishment of a climax forest where the pond once stood. It should be noted that an increase in the variety of species typically occurs during ecosystem development. Also, in the early stages of development, the rate of production by photosynthesis is greater than the rate by which energy is released by community respiration. As the climax stage is reached, however, there is a tendency for the energy that is trapped by the green plants to balance the energy needed to maintain the community in a stable condition.

Figure 14.21 shows an area of abandoned farmland in the first or *weed stage* of succession. Generally this first stage will consist of annuals such as dandelions, but these will rapidly be joined by perennials such as wild strawberries, goldenrod, and clover. The weed stage is followed by the *brush stage* (Figure 14.22), which may include sumacs, willows, thornapple, and red osier dogwood. This stage gives way to the *pioneer forest* (Figure 14.23), consisting of trees such as aspen and birch. The pioneer trees, sometimes called "weed trees," are usually not shade-tolerant. As a result, their saplings will not flourish because of the shade provided by the parent trees. They are therefore replaced by the *climax forest,* a stable community consisting of shade-tolerant trees such as beech, maple, oak, or hickory (Figure 14.24). Box 14.1 (see page 284) presents a summary of a typical developmental pattern that may occur in a cleared area that is allowed to return to the stable, or climax, state.

Summary

In this chapter, we have considered some of the more fundamental principles of ecology. While it has not been possible to consider all of the details of this broad and exceedingly complex science, it is hoped that this brief introduction will help you begin to develop what Dr. Ralph Buchsbaum has called an "ecological viewpoint."

Figure 14.21
The weed stage in an ecological succession involving an abandoned field. The vegetation here consists of both annual and perennial weeds.

Simply stated, one who has an ecological viewpoint recognizes that the world of life has been molded by millions of years of evolution, that it is organized in a most complex fashion, and that it consists of many interdependent parts, each of which is important to the proper functioning of the whole. If you have an ecological viewpoint, you also recognize that tampering with such a complex structure can lead to disastrous results. Man, with his massive and often troublesome brain, is a compulsive manipulator. He tampers with his automobile, with the kitchen clock, and with the environment. Some of the results of that tampering will be considered in the next chapter.

Summary 283

Figure 14.22
In this photograph, the weed stage is giving way to the brush stage.

Figure 14.23
The pioneer forest stage consists of "weed trees" that have a short lifespan and are not shade-tolerant.

Figure 14.24
A climax forest, the final stage in the development of a forest ecosystem. This woods consists mainly of beech and maples together with a few oak and hickory trees.

Box 14.1
Summary of stages in the typical development of a climax forest. (Courtesy of Dr. T. E. Eckert, College at Buffalo, State University of New York.)

ECOLOGICAL SUCCESSION
Secondary Site

↑ Natural Succession—Reforestation
↓ Fire, Forest Clearing, Agriculture, Overgrazing

CLIMAX FOREST: Birch (yellow)—beech—maple—some bosswood—oak—hickory—(chestnut, more southerly), pine—spruce—fir—larch, depending on site (usually more northerly). Mosses, ferns, wildflowers abundant, fewer shrubs; squirrels, black bear, grouse, deer somewhat less abundant.

PIONEER FOREST STAGE: Second growth trees predominant—aspens, ashes, maples (silver in moist areas), birches (grey), cherries, cedars, pines, junipers (young), etc. Wildlife abundant—deer, cottontail, beaver, squirrels, black bear, woodland mice and songbirds, grouse, thrushes, etc.

SHRUB STAGE (brushland): Many perennials persisting from below; shrubs closing in and shading out herbs. Shrubs and young trees—sumac, thornapple, dogwood, willow, wild apple, elderberry, blackberry, arrow-wood, and young saplings of cherry, aspen, and grey birch. Deer increasing, rabbit, goldfinches, sparrows, etc.

GRASSLAND STAGE (old meadow): Grasses dominating are broom sedge, fescues, poverty, orchard, timothy, kentucky blue, quack, and redtop. Perennials—asters and goldenrods abundant; everlasting, strawberries, hawkweeds, fleabanes (daisy and whitetop), black-eyed susan, daisy, beard tongue, joe-pyeweed, boneset, bedstraws. Animals as below plus pheasants, song sparrows, henslow, and grasshopper sparrows, bobolinks, woodchucks, skunks, foxes, hawks, and owls.

PERENNIAL WEED, EARLY GRASSLAND STAGE: Some asters and goldenrod species, sweet clover, other clovers, thistles, cinquefoils, wild carrot, yarrow, mullein, milkweed, hawkweeds, strawberries, wild lettuce, sow thistles, teasel, docks, mats of sod-forming grasses, and scattered bunch grasses (broom sedge). Timothy, orchard, poverty, fescues, and quack. Same animals as below but becoming more abundant (e.g., horned lark). Cottontail, meadowlark, killdeer, upland plover, other ground-nesting birds, and meadow mice.

ANNUAL AND PIONEER WEED STAGE: Ragweeds, lamb's quarters, mustards, smartweeds, peppergrass, shepherd's purse, plantains, horseweed, redroot pigweed, knotgrass, spurges, purslane, and occasional persistent agricultural grasses and clovers and perennial weeds. Horned lark, vesper sparrow; killdeer, field mice, moles, cowbirds, mourning dove, english sparrow, and hunting predators (red fox, sparrow hawk, etc.).

DENUDED STAGE: Cropland operations, aftermath of fire, erosion, removal of topsoil and gravel, roadsides and construction sites, flooding. Few plants, some springing up from perennial rhizomes and taproots. Horned lark, vesper sparrow, moles, field mice, etc.

Questions

1. Summarize the "grand strategy" of life as it relates to solar energy.

2. Discuss the principles of the "trophic pyramid" and its significance in a world of ever-increasing population.

3. Briefly describe the structure and components of the typical ecosystem.

4. Explain what is meant by the "ecological viewpoint." How does this point of view relate to environmental problems?

5. In what ways, if any, has your understanding of the term "ecology" been changed by reading this chapter?

6. Discuss the following statement: Since bacteria cause disease, the best thing for man to do is destroy all bacteria on earth.

7. Discuss what is meant by the term "balance of nature."

8. Describe some of the more important aspects of ecosystem development, or ecological succession.

15

"Subdue the Earth, and Have Dominion..."

Introduction

In man's early beginning he was a part of nature, just another member of the biosphere. As such, he humbly acknowledged nature's power as he fled with other animals before a raging forest fire or huddled in his cave, trembling at the sound and fury of the thunderstorm. He knew the wind's power and felt the warmth from the life-giving sun that daily circled overhead. Primitive man sensed that his own destiny was interwoven with those of the animals and plants with which he shared the earth. And so he respected nature and held it in awe, finding his first gods in the trees, in the antelope, in the streams and lakes, and in the sun itself. Like his fellow predators, he killed only what he needed to survive, and not infrequently he took his turn at being the prey. He lived in harmony with nature.

But evolution had given man a special gift. It was a brain that allowed him to think, to plan, to create, and to build in ways that were far beyond the wildest dreams of his fellow animals. It was probably inevitable that in time, that massive brain would create a new god called Technology. The old gods of nature were swept away and forgotten, and man himself became a god. Now he stood above the ecosphere manipulating, exploiting, and destroying the natural world which had given him birth. It was all in the name of Progress, and if there were moments of doubt, western man could always consult the Book of Genesis to remind himself that, after all, the Supreme Creator *did* tell him to subdue the earth and have dominion over every living thing that moves upon it!

Unfortunately, although man has assumed a god-like position relative to the rest of nature, he has not demonstrated the wisdom that one might expect from a god. His wholesale destruction of the environment and reckless exploitation of natural resources have created problems that seem to defy solution. The air of major cities is unfit to breathe, and lung diseases such as cancer and emphysema are on the rise. The lakes, the streams, and the rivers are polluted with human and industrial wastes. Between 10 and 20 million people die of starvation each year, while many millions more live with ever-present hunger. Still the population continues to grow like a bank account earning compound interest. The earth's stores of fossil fuels are dwindling to a crisis point, while the energy-gobbling affluent nations are demanding more goods, more services, greater production, and ever-higher standards of living. The United States alone uses 30 percent or more of the world's energy resources even though it represents only 6 percent of the total population!

If uncontrolled population growth is not the only cause of our predicament, it is probably the most fundamental. People are the chief polluters of the environment. Therefore, more people means more pollution. If malnutrition and outright starvation are on the increase, it is due in part to the fact that exponential growth of the population must consistently outrun any increases in food production. Also, more people bring more demands on energy resources for both the human machine and the machines of industry. You will recall, for example, that even if the United States were to maintain a so-called "replacement" fertility level, by the year 2020 there will be nearly 300 million energy-hungry Americans.

Unplanned and uncontrolled technological growth must also take its share of the blame. In the spirit of free enterprise, there has been little or no real control over where industries could locate, nor have industries been held responsible for the waste products of manufacturing. As a result the air, the streams, and the lakes have been regarded as convenient and bottomless pits into which vast quantities of wastes could be discharged, rarely with any thought given to the human and environmental consequences.

It is interesting to note that a federal law prohibiting industrial pollution of navigable streams and their tributaries has existed since 1899! It was not enforced largely because few people were concerned about pollution, and besides, it was necessary to encourage industries because they provided jobs and economic growth. Indeed, even in the face of the more obvious facts of environmental deterioration, we continue to be obsessed with increasing the gross national product (GNP) even though it has been shown that unrestricted growth is one of the major environmental culprits.

Water Pollution

Man has always felt at home near the water. Like all other forms of life, his origins trace back to the primeval seas, and he has never totally escaped the watery environment in which life began so many million of years ago. In times past, the lakes and streams abounded with fish and other aquatic life, providing a major source of food for the peoples camped along the shores. Even today there is enchantment in the roar of pounding surf, or in the sound of a stream coursing over rocks and waterfalls, or in the waves lapping the shore of a quiet lake at sunset. Sadly though, the stream all too often reeks with sewage or is milky with the chemical discharge of a nearby factory. The shore of the idyllic lake may be choked with algae and sprinkled with dead fish, and the inviting water often teems with intestinal bacteria from human wastes.

With the growth of modern industrial cities, our natural bodies of water became dumps for municipal sewage and the waste products of industry. Pesticides such as DDT, together with nitrates and phosphates from artificial fertilizers, also find their way into streams and rivers as part of the runoff of rainfall from farmlands.

When untreated or only partially treated sewage is added to a stream, it is attacked by the bacteria that act as decomposers in the ecosystem. This decomposition process requires oxygen, which is obtained from the supply of dissolved oxygen that is normally present in the water. The amount of oxygen taken for these decomposition activities is measured in terms of the *biological oxygen demand,* or BOD. As the BOD increases, the dissolved oxygen is reduced below the point at which game fish such as trout or pickerel can survive.

Water polluted by sewage may also contain infectious organisms carried by human intestinal wastes, including the bacteria and viruses that cause such serious illnesses as polio, cholera, typhoid fever, and hepatitis. In the summer of 1973, for example, a major cholera epidemic in Italy was traced to shellfish that were gathered from water polluted with human wastes. In addition to human wastes and everything else that goes down the drain, municipal sewage contains a relatively new kind of pollutant in the form of the phosphate detergents that have come into popular use along with automatic laundry and dishwashers.

With properly constructed municipal sewage treatment plants, up to 90 percent of the major organic pollutants *can* be removed. Unfortunately, no more than 50 percent of all existing communities have treatment facilities adequate for their needs, and a shocking 25 percent of all communities simply discharge raw sewage directly into nearby streams! Meanwhile, the cost to the taxpayer of high-quality sewage treatment keeps escalating in response to the ever-increasing demands of continued population growth.

In the literature of pollution, the story of Lake Erie has been told and retold as a classic example of the death of a lake at the hands of man, the master polluter. To repeat what was said earlier, Lake Erie

Figure 15.1
Map of the Great Lakes showing principal cities.

is far from dead, but its environment and the forms of life that inhabit it have been drastically changed. Lake Erie is part of the Great Lakes system (Figure 15.1), which was formed over 12,000 years ago by the giant glacier that covered a large part of North America. Normally, a lake or pond goes through a natural aging process as it becomes shallower due to a buildup of soil deposits and organic materials. Ultimately, over a period of thousands of years, the lake basin is filled, forming a swamp or bog. This aging or enrichment process is called *eutrophication*. Lake Erie, however, is a victim of what has been called "cultural eutrophication," because the natural aging process has been rapidly accelerated by man and his activities. In fact, it has been estimated that the lake has aged at least 1500 years during the past 50 years as a result of the sewage, industrial wastes, and farmland runoff that have poured into its waters from Cleveland, Buffalo, and the numerous smaller communities and farms that line its shores (see Figure 15.1).

Aquatic plants in the lake are heavily fertilized with nitrates and phosphates carried by sewage and farmland runoff. This abnormally high concentration of fertilizers produces "algal blooms" (Figure 15.2) far greater than would normally occur. As these vast quantities of plants die and then decay through the activities of the bacterial decomposers, the BOD drastically increases and robs the water of dissolved oxygen.

As in most deep lakes, summer heating of Lake Erie's surface waters produces a *temperature stratification* (Figure 15.3). There is a warmer layer on top called the *epilimnion* and a colder, denser layer on the bottom known as the *hypolimnion*. Separating the two layers is a third layer called the *thermocline*,

Figure 15.2
Algal blooms on a pond. The pond is fed by a creek that is polluted by sewage.

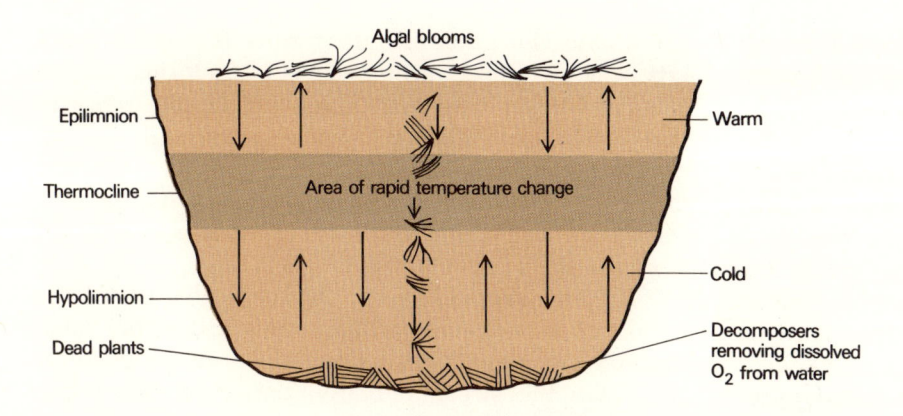

Figure 15.3
Formation of a thermocline in a deep lake. The plant material falls to the bottom where bacteria take oxygen from the water as part of the decomposition process. Because the thermocline prevents circulation of oxygen from above, the oxygen in the deeper levels of the lake is depleted.

which is an area of rapid temperature change. This stratification effect prevents vertical mixing of oxygen and nutrients, and the hypolimnion becomes practically devoid of oxygen due to the action of the decomposers on masses of decaying water plants. Since oxygen is a critical limiting factor for many species of edible fish, their numbers have rapidly dwindled, crippling the lake's once-flourishing fishing industry. Meanwhile the numbers of "trash" fish, so called because most people regard them as unappetizing, have increased due to their ability to adapt to lower oxygen levels and because of the removal of competition from other species.

Pollution has also reduced the number of aquatic insects that were important links in the lake's food chains. For example, not so many years ago the swarms of mayflies on a spring evening were so thick that they had to be literally shoveled from the sidewalks by lakeside residents. Now they are all but gone, good riddance perhaps from the human point of view, but also a loss of an important source of food for the lake's fish population.

Man himself is not unaffected by the changes in Lake Erie. Those who would spend a quiet afternoon fishing must wait for an infrequent tug on the line that usually brings a carp or sheepshead, not the yellow pike or whitefish of another day. Many people who have seen and smelled the water would hesitate to eat the fish anyway. Swimming presents similar problems; the water along the beaches must be constantly checked for certain kinds of bacteria that indicate the presence of human intestinal wastes. The lake is also a major source of drinking water for most communities along its shores, but water with reduced oxygen content tastes and smells bad even though it can be rendered safe by adding chlorine to reduce the number of disease-causing bacteria.

From Lake Erie, the Niagara River flows past Buffalo north to Lake Ontario. The polluted water sparkles with deceptive beauty as it rushes toward

"Subdue the earth, and have dominion..."

the brink of Niagara Falls, where tourists from all over the world view the mighty cataract while breathing the stench from the chemical plants that line the upper river (Figure 15.4). East of Niagara Falls, the Eighteen-Mile Creek originates south of Lake Ontario and winds its way to a small lakeside village that was once a thriving summer resort. Along the creek, arrowheads are turned up by farmers cultivating the fruit orchards, recalling a time when the Indian camped on its banks and fished the bounty that the creek offered.

The creek is formed by two tributaries. One of these tributaries runs through farmland and is muddy with the soil that runs off the land with each rain, carrying fertilizers and pesticides into the stream (Figure 15.5). The other tributary runs through a small city where it picks up the wastes of factories and the sewage from a treatment plant which, like those of most cities, has difficulty keeping up with the increasing demands made upon it. As the tributary flows past the chemical plant shown in Figure 15.6, its waters turn milky and acid. One looks in vain for a sign of life (Figure 15.7).

Prior to World War II, the main channel of the creek ran through farmland, but now residential housing solidly lines its banks and raw sewage finds its way into the stream. Finally, the creek flows into Lake Ontario at the village which in the summers of 40 years ago saw thousands of people crowd its beaches and fill its hotels and cottages. Now the ho-

Figure 15.4
Factories crowded together along the Niagara River. These factories take advantage of the electrical power generated by the force of the rushing water.

Figure 15.5 (above right)
The east branch of the Eighteen-Mile Creek carries fertilizers and pesticides from the farmlands through which it flows.

Figure 15.6
A chemical plant located on the west branch of the Eighteen-Mile Creek.

Figure 15.7
The stream flowing by the plant in Figure 15.6 is highly acidic. It shows no signs of life except for the vegetation that lines its banks.

Figure 15.9
The "Beach Closed" sign has become a permanent part of the scenery. The level of intestinal bacteria makes the water unsafe for swimmers.

Figure 15.8
Midsummer at the resort's midway. On summer days in years past, this resort village was crowded with cottage residents and visitors from nearby cities.

Figure 15.10
The man in the photograph recalls a time when the beach was crowded with bathers.

tels have decayed and disappeared, the cottages are in disrepair, and the buildings of the once-busy midway are closed and silent, their paint peeling off unnoticed (Figure 15.8). Weeds and shrubs are creeping up on the beaches, which are littered with dead fish and green with the blooms of algae that wash up on the shore. The "Beach Closed" signs warn that disease from human waste may lurk in the water, and piers from which small boys once took strings of fish are cracked and broken, victims of the winter storms that lash the lake into a fury which rivals the North Atlantic at its worst (Figure 15.9). The man in Figure 15.10, a survivor of the resort's better days, stands on the deserted beach and ponders the results of man's carelessness with the precious legacy of our natural environment.

Lake Erie and the Eighteen-Mile Creek are but two of the many examples of water pollution that exist across the United States and the world at large. In recent years, the demands of environmentalist groups have brought signs of progress, but economic considerations and political bickering have blunted most of the attacks which have been made on the root causes of water pollution. The private citizen must become involved. He must join with others to demand better sewage treatment in his own community and to insist on laws that will make industry responsible for proper disposal of wastes. Only a concentrated effort by a government prodded by a concerned citizenry will reduce the pollution of our waters. Even then we cannot hope to restore the Lake Erie's of the world to what they once were, to a state that was slowly created over thousands of years of evolution.

Air Pollution

The Surgeon General Has Determined That Cigarette Smoking Is Dangerous to Your Health is the sinister warning stamped on each of the millions of packs sold daily in the United States alone. Despite the predictable protests of innocence from the tobacco companies, there is overwhelming evidence that the Surgeon General is entirely correct. In spite of this, however, the consumption of cigarettes increased during 1973 by 4 percent over the previous year!

Nonsmokers definitely agree with the Surgeon General and react with appropriate disgust when the person in the next seat on an airplane lights up. Yet after landing at the airport, the nonsmoker gets into his car, a prolific source of air pollution, and breathes the pollutants spewing from the factory smokestacks as he drives through the city. If reminded of this, our nonsmoker might well reply that cigarettes are not essential to man's welfare but factories and automobiles are. His argument has merit, since for many people the automobile is not a mere luxury but a necessary mode of transportation. Besides, factories mean jobs, they pay significant taxes, and they produce the necessities and luxuries that we demand. Furthermore, while air pollution is admittedly unsightly and

Figure 15.11 A temperature inversion. In (a), a warm air mass has moved in over a relatively cool air mass. This forms a "ceiling" which prevents upward circulation of air pollutants. In (b), the normal temperature gradient has returned.

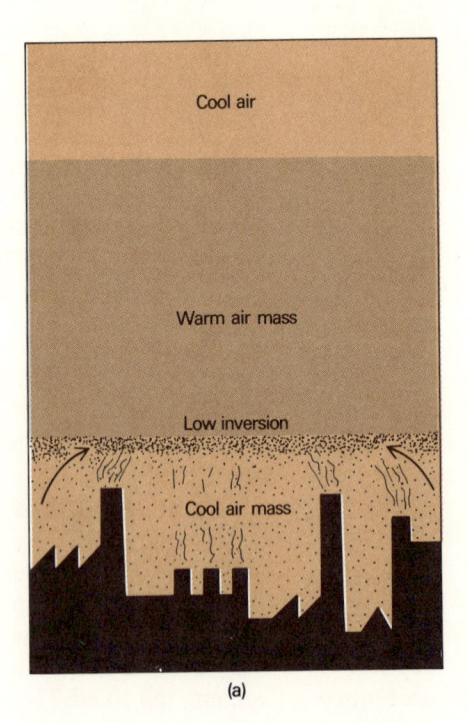

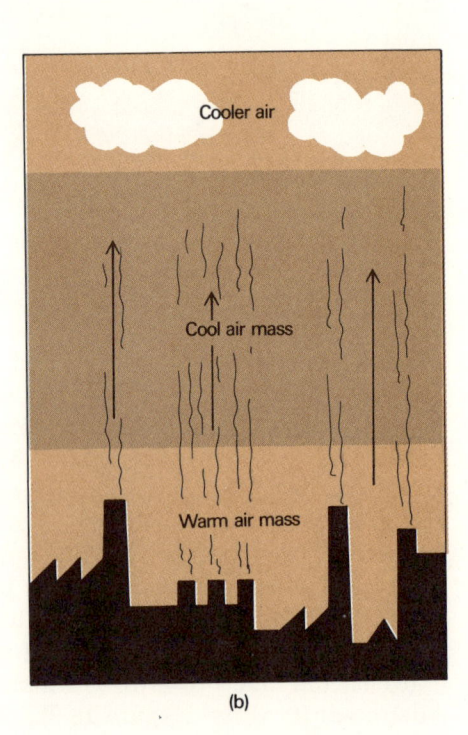

odorous, we have, after all, been living with it for a long time without apparent harm. Or have we?

In 1948, a temperature inversion (Figure 15.11) settled over the small city of Donora, Pennsylvania, an industrial town that contains a steel mill, a sulfuric acid plant, a zinc production facility, and several other factories. A fog began to form, becoming thicker and darker as the factory stacks continued to pour out on the heavy air. When people walked on the streets, they left footprints in the dirt that settled from the air. Before a change in weather conditions brought relief, almost 6000 of the 14,000 residents of Donora became ill, and 17 persons died as a direct result of the heavy concentration of pollutants.

What happened in Donora was a minor tragedy compared to what happened in London in December, 1952. The weather conditions over that city created a concentration of pollution that caused *4000 excess deaths over a four-day period*. During the following two months, there were 8000 more excess deaths, most of them resulting from complications affecting existing respiratory conditions. This situation has been repeated on a smaller but still significant scale in other cities such as New York, where 400 excess deaths were reported during a temperature inversion in 1963, and 80 excess deaths occurred again in 1968.

It is apparent that air pollution is not merely an unsightly and smelly nuisance. Air pollution kills people, sometimes dramatically as in Donora and London, but more often its effects on human health are less obvious. The death certificate does not read "air pollution," but rather lists "emphysema" or "lung cancer."

There is now little doubt that air pollution is largely to blame for an alarming upswing in the number of cases of pulmonary disease. Emphysema, an extremely disabling and ultimately fatal disorder characterized by distention and destruction of the lung's air sacs, is increasing rapidly in the United States. Fully twice as many cases are reported from cities than from rural areas. The same can be said of lung cancer. While cigarettes have been definitely linked with lung cancer, general air pollution cannot be ruled out as an important cause.

Air pollution is also believed to contribute significantly to other respiratory problems such as pneumonia, bronchial asthma, and chronic bronchitis. Chronic bronchitis is characterized by a constant secretion of excessive mucous and a thickening of the bronchial tubes that lead to the air sacs of the lung. It has been reported that chronic bronchitis contributes to as much as ten percent of all deaths in Great Britain.

Air pollutants may also kill or stunt the growth of plant life, affecting crop productivity and destroying residential trees and shrubbery. In the early 1900's, a copper smelting plant in Tennessee added 40 tons of sulfur dioxide to the air every day for a period of years. Upwards of 7000 acres of vegetation were destroyed, and today's traveler in that area can easily imagine himself on Mars as he crosses an eroded and barren red desert that extends as far as the eye can see (Figure 15.12).

Air pollutants can be roughly divided into two types consisting of (1) visible particles such as smoke, and (2) gaseous materials such as carbon monoxide, sulfur dioxide, and hydrocarbons. Most people associate air pollution with smoke, doubtless because it is easily seen, but smoke actually represents a relatively small percentage of the total problem and is not nearly as dangerous as some of the invisible products that come from factory stacks and automobile tailpipes.

At least 50 percent of all air pollution is created by the industrial use of fossil fuels, particularly coal and oil. Both coal and oil contain sulfur in varying concentrations, and the combustion of these fuels produces gaseous sulfur compounds that are invisible, toxic, and highly corrosive. One of these, hydrogen sulfide (H_2S), is toxic at even relatively low concentrations, and in water forms a weak acid. Perhaps the most important of all air pollutants, however, are sulfur dioxide (SO_2) and sulfur trioxide (SO_3). Sulfur dioxide is a highly irritating gas, and sulfur trioxide

Figure 15.12
The Tennessee Copper Basin at Ducktown, Tennessee. The smelter fumes destroyed over 100 square miles of vegetation. (Courtesy of the U.S. Department of Agriculture.)

forms sulfuric acid (H_2SO_4) when mixed with water. Air pollutants are often carried in finely dispersed water droplets called *aerosols*, which are similar to hairspray. These water droplets may carry sulfuric acid, which is highly corrosive and irritating to the respiratory passages and which also attacks stone facings on buildings, causing them to deteriorate much more rapidly than they would under normal weathering conditions. It is estimated that power plants alone release *12 million tons* of sulfur dioxide into the atmosphere annually (Figure 15.13)!

There has been a great deal written about the long-standing love affair between the American and his automobile. This famous romance has borne fruit in the form of ever-increasing millions of cars, thousands of miles of highways, and junkyards which dot the countryside from coast to coast. Of the 550,000 tons of air pollutants released into the air every day in the United States alone, approximately 50 percent are produced by automobiles.

Individually, the automobile is not a significant source of pollution. The problem stems from the fact that so many cars are concentrated in certain areas. The most famous case in point is the Los Angeles area, where there are approximately 3 million cars burning almost *10 million* gallons of gasoline per day! The products of this combustion include carbon monoxide, nitrogen dioxide, and hydrocarbons, as well as lead and other substances that are part of the additives which are necessary for top performance of today's engines. Carbon monoxide (CO) is a colorless, odorless gas that combines with the blood hemoglobin, reducing the oxygen-carrying capacity of the blood. In sufficient concentrations, carbon monoxide can cause death from asphyxiation, or lack of oxygen. Nitrogen dioxide (NO_2) is a brownish yellow, extremely irritating gas that gives smog its hazy yellowish color. Hydrocarbons are compounds of hydrogen and carbon that result from the incomplete burning of gasoline, and they are a major contributor to the smog that often forms where there are high concentrations of automobiles.

Los Angeles is located in a natural basin surrounded on three sides by mountain ranges that prevent circulation of the air. In addition, a stagnant high pressure system dominates the area's weather during most of the year. Pollutants from automobile exhausts gather in the heavy air, and the famous California sunshine provides ultraviolet radiation that initiates a series of reactions between the pollutants and the atmosphere. The result is an irritating yellowish haze called photochemical smog, which at times becomes almost thick enough to blot out the sun. Los Angeles is typical of the way we have built our society around the automobile, a device that requires billions of gallons of a dwindling supply of gasoline, a machine that

Figure 15.13
A coal-fired power plant. Heat from burning coal is used to generate steam, which in turn drives the turbines and generators that produce electrical power.

has demanded that more and more food-producing land be converted to the concrete and asphalt of superhighways.

Unfortunately, there does not seem to be a practical substitute for the automobile in sight at the moment, but there is currently an attempt to reduce the level of automobile pollution by mandating the installation of devices that control emissions from the crankcase, tailpipe, and carburetor. Another device reduces evaporation of hydrocarbons from the gas tank. It should be emphasized, however, that these so-called "antipollution" controls only reduce the emission of pollutants; they do not by any means eliminate them. Unfortunately, these antipollution devices add to the cost of the automobile to the consumer and also increase gasoline consumption in the face of diminishing oil supplies.

There are a number of difficulties inherent in the control of air pollution. For one thing, air pollutants drift with the prevailing winds, often traveling across state lines or even national boundaries. A locality that may have an effective pollution abatement program may therefore still be at the mercy of another area some distance away. Also, air pollution is often intimately tied to the economics of the polluted area. This is especially true of "one-factory" towns, where people are hesitant to demand pollution abatement programs out of fear that the industry on which they depend for jobs may move to a more tolerant community. Finally, as we will see in a later section, increasing concern over dwindling energy resources may lead to a partial retreat from the gains that have been made in recent years, especially in the area of federal controls.

If we recall the lessons of London and Donora, and if we consider the increasing prevalence of disabling and fatal respiratory diseases, it would seem logical that the only sensible choice would be to strictly enforce antipollution laws and to strengthen the laws that now exist. Pollutants such as sulfur dioxide *can* be eliminated from factory stacks; automobile engines that will reduce pollution *can* be developed.

On the other hand, pollution abatement programs are costly, and the added costs of manufacturing are inevitably passed on to the consumer in the form of higher prices for goods and services. This in turn reduces the consumer's buying power and sales fall off. When industries cannot sell their products, they must lay off employees and unemployment increases. A case in point involves an industry which used a material in the manufacturing process that was shown to expose the employees to a high cancer risk. The industry could not afford to "clean up" the process and closed down. It is interesting to note that the employees almost unanimously took the position that if their choice was between a safe environment or a job, they would choose the job! There are, as usual, no simple answers to big problems. How do we strike that delicate balance between a healthy environment and a healthy economy? So far, no one has come forth with a satisfactory answer.

Pesticides

Of all our fellow travelers on this planet, insects and their relatives are probably the most universally unloved. People often wonder at the perverse whim of nature that led to the creation of those creepy, crawly creatures that bite us, sting us, invade our picnics, destroy our crops, spread disease, and drive us indoors on a hot summer night.

There is no doubt that insects take their share of a food supply that is already inadequate to feed the growing billions of human beings on earth. It is estimated, for example, that insects annually destroy as much as 10 percent of the crops in the United States, and that 30 percent or more of the crops of underdeveloped countries fall prey to insect pests. It is also true that some insect species carry diseases, including such mass killers as malaria, typhus, and the bubonic plague, which during the Middle Ages wiped out a large segment of Europe's population.

Shortly after World War II, a new and powerful group of insecticides was introduced. These insecticides, known as *chlorinated hydrocarbons,* were enthusiastically greeted as tools with which man could finally rid the world of his ancient enemies and tormenters. Accordingly, these new insecticides, especially DDT, were put to wide use with considerable success that ranged from the elimination of flies in a cowbarn to a significant reduction in the incidence of malaria, a mosquito-borne disease that had long been the world's number-one killer of human beings.

Unfortunately, it soon became clear that DDT was bringing new problems along with its beneficial effects. For one thing, DDT is a "broad-spectrum" insecticide; that is, it is a powerful poison that kills other organisms in addition to the one it is meant to control. To put it another way, it is not *selective;* we cannot use it to zero in on a "target" organism because it is also capable of killing other insects, birds, fish, mammals, and even man himself. The situation is a little like the farmer putting out rat poison only to find that the dog, the cat, and the chickens have been killed along with the rats!

If we look at the situation from the "ecological point of view" that was discussed in Chapter 14, we can see that man's frantic efforts to eliminate certain insect pests may be yet another example of unwise tampering with the ecosystem's balance. Insects, like all forms of life, have specific functions according to the ecological niches they occupy, and certainly not all insects are mortal enemies of man. Bees, for example, are essential to the life cycles of flowering plants, since in the course of gathering nectar they accidentally transfer pollen from the male parts of the flower to the female part, thus ensuring fertilization and formation of seeds and fruit. Any fruit grower knows that an apple or peach crop depends on the bees "working" in the spring when the trees are in blossom. Other insects are important links in food chains—recall, for example, the case of the mayfly and its association with Lake Erie. Finally, insects prey on other insects, and one unfortunate aspect of a broad-spectrum insecticide such as DDT is the destruction of predator insects along with the target pest.

A major problem with DDT is its tendency to persist in the environment for long periods of time. It decomposes very slowly—it takes about 15 years for one pound of DDT to break down to one-half pound, and another 15 years for the remaining half-pound to decompose to a quarter-pound, and so on. Thus as vast quantities of DDT and other chlorinated hydrocarbons are added to the environment by intensive and often carelessly planned spraying programs, the total amount of these poisonous chemicals circulating in the ecosphere has reached a point where it is estimated in terms of millions of tons! Furthermore, DDT is a fat-soluble compound, and it accumulates in the stored body fat of all animals including man. This accumulation then becomes increasingly concentrated as it is passed up through the food chain, due to the fact that fewer organisms are found on each higher trophic level.

This effect of concentration by the food chain was dramatically illustrated during the mid-1950's by the mystery of the dying robins at the Michigan State University campus. In the early 1930's, a fungus that attacks the Dutch elm tree was brought into the United States from Europe. An insect called the elm bark beetle carried the fungus from tree to tree, spreading its deadly effect slowly but surely so that now the countryside throughout much of the United States is dotted with the rotting skeletons of the huge and stately elms that once gave summer shade to city streets and village greens. In an effort to save the elms on the Michigan State Campus, the trees were sprayed with DDT to kill the elm bark beetle and thus halt the spread of the fungus. In the autumn the leaves fell, carrying DDT with them as they rotted into the soil. Earthworms, turning over the soil and eating the decaying leaves, absorbed the DDT into their own tissues, thus concentrating it and setting a death trap for the robins returning in the spring. Large numbers of dead and dying robins were found on the campus. Further investigation showed fewer nests than normal were being built and fewer than the usual number of young robins were seen as the summer wore on. Also there were instances of birds building nests in which no eggs were laid and still other cases where those eggs that were laid did not hatch. This apparent reduction in fertility was further supported by the discovery that DDT accumulated in the testes and ovaries. The mystery was solved. The earthworms had concentrated the DDT at their level in the food pyramid and the robins, with their huge appetites for earthworms, had concentrated the DDT still further to a point where in many cases it was lethal!

The consequences of the persistent, heavy concentrations of DDT in the environment are already apparent. It is estimated, for example, that DDT can be found in the body fat of human beings to an average extent of 12 parts per million. This may not sound like much, but poisons as powerful as DDT can be injurious even in extremely small concentrations. Indeed it is suspected, though not proven, that some cases of liver cancer may be associated with accumulations of DDT in that organ.

Laboratory experiments have supplied strong evidence that low concentrations of DDT in the diets of birds cause the production of thin, weak eggshells, and it is feared that the consequent loss of fertility is contributing to a decline in the populations of several bird species. Also, DDT poses a considerable threat to those species of fish in which the emerging young live for a time on concentrated fat stored in the eggs. In one case alone, this problem apparently led to the death of almost a million hatching salmon.

From the foregoing examples, it is clear that if DDT and its chemical relatives are to be used at all, they must be used with caution. Many environmentalists insist that they should not be used at all, pointing to their proven disruptive and dangerous effects on the environment and its organisms. Much concern has also been expressed over the persistence of DDT in the environment, and over the fact that it is now circulating in the ecosphere even to the extent that it has been found in the body fat of Antarctic penguins, on a continent where no spraying has even taken place!

We are still left, however, with the problems posed by the fact that man's already insufficient food supply is being further depleted by insect pests and that important diseases are carried by certain insects. Obviously we *do* need some method or methods that will allow us to control certain insects.

The development of new forms of insecticides that do not persist in the environment for long periods of time has been an improvement. This is not a complete answer, however, because these newer insecticides are also "broad spectrum" in their action and therefore do not solve the problem of killing the cat along with the cockroach. Furthermore, we are left with the fact that all chemical insecticides including DDT tend to lose their effectiveness over a period of time due to the development of resistant insect populations. This mechanism of immunity development is an example of evolution in action and will be considered in detail in the next chapter.

The best methods of insect control are those that capitalize on the reproductive processes of a specific pest in place of using chemical poisons. One example of a so-called *biological control* concerns the screwworm, a fly whose larva or wormlike stage attacks cattle, resulting in serious losses. In the case of the screwworm, large numbers of male flies were sterilized and then released to mate with females. Since these matings produced no offspring, it proved an effective method of control. In addition, this approach was specific for the target insect and did not involve the addition of chemical poisons to the environment. Another possibility involves baiting insect traps with substances produced by the female that act as sex attractants to the male. Since these sex attractants are specific for each species, thousands of insects of a single, specific variety can be destroyed without danger of injury to other insect populations in the ecosystem. However, even here we have not answered the question of how the elimination of a given species of insect might affect the "balance" of the ecosystem. Much more research is needed on these and other kinds of biological control, since it is obvious that control of insect pests by indiscriminate use of chemicals is not the answer.

Energy Problems—The Machines

Americans make up only six percent of the world's population, but they gobble over one-third of the available energy. Energy, especially *cheap* energy from oil, coal, and natural gas supplies, has been a major ingredient of the prosperity and the "good life" enjoyed by most Americans. We have used cheap energy to heat and air condition our homes, light our streets and stadiums, operate innumerable labor-saving devices, and run the oversized and overpowered automobiles that clog our streets and pollute the air we breathe. Full employment and affluence is based on growth, and our present rate of growth demands that we *double* our energy consumption every 16 years.

On the other hand, while energy is the chief ingredient of prosperity, it is also the major cause of pollution. It is no accident that the most affluent, industrialized nations are the ones most plagued by polluted air and water. Indeed, it might be said that we are *not* running out of energy sources in the absolute sense; instead, we are beginning to exhaust the environment's capacity to absorb the waste products of energy utilization. It has been said that it costs the earth every time we demand another convenience and that prosperity can be maintained only at increasing destruction, ultimately bringing a lower quality of life for all.

On the other side of the coin, environmentalists have been blamed in part for the energy problems that we now face. They have been accused of exaggerating the problems of pollution, and their opponents point out that clean air and water is not worth the price of unemployment and a reduced standard of living. Environmentalists answer by pointing to the lessons of Donora, London, and Los Angeles, and argue that the only way out of the dilemma is to return to the lifestyle of a simpler time. They also suggest that the energy situation is due in large part to human behavior, including such things as irresponsible attitudes toward childbearing and demands for ever higher standards, all of which is attended by a naive assumption that the earth holds an unlimited supply of cheap energy.

In 1973 when the Arab oil-producing countries placed an embargo on shipments to the United States, Americans and the rest of the world were brought sharply to the realization that energy does not exist as a cheap commodity in endless supply. Among other things, we were made to see that oil supplies cannot provide energy indefinitely and that new sources of energy must be explored.

The problems of electrical power plants provide an excellent example of the energy-environment dilemma. Only a small fraction of power plants in the United States have used water power to turn the turbines that run the generators of electricity. In the past, the great majority of power plants depended on coal to generate steam, which was used to power the turbines. Coal, on the other hand, contains sulfur to varying degrees, and its combustion fills the air with sulfur dioxide. Coal is therefore not very popular with environmentalists, and laws that required pollution abatement procedures were passed. Many power companies therefore turned to cleaner oil and even

Figure 15.14
An area in Montana noted for its rich coal deposits. It was here in 1876 that Custer's Seventh Cavalry was defeated by the Sioux Indians.

cleaner natural gas, rapidly depleting the supply of natural gas which should have been reserved for home heating. Some power companies also turned to the promise of nuclear energy only to find problems with licensing, labor, and environmentalists who fear that nuclear plants may pose dangers as great or greater than pollution from fossil fuels.

What then are the alternatives? For one thing, we are not running out of fossil fuels as rapidly as some believe. For example, it is estimated that we have enough coal in Montana, Wyoming, and the Dakotas to last at least 500 years, which would give us plenty of time to develop alternative sources of energy. In Montana, where Custer and the Seventh Cavalry rode to disaster and immortality at Little Big Horn, lies the richest coal deposit on the North American continent (Figure 15.14). Coal, however, is fraught with problems. Coal is dirty, it involves problems of transportation, and its sulfur content produces air pollution. Secondly, this western coal must be strip-mined, a process in which giant machines remove thousands of acres of fragile plains soil from the coal deposits underneath. The very mention of strip-mining recalls scenes in West Virginia, Kentucky, and Illinois where the land was destroyed and left useless, surrounded by the poverty that now pervades those areas (Figure 15.15).

On the positive side, strip-mining is much safer than deep-mining and the land *can* be reclaimed, although some environmentalists doubt that it can be restored to its original state. At any rate, reclamation is estimated to cost over $6000 per acre, a cost that is certain to be passed on to the consumer. In addition, coal can be cleaned up and the objectionable sulfur removed. Coal can also be liquefied and gasified, but all of these processes must be developed to a practical level. They will cost money, resulting in the inevitable increase in price to the consumer. Experts tend to agree that the western coal fields will be mined in spite of the problems, especially as the demands for energy continue to outrun the supplies.

The offshore oil reserves, which are estimated to contain as much as 130 billion barrels, constitute another likely energy source (Figure 15.16). If recovered at present capability levels, this oil could provide as much as 25 years supply. Unfortunately, it too is expensive to obtain and also involves at least some danger of oil spills that could wreak havoc on marine and bird life (Figure 15.17).

Oil shale, which is an immature form of coal, can be refined into synthetic crude oil. Very extensive oil shale deposits are located in Colorado, Utah, and Wyoming. The process of extraction would be expensive and difficult, however, and would bring the in-

Figure 15.15
An aerial view of the results of extensive strip mining in West Virginia. Note the massive amount of earth that has been moved down the mountainside. (Courtesy of the U.S. Department of Agriculture–Soil Conservation Service.)

Figure 15.16
Aerial photograph showing a spreading oil leak from an offshore oil well at the bottom of the Santa Barbara channel, February 1969. Thousands of gallons of oil polluted nearby shores before the leakage was stopped. (Courtesy of the Environmental Protection Agency, Documerica.)

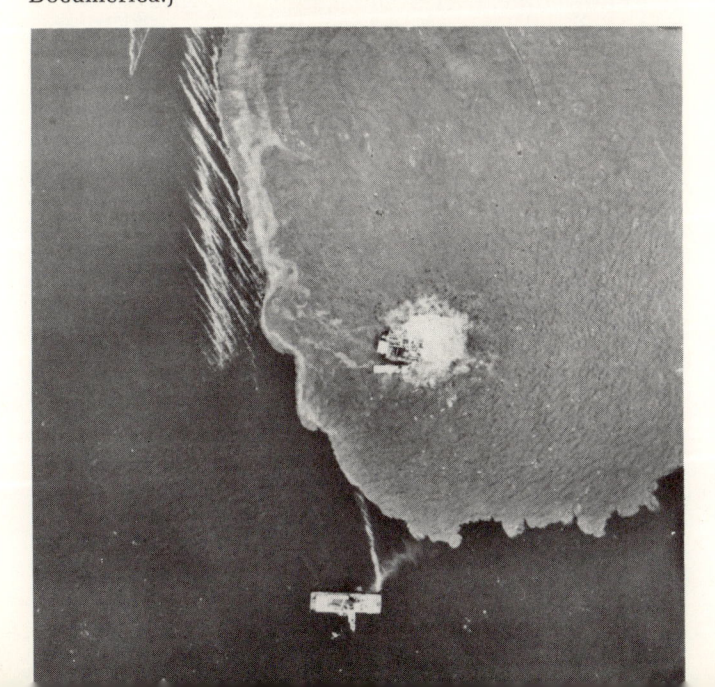

Figure 15.17
Oil-soaked waterfowl found along the New England coast. Oil spills from offshore drilling rigs or from tankers can result in tragedies such as this. (Courtesy of the U.S. Department of the Interior–Sport Fisheries and Wildlife Service.)

Figure 15.18
"Old Faithful," a famous geyser at Yellowstone National Park, Wyoming. Geyser beds such as those at Yellowstone are a potential source of energy to supplement the dwindling supplies of fossil fuels.

evitable environmental headaches. One difficulty is that a large amount of oil shale must be processed to yield a relatively small amount of product. Experts estimate that one ton of rock must be processed to produce between 20 and 40 gallons of oil. The production of 50,000 barrels of oil would involve the disposal of 60,000 tons of waste material. Added to that, there is no doubt that a considerable portion of the scenic beauty as well as the environment of these states would be changed in the process. Also, quite obviously, energy from this source would be far from cheap!

Perhaps the cleanest type of energy would be solar energy derived directly from the sun. The sun would indeed be a limitless source and is regarded by some as the energy source of the future. It seems likely, however, that any hopes for significant contributions from this direction in the near future are overoptimistic. There are numerous difficulties to be overcome, not the least of which is the fact that the sun goes down at night and in some areas is hidden behind cloud cover for long periods of the year. Geothermal energy, which is that derived from the heat of the earth's interior, has received less publicity but is a more practical possibility for the more immediate future. Water heated by molten rock in the tradition of Yellowstone Park's Old Faithful (Figure 15.18) is flashed into steam, which is then used to turn turbines for electrical power. This form of energy was used in Italy as far back as 1904 and has also been successfully used in Iceland. In 1960, energy from the geyser fields of Northern California was harnessed for the production of electricity, and experts believe that other sources of geothermal energy exist in the United States just waiting to be tapped. Outside of the technical barriers, the major problem to be overcome is the restriction of geothermal energy to specific geographical areas. On the other hand, it could be used in those areas where it is available, thus freeing other sources such as fossil fuels for use in other locations.

Figure 15.19
A nuclear power plant. It is expected that this source of power will become increasingly important in the future. (Courtesy of the Yankee Atomic Electric Company.)

In August 1945, an American B-29 bomber named *Enola Gay* (after the pilot's mother) appeared in the skies over Hiroshima, Japan. Within a few seconds, the city and thousands of its inhabitants were destroyed by an explosion that changed the world, ushering in the nuclear age amid the pain and sorrow and suffering of war. The possibilities for peaceful use of the incredible amount of energy released by the atomic bomb has intrigued man from the beginning, and today the nuclear power plant is looked upon as one of the more practical and immediate substitutes for the dwindling oil supplies. In 1973, there were 31 civilian nuclear power plants in operation, and plans were to put at least 900 such plants in operation by the year 2000 (Figure 15.19).

In nuclear fission, certain heavy atoms called uranium-235 are struck by neutron particles, causing the uranium atoms to split into two or more fragments, releasing energy in the process. In a controlled reaction, the heat generated by the uranium fuel is used to convert water to steam. The steam drives turbines, which in turn drive electricity generators. The nuclear reaction thereby substitutes for the fossil fuel that ordinarily provides the necessary heat; the rest of the plant equipment is very much like the conventional type (Figure 15.20).

The essential parts of the nuclear reactor consist of (1) a reactor core containing the uranium-235 fuel, (2) a system of control rods made of a heavy metal that absorbs neutrons and can therefore slow down or stop the reaction when inserted into the core to varying degrees, and (3) a cooling system in which a coolant, usually water, serves both to cool the reactor and to slow down the neutrons so they can be "captured" by the uranium atoms.

Despite its bright promise as a clean, effective source of energy, nuclear power is not without its problems and definitely not without its opponents. First of all, nuclear power plants must be located on a lake or stream from which water is taken to circulate through the structure. The water is then released back to the lake or stream at a considerably higher temperature. The resulting total warming effect on the lake or stream is called *thermal pollution,* and ecologists predict that this will cause severe disturbance of the ecosystem including fishkills caused by oxygen depletion (Figure 15.21). Second, nuclear fuel is radioactive, and radioactivity is known to produce leukemia and other forms of cancer, not to mention unknown and immeasurable numbers of mutations. Advocates of nuclear power plants point out that every precaution is taken to maintain safe levels of radioactivity, but opponents reply that what is a

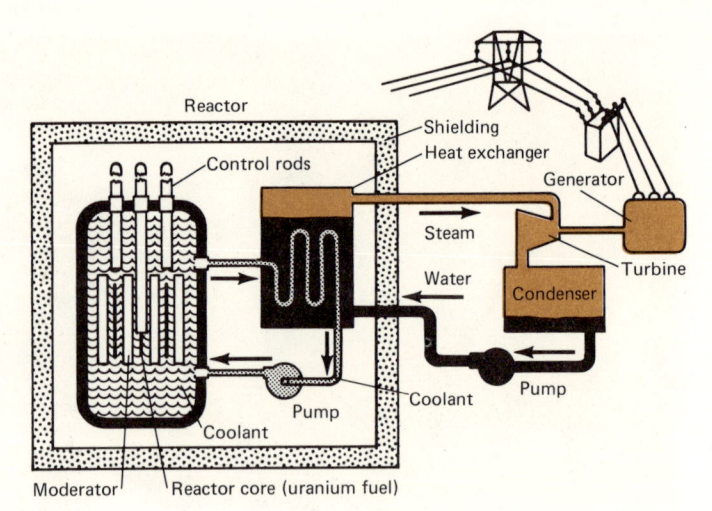

Figure 15.20
Diagram illustrating nuclear power plant operation. The reactor core produces heat to make steam, which in turn drives turbines and electrical power generators. (Atomic Energy Commission.)

"safe level" has not been definitely established and is still debated by the experts.

Another problem involves the nuclear fuel reprocessing plants. After a period of time, the uranium-235 fuel in a nuclear reactor accumulates highly radioactive fission products, requiring its removal and transportation to a reprocessing plant. These waste products of fission must be disposed of; the current method is to bury them underground in storage tanks. Unfortunately, there is no guarantee that these tanks will not break down under the intense heat generated by the waste materials, allowing those waste products to seep into ground water and spread highly radioactive contamination.

There is considerable opposition from residents of areas surrounding proposed nuclear plant locations. Many people recall Hiroshima and envision the reactor as an atomic bomb that could explode at any minute. That a reactor could actually turn into a bomb is

Figure 15.21
Aerial view of a nuclear power plant showing a "thermal plume" extending into a lake. This tendency of nuclear plants to raise the temperature of adjacent waters is of concern to environmentalists. (Photograph courtesy of Dr. Eugene Chermack, College at Oswego, State University of New York.)

unlikely if not impossible, but accidents of other kinds could happen. For example, if the reactor's coolant should be lost through a leak, the core could heat up to the point of causing a "meltdown." Molten uranium could then bore through the bottom of the reactor and into the earth, spreading highly dangerous radioactivity over a large area. Nuclear scientists claim that an accident of this magnitude is highly unlikely, but their estimates of probability tend to be of little comfort to those who live near the site of a proposed reactor. One answer to the "meltdown" possibility is the backup cooling system now being added to reactor design.

It is probable, however, that most of the accidents that might occur would be of the "industrial accident" variety that we already expect in conventional industries such as steel mills. Here the main risks are to the operators, and a nuclear plant is designed to contain the "ordinary" accident within the plant so that people in the surrounding areas will not be affected. At any rate, we will eventually have to balance the likelihood of danger from nuclear plants against the demands for more energy that are made by an expanding economy. As one nuclear scientist points out, we already accept without serious question the annual price of 50,000 human lives that is paid for the privilege of driving automobiles. Such are the choices we must make.

We continue to be plagued by the question of how to balance the needs of the environment against our needs for energy. Environmentalists fear that an energy shortage will almost certainly result in an "environmental retreat" from the antipollution standards they fought so hard to achieve. In point of fact, when the first pinch of the Arab oil embargo in 1973 was felt, one of the first actions taken by the Nixon Administration was to urge relaxation of clean-air laws. Environmentalists also fear that some industries may

"Subdue the earth, and have dominion..."

Figure 15.22
The World Trade Center in New York City illustrates an energy-wasting building design.

take advantage of an energy shortage in order to avoid their responsibilities toward the environment. Quite obviously, adequate supplies of energy will be of no use if we have to wear gas masks on our faces and radioactivity counters around our necks.

A partial solution could involve a significant cutback in our energy demands. We must discard the habits we learned when energy was cheap and plentiful. We must learn to like smaller, less-powerful cars and do without air conditioning. We must insulate our homes and develop efficient forms of mass transit. The World Trade Center in New York City is an example of our energy binge (Figure 15.22). This single building consumes enough energy to power a city of 100,000 people... this is because it was designed with no regard for the cost or availability of energy.

Needless to say, how much energy is used is in part a function of the number of people who use the goods and services that require energy. This presents yet another reason why population control should be the first priority for the human species. The availability of energy, like the food supply, cannot match the population increase. The inevitable result will be increased poverty and less opportunity for upward mobility in our society.

Above all, our problems with energy cannot be solved on a piecemeal, crisis-oriented basis. Money must be spent on extensive programs to develop energy sources that will do minimum damage to the environment. The solutions will require hard decisions, and these decisions will be made largely on the basis of demands made by an informed and concerned public.

Energy Problems—The People

We are fortunate if our individual energy worries are limited to finding enough gasoline to run our automobiles and electricity to power our gadgets. Between one-half and two-thirds of the four billion people on this planet awake each morning to face a continuing energy crisis that is far more fundamental. They are the undernourished and the malnourished, the victims of an enduring hunger that saps the body's strength, robs the brain of its potential, and breaks down the human spirit.

Without energy fuels, the machinery of our civilization will stop. In a similar way, the human body depends upon a continuous supply of nutrients essential to life. We must have carbohydrates, proteins, vitamins, and minerals in quantities sufficient to provide energy, to promote growth, and to participate in the complex reactions that maintain life (Chapter 10).

It is interesting to compare the diet of the average American with the diets of those in the poverty areas of the underdeveloped nations. Americans, for example, consume an average of 186 pounds of meat per capita per year. By comparison, the individual of an underdeveloped area is fortunate if he eats a few pounds of meat during the same time period. The average American eats 314 eggs per year, compared to the average of 8 eggs per person consumed in India. Meat, eggs, poultry, and fish are the major sources of protein in the American diet, but they contribute very little to the diets of people in the underdeveloped areas. Cattle, hogs, and chickens must be fed, and the loss sustained from one trophic level to the next makes meat a luxury that the poor cannot afford. The United States, for example, feeds 80 percent of its grains to livestock, compared to the 0.5 percent that is fed to livestock in India. The poor of the world tend to be "herbivores," going directly to the producers for the bulk of their diets which consist largely of cereals and some legumes such as beans and chickpeas. Since the protein content of cereals is considerably lower than that of meat, cereals must be consumed in greater quantities in order to obtain equivalent protein levels. Unlike animal protein, most cereals are especially deficient in lysine, which is one of the essential amino acids (meaning that it cannot be manufactured by the body but must be taken in as part of the diet). Sorghum cereals that do contain sufficient lysine could help this situation, provided they can be successfully introduced to those areas where they are most needed.

As of now, it is theoretically possible to provide the minimum nutritional needs of all the people on earth, although even this capability will be lost as the population continues its upward spiral. The major problems at this point are concerned with food *distribution*. There are almost insurmountable difficulties that keep surplus food from going to areas where it is needed. Transportation costs are high, the distances are often great, and waste and graft take their toll at each step along the way. Tremendous losses are suffered in storage where food supplies are attacked and spoiled by rats, insects, and fungs growths. It is estimated that the annual storage losses alone could feed a half-billion people!

A more practical solution is to help the underdeveloped countries increase their own food production. Some hope for this possibility has been generated by the "Green Revolution," a term applied to agricultural improvements that include new, high-yield grains and increased production through the use of fertilizers, pesticides, and irrigation. As an example, new varieties of wheat have been developed that can significantly increase the yield per acre. The older varieties did not permit intensive use of fertilizers because they would grow so tall they would fall under the weight of the grain, making harvesting difficult and unproductive. A new variety called Mexican wheat has short, sturdy stalks, a characteristic that allows the use of extra amounts of fertilizers. These new varieties also grow well in latitudes near the equator where many of the underdeveloped countries are located. Dwarf varieties of rice have also been developed that can double the Asian rice crop if carefully managed. These new varieties have the extra advantage of shorter growing cycles, yielding the possibility of an extra annual crop. In the 1968–1969 growing year, the Asian rice crop was increased by 17 million tons.

Fish farming in rice-paddy areas is another possible source of additional food, especially badly needed protein. Fish protein concentrate (FPC) made from trash fish is still another possible source of protein, provided it can gain acceptance. Contrary to popular belief, even people who are bordering on starvation will reject foods that are not to their taste or that may be against religious beliefs or local customs!

Mass programs to fortify existing foods with vitamin and mineral supplements have also been suggested as a way to provide better nutrition for those in poor areas. Programs of this kind existed in the United States as far back as the 1930's when flour was fortified with iron and vitamins. Iodization of salt is also well known in the United States. Iodine, you will recall, is a necessary ingredient of the hormone secreted by the thyroid gland, and if it is in short supply the gland enlarges, causing goiter (Chapter 10). The so-called "common carriers" which include water, salt, wheat, flour, bread, and margarine could be used as vehicles in a fortification program. This approach to combating malnutrition has been tried with considerable success in some areas, but the cost is great and requires heavy government support.

A danger inherent in such programs as the Green Revolution is the tendency for those who oppose population control to see them as justification for such opposition. Actually, given the most concentrated and superhuman efforts, we cannot possibly match the current population doubling rate with anything even approaching sufficient food production. The Green Revolution can be considered only a stopgap measure to avoid widespread famine and give us time to bring population growth under control. The ultimate solution to the problem of malnutrition is the same as that of the problems of energy shortage and pollution. The people problem is the root cause of all, and whether we can or will take the measures necessary to achieve a stable world population is perhaps the single most important question that must be answered in this last quarter of the twentieth century.

Questions

1. It has been said that energy is a major ingredient of both prosperity and pollution. Discuss the social, economic, and environmental implications of this statement.

2. Discuss the following statement: Our present environmental problems are due in part to modern man's failure to have an ecological point of view.

3. Considering that a significant percentage of the world's annual food production is destroyed by insects, how can you justify a ban of DDT in light of the fact that millions of people border on starvation?

4. Opponents of population control point out that (1) man's capability to produce food can be significantly increased, (2) there are vast areas of the earth that are only thinly inhabited, and (3) there is plenty of food for all at the present time. How would you answer these arguments if you were supporting the side of population control?

5. What major steps must be taken now in order to ensure that our great-grandchildren grow up in a world in which the "quality of life" is not reduced?

6. How would you answer the following statement? Unemployment and a lowered standard of living is too high a price to pay for a clean environment.

16

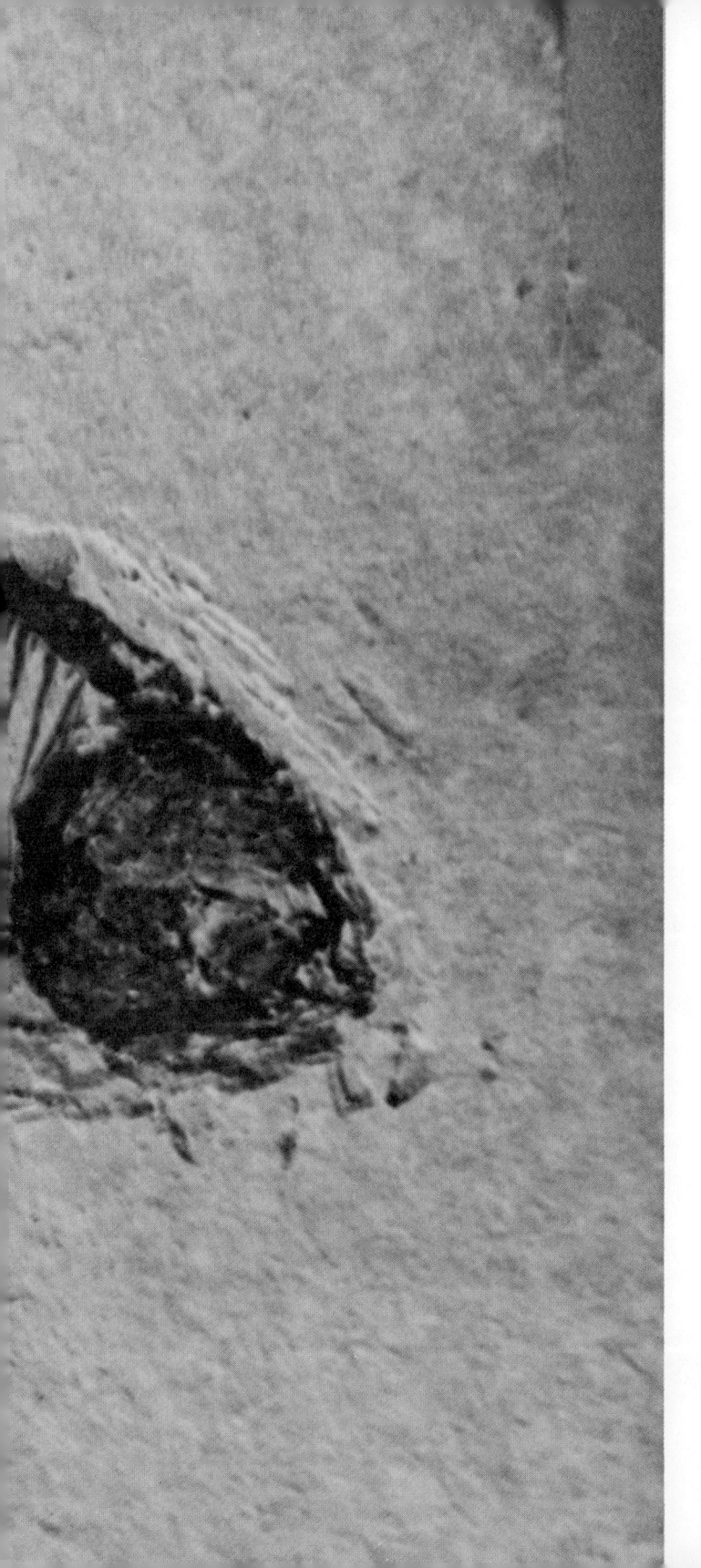

Evolution

The Changing Earth

During the brief moments of a single lifetime, we see little or no evidence of change in the face of the earth or in the plants and animals that share it with us. Lofty mountains have an air of permanence that makes it hard to believe that they were not always there. It is equally hard to believe that the sea once rolled where a bustling city now stands, or that a little over 20,000 years ago a sheet of ice thousands of feet thick covered much of the Northern Hemisphere and all of the British Isles.

"All is change," said the Greek philosopher Heraclitus. "Only change itself is changeless." From the earth's beginning and down through the long ages of its history, change has been a major theme. The climate has fluctuated between periods of cold and

Figure 16.1
The Niagara gorge. The sedimentary rocks exposed in the gorge were laid down layer by layer during the Silurian period (see Box 16.1) which began about 425 million years ago. At that time, this area was covered by a warm, shallow inland sea. Numerous fossils, remains of animals that lived during that period, are found in the rock layers.

warming trends, forming huge continental glaciers that extended down over much of the Northern Hemisphere, leaving a new face on the earth as the temperature rose and they receded northward. Great mountain ranges have been pushed up by the earth's restless crust, only to be eroded away over eons of time by the patient but inexorable forces of water, wind, and ice. The land masses fell and rose again; the seas repeatedly invaded the land and retreated. Each time the oceans left new deposits of sand, mud, and shells which built layer upon layer of sedimentary rock (Figure 16.1), each containing the fossilized remains of the plants and animals of its time (Figure 16.2).

Figure 16.2 ▶
Examples of fossils: (a) the imprint of a fish that lived in Wyoming 50 million years ago, (b) a cast of a trilobite, an organism found in Silurian rocks over 400 million years old, and (c) an ancient rock showing the imprint of a fern.

The changing earth 315

Box 16.1 The geologic time scale.

ERAS	PERIODS	EPOCHS	AQUATIC LIFE	TERRESTRIAL LIFE
(approximate starting dates in millions of years ago)				
Cenozoic 63 ± 2	Quaternary 0.5–3	Recent		*AGE OF MAMMALS* — Man in the New World
		Pleistocene	Periodic glaciation	First men
	Tertiary 63 ± 2	Pliocene	Continental drift continues	Hominids and pongids
		Miocene	All	Monkeys and ancestors of apes
		Oligocene	modern	Adaptive radiation of birds
		Eocene	groups	Modern mammals and herbaceous
		Paleocene	present	angiosperms
Mesozoic 230 ± 10	Cretaceous 135 ± 5		*Still attached are: North America and Northern Europe; Australia and Antarctica* Modern bony fishes; Extinction of ammonites, plesiosaurs, ichthyosaurs	Extinction of dinosaurs, pterosaurs; Rise of woody angiosperms, snakes
	Jurassic 180 ± 5		*AGE OF REPTILES* — *Africa and South America begin to drift apart* Plesiosaurs, ichthyosaurs abundant; Ammonites again abundant; Skates, rays, and bony fishes abundant	Dinosaurs dominant; First lizards: *Archeopteryx*; Insects abundant; First angiosperms
	Triassic 230 ± 10		*Pangaea splits into Laurasia and Gondwanaland* First plesiosaurs, ichthyosaurs; Ammonites abundant at first; Rise of bony fishes	Adaptive radiation of reptiles (thecodonts, therapsids, turtles, crocodiles, first dinosaurs); First mammals

Like chapters in a book of stone, these sedimentary rock layers furnish a history of life on this planet. The deeper rocks thus represent the earlier chapters, and the uppermost lawyers provide evidence of the life forms that existed in more recent times. To be sure, it is a book filled with gaps and uncertainties and is far from complete. Yet the broad outlines of the story of life can be read in the remains of organisms found in the sediments of ancient seas, in the bones of animals trapped in tarpits, in the well-preserved body of a wooly mammoth frozen in the wastes of Siberia, or in a skull fragment that belonged to an apelike ancestor of modern man. It is the story of the slow, random changes of organic evolution, which is the process that produced the many species which exist on today's earth.

Scientists describe the earth's history in terms of the *geologic time scale* (Box 16.1), which is based on the ages and locations of various rock layers. In recent years, the ages of ancient rocks have been esti-

Box 16.1 (continued)

ERAS	PERIODS EPOCHS	AQUATIC LIFE	TERRESTRIAL LIFE
(approximate starting dates in millions of years ago)			
Paleozoic 600 ± 50	**Permian** 280 ± 10	*Appalachian Mountains formed, periodic glaciations and arid climate* Extinction of trilobites, placoderms	Reptiles abundant (cotylosaurs, pelycosaurs) Cycads and conifers, ginkgoes
	Pennsylvanian 310 ± 10 — Carboniferous	*Warm humid climate* Ammonites, bony fishes	First reptiles Coal swamps — AGE OF AMPHIBIANS
	Mississippian 345 ± 10	*Warm humid climate* Adaptive radiation of sharks	Forests of lycopsids, sphenopsids, and seed ferns Amphibians abundant Land snails
	Devonian 405 ± 10 — AGE OF FISHES	*Periodic aridity* Placoderms, cartilaginous and bony fishes Ammonites, nautiloids	Ferns, lycopsids, and sphenopsids First gymnosperms and bryophytes First insects First amphibians
	Silurian 425 ± 10	*Extensive inland seas* Adaptive radiation of ostracoderms, eurypterids	First land plants (psilopsids, lycopsids) Arachnids (scorpions)
	Ordovician 500 ± 10	*Mild climate, inland seas* First vertebrates (ostracoderms) Nautiloids, *Pilina*, other mollusks Trilobites abundant	None
	Cambrian 600 ± 50	*Mild climate, inland seas* Trilobites dominant First eurypterids, crustaceans Mollusks, echinoderms Sponges, cnidarians, annelids Tunicates	None
Precambrian 3000		*Periodic glaciation* Fossils rare but many protistan and invertebrate phyla probably present	None

mated with a high degree of accuracy by a procedure which makes use of the fact that certain radioactive materials change to other materials at a very definite time rate. This process is called *radioactive dating* and has led to an upward revision in the estimate of the earth's age to nearly five billion years.

Note that the geologic time scale is divided into major segments called *eras,* and that these in turn are divided into *periods.* The most ancient rocks are those formed during the Precambrian era. These rocks contain evidence of very simple forms of life but yield no direct fossil evidence of higher organisms. Generally, the evidence suggests that life in the Precambrian era consisted mainly of simpler forms such as bacteria and algae. Other slightly more advanced forms must have also existed, but because of the extreme age of the Precambrian rocks, the fossil record of that era is far from complete.

The Paleozoic era, which began about 600 million years ago, saw the development of major invertebrate groups and the rise of the vertebrates in the forms of fishes, amphibians, and reptiles. In the first part of this era, life multiplied in the sea, which had been its birthplace, and then took the first tentative steps onto the barren rocks of the land. The land offered nothing but an atmosphere, water, and the minerals in the rocks; it was therefore inevitable that the first colonists of the land would be simple forms of green plants with the ability to carry on photosynthesis (see Figure 16.3). As the plants evolved on the land and higher forms appeared, the greening of the once-barren rocks set the stage for the entrance of the insects and amphibians. The amphibians only partially escaped the water, being dependent upon it for part of their life cycle, just as the modern frog must spend part of its life as a fishlike tadpole. By the end of the Paleozoic era, the reptiles had evolved and were spreading over the earth while the amphibians declined. By then the insects were also well established.

The Mesozoic era brought the age of the reptiles, including the famed dinosaurs that ranged from the size of a chicken to the mammoth *Diplodocus* (Fig-

Figure 16.3
The first colonists of the land were simple green plants, which must have been similar to the algae covering the rock in this modern scene.

ure 16.4). The flying reptiles, the *pterysaurs* (Figure 16.5), dominated the skies; the record 51-foot wing span of one variety called *Pteranodon* makes it the largest flying creature of all time.

By the midpoint of the Mesozoic era, the birds and mammals had appeared. The early mammals must have seemed insignificant compared to the giant reptiles that were the current lords of the earth, but their efficient reproductive system and constant body temperature ultimately gave them a dominant position among the world's organisms.

The end of the Mesozoic brought the twilight of the dinosaur gods. They faded from the scene leaving only bones (Figure 16.6) and tracks (Figure 16.7) as reminders of a reign that lasted nearly 100 million

Figure 16.4 (left)
A reconstructed scene depicting the landscape of Colorado, Wyoming, and Utah about 140 million years ago. This was the age of the dinosaurs: Antrodemus (left) was a flesh-eater; Stegosauras (front) was armed with large, bony plates for protection; Diplodocus (right) was a large plant-eater. (Courtesy of the Smithsonian Institution, Photo No. 2526A.)

Figure 16.5 (below)
Restored skeleton of the flying reptile Pteranodon ingens Marsh, found in sedimentary deposits in western Kansas. (Courtesy of the Smithsonian Institution, Photo No. 28138.)

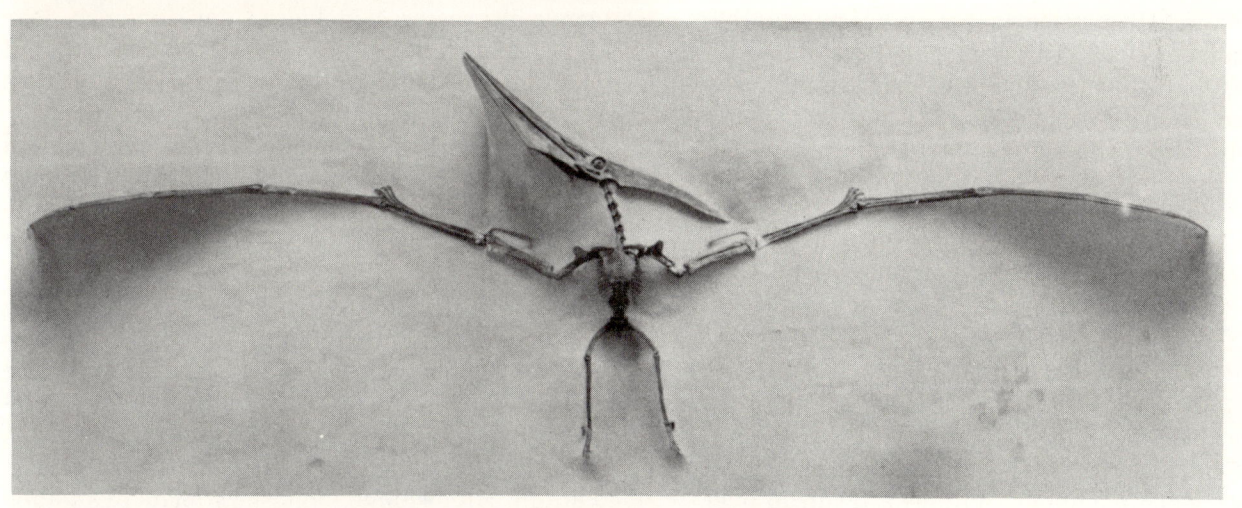

Figure 16.6
Skeleton of the giant dinosaur Diplodocus longus Marsh, reconstructed from sediments found in Utah. (Courtesy of the Smithsonian Institution, Photo No. 43492.)

Figure 16.7
Fossilized dinosaur tracks. (Courtesy of the Smithsonian Institution, Photo No. 75-6949.)

years. No one knows for certain why the dinosaurs became extinct. It may be that they became too big and too cumbersome, with consequent loss of fertility. It may be that the climate changed, and as the world grew colder, the cold-blooded dinosaur's body temperature dropped and its extinction was inevitable. On the other hand, some authorities, notably R. T. Bakker, disagree with the traditional belief that all dinosaurs were cold-blooded. According to Bakker, there is evidence that they were warm-blooded and that their extinction was due to an inability to adapt to the changing land environment because of their great size. Bakker even suggests that dinosaurs are *not* totally extinct but live on in the form of birds, which are their direct descendants.

The Cenozoic era brought the flowering plants, and the birds and mammals continued to spread over the earth, taking the place of the now forgotten dinosaurs. The first primates appeared, which led to the evolution of the monkeys and great apes. Finally during the last part of the Cenozoic, man appeared as yet another offshoot of the primate group.

The fossil record, filled as it is with gaps and uncertainties, presents convincing evidence that the many species now on earth have not been here from the beginning but developed instead by a process of slow change as organisms gave rise to slightly different kinds of organisms over eons of time. On the other hand, the Biblical concept of creation holds that God made all creatures, including man, in the six days of the special creation. According to the Book of Genesis, therefore, the earth and its inhabitants have existed unchanged from the dawn of creation.

For centuries, the concept of *special creation* dominated western thought. Implicit in this concept is the idea that the earth and all its inhabitants were created as a kind of playhouse for man, and that man himself was created in the image of God and put upon earth to rule over all other beings. The Biblical story of creation was further strengthened by Bishop James Ussher (1581–1656), who calculated that the earth and all its inhabitants were created exactly as they are today in six 24-hour days, starting exactly at 9:00 A.M. on the 23rd of October, 4004 B.C.! Bishop Ussher arrived at this astonishingly precise estimate by totaling the ages of the Biblical patriarchs.

Yet there were those who suspected that something was wrong with the idea of an unchanging earth. The discovery of fossils in 1791 by William Smith, an English surveyor, helped create further doubts. Georges Cuvier (1769–1832), a Frenchman, tried to explain the fossil record with his theory of *catastrophism,* which proposed that changes in fossil types from one sedimentary layer to another were evidence that life on earth had been destroyed several times in the past by forces akin to the Biblical flood, with a special creation of new and improved species following each catastrophe. In other words, God had experimented with earth's life, destroying and recreating until he found a group of species that suited Him. Reflecting upon Cuvier's theory, one might wonder somewhat nervously if God is happy now. This theory of catastrophism was opposed by Charles Lyell (1797–1875), an Englishman who proposed his theory of *uniformitarianism*. Stated simply, Lyell's theory proposed that changes that took place on the earth's surface in the past were not of mysterious or supernatural origin and could be explained in terms of the same natural geological processes that can be observed today.

The first major attempt to explain the *process* of evolution was made by Jean Lamarck (1744–1829). Lamarck, a Frenchman, proposed that species changed because they were modified by environment. His most famous illustration of environmental modification was his claim that the long neck of the giraffe had been acquired because generations of giraffes had stretched their necks to reach foliage in the trees (Figure 16.8). Unfortunately, Lamarck's hypothesis was based on the false concept of the inheritance of *acquired characteristics*. If this concept were valid, then a man who works hard to develop large muscles should produce offspring with at least slightly more well-developed muscles than usual. Or to carry the concept to an extreme, a man or woman who loses an arm in an accident might be expected to produce a child without an arm. We now know, of course, that the genetic material which determines that an offspring will have two arms, normal muscle development, a short neck, and so on is carried safely in the sperm or egg cell, unaffected by environmentally induced changes in the parent's body. The only exceptions are mutations in the genetic material which may result from radiation or the action of other mutagenic agents.

Evolution, therefore, remained only an intriguing idea, since despite the suggestive evidence of the fossil record, evolution could not be taken seriously until someone came forward with a logical and convincing explanation of the *process* whereby changes in organisms could have occurred. Finally in 1859, a book called *On the Origin of Species by Means of Natural Selection and the Preservation of Favoured Races in the Struggle for Life* came upon the scene. Its author was Charles Darwin.

Darwinism

Charles Darwin was born in England on February 12, 1809, the same day that Abraham Lincoln was born in

Figure 16.8
Lamarck's giraffes. According to Lamarck's theory, continual stretching to reach food resulted in a gradual lengthening of the neck from one generation to the next.

a rough log cabin on the other side of the Atlantic Ocean. Unlike Lincoln, Darwin came from a wealthy and prominent family, and his early life was that of a typical English gentleman of the time (Figure 16.9). His grandfather, Erasmus Darwin, was a distinguished physician with something of a reputation as an evolutionist. His father, Robert, was also a successful doctor, so it is not surprising that Charles was eventually sent off to study medicine. Fortunately, as it turned out, Darwin's medical career was short-lived. Those were the days before anesthetics, and the sight of a child undergoing an operation so horrified the young Darwin that he fled from the cries of agony that filled the operating room. He then entered Cambridge University where he obediently if unenthusiastically studied for the ministry.

From the time of early boyhood, Darwin's true interest was nature study. He found his natural surroundings an ever-present source of wonder and fascination, and he spent long and happy hours collecting, examining, and classifying rocks, plants, insects, and everything else that failed to escape his collector's eye. Although Darwin himself regarded his passion for nature study as only a hobby, his knowledge so impressed his professors at Cambridge that he was recommended for the unpaid post of ship's naturalist on the H.M.S. Beagle, a 90-foot naval vessel that was to voyage around the world on a surveying mission. After much persuasion, a disappointed Robert Darwin finally gave grudging consent to what he must have felt was at best a frivolous lark, never dreaming that his decision would ultimately lead to his son's becoming one of the giants of human history. Thus it was that on a gray December day in 1831, the Beagle sailed from Plymouth, and Charles Darwin, not quite 23 years old, began a voyage that was to make him miserably seasick and bring him everlasting fame.

The voyage of the Beagle lasted five long years, during which it sailed up and down the coast of South America, to New Zealand, Australia, the Pacific Islands, around the tip of Africa, and finally home to

Figure 16.9
Charles Darwin as a young man. (Courtesy of the American Museum of Natural History.)

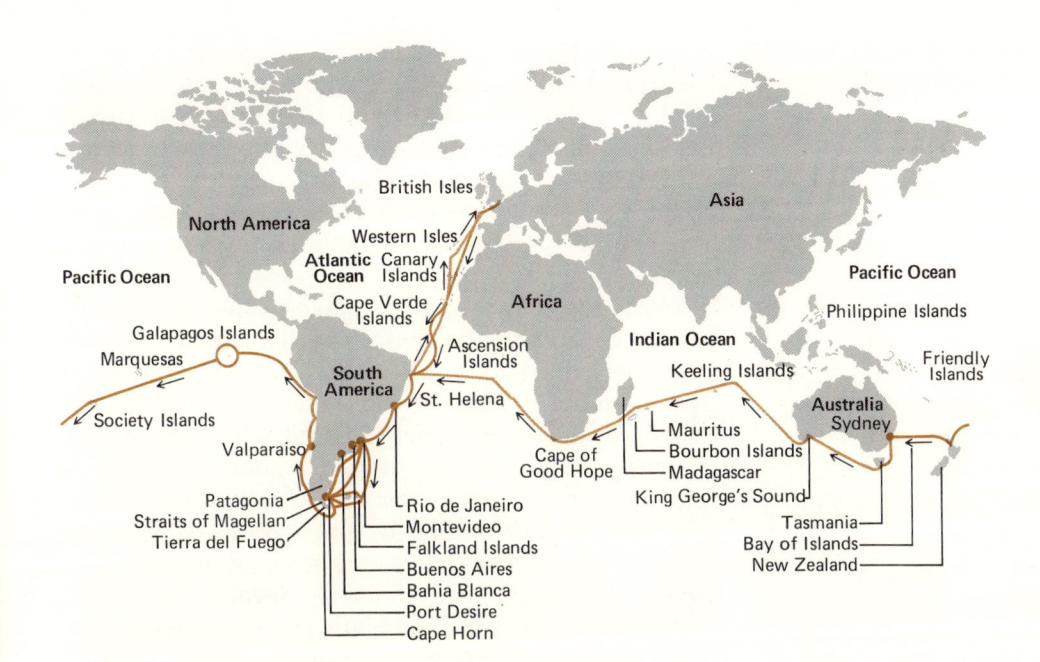

Figure 16.10
Voyage of the H.M.S. Beagle.

Plymouth (Figure 16.10). While the ship's crew carried on the job of charting the coastline, Darwin excitedly observed and recorded the thousands of strange specimens that he collected from the even stranger places visited by the Beagle. He was especially excited by the discovery of seashells at a point 12,000 feet high in the Andes Mountains. How, he asked himself, could this find be reconciled with the belief that the mountains were always there, created by God in the beginning? He reasoned that the shells must have been deposited at a time when the same land was below sea level; the shells were evidence that the face of the earth had changed over what must have been eons of time since its formation.

It is generally believed that Darwin's first glimmerings of his theory of evolution began when the Beagle came to the Galapagos archipelago, a group of bleak, uninviting islands which lie about 500 miles off the coast of South America (Figure 16.11). *Galapagos* is Spanish for *tortoise;* the islands were named for the giant tortoises that roamed over their black, lava-covered surfaces. These islands teemed with other varieties of life, including numerous species of lizards and birds from which Darwin feverishly collected specimen after specimen, carefully labeling each one according to the island from which it came.

Later as he leisurely examined his specimens on board the Beagle, he was struck by the fact that the lizards, birds, and insects he had collected resembled other species found on the South American mainland, and yet they were different in certain ways. Furthermore, members of the same species differed from one island to another. He recalled that one of the residents of the islands had remarked that he could tell by looking at a tortoise *which island it came from.*

Among Darwin's specimens were a number of birds called finches, which he later described so completely that they became known as Darwin's Finches

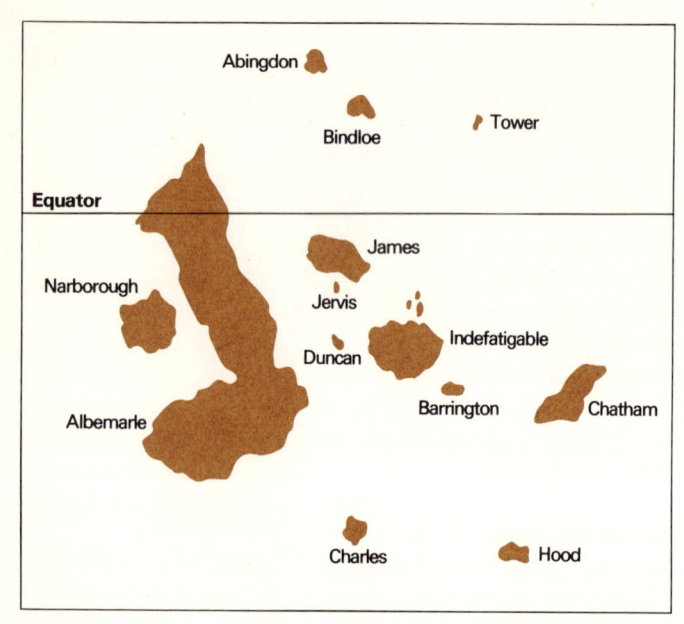

Figure 16.11
The Galapagos Islands. It is said that Darwin first began to formulate his Theory of Natural Selection after visiting these islands.

(Figure 16.12). He noted that the apparent effects of isolation on different islands were even more marked in the finches. On one island the birds had strong beaks adapted for cracking nuts and seeds, while on another the shape of the beak seemed to be more adapted to feeding on fruits and flowers. There was still another variety that used a cactus spine to dig grubs.

Why did the various specimens differ from island to island? Could it be that a species which was isolated on a particular island had somehow changed over a period of time in ways that helped it survive in the specific environment of that island? More importantly, is it possible that this gradual change is universal among organisms, and might it not therefore explain the origins of the various species on earth? In an earlier time this kind of thinking might well have brought torture on a medieval rack; even in the first third of the nineteenth century it was close to heresy. But in Darwin's own words he felt that "Here, both in space and time, we seem to be brought nearer to that great fact—that mystery of mysteries—the first appearance of new beings on this earth."

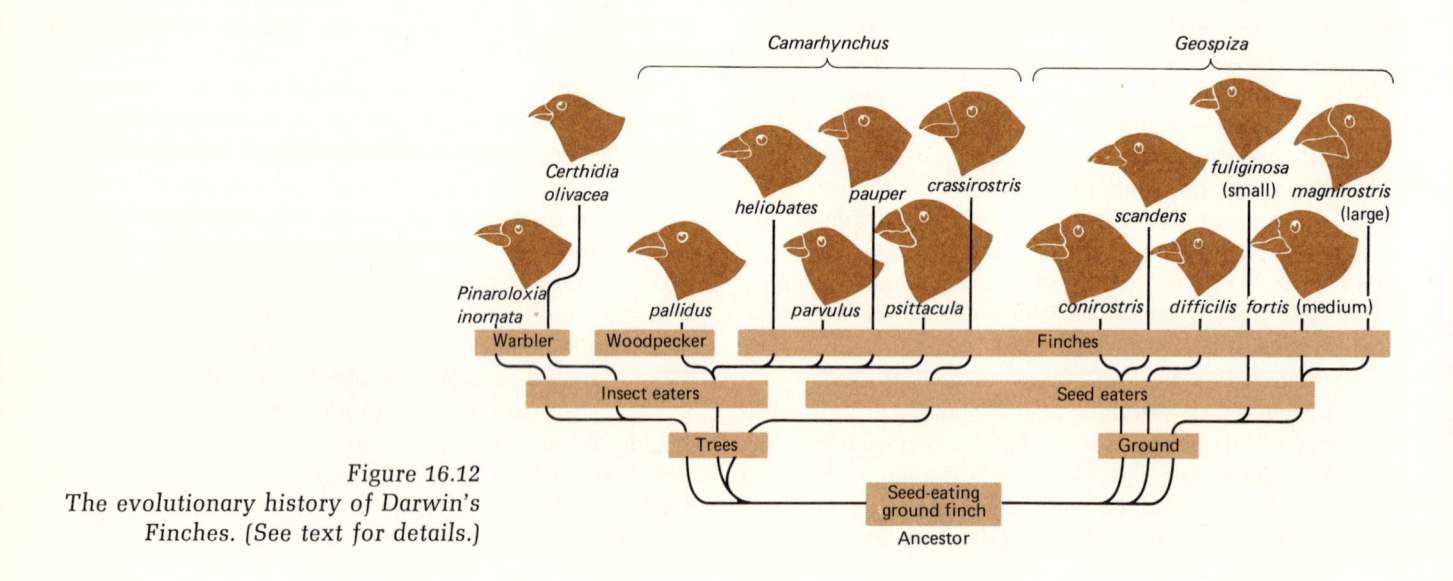

Figure 16.12
The evolutionary history of Darwin's Finches. (See text for details.)

After returning home, Darwin spent years organizing the vast store of data gathered from his observations of plants and animals living in the areas visited by the Beagle. Finally he compiled it all in the form of a book called *On the Origin of Species*. In this book, Darwin proposed that all species now existing on earth are descended from preexisting species. Through a process of slow change, all organisms including man are changed descendants of forms of life that existed in the past. It happened that another Englishman, Alfred Wallace, had intuitively arrived at a theory similar to Darwin's at the same time, but the massive and carefully organized evidence gathered on the long voyage of the Beagle forever established Charles Darwin as the principal founder.

Darwin's theory of natural selection can be briefly summarized as follows.

1. Plants and animals generally produce more offspring than are needed to maintain the species. Familiar examples include the great numbers of winged seeds falling from a maple tree, the masses of frog eggs in a pond, the multiple births that occur in many mammals, and the number of children that will be produced by a fertile human couple if that fertility remains unchecked.

2. In a natural setting, only a portion of the total offspring produced by a given generation survives to maturity. Natural environmental forces such as predators, limited food supplies, weather conditions, disease, and other population-regulating mechanisms take their toll.

3. Those offspring that do survive are likely to include individuals that possess inherited variations which better equip them to be successful in the "struggle for existence." In other words, those that survive will likely be better able to escape from predators, make maximum use of available food, resist disease, and so on. Having survived to sexual maturity, these variant forms will pass their variations on to their offspring, producing a small change from one generation to the next. Over a long period of time these changes accumulate, resulting in noticeable and significant changes in the species itself.

It is perhaps easier to understand how species have changed by looking at the consequences of *artificial selection*. We are familiar, for example, with practices involved in breeding horses. Horses that demonstrate outstanding speed or strength are *selected* for breeding purposes. Meanwhile, those horses that do not exhibit the characteristics the breeder is looking for may find their way to the glue factory or into the dogfood can. Therefore, by permitting only selected individuals to breed and to contribute to the next generation, man produces new varieties of dogs, horses, cattle, wheat, and fruit trees in a relatively short time. To achieve similar results, *natural selection* must operate over a much longer period of time because it is much less efficient. Not all the offspring that survive to produce another generation are variant forms. The two kinds of selection are similar, however, in that both depend upon some outside force to increase the chances that certain individuals will make greater contributions than others to the next generation. In the case of artificial selection, this outside force is man, while natural selection depends upon environmental forces to achieve a similar result.

It is important that we understand that the term "struggle for existence" does not necessarily imply a fight to the death with fang and claw. The "struggle" is more often than not a quiet one involving differences in ability to adapt to the physical and biotic environments of the ecosystem. It may simply mean better camouflage, or a better ability to utilize metabolically produced water, or a higher resistance to disease such as that which the sickle-cell gene provides against malaria. In the same way, we should understand that the phrase "survival of the fittest" does not necessarily mean that only the strongest survive. The insect that looks like just another twig of the branch on which it is resting or the rabbit that turns white to match the winter snows are also "fit" in the true sense of the word, because they are more likely to escape predators and leave offspring. In modern terminology, *fitness* refers to the number of offspring that an individual leaves. Thus, as geneticists are fond of pointing out, George Washington's fitness was zero, since he left no children as far as we know, yet a mentally retarded individual who leaves ten children would be assigned a very high fitness value! Natural selection can therefore be described most simply and most accurately in terms of *differential reproduction*. In other words, evolution depends in part on the fact that different individuals leave different numbers of offspring.

The publication of Darwin's *Origin of Species* produced a controversy that even to this day is not completely stilled. Even with evolution now established as a basic biological principle, there are groups that still object to its being taught in the public schools because they claim that evolution contradicts a divinely inspired story of special creation. The famous Scopes trial, known also as the "Monkey Trial," took place as late as the 1920's when the dismissal of a Tennessee biology teacher for teaching evolution brought together two famous antagonists in the persons of William Jennings Bryan for the prosecution and attorney Clarence Darrow for the defense. Scopes lost, but in the process the cause of evolution gained considerable ground.

Whether evolution is really "antireligion" or not is largely a matter of interpretation. The major organized segments of Judeo-Christianity have largely retreated from a literal interpretation of the Book of Genesis, accepting it as a parable rather than a literal description of creation. It is possible, in fact, to interpret certain passages in the Book of Genesis as being consistent with evolutionary concepts. God said, for example, "Let the waters bring forth abundant life . . . let the Earth bring forth . . . ," and so on. Even the Biblical implication that man was the last to be created is consistent with the fossil record. It should be stressed however, that these can only be individual interpretations. As far as science is concerned, evolution is a random, accidental, and blind process, having no purpose and motivated by nothing other than the natural forces of the environment.

In Darwin's time, the proposal that creation occurred by means of a blind process of evolution was a sacrilege to most people, and the notion that man had descended from apes or apelike creatures was a blow to the human ego. Like the discoveries of Copernicus and Kepler, who showed that the earth was not the center of the universe but merely another speck among billions of other specks, Darwin destroyed man's self-perception as the lord of creation made in the image of the Creator. Like the fish, the insect, or the elephant, he came about as just another product of a random evolutionary process.

The Process of Evolution

Although Darwin supported his Theory of Natural Selection with extensive and persuasive evidence, important questions still remained. Like Lamarck before him, Darwin could not explain *why* changes occurred in organisms and *how* those changes were passed on to succeeding generations. Gregor Mendel's work in genetics was still to come (1866), and evolution could not be cast in the form of a scientific principle until the mechanism of variation could be explained. It is not surprising, therefore, that the modern view of the evolutionary process is firmly rooted in principles of genetics. Today's biologist thinks of the evolutionary process as an interplay among mutations, natural selection, and other factors which result in changes in population *gene pools*.

In Chapter 7 we discussed how gene frequencies could be used to predict hereditary conditions such as

Tay-Sachs disease. At that point we introduced an important principle known as the Hardy-Weinberg Law (A quick review of pages 141 through 143 would now be useful.). The Hardy-Weinberg Law is an important basis of population genetics and therefore of the modern view of evolution.

You will recall that the Hardy-Weinberg Law can be expressed by

$$p^2 + 2pq + q^2 = 1$$

where, in terms of a single pair of genes found in a given population, p^2 represents the frequency of gene combination AA, $2pq$ represents the frequency of Aa, and q^2 represents the frequency of aa.

Let us now consider a very hypothetical population which consists of 36 percent AA combinations, 48 percent Aa combinations, and 16 percent aa combinations. Therefore, in terms of the Hardy-Weinberg expression,

$$p^2\ (AA) = 36\%$$
$$2pq\ (Aa) = 48\%$$
$$q^2\ (aa) = 16\%$$

It therefore follows that

$$p\ (A) = \sqrt{p^2} = \sqrt{0.36} = 0.60$$

and

$$q\ (a) = \sqrt{q^2} = \sqrt{0.16} = 0.40$$

Now, we need to recall that p and q represent the frequencies of genes A and a in our hypothetical population. These frequencies can also be considered as probabilities; in other words, if we now consider our population to be a collection of sperm and egg cells, the probability that any given sperm (or egg) cell selected randomly from the population carries gene A is 0.60 and the probability that it carries gene a is 0.40. If we now assume that the males and females mate *randomly*, and therefore that the sperm and egg cells come together *randomly*, we can predict the resulting sperm-egg combinations in the same way as we predict the results of tossing two coins. Thus,

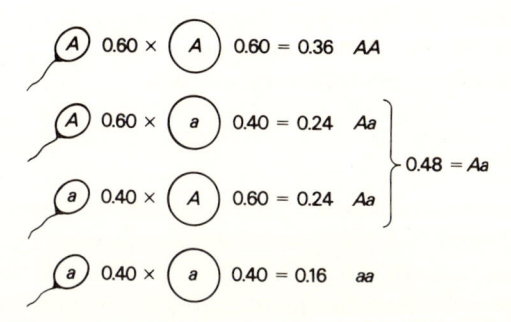

$$A\ 0.60 \times A\ 0.60 = 0.36\ AA$$
$$A\ 0.60 \times a\ 0.40 = 0.24\ Aa$$
$$a\ 0.40 \times A\ 0.60 = 0.24\ Aa$$
$$a\ 0.40 \times a\ 0.40 = 0.16\ aa$$

$$0.48 = Aa$$

From the above we can see that both combination frequencies and gene frequencies have remained unchanged in our population from one generation to the next. In other words, the Hardy-Weinberg Law tells us that a population gene pool should remain stable down through the generations.

If this is so, how then can we explain the changes that take place in evolution? The answer lies in the fact that the Hardy-Weinberg relationships apply in the strict sense only to *theoretical* populations in which:

1. There are no mutations, and all genes present in the gene pool therefore remain forever unchanged in their structures and effects.

2. There is purely random mating between males and females. That is, there are no selective forces operating to prevent all possible combinations of males and females occurring with equal probability.

3. All individuals in the population produce offspring; furthermore, all individuals produce the same numbers of offspring.

4. The population is large.

The fact that some or all of the above conditions do *not* hold true for a given *natural* population at a given time results in a "violation" of the Hardy-Weinberg Law with consequent shifts in gene frequencies. The slow but steady changes in the composition of population gene pools are therefore the essence of the evolutionary process.

Mutation

In Chapter 6 we saw that the action of a gene depends upon a very precise arrangement of certain specific bases in the DNA molecule. You will also recall that a DNA molecule provides an explicit pattern or template for the production of an amino acid chain, thereby determining the structure of enzymes and other protein molecules that are critically important to an organism's genetic makeup. A *mutation* is a random or chance change in the DNA base arrangement that alters the pattern on which a protein molecule is built (Figure 16.13). Most importantly, this altered or *mutant* gene may be passed on to succeeding generations and become part of the population gene pool.

A mutant gene may produce effects that range from negligible to very large. Such conditions as phenylketonuria in man and vestigial wings in fruit flies are examples of the infinite variety of observable effects of mutant genes. Although it is not completely understood why mutations occur, it is known that radiation and certain *mutagenic* substances such as mustard gas may significantly increase the mutation rate. In fact, it is a fear of increased mutation rates that has prompted people in some localities to object to having a nuclear power plant in their midst. Indeed, it has not been shown that these fears are completely groundless despite official assurances to the contrary.

It is known that mutations do occur with surprising regularity. E. P. Volpe (1972) points out, for example, that if man carries at least 25,000 genes, the average number of *new* mutant genes carried by each sperm or egg cell is 0.50. In other words, 500 out of every 1000 sperm or egg cells can be expected to carry a new mutation. At this rate, Volpe calculates that about 75 percent of all offspring in a given human generation contain at least one new mutant gene. It is therefore obvious that these spontaneous and random changes in the basic hereditary material are a primary source of Darwin's "variations."

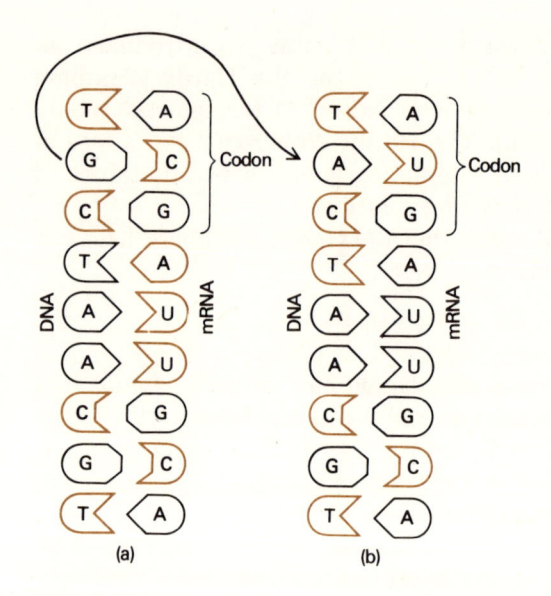

Figure 16.13
The mechanism of mutation. The basic pattern of the DNA in (a) is changed to that in (b) as a result of the substitution of adenine for guanine. This changes the associated mRNA triplet from G-C-A to G-U-A.

Selection

Whether or not a newly mutated gene becomes part of the gene pool depends to a large extent on the selection process. The fact that selection, or nonrandom mating, takes place is one reason why the

Hardy-Weinberg Law does not operate at 100-percent efficiency in a natural population. Males and females follow a choice pattern which reaches the ultimate in a human population, where purely social and cultural factors tend to prevent random mating. In the United States, for example, members of the same religious, racial, ethnic, and socioeconomic backgrounds tend to marry each other more often than not. Also, tall men tend to marry tall women, short men marry short women, and so on. In other species, some males appear to be more sexually attractive to females than others, and in some species the strength of the male determines his success in competition with other males for the favors of the female. Sex appeal is therefore one of the forces that alters the gene pool from one generation to the next.

In addition, it is obvious that a newly mutated gene will not become part of the population gene pool if its carrier does not produce offspring. It is further obvious that in order to produce offspring, the individual must survive to sexual maturity. Genes that tend to provide better adaptation to environmental conditions are therefore said to be "selected for," and those that create effects that reduce adaptation are "selected against." If a newly mutated gene causes the death of its carrier in infancy, this represents the ultimate in *negative selection*. Again we should stress that in the language of modern biology, the term "fitness" refers not to physical characteristics such as strength but to number of offspring produced. You will recall that it is not uncommon to hear the evolutionary process described simply in terms of *differential reproduction,* suggesting that evolution occurs primarily because different individuals produce more or less offspring, and thus contribute more or less of their genes to the gene pool of the next generation.

Sickle-cell anemia provides an excellent example of natural selection. The gene for this condition originated as a mutation in Africa and persisted in the population to a marked extent because its presence afforded a high degree of resistance to malaria, which acted as the environmental force of selection. Since those individuals who possessed the gene for sickle-cell anemia were more likely to escape death from malaria, they were therefore more likely to produce offspring and contribute to the gene pool of the next generation.

The rapid development of DDT-resistant insect strains is another example of how a population gene pool can be changed by an interplay between genetic and environmental factors. Figure 16.14 illustrates the mechanism of this example of "evolution in action." Note that those insects that carry the mutant "DDT-resistant" gene or genes slip through the "environmental screen" provided by the DDT in much greater numbers than their nonmutant brothers. The mutant forms are thus selected by the DDT-screening action to live and therefore to reproduce, much as a dog breeder selects a certain member of a litter over others. In this way, insect populations having a greater proportion of mutant forms are produced with each succeeding generation. This example of *differential reproduction* also explains the development of bacteria strains that are highly resistant to penicillin and other antibiotics.

Probably the most famous example of natural selection in action is provided by the extensive studies on the peppered moth that were carried out in England by Dr. H. B. D. Kettlewell. The peppered moth (*Biston betularia*) exists as two types: one type is a light "pepper" color and the other variety is dark. Breeding experiments have shown that the dark coloration is due to a small number of genes that are produced at a low rate of spontaneous mutation.

Records of naturalists show that the dark form was extremely rare prior to 1850. Then, with the growth of coal-fired industries in certain areas of England, the surrounding countryside became blackened as soot from the factories settled on the rocks, trees, and foliage. Finally by 1950, it was noted that the dark form of the moth predominated in the woodlands around the industrial cities.

330 Evolution

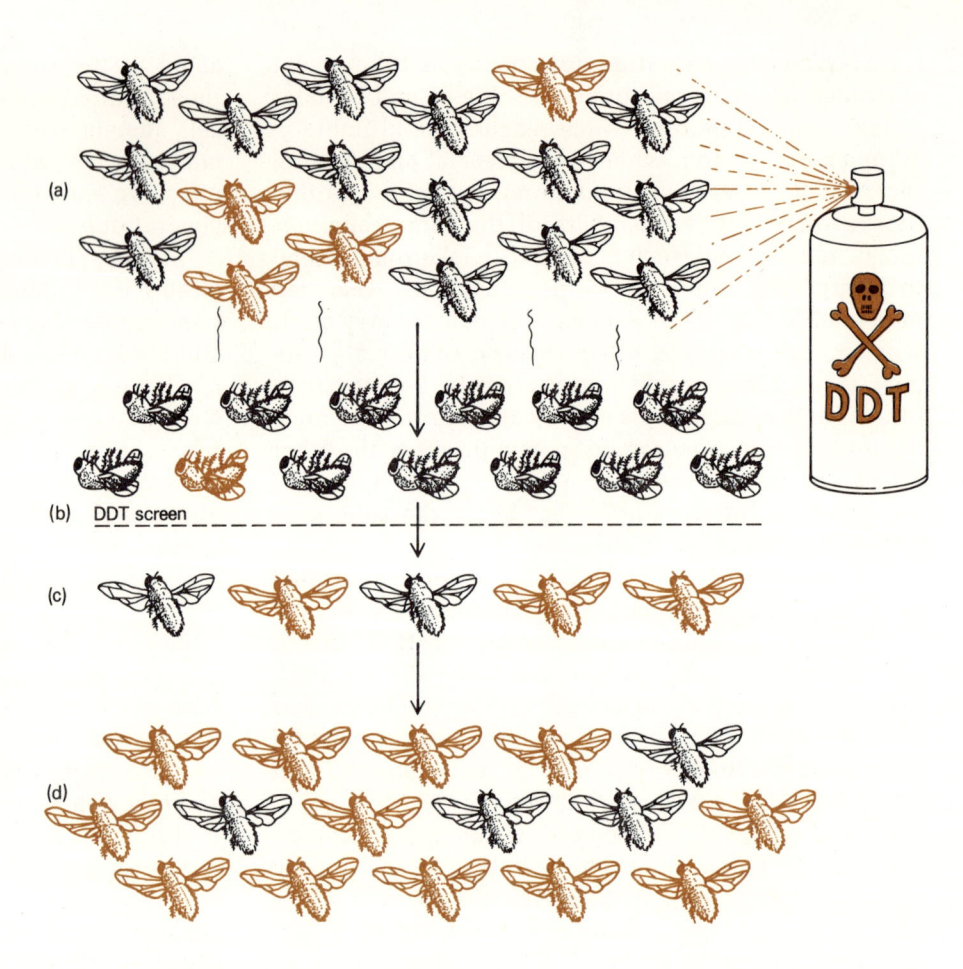

Figure 16.14
The DDT "screen." The solid-color flies are mutant forms that have a high resistance to DDT. Note that a higher proportion of these mutant forms escapes through the DDT "screen," thus contributing a greater proportion of mutant genes to the gene pool of the next generation.

Kettlewell carried out a series of elaborate studies on the relationship between moth coloration and predation by birds. He found that the dark moths blended into the sooty background and were thus much more likely to escape being eaten by birds than the light-colored moths that were adapted to the background coloration of the preindustrial countryside. The mutant genes of the dark moth, originally a disadvantage, had thus become an advantage as the background changed. In the Darwinian sense, the dark moths had become more "fit" than the light variety. In other words, more dark forms than light forms escaped being eaten by birds and therefore lived to reproduce and contribute to the gene pool of the next generation (Figure 16.15).

The foregoing examples might tempt us to think of mutations as arising in *response* to a specific need. It is easy, for example, to slip into the trap of concluding that the DDT-resistant mutations in insects are *caused* by the presence of DDT, or that the presence of malaria *caused* the sickle-cell gene to arise in African populations. We might even be tempted to

Figure 16.15
The black moth blends with the sooty bark of trees in polluted areas but stands out against the background provided by trees in unpolluted areas. The situation is reversed in the case of the pepper-colored variety. (Courtesy of Dr. H. B. D. Kettlewell.)

go further and say that the sickle-cell mutation was ordered by some "higher intelligence" to save the African population from malaria.

First of all, it must be stressed that all mutations are purely random; they occur by chance and should in no way be regarded as arising in response to environmental forces. If the DDT-resistant mutations had not occurred by a lucky accident (lucky for the insects), then the resistant strains would not have developed. We can also say that if it had not been for the presence of the DDT "screen," the resistant strains would not have developed because there would have been no differential reproduction.

This concept of pure chance is the basis of the entire process of evolution that has produced today's life forms, including man. One's own personal philosophy or religious faith may lead to private speculation concerning the role of supernatural forces, but without evidence to the contrary, we must repeat that the *scientific* view must hold evolution to be a blind, random process with no purpose.

The Formation of a Species

It is not difficult to understand how natural selection could produce a population of insects that is resistant to DDT. This is, after all, a minor change and we would be hard put to distinguish between a resistant housefly and one which is not. The formation of new species, however, is a quite different matter and requires a rather large jump in our thinking.

Basically, we can define a *species* as a group of like-appearing organisms that can breed together and produce similar offspring that can also breed together. Dogs and cats certainly belong to different species. Not only are they very different in appearance, but they are also unable to breed with each other. Horses and donkeys also belong to different species, but they are much more closely related. They do breed together but the offspring produced are mules, which are sterile. On the other hand, all groups and races of man breed together and produce fertile offspring. Man, therefore, belongs to a single species called *Homo sapiens.*

In general, a given species remains remarkably stable from generation to generation. Dogs remain dogs and cats remain cats, and we can see little or no change in the human species throughout recorded history. Species stability is based on the Hardy-Weinberg Law, which dictates that gene frequencies in a given population will not change from one generation to the next unless mutations and natural (or artificial) selection are present. Even with these evolutionary forces present, however, there is a *tendency* for a given population gene pool to remain unchanged. Evolution therefore proceeds slowly, and it is little wonder that millions of years are required for significant changes to occur.

The evolution of a new species is called *speciation*. The basic mechanism of speciation involves the operation of natural selection in reproductively isolated populations. *Reproductive isolation* means that no gene flow takes place between populations and that evolution therefore proceeds independently within each separate population. Since evolution is a blind, random process, it follows that the isolated populations will change in different ways, especially if they are subjected to different environmental forces.

Figure 16.16 shows a hypothetical island which contains a single population of bird species A. Since there is no isolation, all males of species A are theoretically free to mate with all females and there is free gene flow throughout the population. As species A slowly changes as a result of evolutionary processes, we would therefore expect that those changes will affect the entire population more or less uniformly. Suppose, however, that the island is divided by geologic processes into two islands. Assuming that species A cannot swim or fly between the islands, we now have two populations of species A that are geographically isolated and therefore *reproductively* isolated from each other. Evolution therefore proceeds independently within each of the two populations, and after a long period of time there are two new species, B and C, both of which are descended from

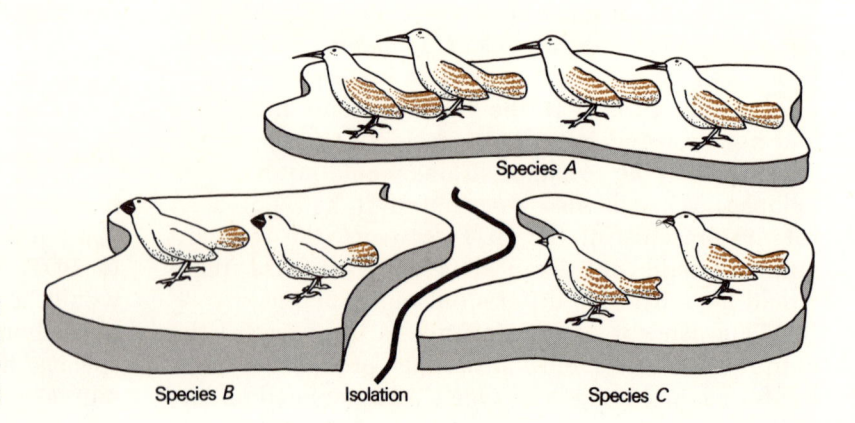

Figure 16.16
The evolution of bird species A into species B and C as a result of isolation. (See text for details.)

A. By this time they are no longer able to breed with each other even if brought into contact.

There are several ways by which populations may be reproductively isolated. They may, for example, develop sex-organ differences that make interbreeding mechanically impossible. Or they may develop different breeding seasons that will also effectively isolate them reproductively even if they occupy the same geographic area. Differences in mating behavior may arise, involving such things as mating calls or scent glands that act as sex attractants. In plants and in certain animals such as fruit flies and salamanders, a condition known as polyploidy may arise spontaneously and produce "instant" reproductive isolation (Figure 16.17). *Polyploidy* is a condition in which the organism has multiple sets of chromosomes, and when crossed with the parent species, the offspring are sterile.

Generally, geographic isolation is the first step in the development of permanent reproductive isolation between populations. Sometimes geographic isolation is due to the size of the area over which a certain species is distributed. The common leopard frog, for example, is found over an area extending from Vermont to Mexico, forming a *cline* (Figure 16.18). It has been shown that the degree of sterility between groups of leopard frogs in the cline is proportional to the distances between the groups.

Figure 16.17
Polyploidy in a lily. The different statures of the two plants are due to a difference in chromosome number. (Courtesy of the U.S. Department of Agriculture.)

334   Evolution

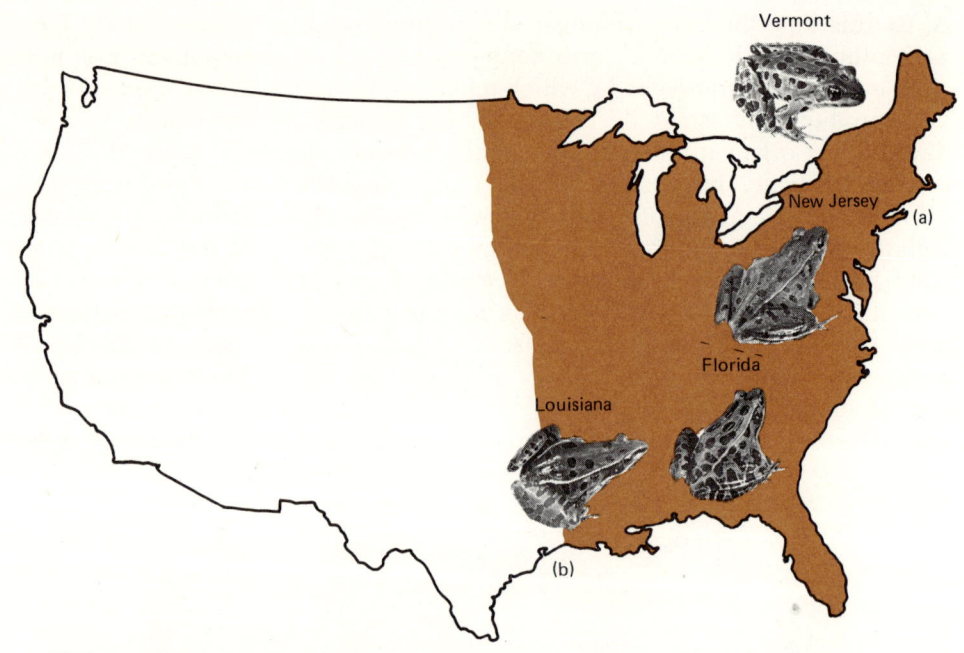

Figure 16.18
Diagram illustrating a cline in the leopard frog. Due to isolation caused by distance, those frogs at (a) have evolved differently than those at (b). Laboratory experiments have shown that the offspring from matings between frogs at (a) and (b) are sterile. There are also differences in ability to adapt to temperature ranges from (a) to (b).

Species formation, or speciation, is often described in terms of *adaptive radiation* (Figure 16.19). When populations of a given species migrate, they not only become geographically and reproductively isolated, but they also migrate into differing environments. As the different populations adapt to the selective forces in their individual environments, they evolve in different ways, ultimately producing permanent reproductive isolation and speciation.

Genetic drift may also play a part in species formation, but its role is probably minor as compared to adaptive radiation. Genetic drift refers to the fact that when a population is very small, certain genes may be accidentally lost and others increased purely by chance as one generation gives rise to another. This is called *nonadaptive evolution,* since chance alone and not selection pressures determine changes in gene frequencies. The *founder principle* is a special case of genetic drift and may operate when a small group breaks off from a main population and migrates to a

The formation of a species 335

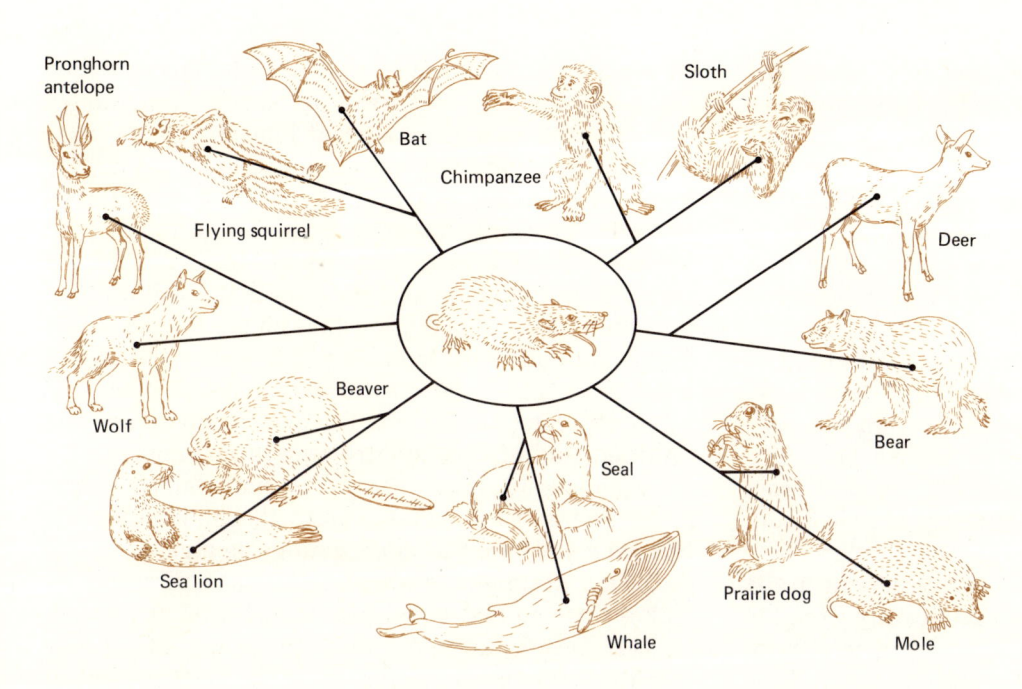

Figure 16.19
Adaptive radiation in mammals. Beginning with some primitive mammal, a variety of species have evolved to fill different ecological niches.

Figure 16.20
Genetic drift. Note the relative frequencies of blood types in the parent population. By chance alone, the relative frequencies in the emigrant population differ from those in the parent population.

different geographic area. In this situation, chance alone may determine that certain genes may be taken along in greater or lesser frequency than they occur in the parent population (Figure 16.20).

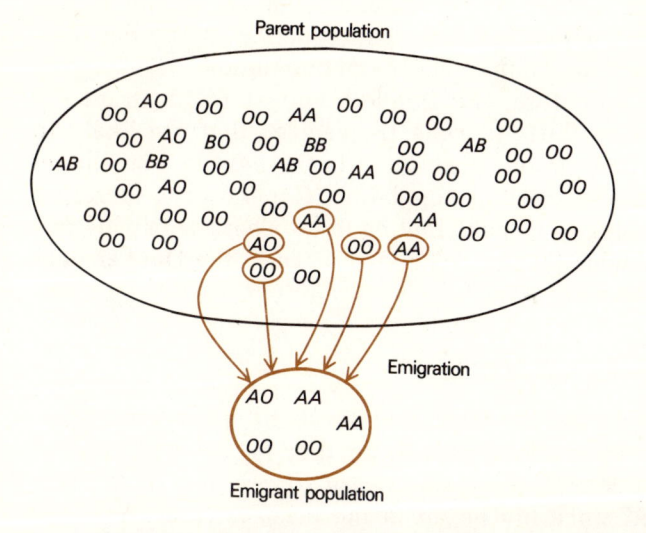

The American Indian, for example, is believed to be descended from Asian groups that migrated to North America over a land bridge that once extended across the Bering Strait. It is noted that American Indians have no gene for type B blood, although this gene is widespread in Asia. One might therefore speculate that by chance alone the "founder" group had no type B genes, or the few that did failed to leave offspring. On the other hand, most American Indians have type O blood; this means that the genes for type A *and* type B are both missing. Most evolutionists therefore suspect that natural selection must have played a part in the loss of these genes, since it is highly unlikely that both types would have been left behind by chance alone.

In the early eighteenth century, a small group of members of a religious sect called the Dunkers migrated from Germany and settled in eastern Pennsylvania. The Dunkers form what is known as a "religious isolate." They are genetically isolated from the main population by rigid customs which forbid members to marry outside the group. Studies have shown that the frequencies of certain genetic traits found among the Dunkers differ significantly from the frequencies of those same traits in the general populations of both the United States and Germany. Type O blood, for example, is relatively rare among the Dunkers. Type A occurs most frequently, while the gene for type B appears to be almost entirely lost. Other traits such as ear-lobe patterns and the presence of hair on the middle segments of the fingers occur with a much lower frequency among the Dunkers than among the general population.

Therefore, the Dunkers appear to illustrate the chance shifts in gene frequencies that are likely to occur from one generation to the next in a small, genetically isolated population. Genetic drift therefore appears to be contributing to the changes in gene frequencies (evolution) taking place in the Dunker population.

The Arrival of Man

From a strictly biological viewpoint, man is no different from other animals in that he is an accidental creation of the evolutionary forces that have operated since life began in the primeval seas. Yet it is difficult to say that man is "just another animal." He alone possesses that unmatched intelligence which has given him mastery over the earth, even though he has yet to become master of himself.

Man is a study in contradictions. From his mighty brain and restless energy have come great civilizations. He has built the great pyramids of Egypt with the sacrifice of untold numbers of his own kind. He erects giant towers of shining steel that reach upward into the clouds of pollution that choke his cities. He builds churches and temples to his gods and talks of fellowship and brotherhood while he perfects his weapons of terror. He has traveled to the moon and returned, an unbelievable triumph of technology.

Man is kind, generous, and sentimental. He weeps at the death of a loved one and will risk his own life to save a stray dog from drowning. He gives freely to programs designed to alleviate human suffering. He makes laws so that equal justice might be had by all. He is civilized. But under that thin veneer of civilization that covers his origins, he is the most violent of all animals on earth. No beast of the jungle can match him in ruthless and senseless savagery. He uses his cunning brain to prey upon the earth, upon his fellow creatures, and upon himself. He has refined the art of killing his own kind to the ultimate degree with a weapon that is capable of unleashing a final flood of destruction upon the planet that gave him birth.

Man is special. He is unique. Yet his origins are as humble as the animals that he views with amusement on a Sunday afternoon at the zoo. The full story of man's evolution will never be told; too many details are buried in the dim past. From the scattered bits of evidence in the form of bones and tools unearthed from the geologic strata, scientists have pieced together a history of man's development. However, much of it is open to argument and subject to change in the future. What follows is a brief summary of that history.

Somewhere in the dim past, probably about 75 million years ago, a new group of mammals called the *primates* arose. They began with small, insect-eating animals called *prosimians,* which were similar in appearance to the modern shrew (Figure 16.21). Figure 16.22 shows the family tree that grew from the original primate stock. Note that there are two main branches. The first, called the *Prosimii*, led to the modern tree shrews, lemurs, and tarsiers. The second branch, the *Anthropoidea* (meaning "manlike"), gave rise to the monkeys, the great apes, and man. Another look at Figure 16.22 shows that the branch that produced man and the apes divided approximately 25 million years ago to give rise to man and to our first cousins, the apes.

Figure 16.21
A modern squirrel shrew. (Courtesy of the American Museum of Natural History.)

Figure 16.22
Diagram showing the primate family tree arising from original primate stock. Note that man is included with the anthropoids, meaning "manlike." (Adapted from E. P. Volpe, Understanding Evolution, 2nd ed. Dubuque, Iowa: Wm. C. Brown, 1970.)

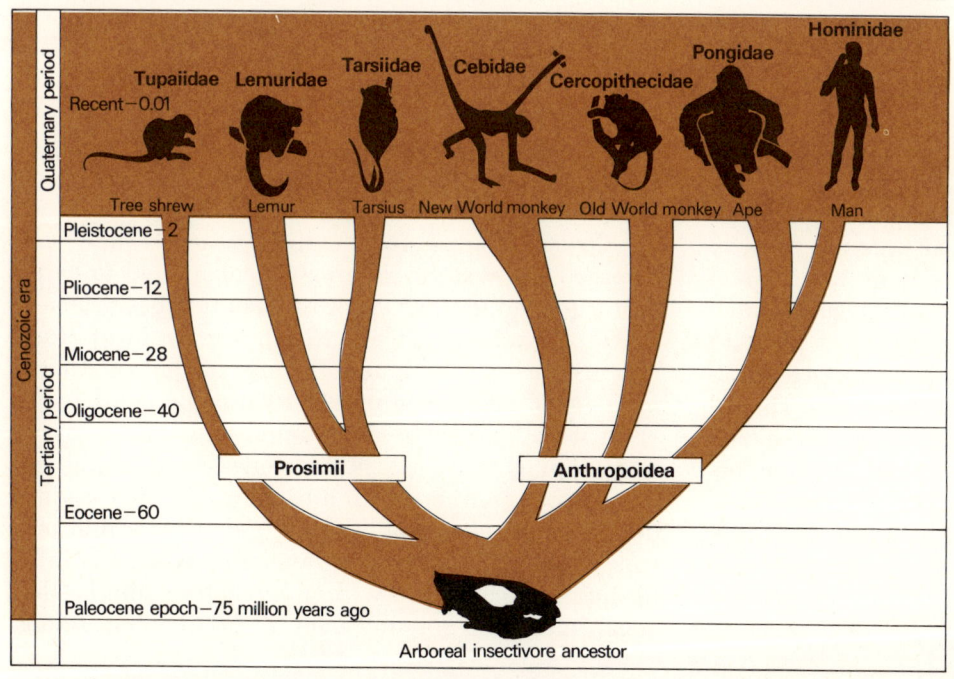

In the 1930's, the fossil remains of *Proconsul africanus* were discovered in Kenya, Africa. *Proconsul* is believed to have existed about 25 million years ago, and it has been suggested that it is closely related to the stock that branched by adaptive radiation into the lines leading to the apes and to man. *Proconsul* was about three feet tall with huge canine teeth, receded forehead, and covered with hair. It was evidently able to stand erect but preferred to walk on all fours.

Proconsul may have been among the first of the primates to venture from the trees, probably because the climate of East Africa at that time was causing the rain forests to dwindle, forming open grasslands. The primates were largely tree dwellers, and their adaptation to their arboreal existence included the development of flat nails instead of claws, grasping fingers and toes, and eyes placed in front of the head to provide the depth perception so useful to an animal that swings from branch to branch. These same features we can recognize in ourselves, and when *Proconsul* was forced to leave the safety of the trees and venture across the grassy African veldt, it may well have been one of the major events in the development of man.

One of man's earliest forerunners was *Australopithecus africanus*, a man-ape that lived on the plains of East Africa about 5 million years ago (Figure 16.23). *Australopithecus* was about four feet tall with a manlike body and apelike head, and he was not a likely candidate for survival. Practically defenseless against the stronger and swifter predators that surrounded him, he ventured onto the dangerous open plain in search of food, competing with other scavengers for the scraps left by the more skillful predators. He did have a few advantages: he walked erect, his hands were free, and he had depth perception. Most important, he had a brain capacity of about 600 cubic centimeters, more than a third that of modern man. This developing brain helped him learn to use stones and sticks as simple tools. *Australopithecus* was not yet a man, but he was a beginning.

Australopithecus africanus

Figure 16.23
Australopithecus africanus *was one of the earliest ancestors of modern man.*

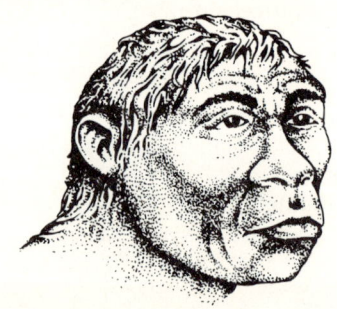

Homo erectus

Figure 16.24
Homo erectus *represents a stage in the development of man that is well advanced over* Australopithecus.

Over the next four million years, a new man emerged. He was *Homo erectus*, considered to be the first true man that we know of from the fossil remains that are available (Figure 16.24). The brain of *Homo erectus* was nearly double the size of that of *Australopithecus*, measuring nearly 1000 cubic centimeters. With it he learned to make crude tools and weapons by chipping sharp edges on stone. He captured the use of fire, and he set up a social order that divided duties within the group. *Homo erectus* saw that man was weak when alone but invincible when his intelligence was combined with group action, and he thus began man's domination over all other living things.

Figure 16.25
Neanderthal man. (Courtesy of the Field Museum of Natural History, Chicago, Illinois.)

It was at this point that man's brain became the ultimate weapon of the hunter, and like his descendents, *Homo erectus* came to fear only his own kind.

The climate changed, food became scarce, and *Homo erectus* migrated from the tropical cradle of man, driven by hunger and possibly the pressures brought to bear by his own increasing numbers. He spread to Java, China, and Europe, and then he disappeared.

He was replaced about 100,000 years ago by *Neanderthal man*, a brutish individual only about five feet tall but extremely strong and muscular (Figure 16.25). His brain equaled that of modern man in size, and he made sophisticated tools along with a growing array of weapons that made him the greatest hunter the earth had ever seen. Neanderthal man became the true "cave man," seeking out caves as shelter against the glacial cold. He made clothes of animal skins and built crude shelters as he migrated northward. Then he too vanished, to be replaced by *Cro-Magnon man*, who was the first of our kind that today we call *Homo sapiens*, meaning "man the intelligent, man the wise." Why Neanderthal man disappeared is one of the great mysteries of the evolution story. He may have represented a "dead end" in man's evolution, dying out because of an inability to further adapt to a changing world. Or he may simply have been absorbed by the superior Cro-Magnon man. No one knows for sure.

It was only about 40,000 years ago that man began his dominant role in earth's affairs. It is indeed an incredibly short time to go from caves to the moon, but our pride in our accomplishments should not cause us to forget that the dinosaurs lasted for 100 million years. There is cause to wonder if the unrestricted fertility and the hydrogen bombs of "man the wise" will allow him to succeed half as well as those former lords of the earth whose bones now grace our museums.

The Races of Man

Throughout mankind's recorded history, there have been repeated attempts by groups of human beings to subjugate and even destroy other groups because they were "different." During the Hitler Regime, the Nazis and their collaborators "eliminated" an estimated ten million people. Many of the victims were Germans who opposed the Nazi policies, but they also included six million Jews who were killed because they belonged to an "inferior race."

Long before that, the white man in America operated on the theory that "the only good Indian was a dead Indian." With his superior weapons and numbers, he slaughtered many American Indians and herded the survivors onto reservations where today their descendants live in varying degrees of poverty. Prior to the Civil War, black people were sold on the auction block as subhuman beasts of burden for the white man. The fruits of those years of slavery are still with us in the ghettos of the cities and in the sharecropper's shack of the rural South.

It is small wonder, therefore, that the word *race* evokes deep emotions. In America especially, we associate race with *racism,* a word that conjures up black-white differences and the problem of race relations. Neither is it surprising that following the Nazi holocaust, it was popular for awhile to deny the very existence of races, and that there were those who said "there is no such thing as race" when they really meant that there is no such thing as a *pure* race. The various races of man do exist, just as different breeds of dogs or horses undeniably exist, and to say that they do not will hardly make that fact go away. There are those who interpret the phrase "all men are created equal" as meaning that no differences exist among human beings. The differences, however, are obvious, not only between racial groups but within racial groups and even among members of the same family. The differences between blacks and whites in terms of certain characteristics are obvious, but to be different does not imply a lack of equality. Part of the problem rests in our tendency to stereotype racial and ethnic groups. We tend to think of blacks, Italians, the Irish, and white Anglo-Saxon Protestants (WASP's) in terms of classic stereotypes instead of realizing that within each of these and other groups there are wide variations. Not all Italians love spaghetti, not all Irishmen have red hair and drink whiskey, and not all WASP's are bank presidents. On the other hand, many if not most WASP's *do* love spaghetti, and some Italians are bank presidents. Not all blacks are on welfare, though some are, but then some WASP's, Italians, and Irishmen are also on welfare. And so it goes.

Races are groups within a single species, and all human races on earth today belong to the species *Homo sapiens*. A race may be defined as a breeding population having a gene pool that is different from other gene pools that are almost but not completely isolated from one another.

As man spread across the earth, it was inevitable that various groups would become isolated from one another by geographic barriers and distance. The three major forces of evolution were therefore present: (1) isolation, (2) genetic variation, and (3) environmental selection pressure. In the different environments occupied by the migrant groups, certain gene combinations and frequencies were favored over others, allowing successful adaptation to the environmental demands. You will recognize the situation we have described as another example of *adaptive radiation*; in fact, it can possibly be assumed that in some instances, race formation was a stage in the development of new species. We can also assume, however, that modern modes of transportation and communication have halted any speciation process that may have been going on. All human beings will undoubtedly continue to belong to a single species.

A study done several years ago on the English sparrow provides an excellent example of race formation in action. The English sparrow was introduced to the east coast of the United States in 1852. From there it spread to California, to Mexico, and north to Canada. Different geographical populations already exhibit differences in color, wing length, bill length, and body weight. The fact that these racial groups developed in approximately 100 sparrow generations shows how rapidly geographic isolation can produce significant differences in populations. If we assume that the human generation time is 30 years, then significant changes could have taken place in isolated human groups in 3000 years or less.

Although man is a compulsive classifier, there is no real agreement concerning the number of racial groups within *Homo sapiens*. For one thing, races are much more difficult to classify than species, since species by definition do not interbreed and are therefore reproductively isolated. Races, on the other hand, do interbreed and there is constant gene flow between neighboring geographic populations. Racial boundaries are consequently blurred. In addition it must be recognized that despite our tendency to invent stereotypes, more variation occurs *within* racial groups than *between* them. Finally, we must be careful not to confuse racial groups with cultural or ethnic groups. Jews, for example, do not constitute a race, the Nazis to the contrary. They represent instead a group that is bound together by a common religion and by other cultural traditions. The Jewish population presently found in the state of Israel demonstrates very significant differences between Jews from North Africa and those from Eastern Europe. Italians, Greeks, Swedes, and so on represent ethnic groups separated by national boundaries, language, and customs. They all belong to the same race, the *Caucasoids*, which illustrates the wide differences found within a single racial group.

The great majority of people on earth can be divided into (1) the Mongoloids, (2) the Caucasoids, (3) the Negroids, and (4) the masses of people living on the Indian subcontinent. Among the numerous smaller groups classified by anthropologists are the aborigines of Australia and New Guinea and the various peoples of the Pacific Islands, including the Polynesians native to Hawaii.

The *Mongoloids* comprise a large group that includes the Chinese, Japanese, Southeast Asians, Tibetans, and the American Indians among others (Figure 16.26). Mongoloids are traditionally thought of as the "yellow race," but their skin color actually ranges from nearly white through yellow and brown. The lighter-skinned Mongoloids are found in northern latitudes, and the darker skinned varieties are associated with tropical climates. They have very little body hair (consider the American Indian), and the hair on the head tends to be black and coarse. An easily recognized feature is the *epicanthic fold* on the upper eyelid which gives the eyes a characteristic almond shape.

The *Caucasoids* are generally thought of as "white," but actually they exhibit a wide range of skin color from the very light Scandinavian to the very dark Arab. Hair color is extremely variable, ranging from blonde to black. Caucasoids may have abundant body hair, and the males tend to have full beards (Figure 16.27).

In the *Negroids*, skin color ranges from deep brown to black. The hair is usually tightly curled, and full lips and broad noses predominate. Body hair tends to be sparse. The major Negroid groups include black Africans, North American blacks, and South American blacks (Figure 16.28). It should be empha-

Figure 16.26
(a) Chief Thundercloud, an American Indian and member of the Mongoloid race. (Courtesy of the Peabody Museum, Harvard University.) (b) Hideki Yukawa, Nobel Prize Laureate in Physics, 1949. (Courtesy Nobel Foundation.)

Figure 16.27
(a) A young Caucasoid woman. (b) Linus Carl Pauling, Nobel Prize Laureate in Peace, 1962. (Courtesy Nobel Foundation.)

Figure 16.28
(a) A member of the Negroid race. (Photo by Elizabeth Hamlin, Stock Boston.) (b) Martin Luther King, Jr., Nobel Prize Laureate in Peace, 1964. (Courtesy Nobel Foundation.)

sized that American blacks represent a gene pool that is different from that of their African counterparts. Interbreeding between blacks and whites on both American continents has resulted in *hybridization*, or a mixing of two races. In fact, it is estimated that 20 percent or more of the gene pool of blacks in the United States consists of Caucasoid genes. This accounts for the tremendous variation within the American black population.

When we think of races, we tend to think of skin color as the primary distinguishing trait. Actually skin color is only one of the many traits in which racial *populations* differ. In his excellent book *Race and Races,* Richard A. Goldsby points out that if 100 each of Caucasians, black Africans, and Chinese were put into separate rooms, he could *sight unseen* determine which room held the Caucasians, which contained the blacks, and which held the Chinese. He could do this provided he had a sample from each individual of (1) ear wax, (2) blood, (3) urine, and (4) information as to whether or not the individuals could taste a chemical called PTC. Therefore, as Goldsby points out, racial differences are more than skin deep.

On the other hand, since so many of us seem to be obsessed with skin color as the primary racial trait, it might be well to consider the origins of the spectrum of colors that characterizes the human species. First of all, it is obvious that different shades of skin color are associated with latitude, that is, distance north or south of the earth's equator. In other words, people who are native to the northern regions of the world are generally light-skinned, and those found in tropical areas typically are heavily pigmented. Biologists have good reason to believe that skin pigmentation is closely associated with the production of vitamin D. Vitamin D is called the "sunshine vitamin" because it is produced in the sublayer of the skin when ultraviolet radiation is present.

Vitamin D is an important factor in controlling the body's utilization of calcium. Too little vitamin D causes rickets, a disease in which the bones are soft, resulting in bowed legs and spinal deformities. Too much vitamin D, on the other hand, produces brittle, easily broken bones and other problems such as kidney stones. Unlike some vitamins, therefore, the quantity of vitamin D in the body must be controlled within relatively narrow limits. In the evolution of races, therefore, heavy pigmentation was a necessary adaptive trait in tropical areas due to the necessity to filter out a portion of the intense ultraviolet radiation and thereby prevent the synthesis of *excessive* amounts of vitamin D. As early man migrated to northern climates where the sunlight is much less intense, a lightly pigmented skin was equally adaptive, since it is important to allow enough ultraviolet to penetrate the outer layer of skin in order to synthesize *enough* vitamin D.

From the foregoing it should be immediately obvious that black children in the northern cities of the United States are poorly adapted to the less intense sunlight and are especially susceptible to vitamin D deficiency. It is therefore important that they be given a vitamin D supplement, especially during the winter months.

Although it is true that skin color and other physical features have been most emphasized as distinguishing racial characteristics, such traits are actually superficial and their importance lies mainly in the eye of the beholder. During recent years, there have been at least some changes in social attitudes which hold promise that such superficial differences will be much less important to future generations. Of much greater importance in our culture are those traits that enable the individual of any racial group to achieve maximum "success" and quality of life. Chief among these is that indefinable something called "intelligence."

One of the more controversial issues of recent times concerns the question of genetically determined interracial differences with respect to intelligence. The results of a large number of IQ tests given over a period of time indicates that on the average, black Americans score about 15 points lower than white Americans. Remember we said "on the average." As with other traits, there is considerable variation in IQ test performance *within* racial groups, and many blacks score higher than many whites. A. R. Jensen, a psychologist, and W. Shockley, a Nobel Prize winner in physics, stirred up an emotional controversy by suggesting that the difference in average IQ between blacks and whites is due largely to inheritance. To put it more bluntly, the suggestion is that blacks are genetically inferior to whites as far as that all-important commodity of intelligence is concerned.

First of all, intelligence is a trait that must have a high degree of genetic influence, since intelligence, even as we imperfectly understand it, must be largely a function of the brain and since the brain is constructed under the direction of genes. It is therefore difficult to accept the extreme view of some environmentalists who claim that intelligence is determined wholly or largely by environmental influences. On the other hand, there are far too many problems involved with the Jensen-Shockley point of view to permit us to accept that at face value either.

"Intelligence" is a vague term that has different meanings to different people under different circumstances. It may refer to an ability to get an "A" in calculus or even in biology, or it may imply a talent in art or music. It may even imply "social intelligence," which is an ability to interact well with other people. It is therefore not surprising that intelligence is extremely difficult—some say impossible—to measure, and that IQ tests are often regarded with suspicion. This is especially true when one considers that IQ tests are for the most part "culturally biased" in the direction of white culture and experience rather than toward the kind of experience of most American blacks. To illustrate cultural bias, suppose we were to ask a child of any race brought up on the streets of New York City to tell us the best time of year to plant corn. If he fails to give the correct answer, do we assume that he is lacking in intelligence?

Motivation and self-confidence play a large role in taking any kind of examination. Therefore it seems logical to consider the fact that black children are the product of a long history of oppression and discrimination. If they feel that their opportunities are limited, as must often be the case, a "what's the use" attitude may very likely affect motivation and significantly lower test scores. The question of nutrition also clouds the picture. Large numbers of black Americans live in ghettos at poverty levels, and lower achievement on IQ tests may be due to poor nutrition, both prenatal and early childhood.

Jensen points out that American Indians, often more economically disadvantaged than blacks, still score significantly higher than blacks. However, R. A. Goldsby makes the point that a very large percentage of American Indians attend white schools, whereas black children, especially those living in the large cities, tend to be segregated into all-black schools. In fact, it has been shown that blacks who attend white schools score higher than those who attend all-black schools. This would suggest that there is some merit to proposals to bus children to different neighborhoods in order to break down de facto school segregation, but so far such proposals have met with resistance from white parents and even from some black parents.

From a purely scientific standpoint, we must conclude that the answer to the question of interracial genetic differences in intelligence is simply not available at this time. Furthermore, it is likely that we will never know unless valid measuring instruments are developed and obvious environmental influences are eliminated. In the meantime, Goldsby's recommendation appears logical—to continue to improve social and educational programs aimed at expanding opportunities for all racial minorities.

We are truly more alike than we are different. We had a common beginning in that East African Garden of Eden, and we all live on the same family tree with *Proconsul, Australopithecus, Homo erectus,* and Neanderthal man. In the end we will all share the same destiny, be it good or bad.

Questions

1. Describe the Kettlewell experiments with the peppered moth and state their significance.

2. Explain the basic process involved in Darwin's principle of "survival of the fittest." What does "fitness" mean in this case?

3. Discuss Lamarck's theory of evolution. What is the basic error in this theory?

4. Explain why the stopover at the Galapagos Islands was such an important event in the voyage of the Beagle.

5. Explain the mechanism that has reduced the effectiveness of DDT and penicillin.

6. Describe the Hardy-Weinberg Law and explain its significance.

7. Discuss the principle of evolution as an interaction between mutations and selection.

8. What factors contribute to the high frequency of the sickle-cell gene in African populations? How do these factors contribute to the frequency of this gene?

9. Explain how the various races of man were formed. What do we mean when we say that all groups of man belong to the same species?

10. Discuss the following statement: Being different is not necessarily the same as being unequal.

17

Epilogue

> If we are going to develop a civilization broadly and soundly based upon scientific foundations—and we can hardly escape that now—every citizen, every man on the street, must learn what science truly is, and what risks and what quandaries, as well as what magnificent gifts, are engendered by the powers that grow out of scientific discovery.
>
> —Bentley Glass

At the beginning of this book, you were promised that modern biology involves a great deal more than rusty scalpels and dead cats. It is hoped that by now you are convinced that the principles of biology are directly related to some of the most fundamental problems and ethical issues that affect all humankind.

What of the future? It has been said that those who would pretend to know the future are either fools or knaves, but some things are at least predictable from what is going on at the present time. It appears likely, for example, that the horrendous projections of future population growth do not bode well for those already in the grip of famine or for millions more that stand on the brink of starvation. If something is not done about population growth and the distribution of food, malnutrition will continue to run rampant through the poor areas of the world, stunting the bodies and minds and breaking the spirits of the children in whom the hope of the next generation resides.

The great ecology crusade of the 1960's has apparently been supplanted by a new concern over energy and the economic disaster that dwindling energy supplies could bring. Even so, it left us with a belated awareness of the intricate workings of the ecological machinery, and man *may* at long last be learning that he must be careful how he tinkers with that machinery.

In the waning years of the twentieth century, one of our most difficult tasks will be to come to grips with the dilemmas and decisions that are already being thrust upon us by advances in scientific knowledge. Are we prepared, for example, to use our knowledge of birth control to reduce birth rates in a world where millions border on starvation? Are we prepared to face up to the issues that advances in genetics and embryology are sure to bring before us in the years to come?

Bitter arguments already rage over the possibilities of *in vitro* (test-tube) fertilization of human egg cells. Scientists are already able to remove an egg cell from a woman with an instrument called a laparoscope. The egg cell can then be mixed with her husband's sperm and fertilized. The resulting blastocyst can then be implanted in her uterus for further development. For women who are sterile due to blocked or poorly developed fallopian tubes, this method offers the only way to have a baby. If this can be done, how about implanting the blastocyst in the uterus of another woman, who will then carry the baby to full term and turn it over to the "natural" mother after it is born? Thus we have the possibility of "surrogate" mothers hired for a price to go through pregnancy and childbirth for other women who may be too busy or just plain reluctant to go through it themselves! But wait! Who *is* the "natural" mother in this situation? Is it the donor of the egg cell or the woman who housed the developing fetus within her body for nine months? What happens if she refuses to hand the baby over after it is born? What, then, are the legal entanglements, not to mention the ethical problems, that could arise?

Of even more concern is the fact that the technical capability to create "artificial wombs" is well within our grasp. This immediately brings to mind the nightmares described in Aldous Huxley's 1932 novel *Brave New World,* in which the elite "predestinators" created test-tube babies on an assembly line, giving each baby a set of physical and mental characteristics that would ensure its maximum contribution to the State. In 1932 this was science fiction; opponents of the artificial womb are afraid that it could become a reality. Supporters answer by pointing out that no way is in sight by which the genetic makeup of babies, test-tube or otherwise, can be controlled. Besides, they argue, dictators have never needed sophisticated methods to oppress whole peoples; the old-fashioned methods are quite adequate.

Opponents argue further that the development of babies *in vitro* would be immoral, destroy the family concept, and constitute interference by man in the "miraculous" process by which life is created. Proponents argue, on the other hand, that man is not creating life but only introducing a live sperm to a live egg cell. They further point out that fetal growth *in vitro* could be monitored, possibly even by computers, and the fetus could be aborted if there were evidence of malformation. *In vitro* development could also be a boon to women who continually miscarry, since their children could develop safely in the artificial womb. Finally, it has been seriously suggested that the artificial womb could spell the end of the myth that women must forever suffer the painful punishment of childbirth because Eve stole an apple, an offense that today would unlikely bring more than a stern warning from a kindly judge.

If growing babies in artificial wombs is filled with potential problems, consider this possibility. Suppose that we remove an egg cell from a woman's ovary and then destroy the nucleus of the cell by irradiation. Now we remove one of your body cells and transfer its nucleus into the egg cell. Keep in mind that this nucleus from your body cell contains the special combinations of 23 pairs of chromosomes that make you unique. Now we put the egg cell into an artificial womb and let it develop into a baby. That baby, by virtue of its genetic makeup, is a carbon copy of you! Science fiction? Only in part, because this procedure, called *cloning* (Greek: "throng"), has already been done successfully with frogs. Will scientists actually develop ways to successfully clone human beings? If they do, and some consider this a definite possibility within the next fifty years, what will be the consequences (social, legal, moral) of a biological xeroxing process by which practically unlimited carbon copies of any individual can be made?

What about changing the genetic makeup of an individual by altering the genes in the egg or sperm cell from which he or she is formed? Some work has already been done in this area, but geneticists disagree on whether it will ever be a practical reality. Even without such capabilities, however, we have already begun to assume a degree of control over our own evolutionary process merely through our ability to "cure" people with inherited conditions such as PKU, diabetes, or retinoblastoma who would otherwise have not been able to contribute to the gene pool of the next generation. Hemophilia, for example, can now be controlled by frequent injections of antihemophilic hemoglobin; with successful treatment, it is estimated that the frequency of hemophilia may double in as few as four generations.

Man's knowledge of his biological self is increasing by leaps and bounds. The mere acquisition of knowledge, however, is not enough. He must also have the *wisdom* to use that knowledge to better his own condition and to preserve the natural environment which he has shared with his fellow travelers down the long path of evolution. If knowledge is to be gained simply for its own sake, if it is not used to alleviate human suffering and to stop the creeping malnutrition that saps the strength of millions, and if it does not give future generations a better world than we have now, then from the human point of view our age of scientific enlightenment will be little better than the Dark Ages.

As we said in the beginning, the enormous ethical issues that are raised by population problems, environmental deterioration, abortion, and the implications of the new genetics are far too important to be left to the biologists to solve. Everyone, including you, must participate in decisions which will ensure that our knowledge is used to maximum advantage. In some cases this may require that we apply moral and ethical criteria that are different from those that fit the situations of years past.

Man has come a long, long way since *Australopithecus* set out on that great and perilous adventure across the African veldt. He has dared to steal knowledge from the gods, but if like Prometheus he is punished, it will be through his own misuse of that knowledge. He can use that mighty brain that is both his curse and his blessing to rise far above what he has ever been, or the dream can end in a cataclysm of famine and nuclear horror. Which shall it be?

Glossary

The following glossary has two purposes. First, it provides the student with brief definitions of common biological terms used in this text. Second, it provides the student with an aid to pronunciation. Phonetic symbols and phonetic spelling are kept as simple as possible, indicating only the "long" sound of a given vowel. Actually, the major problem encountered in pronouncing unfamiliar words is in knowing which syllable to accent.

The student is cautioned that a glossary is not a source of in-depth understanding of a word or concept. When this is required, the index should be used to locate the page or pages on which the term or concept is discussed.

Abortifacient (ah-bōr-ti-fa′-shent). A drug or other chemical that causes premature explusion of the embryo or fetus.

Abortion (ah-bōr′-shun). The expulsion of an embryo or fetus from the uterus before it has reached full term.

Absorption (ab-sōrp′-shun). The passage of food nutrients through the wall of the stomach and small intestine into the bloodstream or lymphatic system.

Acetylcholine (a-sē-til-kō′-lēn). A chemical secreted at a synapse that carries the nerve impulse across the synaptic space. A neurotransmitter.

Acrosome (ak′-roh-sōm). The caplike structure located on the head of a sperm cell.

Adaptive radiation (ah-dap′-tiv rā-di-ā′-shun). Evolution of a single group of organisms into several lines due to the selection factors of different environments.

Adenine (ad'-e-nēn). A purine compound that is part of the structures of DNA, RNA, and ATP.

Adenosine diphosphate (ADP) (ah-den'-o-sēn dī-fos'-fāt). A compound which is converted to adenosine triphosphate (ATP) in the process of storing energy derived from respiration.

Adenosine triphosphate (ATP) (ah-den'-o-sēn trī-fos'-fāt). A compound used as a source of energy for cell activities.

Albinism (al'-bi-nizm). A condition characterized by insufficient pigment in the skin, hair, and eyes.

Alveoli (al-vē'-o-lē). Small air sacs making up the lung.

Amino acid (ah-mē'-nō). An organic acid containing the amine group (NH$_2$) and the carboxyl group (COOH). They are the constituents of proteins.

Ammonia (ah-mō'-nē-ah). A gaseous compound of nitrogen and hydrogen (NH$_3$).

Amniocentesis (am-nē-ō-sen-tē'-sis). A technique by which amniotic fluid is withdrawn from the pregnant uterus. May be used in the prenatal diagnosis of certain genetic defects.

Amnion (am'-nē-on). The membrane enclosing the fluid in which the fetus is immersed prior to birth.

Amniotic fluid (am-nē-oh'-tik). The fluid in which the fetus is immersed prior to birth.

Amoeba (ah-mē'-bah). An organism consisting of a single cell; a member of the kingdom *Protista*.

Amphibian (am-fib'-ē-an). A member of a group of animals that includes frogs and salamanders; spends part of its life cycle in the water.

Amygdala (ah-mig'-dah-lah). A brain structure that is believed to be involved with certain emotional responses.

Anaerobic respiration (an-ah-rō'-bik res-pi-rā'-shun). The release of energy from food materials without the involvement of oxygen.

Anemia (ah-nē'-mē-ah). A general term pertaining to a number of conditions in which insufficient oxygen is carried by the blood to the cells of the body.

Annelid (an'-e-lid). One of a group of worms characterized by segmented bodies. The earthworm is included in this group.

Aneuploidy (an'-yū-ploy-dē). The state of having more or less than the normal diploid number of chromosomes.

Aneurysm (an'-yū-rizm). A sac formed by dilation of the walls of an artery or vein.

Antibody (an'-tē-bo-dē). A substance produced by the body in response to the presence of a foreign protein, or antigen.

Antigen (an'-ti-jen). A substance which when injected into an animal body causes production of antibodies.

Artificial insemination (ar-ti-fi'-shal in-sem-i-nā'-shun). Introduction of seminal fluid into the vagina by artificial means.

Arthropod (ar'-thrō-pod). The term means "jointed legs." A member of one of the largest groups of organisms which includes insects, crayfish, lobster, crabs, and shrimp, among others.

Ascorbic acid (as-kōr'-bik). One of the vitamins (C), the deficiency of which results in a disease called *scurvy*.

Asexual (ā-sex'-ū-ahl). A form of reproduction that requires only one type of parent cell.

Atherosclerosis (ath-er-ō-skler-ō'-sis). A condition in which fatty materials are deposited on the inner wall of an artery.

Australopithecus (aw-stra-lō-pi'-the-kus). An early ancestor of modern man having some manlike and some apelike features.

Autosome (aw'-tō-sōm). A chromosome other than the X and Y chromosomes.

Autotroph (aw'-tō-trōf). The term means "self-feeder." An organism such as a green plant that makes food from environmental materials.

Barbiturate (bar-bit'-ū-rat). One of a class of depressant drugs which includes phenobarbitol.

Bile (bīl). A fluid secreted by the liver. It participates in the digestion of fats.

Biosphere (bī'-ō-sfēr). The sum total of living organisms on earth.

Blastocyst (blas'-tō-sist). The hollow-ball stage of the early embryo.

Brachydactyly (brak-ē-dak′-ti-lē). A dominant inherited condition that involves an abnormal shortness of fingers and toes.

Bryophyte (brī′-o-fīt). A member of a group of plants that includes the mosses and liverworts.

Carbohydrate (kar-bō-hī′-drāt). One of the basic food nutrients; it includes sugars and starches.

Castration (kas-trā′-shun). Removal of the gonads; usually refers to the removal of the male testes.

Catalyst (kat′-ah-list). A substance which changes the speed of a chemical reaction.

Caucasian (kaw-kās′-ē-an). A person generally belonging to the "white" racial group. However, "white" describes a wide range of skin colors and types.

Cell (sel). The basic unit of structure and function in the living organism.

Cellulose (sel′-yū-lōs). A carbohydrate which forms the skeleton of most plant structures and of plant cells.

Centriole (sen′-trē-ōl). A cell structure involved in cell division.

Centromere (sen′-trō-mēr). The region of attachment between two chromatids; observable in chromosomes caught at the metaphase stage of cell division.

Cerebellum (ser-e-bel′-um). The brain structure that is primarily responsible for balance and coordination.

Cerebrum (ser′-e-brum). A portion of the brain that is especially well developed in the human. It controls many voluntary functions and is the seat of learning and memory.

Cervix (ser′-viks). The necklike structure at the opening of the uterus.

Chlorophyll (klōr′-ō-fil). The green substance in plant cells which "traps" light energy during the process of photosynthesis.

Chloroplast (klōr′-ō-plast). One of many small green bodies found in plant cells; it contains chlorophyll.

Cholesterol (kō-les′-ter-ol). A member of the nutrient group called *lipids*. Found in brain tissue, egg yolk, and bile.

Cholinesterase (kō-lin-es′-ter-āz). An enzyme found in the nervous system. It functions to break down acetylcholine to stop nerve impulses from crossing the synaptic space.

Chordate (kor′-dāt). One of a group of animals characterized by the presence of a structure called a *notochord* at some time during embryological development.

Chorionic gonadotropin (kōr-ē-on′-ik gō-nahd-ō-trō′-pin). A hormone released from the embryonic membranes; maintains the corpus luteum during early pregnancy.

Chromatid (krō′-mah-tid). One of the two spiral filaments making up a chromosome which separates in cell division. Each goes to a different pole of the dividing cell.

Chromosome (krō′-mo-sōm). One of the rod-shaped bodies in the nucleus of a cell that contain the genes or hereditary factors. They are constant in number for each species.

Cilia (sil′-ē-ah). Tiny hairlike structures arising from the surface of a cell.

Cleavage (klē′-vij). A series of regular mitotic divisions of the fertilized egg cell.

Clitoris (kli′-tor-is). A small body of erectile tissue; part of the external female sex organs.

Coelenterate (se-len′-ter-āt). One of a group of invertebrates characterized by a saclike body; includes jellyfish, hydra, and others.

Coitus interruptus (kō′-ī-tus in-ter-rup′-tus). Withdrawal of the penis from the vagina prior to ejaculation; usually done for contraceptive purposes.

Colon (kō′-lon). That part of the large intestine which extends from the cecum to the rectum.

Commensalism (ko-men′-sa-lizm). A relationship in which an organism lives with a host and the host is neither helped nor harmed by the association.

Condom (kon′-dom). A sheath worn on the penis during intercourse to prevent infection or fertilization.

Contraception (kon-tra-sep′-shun). The prevention of pregnancy.

Coronary thrombosis (kor′-o-na-rē throm-bō′-sis). Heart attack caused by blockage of one or more of the arteries that supply the heart muscle with blood.

Corpus callosum (kōr′-pus kah-loh′-sum). A mass of fibers connecting the left and right cerebral hemispheres.

Corpus luteum (kōr′-pus lū′-tē-um). A mass of tissue in the ovary which secretes progesterone.

Covalence (kō′-vā-lents). A form of chemical bonding in which electrons are shared between atoms.

Cowper's gland (kow′-perz). A small gland (paired) located near the prostate. It adds small amounts of secretion to the seminal fluid.

Cri du chat (krē′-dū-shah). The "cry of the cat" syndrome; caused by deletion of part of the short arm of chromosome 5.

Cristae (kris′-tē). Projections leading into the interior of the mitochondrion.

Cryptorchidism (kript-or′-ki-dizm). A condition characterized by failure of the testes to descend into the scrotum.

Cytoplasm (sī′-tō-plah-zum). The protoplasm of a cell exclusive of the nucleus.

Cytosine (sī′-tō-sēn). One of the bases found in DNA and RNA.

Decomposer (dē-kom-pō′-zer). One of a group of organisms, such as bacteria, that breaks down dead organic materials in the ecosystem.

Deoxyribose (dē-aks-ē-rī′-bōz). A five-carbon sugar found in the structure of deoxyribonucleic acid (DNA).

Diaphragm (dī′-ah-fram). A device of molded rubber or other soft plastic material placed over the cervix to prevent the entrance of sperm cells.

Diatom (dī′-ah-tom). Any unicellular, microscopic form of algae having a wall of silica.

Diethylstilbestrol (DES) (dī-eth-il-stil′-bes-trol). A synthetic hormone sometimes given after intercourse to prevent pregnancy.

Digestion (dī-jest′-yun). The process of converting food into materials that can be absorbed and assimilated.

Deoxyribonucleic acid (DNA) (dē-oxy-rī-bō-nū-klē′-ik). The hereditary substance found in the chromosomes of cells.

Dominant (dohm′-i-nent). In genetics, refers to a gene that exerts its full effect regardless of the effect of its partner.

Dopamine (dō′-pah-mēn). A chemical neurotransmitter in the brain. An insufficiency of this substance is associated with Parkinson's disease.

Down's syndrome (Downz sin′-drōm). Commonly known as "mongolism," a condition of mental retardation caused by the presence of a triplet at chromosome 21.

Echinoderm (ē-kīn′-ō-derm). One of a large group of invertebrates characterized by radial symmetry. This group includes starfish and sea urchins, among others.

Ecology (ē-kol′-oh-jē). The study of the interrelationships among living organisms and between organisms and their physical environments.

Ecosphere (ē′-kō-sfēr). The sum total of all living organisms on earth together with the physical environment.

Ecosystem (ē′-kō-sis-tem). The sum total of living organisms, physical environmental factors, and their interrelationships occurring in a given area such as a pond or woods.

Ectoparasite (ek-tō-par′-ah-sīt). A parasite that lives attached to the outside of the host's body.

Embryo (em′-brē-ō). The early stage of development of an organism. In the human, this stage is generally considered to include the period from conception to the end of the first eight weeks.

Emphysema (em-fi-sē′-mah). A lung disease characterized by loss of elasticity of the air sacs.

Endometrium (en-dō-mē′-trē-um). The lining of the uterus, the thickness and structure of which vary with the phases of the menstrual cycle.

Endocrine (en′-dō-krin). Refers to glands that secrete substances called hormones directly into the bloodstream.

Endoparasite (en-dō-par′-ah-sīt). A parasite that lives in the host's internal organs.

Endoplasmic reticulum (en-dō-plaz′-mik re-tik′-yū-lum). A system of tortuous pathways in the cytoplasm of a cell.

Enzyme (en′-zīm). A protein compound that catalyzes chemical reactions which take place in the cell.

Epididymis (ep-i-did′-i-mis). A coiled structure attached to the upper part of the testis in which mature sperm cells are stored.

Epilimnion (e-pi-lim′-ne-on). The top layer of water in a deep lake that has undergone stratification.

Epinephrine (ep-i-nef′-rin). A chemical neurotransmitter in the brain; also the hormone produced by the medulla of the adrenal gland.

Erythroblastosis (ē-rith-rō-blas-tō′-sis). A condition involving the destruction of fetal red blood cells due to Rh incompatibility.

Estrogen (es′-trō-jen). A major female sex hormone; responsible for the development and maintenance of secondary sex characteristics.

Eucaryote (yū-kar′-ē-ōt). An organism in which the cells contain highly specialized organelles such as nuclei and mitochondria.

Eutrophication (yū-trōf-i-kā′-shun). The aging process of a lake. Increased nutrients produce algal blooms which decompose, thus reducing the oxygen content of the lake.

Evolution (eh-vō-lū′-shun). A slow process of change in the genetic makeup of a population.

Fallopian tube (fah-lō′-pē-an). Also called the *oviduct*, the tube that leads from the ovary to the uterus.

Fertilization (fer-ti-li-zā′-shun). The fusion of a sperm cell nucleus with an egg cell nucleus to form a zygote, which is the first stage in the development of a new individual.

Fetus (fē-′tus). In the human, the term is used to describe the developing individual from the eighth week to birth.

Fission (fish′-un). A form of asexual reproduction in which the cell simply divides into two equal parts.

Flagellum (flah-jel′-um). The taillike structure by which a sperm cell propels itself through the seminal and uterine fluids.

Follicle stimulating hormone (FSH). A hormone that stimulates the growth and maturation of follicles in the ovary and the formation of sperm cells in the testis.

Fructose (fruk′-tōs). A simple sugar found in sweet fruits; forms part of the sucrose molecule.

Fungi (fun′-jī). A group of plantlike organisms that do not possess chlorophyll.

Galactose (gah-lak′-tōs). A simple sugar that makes up part of the molecule of lactose, or milk sugar.

Galactosemia (gah-lak-tō-sē′-mē-ah). An hereditary disorder of galactose metabolism characterized by vomiting, diarrhea, jaundice, poor weight gain, and malnutrition in early infancy.

Glial cells (glē′-al). Brain cells that serve as connective tissue; their full significance is not completely understood.

Glucose (glū′-kōs). A simple sugar that is an important source of energy for the living organism.

Glycogen (glī′-kō-jen). The chief carbohydrate storage material in animals. It is formed by and largely stored in the liver. It is also called *animal starch*.

Glycolysis (glī-kohl′-i-sis). The breaking down of sugars into simpler compounds in the process of respiration.

Golgi complex (gōl′-jē). A type of cell organelle that is believed to participate in secretion.

Gonadotropic (gō-nahd-ah-trō′-pik). Refers to the stimulation of the testes or ovaries.

Gonad (gō′-nad). A sex gland, consisting of the ovary in the female and the testis in the male.

Gonorrhea (gon-ō-rē′-ah). A disease characterized by contagious inflammation of the genital mucous membrane; transmitted mainly by sexual intercourse.

Guanine (gwan′-ēn). One of the purine bases found in the structures of DNA and RNA.

Hallucinogen (hah-lū′-si-nō-jen). A drug, such as LSD, which changes the perception of reality, producing bizarre visions or hallucinations.

Haploid (hap′-loyd). Refers to a cell having only one member of each chromosome pair, as in a sperm or egg cell.

Hemoglobin (hē′-mah-glō-bin). The iron compound found in red blood cells which functions as an oxygen carrier.

Hemophilia (hē-mō-fil′-ē-ah). An hereditary condition characterized by deep-tissue bleeding.

Herbivore (er′-bi-vōr). An animal that subsists wholly on plant food.

Heterotroph (het′-er-ō-trŏf). The term means "other feeder." An organism that does not carry on photosynthesis and depends upon the producers for its food.

Hexosaminidase-A (hek-sos-a′-min-i-dāz). An enzyme involved in the metabolism of fats. A deficiency of this enzyme results in Tay-Sachs disease.

Hippocampus (hip-ō-kam′-pus). One of the "lower" brain structures; believed to be associated with functions involving memory.

Homo erectus (Hō′-mō ē-rek′-tus). A fossil that represents an early stage in the development of man, having more distinct manlike features than *Australopithecus*.

Homo sapiens (Hō′-mō sā′-pē-ans). The modern human species; means "man the wise."

Hormone (hor′-mōn). A chemical substance secreted by an endocrine gland.

Hyaluronidase (hī-al-yū-ron′-i-dāz). An enzyme secreted by sperm cells evidently to assist in penetrating the substance surrounding the egg cell.

Hyperkinetic (hī-per-ki-net′-ik). Refers to the condition of being abnormally active; often applied to children having minimal brain damage (MBD).

Hypertonic (hī-per-tohn′-ik). Having a greater concentration of dissolved particles than a solution on the other side of a membrane. Water will move through the membrane toward the *hypertonic* solution.

Hypolimnion (hī-pō-lim-nē-on). The bottom layer of water in a deep lake that has undergone stratification.

Hypothalamus (hī-pō-thahl′-uh-mus). A region of the brain containing various "lower" centers such as those for temperature and appetite control.

Hypotonic (hī-pō-tohn′-ik). Having a lesser concentration of dissolved particles than a solution on the other side of a membrane. Water will move through the membrane in a direction away from the *hypotonic* solution.

Implantation (im-plan-tā′-shun). Attachment of the blastocyst to the lining of the uterus and its embedding in the endometrium, occurring six or seven days after fertilization of the egg.

Infanticide (in-fan′-ti-sīd). The taking of the life of an infant.

Inorganic (in-ohr-gan′-ik). Pertaining to substances that are not produced by living tissue.

Intrauterine device (IUD) (in-tra-yū′-ter-in). A plastic object inserted in the uterus for the purpose of preventing pregnancy.

Isotonic (ī-sō-tohn′-ik). Having the same concentration of dissolved particles as a solution on the other side of a membrane. There is no net movement of water molecules between the two solutions.

Karyotype (kehr′-ē-ō-tīp). A systemized arrangement typical of an individual or a species of the chromosomes of a single cell.

Klinefelter's syndrome (Klīn′-fel-terz sin′-drōm). A condition characterized by the presence of small testes; associated with an abnormality of the sex chromosomes (XXY).

Kwashiorkor (kwash-ē-ōr′-kōr). A syndrome produced by severe protein deficiency with characteristic changes in pigmentation of the skin and hair.

Labia majora (la′-bē-uh mah-johr′-uh). The large folds of skin found on the external female genitals.

Labia minora (la′-bē-uh mī-nohr-uh). Small folds of skin located on either side of the vaginal opening.

Lactation (lak-tā′-shun). The secretion of milk.

Lactose (lak′-tōs). A double sugar consisting of galactose and glucose; found in milk.

Legume (leg′-yūm). One of a group of plants including clover, beans, and alfalfa. The roots contain small nodules of nitrogen-fixing bacteria.

Leukemia (lū-kē′-mē-ah). A disease of the blood-forming organs characterized by a marked increase in the number of leukocytes in the blood, together with the enlargement of the spleen, lymphatic glands, and bone marrow.

Leydig, cells of (Lī′-dig). Testicular cells that secrete the male hormone testosterone.

Limbic system (lim′-bik). A "lower" group of structures of the brain which is apparently involved with emotional responses.

Lipid (lip′-id). One of a group of nutrients that includes triglycerides, phospholipids, and cholesterol.

Luteinizing hormone (LH) (lū'-ten-īz-ing). A pituitary hormone that participates in ovulation. It also stimulates the testes to produce the male hormone.

Lysosome (lī'-soh-sōm). A cell organelle that contains enzymes that destroy the cell when released.

Malnutrition (mal-nū-tri'-shun). A condition resulting from a deficiency of one or more of the basic nutrients.

Maltose (mawl'-tōs). A double sugar resulting from the combination of two glucose molecules.

Mammary gland (ma'-mer-ē). A specialized gland which in the female mammal produces milk for the young.

Marasmus (mah-raz'-mus). A disease of malnutrition caused by a protein-calorie deficiency.

Marsupial (mar-sū'-pē-al). One of a class of mammals characterized by the possession of an abdominal pouch in which the young are carried for some time after birth.

Mechanistic (meh-ka-nis'-tik). Pertaining to the concept that life can be explained in terms of basic physical and chemical principles.

Medulla (me-dul'-lah). The brain stem; contains vital control centers affecting respiration, heart rate, and blood pressure.

Meiosis (mī-ō'-sis). A form of cell division which produces sex cells having one-half the species chromosome number.

Melanin (mel'-ah-nin). A dark pigment which contributes to the color of the skin, hair, and iris of the eye.

Menopause (men'-ō-pawz). Cessation of menstruation in the human female usually between the ages of 45 and 50.

Menstruation (men-strū-ā'-shun). The cyclic uterine bleeding which recurs at approximately four-week intervals in the human female.

Metabolism (me-tab'-ō-lizm). The sum of physical and chemical processes by which living organisms are produced and maintained, and the transformation by which energy is made available to the organism.

Metaphase (met'-ah-fāz). The middle stage of mitosis in which the lengthwise separation of the chromosomes occurs.

Methadone (me'-tha-dōn). A drug used in the treatment of drug addiction; also used as a substitute for heroin.

Miscarriage (mis'-kar-age). The premature expulsion of the products of conception from the uterus.

Mitochondrion (mī-to-kon'-drē-on). A cell organelle in which the major energy-releasing reactions take place.

Mitosis (mī-tō'-sis). A form of cell division normally characterized by exact duplication of chromosomes and chromosome number.

Mollusc (mol'-lusk). A member of a group of invertebrates that includes clams and oysters, among others.

Monera (mo-ne'-rah). A major taxonomic category consisting of bacteria and blue-green algae; characterized by the absence of highly organized cell organelles.

Monosaccharide (mon-ō-sak'-ah-rīd). A simple sugar such as glucose, fructose, or galactose.

Mutation (myū-tā'-shun). A stable change in gene structure that may be passed on to succeeding generations.

Mutualism (myū'-chū-a-lizm). The living together of two organisms in a close and mutually helpful relationship.

Neanderthal man (Nē-an'-der-thahl). A fossil human form, relatively advanced, that disappeared about 50,000 years ago.

Neuron (nū'-ron). A nerve cell; regarded as a structural unit of the nervous system.

Neurotransmitter (nū-rō-tranz-mi'-ter). A chemical secreted by nerve-cell endings that carries the nerve impulse across the synaptic space.

Nondisjunction (non-dis-junk'-shun). Failure of a pair of chromosomes to separate at meiosis, so that both members of the pair are carried to the same daughter nucleus and the other daughter cell is lacking that particular chromosome.

Norepinephrine (nor-ep-i-nef'-rin). A chemical which acts as a neurotransmitter in the brain.

Notochord (nō'-tō-kōrd). A structure that is found in the early embryological stage of vertebrates and is found in the adult stage of other chordates such as amphioxus.

Nucleic acid (nū-klē'-ik). A compound that is composed of nucleotides joined together. DNA and RNA are especially important ones.

Nucleolus (nū-klē'-o-lus). A dense body within the cell nucleus; composed largely of RNA.

Nucleotide (nū'-klē-ō-tīd). A molecule consisting of a five-carbon sugar, a phosphate group, and a base. A fundamental unit of DNA or RNA.

Nucleus (nū'-klē-us). A structure found in most living cells; contains the hereditary material.

Oögenesis (ō-ah-jen'-e-sis). The process of the origin and development of the ovum, or egg cell.

Orchiditis (ōr-ki-dī'-tis). Inflammation of a testis.

Organelle (ōr-gan-el'). Organized subcellular structures such as mitochondria and lysosomes; found in all cells except the *Monera*.

Orgasm (ōr'-gazm). The point of climax of sexual excitement involving the release of sexual tensions.

Osmosis (ohs-mō'-sis). The passage of water from the lesser to the greater concentration when two solutions are separated by a membrane which selectively prevents the passage of solute molecules but is permeable to water.

Ostium (ohs'-tē-um). The structure located at the ovarian end of the fallopian tube that picks up the egg cell following its release from the ovary.

Ovary (ō'-vah-rē). The female gonad, or sex gland, from which egg cells and hormones are released.

Ovulation (ov-yū-lā'-shun). The discharge of a mature egg cell from a follicle of the ovary.

Oxygen (ohk'-si-jen). A gaseous element existing free in the air. One of the four principal elements of living tissue.

Oxytocin (ohk-sē-tō'-sin). A hormone secreted by the hypothalamus and stored in the posterior lobe of the pituitary. Causes contractions of uterine muscles and the release of milk from the mammary glands.

Pancreas (pan'-krē-us). A large elongated gland located behind the stomach; secretes digestive enzymes and the hormone insulin.

Paramecium (par-ah-mē'-sē-um). A protozoan and member of the kingdom *Protista*.

Pathogen (path'-ō-jen). Generally, a microorganism that causes disease.

Parasite (par'-ah-sīt). An organism that obtains nutrients from another organism, called the *host*, and which harms the host to some extent (depending on the kind of parasite).

Penis (pē'-nis). The male organ of copulation.

Phenylalanine hydroxylase (fen-il-ā'-lah-nēn hī-drok'-si-lāz). An enzyme that aids in the conversion of phenylalanine to tyrosine. Lack of this enzyme results in phenylketonuria (PKU).

Phenylketonuria (PKU) (fen-il-kē-toh-nūr'-ē-ah). A form of mental retardation inherited through a recessive gene. Involves the inability to convert phenylalanine to tyrosine.

Phosphoglyceraldehyde (fos-fō-glis-er-al'-de-hīd). One of the products of photosynthesis.

Phospholipid (fos-fō-lip'-id). One of a class of lipids; an important structural component of the cell membrane.

Photosynthesis (fō-tō-sin'-the-sis). The process of formation of carbohydrates from carbon dioxide and water in the presence of light and chlorophyll.

Phylogeny (fī-loj'-e-nē). The study of the evolutionary history and interrelationships among groups of organisms.

Phylum (fī'-lum). A major taxonomic division of a kingdom.

Phytoplankton (fī-tō-plank'-ton). Simple, free-floating, aquatic plants; often the basic producers in an aquatic ecosystem.

Pituitary (pi-tū'-i-tar-ē). An endocrine gland located at the base of the brain.

Placenta (plah-sen'-tah). The organ of pregnancy that provides for the exchange of gaseous, nutritive, and hormonal substances between the maternal and fetal bloodstreams.

Polypeptide (pol-ē-pep'-tīd). One of a group of amino acids bonded together by a peptide linkage.

Polyploidy (pol'-ē-plōy-dē). The state of having more than two full sets of homologous chromosomes.

Polysaccharide (pol-ē-sak'-ah-rīd). A large carbohydrate molecule, such as starch or cellulose, made up of simple sugars.

Procaryote (prō-kar'-ē-ōt). An organism that does not possess well-organized subcellular structures such as mitochondria and nuclei.

Progesterone (prō-jes'-ter-ōn). The hormone produced by the corpus luteum; prepares the uterus for the reception and development of the fertilized ovum.

Prolactin (prō-lak'-tin). A pituitary hormone which stimulates the mammary glands to produce milk.

Prostaglandin (pros-tah-glan'-din). A substance found in seminal fluid; produces strong contractions of smooth muscle.

Prostate (pro'-stāt). A gland located in the male reproductive system; secretes seminal fluid.

Protein (prō'-tēn). An important structural component of living organisms; an organic compound composed of amino acids.

Protist (prō'-tist). A member of a large taxonomic group of simple organisms which includes the protozoa and most types of algae.

Psychosurgery (si-kō-ser'-jer-ē). Brain surgery performed for the relief of mental and emotional disturbances.

Pterysaur (ter'-i-sōr). A member of a group of flying reptiles that lived during the Mesozoic era.

Puberty (pyū'-ber-tē). The stage of development at which the sex organs begin to function and the secondary sex characteristics begin to develop.

Purine (pyū'-rēn). A class of organic compounds that includes adenine and guanine; important constituents of DNA and RNA.

Pyrimidine (pi-rim'-i-dēn). A class of organic compounds that includes cytosine and thymine (or uracil); important constituents of DNA and RNA.

Recessive (rē-ses'-iv). In genetics, refers to a gene that does not exert its influence in the presence of a dominant partner.

Retinoblastoma (re-ti-nō-blas-tō'-mah). A tumor of the eye inherited through a dominant gene.

Ribonucleic acid (RNA) (rī-bō-nū-klē'-ik). A molecule similar to DNA, but with ribose sugar and uracil in place of deoxyribose sugar and thymine, respectively.

Ribosome (rī'-boh-sōm). One of many small granules in the endoplasmic reticulum which contain RNA and are the sites of protein synthesis.

Rubella (rū-bel'-lah). A virus disease, also called German measles, which causes birth defects when the mother is exposed in early pregnancy.

Salpingitis (sal-pin-jī'-tis). A condition characterized by inflammation of the fallopian tube.

Saprophyte (sap'-rō-fīt). A plant organism that lives on dead or decaying organic matter, obtaining nutrients by absorption.

Schizophrenia (skit-sah-frē'-nē-ah). A form of mental illness.

Scrotum (skrō'-tum). The pouch which contains the testes and their accessory organs.

Seminiferous tubule (se-min-if'-er-us tūb-yūle). A long, coiled tubule located in the testis; the site of sperm-cell production.

Spirogyra (spī-rō-jī'-rah). A filamentous green alga.

Spontaneous generation (spon-tā'-nē-us gen-er-ā'-shun). The concept that a living organism can arise from inorganic materials.

Starch. A group of carbohydrates composed of large molecules consisting of linked simple sugars.

Sucrose (sū'-krōs). A double sugar consisting of a molecule of glucose and a molecule of fructose.

Symbiosis (sim-bī-ō'-sis). The living together of two organisms in a close relationship. Such a relationship may involve commensalism, mutualism, or parasitism.

Synapse (sī'-naps). A region where the nerve impulse is transmitted from a neuron to an adjacent neuron.

Syndrome (sin'-drōm). A general set of symptoms which is characteristic of a specific condition, such as *Down's syndrome*.

Taxonomy (taks-on'-o-mē). A branch of biology which deals with the classification of living organisms.

Tay-Sachs disease (Tā-saks'). A recessive, inherited condition found largely in Ashkenazi Jews involving deterioration of the nervous system.

Teratogen (ter-a'-toh-jen). An agent that causes physical defects in a developing embryo; for example, rubella virus or thalidomide.

Testis (tes'-tis). The male gonad; an egg-shaped gland located in the scrotum.

Testosterone (tes-tos'-ter-ōn). The principal male hormone; produced in the testes.

Tetrahydrocannibinol (tet-rah-hī-dro-kan-i'-bin-ohl). The active chemical found in marijuana.

Thalamus (thal'-ah-mus). The brain structure that transmits sensory stimuli to the cerebrum.

Thalidomide (thah-lid'-ō-mīd). A drug that was discovered to be the cause of serious congenital abnormalities in the fetus; used commonly in Europe as a sedative in the 1960's.

Thermocline (ther'-mō-klīn). A water layer that separates the epilimnion and the hypolimnion in a deep lake that has undergone stratification.

Tracheophyte (trā'-kē-ō-fīt). A plant organism characterized by the presence of conducting tissues in the roots, stems, and leaves.

Translocation (trans-lō-kā'-shun). In genetics, the shifting of a segment of one chromosome to another chromosome.

Triglyceride (trī-glis-er-īd). One of a group of lipids with molecules consisting of glycerol and fatty acids.

Tropic (trō'-pik). Pertains to action brought about by specific stimuli; for example, *gonadotropic* means "stimulating the gonads."

Turner's syndrome. A condition found in human females who have only one X chromosome; involves retarded growth and sexual development.

Umbilical cord (um-bil'-i-kal). The structure that connects the fetus to the placenta.

Uracil (yū'-rah-sil). A pyrimidine base found in ribonucleic acid (RNA).

Urea (yū-rē'-ah). A nitrogenous compound produced in the liver; contains waste products of amino-acid metabolism.

Urethra (yū-rē'-thrah). The canal that carries urine from the bladder to the exterior of the body. In the male, it carries sperm cells from the vas deferens to the outside.

Uterus (yū'-ter-us). The hollow, muscular, female organ in which the fetus develops.

Vagina (va-jī'-nah). In the female, the canal that extends from the vulva to the cervix of the uterus.

Vascular (vas'-kew-lahr). Pertaining to the blood circulating system.

Vas deferens (vahz def'-er-ens). Also called the *ductus deferens,* a paired duct that conducts sperm cells from the testes to the urethra.

Vasectomy (va-sek'-to-mē). A method of sterilization in the male. The vas deferens is cut and tied to prevent the ejaculation of sperm cells.

Vertebrate (ver'-te-brāt). A member of a group of animals characterized by the presence of a backbone (or notochord) in the adult stage.

Vitalism (vī'-ta-lizm). The concept that biological activities are directed by a supernatural force.

Vitamin (vī'-tah-min). A general term for a number of substances that are necessary in small amounts for assisting metabolic processes.

Vulva (vul'-vah). The external female reproductive organs.

Zona pellucida (zō'-nah pel-lū'-si-dah). A transparent noncellular layer surrounding the egg cell.

Zygote (zī'-gōt). The cell that results from the fusion of a sperm cell and an egg cell, i.e., the fertilized egg cell.

Annotated Reading List

The following books were selected primarily for their nontechnical presentation and lucid writing style. Most of the books on this list are directed to the reader who has little or no formal background in biology. They are separated into subject-matter areas for the convenience of the student who may wish to pursue further study on a specific topic.

General Issues

Augenstein, Leroy, *Come, Let Us Play God.* New York: Harper & Row, 1969.

The author discusses the "biological revolution" and the new dilemmas produced by increasing knowledge and sophisticated technology. The topics range from organ transplants to birth defects, as Dr. Augenstein examines (with case histories) the possible consequences of man's increasing ability to control his own destiny.

Hardin, Garrett, *Stalking The Wild Taboo.* Los Altos, Calif.: William Kaufmann, 1973.

The author describes how traditional taboos found in human society often prevent us from dealing directly and rationally with problems such as those involving abortion, overpopulation, and the environment.

Human Reproduction

Bender, Stephen J., and Stanford Fellers, *Contraception, By Choice or By Chance.* Dubuque, Iowa: Wm. C. Brown, 1972.

This book is dedicated by the authors to "all those who have been affected by an unwanted pregnancy." They describe the modern methods of contraception and discuss abortion as a method of birth control.

Bender, Stephen J., *Venereal Disease,* 2nd ed. Dubuque, Iowa: Wm. C. Brown, 1975.

Written in nontechnical language, this book is a knowledgeable discussion of the major venereal diseases, including their description, treatment, and prevention.

Hardin, Garrett, *Birth Control.* New York: Western Publishing Co., 1970.

An excellent, highly readable book in which the author describes the various methods of birth control and examines the issues that surround contraception and abortion.

Katchadourian, H. A., and D. T. Lunde, *Fundamentals of Human Sexuality,* 2nd ed. New York: Holt, Rinehart and Winston, 1975.

A detailed but readable book on human reproduction. This book covers practically very facet of human sexuality and its related problems.

Pengelley, Eric T., *Sex and Human Life.* Reading, Mass.: Addison-Wesley, 1974.

An excellent treatment of human sexuality with in-depth but understandable discussions of the biology of human reproduction. A useful book for those who want to understand the vast biological, psychological, social, and political aspects of human sexuality.

Human Genetics

Bresler, Jack B., *Genetics and Society.* Reading, Mass.: Addison-Wesley, 1973.

An introduction to sociogenetics through a series of readings from journals in the fields of genetics, sociology, medicine, psychiatry, and anthropology. Some knowledge of elementary genetics is assumed.

Fletcher, Joseph, *The Ethics of Genetic Control.* Garden City, N. Y.: Anchor Press/Doubleday, 1974.

In this highly readable, nontechnical book, the author raises ethical problems related to the new genetics, and he also attempts some answers.

Ingle, Dwight J., *Who Should Have Children?* New York: Bobbs-Merrill, 1973.

The author maintains that society has a clearly defined right to exercise intelligent control over its future. This is a highly provocative book. Not everyone will agree with the author's ideas, but it is well worth reading.

Volpe, E. Peter, *Human Heredity and Birth Defects.* New York: Bobbs-Merrill, 1971.

This is a simple, well-written, and excellent discussion of some of the important birth defects. The discussion builds upon basic principles and is presented in an understandable fashion.

Winchester, A. M., *Human Genetics.* Columbus, Ohio: Charles E. Merrill, 1971.

This book presents a simple approach to the basic principles of human genetics. Major types of birth defects are also discussed.

Psychobiology

Girdano, D. D., and D. A. Girdano, *Drugs, A Factual Account.* Reading, Mass.: Addison-Wesley, 1973.

The authors relate the effects of drugs to the basic physiology of the nervous system. The psychological and social factors that underlie drug use are discussed, as are the specific effects and problems associated with the various groups of drugs most subject to abuse.

Hubbard, John I., *The Biological Basis of Mental Activity.* Reading, Mass.: Addison-Wesley, 1975.

For the reader who wishes to further explore the structure and function of the human brain, this book discusses the neural basis for emotion, learning, and memory. Sleep, dreams, and the question of the link between brain and mind are also discussed.

Klemm, W. R., *Science, The Brain, and Our Future.* New York: Bobbs-Merrill, 1972.

In simple and highly readable language, the author presents major ideas concerning brain function, describes the results of research on the brain, and discusses the social significance of new knowledge of brain functions.

Energy and the Environment

Brown, Theodore L., *Energy and the Environment.* Columbus, Ohio: Charles E. Merrill, 1971.

A short, interesting account of the nature of energy and the relationship of energy to environmental pollution.

Ingles, David R., *Nuclear Energy, Its Physics and Its Social Challenge.* Reading, Mass.: Addison-Wesley, 1973.

An examination of the nature of nuclear energy, its promise, and its problems.

Kormondy, Edward J., *Concepts of Ecology.* Englewood Cliffs, N. J.: Prentice-Hall, 1969.

Discusses the structure and function of ecosystems, energy flow, nutrient cycling, and population growth. Emphasizes modern concepts of ecology.

Wagner, Richard H., *Environment and Man,* 2nd ed. New York: W. W. Norton, 1974.

This book is a complete account of the environmental and energy problems that currently plague the earth and mankind. It is nontechnical and written in a style that is easily understood by the reader with no background in biology.

Evolution

Orgel, L. E., *The Origin of Life: Molecules and Natural Selection.* New York: John Wiley, 1973.

An excellent account of the way life probably originated on earth. The author takes the reader through a step-by-step development of life, beginning with the basic materials provided by the ancient earth and atmosphere.

Gastonguay, Paul R., *Evolution for Everyone.* New York: Bobbs-Merrill, 1974.

A nontechnical description of the principles and processes of evolution, written in a highly entertaining style.

Volpe, E. Peter, *Understanding Evolution,* 2nd ed. Dubuque, Iowa: Wm. C. Brown, 1970.

A readable account of evolution that takes the student step-by-step through the basic principles that underlie the evolutionary process.

Goldsby, Richard A., *Race and Races.* New York: Macmillan, 1971.

An excellent, highly readable examination of the meaning and reality of race, presented within a framework of basic biological principles. The author shows that racial groups are indeed different in certain biological aspects, but he also emphasizes the wide variations within as well as among racial groups.

Index

Abortifacient, 105
Abortion, and birth control, 108
   of defective fetuses, 173
     legal aspects of, 110
     methods of, 108
Acetylcholine, 217, 218
Acquired characteristics, inheritance of, **321**
Acrosome, 72
Activation energy, 115
   barrier, **116***
Active transport, 49
Adaptive radiation, 334, **335**
Adenine, **120**
Adenosine triphosphate (ATP), 42, 43
Air pollution, 294–298
   sources of, 295–297

---
* Page numbers in boldface type refer to illustration appearing on that page.

   types of pollutants, 295
Albinism, **139, 140**
Alcohol, 233, 234
Algae, blue-green, 242, **243**
   brown, 246
   green, 246, **247**
   red, 246
Algal blooms, **290**
Alimentary canal, 183, **184**
Amino acids, 13, 117
Amniocentesis, 91, 170–174
   and abortion, 173
   procedure, **172**
Amnion, 91
Amniotic fluid, 91
Amniotic sac, 91
Amphetamines, 230, 234
Amphibians, 258, **259**
Amphioxus, **257**
Amygdala, 227, 229, 231

Anemia, 190
   sickle-cell; *see* Sickle-cell anemia
Aneurysm, 196
Animal kingdom, 252–261
Annelids, 254, **255**
Anthropoidea, 337
Arthropods, **256**
Artificial insemination, 86
Asexual reproduction, 62–65
   budding, 62
   fission, 62
   fragmentation, 64
   grafting, **64,** 65
   stolons, **64**
   tubers, 64
Atherosclerosis, **196,** 197
Atmosphere, oxidizing, 17
   primitive, 16
   reducing, 17
Atomic structure, 9, **10**

365

Australopithecus africanus, **338**
Autosomes, 54
Axon, 216

Bacteria, 243, **244**
"Balance of nature," 274, 275
Barbiturates, 234
Barr body, **166**
Barr, Murray L., 166
Beagle, voyage of, 322, **323**
Behavior control, 229–232
    hyperkinetic children, 230
    psychosurgery, 229, 231
    schizophrenia, 230
Beriberi, 190
Biological oxygen demand (BOD), 289
Biosphere, 267
Birds, **260**
Birth, **93**
Birth control, 97–111
    methods of, 99
Birth defects, environmental, 170, 171
    rubella, 170
    thalidomide, 170
Blastocyst, 87
Blood types, 149, 150
Brachydactyly, **138**, 139
Brain, human, 219–228
    cerebellum, **225**
    cerebrum, 220–225
    general structure of, **220**
    hypothalamus, 227, 228
    limbic system, 227, **228**
    medulla oblongata, 225
    pons, 225
    reticular activating system (RAS), 226
    stem, 225, 226
    thalamus, 227
    weight of, 219
Bryophytes, 246, **247**
    liverworts, 246, **247**
    mosses, 246, **247**
Buchsbaum, Ralph, 282

Caffeine, 235
Carbohydrates, 11, 178
Carbon cycle, **277**
Carnivora, 240

Carrier identification, 145
    in galactosemia, 131
    in phenylketonuria, 131
    in Tay-Sachs disease, 131
Castration, 73
Catastrophism, 320, 321
Caucasoid race, 341, **342**
Cell, 31–62
    division, 55
    eucaryotic, 242
    evolution of, 24, 25
    glial, 216
    membrane, 33
    organelles, 35, 36, 52
    procaryotic, 242
    structure, **34**
    study of, 32
Cells of Leydig, 73
Centromere, 54
    positions of, 54, 55
Cerebrum, functions of, 220–225
    frontal lobes, 223
    memory, 223–225
    motor activities, **222**
Cervix, 76
Chlorophyll, 36
Chlorpromazine, 230
Cholesterol, 180, **196**
Cholinesterase, 218
Chorionic gonadotropin, 87
Chromatids, 54
Chromosome deletions, 169, 170
    cri du chat, **169**
    Philadelphia chromosome, 170
Chromosome translocation, 167, 168
    in Down's syndrome, **168**
Chromosomes, 51–55
    autosomes, 54
Classification, 239–261
    categories; see Taxonomic categories
    and evolutionary patterns, 241
Cleavage, 86
Clitoris, 76
Cloning, 349
Coacervates, 21
    sensory functions, **222**
Coal deposits, **302**
Coelenterates, **253, 254**

Coenzymes, 117
Coitus interruptus, 99
Commensalism, 276
Community, ecological, 268
Compound interest, 201
Comstock, Anthony, 98
Comstock Law, 98
Condom, **100**
Consumers, 269
Contraception, 99
Contraceptive foams, 101
Contraceptives, oral, 103
Corpus luteum, 84, 87
Covalent bonding, 10
Cowper's gland, 74, **75**
Crick, Francis E., 4, 119
Cro-Magnon man, 339
Cryptorchidism, 71
Cuvier, Georges, 320
Cycles, ecological, 277–279
Cytosine, **120**

Darwin, Charles, 241, 321, **322**
Darwin's Finches, 323, **324**
Darwinism, 321
Deamination, 187
Decomposers, **269**
DDT, 299, 300
Deletions, chromosome; see Chromosome deletions
Delgado, Jose M. R., 231
Dendrites, 216
Deoxyribose nucleic acid; see DNA
Deoxyribose sugar, **120**
Development, embryological, 91
Diaphragm, 100, **101**
Diethylstilbestrol (DES), 107
Differential reproduction, 326
Diffusion, 45
    facilitated, 46
    gradient, 46
Dinosaurs, 318, **319**, 320
Disaccharide, 11
DNA, 119–122
    replication of, 122
    structure of, **121, 122**
    structures of components in, **120**
Donora, Pennsylvania, 295
Dopamine, 217, 218

Douche, post-coital, 99
Down's syndrome, 160–162
   karyotype of, 162
   nondisjunction, 160
   physical characteristics of, 160, **161**
Drugs, 232–237
   alcohol, 233, 234
   amphetamines, 234
   barbiturates, 234
   caffeine, 235
   depressants, 233
   hallucinogenic, 235, 236
   heroin, 237
   LSD, 235
   marijuana, 236
   nicotine, 235
   opiates, 236
   stimulants, 234, 235
   tolerance to, 232, 233
"Drumstick," **167**
Dunkers, 336
Duodenum, 184

Earth, origin of, 14–16
Eccles, John, 214
Echinoderms, **257**
Ecological succession; *see* Succession, ecological
Ecology, 263–284
   definition of, 265–267
Ecosphere, 267
Ecosystem, 267–270
   circulation of materials in, **277, 278,** 279
   community, 268
   consumers, 269
   decomposers, **269**
   development; *see* Succession
   energy flow, 270–272
   negative feedback, 274, **275**
   niche, 268–270
   physical factors of, 273, 274
   population, 267
   producers, 268, 269
   scavengers, 269
Egg cell, life span of, 85
Ehrlich, Paul, 208
Eighteen-Mile Creek, 292
Ejaculation, 74

Embryo, 88
Embryological development, 90–93
Endergonic reaction, 116
Endometrium, 76
Endoplasmic reticulum, 35, 51
Energy, 37
   activation, 115
   of chemical reactions, 115
   in an ecosystem, 270–272
   and the environment, 301–309
   and food sources, 309, 310
   geothermal, **304**
   kinetic, 37
   potential, 37
   solar, 263, **264, 265**
   sources of, 301–306
Energy needs, human, 309, 310
Entropy, 37
Enzymes, 116
   function of, **118**
Epididymis, 73, **75**
Epinephrine, 217
Erythroblastosis, 151–152
Estrogen, 81, 82, 84
Estrus, 81
Eucaryotes, 242, 244
Eutrophication, 290
Evergreens, **249**
Evolution, 313–340
   acquired characteristics, **321**
   adaptive radiation, 334, **335**
   catastrophism, 320, 321
   differential reproduction, 326
   genetic drift, 334, **335**
   Hardy-Weinberg Law, 327, 328
   human, 336–340
   mutations, 328
   natural selection, 325, 326, 328, 329
   reproductive isolation, **332, 333, 334**
   species formation, 331–336
   uniformitarianism, 321
Exergonic reactions, 116

Fallopian tube, 76
Fats, 11, 179, 184
Female reproductive system, 76, **77**
Ferns, **248, 249**
Fetus, definition of, 88
Finches, Darwin's, 323, **324**

Fish, 257, **258**
Fitness, 326
Flatworms, 254, **255**
Fluid, amniotic, 91, 171
Foams, contraceptive, 101
Folic acid, 182, 193
Follicle stimulating hormone (FSH), 72, 81
Food chain, 270
Food web, 270, **271**
Fossils, **314, 315,** 316
Fructose, 11, 185
Fungi, 250–252
   club, **251**
   sac, 252

Galactose, 11, 127
Galactosemia, 127, 130
Galapagos Islands, 323, **324**
Gaylin, Willard M., 231
Gene frequencies, 141–145
   and DNA, 113–131
   dominant, 137
   importance of, 114
   recessive, 139
Genes, 113
Genetic drift, 334, **335**
   American Indian, 335
   Dunkers, 336
Genetics, Mendelian, 133–140
Geologic time scale, **316, 317**
Geothermal energy, **304**
Gland, Cowper's, 74, 75
   pituitary, 68, **69,** 81, 83, 94
   prostate, 74, **75**
Glands, endocrine, 68
   sex, 68
Glass, Bentley, 347
Glial cells, 216
Glucose, 11, 185–187
Glycerine (glycerol), 12, 185
Goldsby, Richard A., 343, 345
Golgi complex, **35,** 36
Gonadotropic releasing factors (GRF), 68, **69**
Great Lakes, **289**
Green revolution, 310
Guanine, 120
Guthrie test, 130

Haldane, J. B. S., 19
Hapsburg lip, 137
Hardy-Weinberg Law, 142, 327, 328
Hemoglobin, 126
Hemophilia, 152, **153**
Heroin, 236, 237
Hexosaminidase-A, 127, 145
Hippocampus, 229
Homo erectus, **338**
Hormones, 68
  chorionic gonadotropin, 87
  estrogen, 81, 82, 84
  follicle stimulating hormone (FSH), 81
  gonadotropic, 68
  luteinizing (LH), 83
  luteotropic (LTH), 84
  oxytocin, 94
  progesterone, 81, 82
  prolactin, 94
  testosterone, 70, 73
Horsetails, **248**
Human evolution, 336, 340
Human nutrition; *see* Nutrition, human
Human sexuality, 67, 68
Hyaluronidase, 72
Hyden, Halger, 224
Hymen, 76
Hypertonic solution, **48**
Hypothalamus, 68, **69**
  feeding center, 228
  "pleasure center," 228
  in temperature control, 228
Hypotonic solution, **48**

Implantation, 87, **88**
Intelligence and race, 344
Interstitial cell stimulating hormone (ICSH), 73
Intrauterine device (IUD), **105**
I.Q., 344, 345
Isolation, reproductive, **332, 333, 334**
Isotonic solution, **48**

Jensen, A. R., 344
John, E. Roy, 225
Johnson, V., 78

Karyotype, 53, 136
  human female, **136**
  human male, **136**
Kelsey, Frances O., 170
Kettlewell, H. B. D., 329
Kinsey, A. C., 80
Kistner, Robert W., 105
Kwashiorkor, 189, **190**

Labia majora, 76
Labia minora, 76
Laborit, Henri, 230
Lactation, **94**
Lactose, 12, 130, 185
Lake Erie, 289–291
  temperature stratification in, 290, **291**
Lake Ontario, 291, 292
Lamark, Jean, 321
Lashley, Karl, 225
L-dopa, 218
LH, 72, 73, 83, 85
Lichen, **276**
Life, definition of, 9
  major elements of, 9
  major molecules of, 11–14
  nature and origins of, 7–29
Limbic system, 227, **228**
Linnaeus, Carl, 240
Lipids, 12, 179
Lipofuchsin, 224
Little, Rev. William John Knox, 98
Liver, synthesis of urea, 184
Liverworts, 246, **247**
LSD, 235
Luteinizing hormone (LH), 72, 73, 83, 85
Luteotropic hormone (LTH), 84
Lyell, Charles, 321
Lysine, 309
Lysosome, 35

Male reproductive system, **70**, 70–76
Mammals, 260, **261**
Marasmus, 190
Marijuana, 236
Masters, W., 78
Mechanistic Theory of Life, 8
Medulla oblongata, 225, 226

Meiosis, 58, **59**
Membrane, cell, 33, 45
Memory, 223–225
Mendel, Gregor, 326
Menopause, 81
Menstrual cycle, 80-84
Menstruation, 81
Metabolic block disorders, 126–131
  albinism, 139
  galactosemia, 127, 130
  phenylketonuria, 127, 128, 148
  Tay-Sachs disease, 127, 130, 143
"Metabolic mill," **188**
Metabolism, 37, 188
Migraine, **141**
Miller, S. L., 19
Minimal brain dysfunction (MBD), 230
Mitochondrion, 36
Mitosis, 55, **56,** 57
  anaphase, 57
  metaphase, 55
  prophase, 55
  telophase, 57
Molds, 252
Molluscs, 254
Monera, 242, 243
"Mongolism"; *see* Down's syndrome
Mongoloid race, 341, **342**
Monosaccharide, 11
Mosses, 246, **247**
Multiple sclerosis, 216
Mutagenic, 328
Mutations, **328**
Mutualism, 276
Myasthenia gravis, 219

NADP, 41
Natural selection, 325, 326, 328, 329
Neanderthal man, **339**
Negative feedback, **68,** 274, **275**
Negroid race, 341, **343**
Neostigmine, 219
Nerve, 216
Nerve impulse, 216, **217**
Neuromuscular junction, 218, **219**
Neuron, 191, **215,** 216
  axons, 216
  dendrites, 216

Neurotransmitter, 217, 218
Niagara river, **292**
Niche, ecological, 268–270
Nicotine, 235
Nitrogen cycle, 278, 279
Nondisjunction, 159, **160**
Norepinephrine, 217
Notochord, **257**
Nuclear power, **305, 306, 307**
Nucleic acids, 22–25, 120
    duplication of, 23
    structure of, 22
Nucleolus, 51
Nucleotides, 21, **22**, 121
Nutrition, human, 177–196
    and aging, 192, 193
    and brain development, 191, 192
    carbohydrate metabolism, 186
    classes of nutrients, 178–182
    digestion and absorption, 182–186
    lipid metabolism, 186, 187
    malnutrition, 189, 190
    obesity, 195, 196
    and the "Pill," 193
    in pregnancy, 193, 194
    protein metabolism, 187

Oil spills, **303**
Oögenesis, **61**
Oparin, A. I., 19
Oral contraceptives, 103
Orchiditis, 71
Orgasm, 79, 80
Orgel, L. E., 24
Origin of Species, 321
Ornstein, Robert, 221
Osmosis, **47**
Osteoporosis, 193
Ostium, 76
Ovaries, 76
Ovulation, 83, 101, 102
Oxytocin, 94
Ozone, 17

Pancreas, 184
Parasitism, 276, 277
Parkinson's disease, 218
Pasteur, Louis, 27
Penis, **70**, 74, **75**

Peppered moth, selection in, 329, 330, **331**
Peptide linkage, 14
Pesticides, 298–301
    concentration in food chains, 299, 300
    DDT, 299, 300
Phenothiazine, 230
Phenylalanine hydroxylase, 128
Phenylketonuria (PKU), 127, 128, 148
    Guthrie test, 130
Phosphoglyceraldehyde (PGAL), 42
Phosphorus cycle, 279
Photosynthesis, 38–42
    dark reactions, **41**
    light phase, **40**
Phytoplankton, **268**, 269
Pituitary gland, 68, **69**
    anterior lobe, 68
    posterior lobe, 68
Placenta, 88, **90**
Plant kingdom, 246–250
Plants, flowering, 249–251
    reproduction in, 249, **250, 251**
Pollution, 2, 287–298
    of air, 294–298
    with pesticides, 298
    thermal, 306, **307**
    of water, 288–294
Polygenic inheritance, 154, **155**
Polynucleotides, 22, 121
Polypeptides, 14, 125
Polyploidy, **333**
Pons, 225
Population, ecological, 267
Population genetics, 143
Population growth, 199–211
    birth rate, 201
    causes, 203–206
    consequences of, 208–211
    death rate, 201
    formulas, 201–203
    rate of increase, 201
    in India, 207
    in Latin America, 207
    in the United States, 206
Population Reference Bureau, 199
Post-coital douche, 99
Prefrontal lobotomy, 229

Pregnancy, 86–93
    and nutrition, 92, 193, 194
Prenatal diagnosis, 171–174
Primates, 240, 336
Probability, 133–135
    addition rule, 135
    product rule, 134
Procaryotes, 242
Proconsul africanus, 338
Producers, 268, 269
Progesterone, 81, 82, 84, 87
Prolactin, 94
Prosimians, 336
Prostate gland, 74, **75**
Protein synthesis, **124,** 124–126
Proteins, 12, 180, 187, 190
Protists, 244, **245,** 246
Psychobiology, 213–237
Psychosurgery, 229, 231
Puberty, 70, 81
Purines, 120
Pyramid, trophic, 270, 271, **272**
Pyrimidines, **120**

Races of man, 340–345
    Caucasoids, 341, **342**
    Mongoloids, 341, **342**
    Negroids, 341, **343**
RAS, 226
REM sleep, 227
Reproduction, 62–66
    asexual, 62–65
    human, 67–95
    sexual, 65
Reproductive isolation, **332, 333, 334**
Reptiles, **259**
Respiration, 25, 26, 42
Reticular activating system (RAS), 226
Rh factor, 151, **152**
Rhythm method, contraception, 101, **102**
Ribonucleic acid; see RNA
Ribose sugar, 123
Ribosomes, 35
RNA, 123–126
Rolando, fissure of, **222**
Roundworms, 254, **255**
Rubella, 90

Sanger, Margaret, 98
Scavengers, 269
Schizophrenia, 230
Selection, artificial, 325
   natural, 325–329
Seminal vesicles, 74, **75**
Serotonin, 217
Sex chromatin, 166–167
Sex chromosome abnormalities, **163**, 163–167
   Klinefelter's syndrome, 163, **164**
   Turner's syndrome, 164, **165**
   XXX combination, 163
   XYY syndrome, 164, 165
Sex determination, 135
Sex linkage, 152
Sex ratio, 136
Sexual intercourse, 78–80
   excitement phase, 79
   orgasmic phase, 79
   orgasmic platform, 79
   plateau phase, 79
   resolution phase, 80
Sexual reproduction, 65
Shockley, W., 344
Sickle-cell anemia, 126, 146, **147**
Skin color, origin, 343, 344
Sleep, 227
Solar energy, 263, **264, 265**
Species, 240, 332
Species formation, 331–336
Sperm cell, **73**
Spermatogenesis, 59–62, 72
Sperry, Robert, 221
Sponges, **252**
Spontaneous generation, 27, 28
Sterilization, 106
Strip mining, **302**
Stomata, 41
Substratum, 273
Succession, climax stages of, 282, **283**
   denuded field, 282, **283, 284**
   ecological, 279–282
   pioneer stage, 279
   pond, **281,** 282

rock, 279, **280**
Sweeney, Robert A., 266
Symbiosis, 276–277
   commensalism, 276
   mutualism, 276
   parasitism, 276, 277
Synapse, 217, **218,** 219
Syndrome, 160

Taxonomic categories, 240–241
   class, 240
   family, 240
   genus, 240
   kingdom, 240
   order, 240
   phylum, 240
   species, 240
Taxonomy, 240
Tay-Sachs disease, 127, 143
Temperature inversion, **294,** 295
   Donora, Pennsylvania, 295
   London, 295
Testes, **71**
Testosterone, 73
Thalamus, 227
Thalidomide, 91
Thermal pollution, 306, **307**
Thermocline, 290, **291**
Thermodynamics, First Law of, 37
   Second Law of, 37
Thiamine, 190
Thymine, **120**
Tracheophytes, **248, 249,** 250
   ferns, **248, 249**
   flowering plants, 249
   horsetails, **248**
Transamination, 187
Translocation, chromosomes, 167–168
   Down's syndrome, **168,** 169
Triglycerides, 185
Trophic pyramid, 270, 271, **272**
Trophoblast, 87
Tubal ligation, 106, **107**
Twins, fraternal, 86
   identical, 86

Tyrosine, 128

Umbilical cord, 89
Ungar, Georges, 224
Uniformitarianism, 321
Uracil, 123
Uranium–235, 306
Urea, 187
Urethra, 73–75
Ussher, Bishop James, 320
Uterus, 76, **77,** 91, 93

Vas deferens, 73, **75**
Vasectomy, 106, **107**
Vertebrates, 257–260
   amphibians, 258, **259**
   birds, **260**
   fish, 257, **258**
   mammals, 260, **261**
   reptiles, **259**
Villi, intestinal, **185**
Vitalism, 8
Vitamins, 182, **183**
   C, 182
   D, 344
   E, 182
Volpe, E. Peter, 328
Vulva, 77

Waldheim, Kurt, 200
Water, importance of to life, 18
   in nutrition, 182
Water cycle, **278**
Watson, James D., 4, 119
Whittaker, R. H., 242
Wohler, Friedrich, 8

X-linked genes, 152–154
   color blindness, 154
   hemophilia, 152, **153**
   ichthyosis, 154
   muscular dystrophy, 154

Zona pellucida, 86